AF520893

Herbal Resources of India and Nepal

(Tonic, Memory Boosters and Mood Swing)

S.K. Sood
Rakesh Thakur
Department of Biosciences
Himachal Pradesh University
Shimla – 171005, H.P.

Published by:

Scientific Publishers (India)
5 A, New Pali Road, P.O. Box 91
Jodhpur 342 001 (India)

Branch Office
Scientific Publishers (India)
4806/24, Ansari Road, Daryaganj
New Delhi - 110 002 (India)

E-mail: info@scientificpub.com
Website: www.scientificpub.com

Print : 2021

ISBN: 978-81-7233-927-2

Typeset by Rajesh Ojha
Printed in India

Preface

Since time immemorial, plants have played an integral part in the development of human civilization. Undoubtedly, India holds a prominent position in utilization of its time-tested traditional herbal knowledge for overall well-being of mankind. Unfortunately, most of the age-old plants based traditional practices developed over years of quest through trials and errors are vanishing consequent to the assault of modern civilization into tribal areas and changes in sustenance economy, and needs to be popularized with the present and future generations. For doing so, one of the main tasks for the scientific community is to develop sound database on their ethnobotanical prudence. Viewed in this context, the preparation of this comprehensive invaluable manual has been necessitated for which no efforts have been made till date with all earnest efforts to provide up-to-date information on botanical names, synonyms, family, English, Hindi and Sanskrit names, distribution, reproductive cycle, part/s used, folk use(s) and active constituents of tonic, memory boosters and mood swing promising herbal resources of India and Nepal, the need for which has long been felt by biotechnologists, phytochemists, nutritionists, food scientists, pharmacologists, students and researchers of these disciplines and laymen alike in today's stressful and competitive world. Such a valuable baseline data offers tremendous scope for its effective bio-prospecting and hopefully, be put to use for welfare of mankind all over the world for its commercial exploitation. No institutional library in India and abroad should be without it.

We are indebted to Prof. A.D.N. Bajpayee, Vice-Chancellor, Himachal Pradesh University, Shimla, Prof. T.N. Lakhanpal and Prof. V.K. Mattu for their encouragement, support and well wishes. We thank Dr Kuldeep Dogra, Dr Dheeraj Rawat, Dr Suresh Kumar, Dr Romita Sharma, Mr Ved Prakash, Ms

Chandrika, Ms Parul, Ms Anjana, Ms Indu Chauhan, Mr Ram Kishan, Ms Poonam and Mr Bal Krishan for their ungrudging help in various ways. Words of gratitude are expressed to the publisher for bringing out the book in its present form. Thanks are also due to authors of books, papers and figures listed in the treatise. Finally, we thank the members of our families for bearing the opportunity cost.

The work does not pretend to be complete; suggestions for improving it, if any, are most welcome.

S.K. Sood
Rakesh Thakur

Abbreviations

Al	Aluminium	**Me**	Methyl
Alc.	Alcohol	**mg**	Milligram
α	Alpha	**Mg**	Magnesium
&	And	**ml**	Millilitre
A.P.	Andhra Pradesh	**Mn**	Manganese
As	Arsenic	**M.P.**	Madhya Pradesh
β	Beta	**N.**	North/Northern
C.	Central	**Na**	Sodium
Ca	Calcium	**N.E.**	North East
Cl	Chlorine	**N.W.**	North West
CNS	Central Nervous System	**OH**	Hydroxy
Co	Cobalt	**P**	Phosphorous
Cu	Copper	**%**	Percentage
Δ	Delta	**Ψ**	Psi
δ	Delta	**S**	Sulphur
E.	East/Eastern	**S.**	South/Southern
Eng.	English	**Sans.**	Sanskrit
et al.	*et alii* (and other authors)	**S.E.**	South East
etc.	*et cetera* (so on)	**Sp.**	Species
EtOH	Ethanol	**Si**	Silicon
Fe	Iron	**Sub sp.**	Sub Species
γ	Gamma	**Syn.**	Synonymous
g	Gram	**tsp.**	Tea Spoon
I	Iodine	**U.P.**	Uttar Pradesh
i.e.	*id est.* (that is)	**U.S.**	United States
in vitro	In Glass	**var.**	Variety
in vivo	In Living Tissue	**viz.**	*Vide licet* (namely)
J. & K.	Jammu & Kashmir	**Vol. (s)**	Volume (s)
K	Potassium	**W.**	West/Western
m	Metre	**W.H.O.**	World Health Organisation

Contents

Introduction

Man and Traditional Medicine

Herbs have been highly valued and used regularly for thousands of years as the medicine of the masses (Klemens, 2006). Undoubtedly, India with about 68 million people belonging to 227 ethnic groups and 573 tribal communities derived from six racial stocks in the country (Pushpgandhan, 1994) and over 6,000 of 15,000 species of higher plants being used in traditional, folk and herbal medicine, is regarded as a gold mine of well-recorded and traditionally well practiced knowledge of herbal medicine (Dhamija *et al.*, 2011; Jeyapardha *et al.*, 2011). Herbal drugs also constitute a major share of all the officially recognised systems of health in India viz. Ayurveda, Unani, Siddha, Homeopathy and Naturopathy, except Allopathy. More than 70% of India's 1.1 billion population still use these non-allopathic systems of medicine (Vaidya & Devasagayam, 2007). World Health Organization (WHO) has made an attempt to identify all medicinal plants used globally and listed more than 20,000 species. NAPRALERT database documents ethnomedicinal uses alone for 9,200 of 33,000 species of monocots, dicots, gymnosperms, pteridophytes, bryophytes and lichens, suggesting thereby that 28% of plants on earth have been used ethnomedicinally (Farnsworth & Soejarto, 1991). Also, from the export-import Bank reports, it is evident that the global trade of plant-derived and plant originated products is around US $60 billion (with growth of 7% per annum) where India holds stake of US $1 billion (Raskin *et al.*, 2002; Mathur, 2003) which is expected to reach 3 trillion US $ by the end of 2015. WHO has estimated that at least 80% of the population globally relies on traditional medicine to meet their primary health care needs (Bannerman, 1982; WHO, 2000). Nevertheless, use of traditional medicine has changed dramatically over the years due to its affordability, availability, accessibility and acceptability (WHO, 2008).

Psychoactive Plants

References to the use of psychoactive plants (mind swings) in literature dates to prehistoric times, and these are still the most widely used group of pharmacological agents. Infact, psychoactive chemical substances affect the central nervous system mainly by influencing the release of neurotransmitters or mimicking their actions. Based on their effects, they can be classified as stimulants (excite and enhance mental alertness and physical activity; reduce fatigue and suppress hunger), depressants (dull mental awareness, reduce physical performance, and often induce sleep or a trance-like state) and hallucinogens (produce changes in perception, thought and mood, often inducing a dream-like state (Levetin & McMohan, 2003). Different species and cultivars of psychoactive plants, as well as varying admixtures and doses, are used to obtain different intensities and qualities of psychoactive experience, depending upon the individual's goals (Shephard, 1998).

Memory Boosters

Memory (cognition) is the ability of the brain to encode, store and retrieve information (Suba *et al.*, 2002). Factors like aging and Alzheimer's disease (AD) leading to memory loss has emerged as a major concern of modern scientists. The term memory enhancers refers to a broad range of substances including drugs, nutrients and herbs that improve impaired human cognitive abilities (the function and capacities of the brain). The Indian system of medicine is replete with medicinal plants claimed to promote learning, memory and intelligence (Satyavati, 1993). Consumption of plants like *Albizzia lebbeck, Azadirachta indica, Bacopa monniera, Desmodium gangesicum, Ginseng, Hypericum perforatum, Ocimum sanctum, Vitis vinifera, Withania somnifera* is believed to enhance memory and intelligence (Shivakumar *et al.*, 2011). Information on these aspects recorded in traditional system would certainly help in the management of cognitive disorders.

Tonic Plants

Medicines or agent that strengthen and invigorates or increase the feeling of wellbeing are referred to as tonics. The earliest known usage of tonic in English dates from the 17th century. Generally, tonics do not over-stimulate the body. They are bidirectional, capable of both increasing and decreasing the activity of body processes. The regular use of botanicals by traditional societies is an important aspect of nutrition and needs to be exploited for its practical application for overall well being of mankind.

Overview of Current Status of Work

The earliest credit to document the scientific study of Indian indigenous drugs goes to Jones (1799) who recorded their use in his book entitled *Botanical Observations on Select Plants,* followed by other valuable contributions of the 19th century, viz., *A Catalogue of Indian Medicinal Plants and Drugs* (Fleming, 1810), *Materia Medica of Hindoostan* (Ainslie, 1813), *Flora Indica* (Roxburgh, 1832), *An Essay on the Antiquity of Hindu Medicine* (Royle, 1837), *The Bengal Dispensatory* (the first ever book dealing exclusively with the properties and use of medicinal plants; O'Shaugnessy, 1841), *Pharmacographia* (Fluckier & Handburg, 1874), *Materia Medica of the Hindus* (Dutta, 1877), *Vegetable Materia Medica of Western India* (Dymock,

1883), *The Bombay Materia Medica and Their Therapeutics* (Khory, 1887), *Indegenous Drugs of India* (Dey, 1896), *Pharmacographia Indica* (Dymock *et al.*, 1889-1893), and *Dictionary of Economic Products of India* (Watt, 1889-1896). But the study of indigenous drugs gained momentum with the publication of Chopra's book *Indigenous Drugs of India* in 1933.

Other works which added to the wealth of literature on medicinal plants pertain to compilations like *Indian Materia Medica* (Nadkarni, 1926), *Medicinal and Economic Aspects of Some Indian Medicinal Plants* (Chopra, 1932), *Pharmacognosy of Indian Root and Rhizome Drugs* (Dutta, 1949), *Pharmacognosy of Indian Leaf Drugs* (Datta & Mukerji, 1952), *Review of Work on Indian Medicinal Plants* (Chopra & Chopra, 1955), *Medicinal Plants of India* (Satyavati *et al.,* 1976), *Herbal Drugs in Indian Pharmaceutical Industry* (Kapoor, 1979), *Marketing Directory of Medicinal Plants* (Kumar, 1979), *Cultivation and Utilisation of Medicinal Plants* (Atal, 1982), *Drug Plants of Indian: Root Drugs* (Agarwal, 1985), *Medicinal Chemistry Research in India* (Singh, 1985), *Current Research on Medicinal Plants* (Dhawan, 1986), *Drug Plant Sources of Central India* (Srivastava, 1989), *Medicinal Plants of India* (Jain & De Filipps, 1991), *Recent Advances in Medicinal, Aromatic and Spice Crops* (Raychaudhari, 1991), *Glossary of Indian Medicinal Plants with Active Principles* (Asolkar *et al.*, 1992), *History of Indian Medicine* (Mukhopadhyaya, 1992), *Production Technology of Medicinal and Aromatic Crops* (Farooqi, 1993), *Tropical Indian Medicinal Plants Propagation Methods* (Ved, 2000), and *Cultivation Practices of Some Commercially Important Medicinal Plants* (Rawat, 2002). Not only this, a large number of medicinal flora of different states of the country, namely Arid zone (Chopra, 1960), A.P. (Raghunathan, 1976; Anonymous, 1999), Garhwal Himalayas (Uniyal, 1989), Gwalior (Singh, 1984), H.P. (Chauhan, 1999), Kachchh – Gujarat (Goyal, 1999), Kerala (Anonymous, 1981; Nambiar, 1985), M.P. (Tiwari, 1998), Nilgiris (Krishnamurthy, 1953), North East India (Kumar, 2002), Patalkot (Saxena, 1971), Saharanpur (Ahuja, 1965), Sikkim Himalayas (Biswas & Chopra, 1982; Rai, 1994), S. India (Sundararaj, 1959), Uttarakhand Himalayas (Uniyal, 1997) and research publications in the ethnobotanical (Jain *et al.*, 1984; Jain, 2002) and floristic studies (Samant *et al.*, 1996, 2001; Dhar *et al.*, 1997, 1998; Purohit & Samant, 1995; Aswal & Mehrotra, 1994; Atkinson, 1982; Pande *et al.*, 1990) have also enriched our knowledge on the subject. More importantly, two directories – *Herbal Drugs Used in Pharmaceutical Industry* (Kapoor & Mitra, 1979), and *Directory of Crude Drugs and Aromatic Plants, Dealers, Producers and Exporters in India* (Shah & Virmani, 1982); two bibliographies, viz., *Bibliography on Investigated Indian Medicinal Plants* (1950-1975) (Iyenger, 1977) and *Bibliography on Pharmacognosy of Medicinal Plants* (Mitra, 1985); and review article of Mehra *et al.*, (1969) on *Research in Pharmacognosy in India* and that of Mitra & Jain (1991) provided a brief history and evolution of work on indigenous drugs, and establishment of various medicinal centres in India whereas that of Prakash (1998) highlights the current status of knowledge on Indian Medicinal Plants in relation to their use in Ayurveda, ethnomedicine and modern medicine, and also on the efforts of researchers towards the development of modern medicines.

In the last half century, psychoactive plants have attracted worldwide attention as evident from notable publications of Wasson (1967) – *Soma–Divine Mushroom of Immortality*, Furst (1972, 1976) – *Flesh of the Gods, Hallucinogens and Culture*, Merlin (1972, 1984) – *Man and Marijuana, On the Trail of Ancient Opium Poppy*, Schulte's & Hoffmann (1980, 1987) – *Botany and Chemistry of Hallucinogens, Plants of the Gods* and Wilbert (1988) – *Tobacco Shamanism in South America*. From Indian perspective however, Jain (2009) provided valuable information on psycho-active properties of over 300 plants in his book entitled *Plants Affecting Human Mind* (Psychoactive Plants). Besides, contributions like Vegetable psychotropic drugs of India (Singh & Trivedi, 1978), therapeutic role of ayurvedic herb in mental disorders (Upadhyay, 1986), the pharmacological basis of herbal drugs acting on CNS (Dandiya, 1990), leads from Ayurveda on medicinal plants acting on the nervous system (Satyavati, 1993), some potential psychoactive plants of India (Shukla *et al.*, 1997), and reviews on antidepressant plants (Dhingra & Sharma, 2006; Dhamija *et al.*, 2011) and CNS disorders (Kumar, 2006; Husain *et al.*, 2007) have further enriched our knowledge on the subject.

For aspects like memory boosters, Bhowmik (2010) dealt with traditional Indian memory enhancers and their medicinal importance whereas Miland & Nirmal (2004) published information on animal models for testing memory. Also, information on memory enhancing properties of *Clitorea ternatea* (Rai *et al.*, 2000), Abana – an Indian ayurvedic poly-herbal formulation (Parle & Vasudevan, 2006), *Convolvulus pleuricaulis* (Nahata *et al.*, 2008; Sethiya *et al.*, 2009) and *Bacopa monnieri* (Dubey *et al.*, 1994; Sudkarani *et al.*, 2011) and memory enhancers (Bharkatiya *et al.*, 2006) have been provided.

Comparatively, published literature pertaining to Indian tonic plants is scanty. Rai (1987) described tonic plants of Patalkot and Tamiya (Distt. Chhindwara) M.P. Work of Perri (2009) exclusively dealt with tonic plants for pregnancy. Information pertaining to the use of *Chlorophytum borivilianum* and *C. arundinaceum* as vital tonics in Indian system of medicine (Phurailatpam *et al.*, 2009) and *Centella asiatica*: a brain tonic at our disposal (Kumar, 2011) have also been described.

From the above foregoing, it is evident that no attempt has been made so far to compile systematically widely scattered large mass of Indian data on plants used for mood swings, tonics and boosting memory. The present work is an earnest attempt in this direction, and it is hoped that this attempt though belated, will fill the gap adequately.

Enumeration

For collecting data of interest on tonic, memory boosters and mood swing herbal resources of India and Nepal, scrutiny of outdated classical material, viz., *Vegetable Materia Medica of Western India* (Dymock, 1883), *A Dictionary of Economic Products* (Watt, 1889-1896), *Indigenous Drugs of India* (Dey,

1896), *Materia Medica of Indian and Their Therapeutics* (Katark, 1903), *Vegetable Drugs of India* (Sanyal, 1984), *Materia Medica* (Ainslie, 1926), *Indian Materia Medica* (Nadkarni, 1926), *Commercial Drugs of India* (Dutt, 1928), *Vegetable Drugs of India* (Robert, 1932), *Pharmacopoeia Indica* (Bose, 1932), *Indigenous Drugs of India* (Chopra, 1933), *Indian Medicinal Plants* (Kirtikar & Basu, 1935), *The Wealth of India: A Dictionary of Indian Raw Materials and Industrial Products* (Anonymous, 1948-1976), *Pharmacognosy of Indian Root and Rhizome Drugs* (Datta, 1949), *Glossary of Indian Medicinal Plants* (Chopra *et al.*, 1956), *Medicinal Plants of Arid Zone* (Chopra, 1960), *Supplement to Glossary of Indian Medicinal Plants* (Chopra *et al.*, 1974), *Dictionary of Economic Products of India* (Maheshwari & Singh, 1965), *Medicinal Plants of India & Pakistan* (Dastur, 1970); scarcely available standard works (Atkinson, 1982; Singh *et al.*, 1983; Dutta, 1985; Agarwal, 1986; Ambasta, 1986; Singh, 1996; Uphof, 1986; Nayar *et al.*, 1989; Thakur *et al.*, 1989; Jain, 1991; Jain & De Filipps, 1991; Jain *et al.*, 1991; Chatterjee & Pakrashi, 1997; Ambasta *et al.*, 1992; Hussain, 1992; Saklani & Jain, 1994; Warrier *et al.* 1994-1996; Pandey, 1995; Maheshwari, 1996; Rao & Henry, 1996; Varghese, 1996; Kaul, 1997; Sharma & Sood, 1997; Vedavathy *et al.*, 1997; Rawat & Chaudhary, 1998; Singh & Pandey, 1998; Chauhan, 1999; Kumar, 1999; Joshi, 2000; Maheshwari, 2000; Bhattacharjee, 2001; Sood *et al.*, 2001; Kumar, 2002; Caius, 2003; Sood & Thakur, 2004; Sood *et al.* 2005 a-c; Prajapati *et al.*, 2006; and a large number of research papers published in various journals on economic botany and ethnobotany, was undertaken as per the standard methodology outlined by Schultes (1960), Jain (1987, 1989) and Jain & Mudgal (1999). Wherever, possible, data gathered from own field ethnobotanical investigations were also included; and such plants have been marked with asterisk under the present observations. Besides, the information on English, Hindi and Sanskrit names, distribution and active constituents was largely based upon Hooker (1872-1897), Rastogi & Mehrotra (1998, 1999a, b, 2001, 2002) and Asolkar *et al.* (1992). The relevant informations are recorded in specific proforma, designed by modifying that of Saklani & Jain (1994).

Altogether systematic enumeration of 1000 species of plants possessing tonic, mood swings and memory boosting properties has been presented alphabetically with updated botanical names (nomenclature based upon Bennet, 1986) and synonyms (wherever necessary), followed by the name of the family, English, Hindi and Sanskrit names, distribution, part (s) used, utilization, active constituents and relevant references, colour photographs and 9 tables have been also included to explain the textural matter. Appendices include index to families, genera and species of plants used for mood swings, tonic purposes and boosting memory (Appendix I), index to families of mood swings, tonic and memory boosters (Appendix II), index to plant part/s used for tonic purposes (Appendix III), index to plant part/s used for mood swings (Appendix IV) and index to plant part/s used as memory boosters (Appendix V).

Psychoactive, Memory Boosters and Tonic Plants of Economic Significance

1. ***Abelmoschus esculentus*** (L.) Moench (Pl. 1A)

   Syn.: Hibiscus esculentus L.; *H. longifolius* Roxb.

   Family: Malvaceae.

   English, Hindi and Sanskrit Names: **Eng. :** Gumbo, Lady's Finger, Okra; **Hindi :** Bhindi, Bhindi-tori, Katavandai, Ramturai; **Sans. :** Asrapatraka, Bendokaya, Dindish, Gandhamula, Pichhila.

   Distribution : Throughout India, cultivated in areas upto 1,200m.

   Parts Used : Seeds.

   Flowering & Fruiting : June – December.

   Folk Use : Seeds used as stimulant.

   Active Constituents : Gossypol, thiamine, riboflavin, α-tocopherol, myristic, palmitic, palmitoleic, oleic, linolenic acids and tocopherols characterized from *seeds.*

   References : Farooq (2005), Kaushik & Dhiman (2000), Kirtikar & Basu (1984), Prajapati *et al.* (2006), Singh & Srivastava (2000).

2. ***Abelmoschus moschatus*** L.

   Syn.: A. rugosus Wight. & Arn.; *Hibiscus abelmoschus* L.

   Family: Malvaceae.

   English, Hindi and Sanskrit Names: **Eng. :** Ambrette Plant, Musk Mallow; **Hindi :** Latakasturi, Muskdana; **Sans. :** Latakasturikam.

   Distribution : Cultivated throughout India.

   Parts Used : Seeds.

   Flowering & Fruiting : Flowering: August – November; Fruiting: October – January.

   Folk Use : Seeds used as a tonic and stimulant.

   Active Constituents : Seeds rich in β-sitosterol and its glucoside, campesterol, stigmasterol, cholesterol and ergosterol.

References : Ambasta (1986), Dwivedi (2007), Kirtikar & Basu (1984), Kumar & Lal (2003), Prajapati *et al.* (2006), Sharma (2004), Singh *et al.* (2007).

3. ***Abies spectabilis*** (D. Don) Spach.

Syn.: A. densa Griffith; *A. webbiana* Lindl.; *Pinus tinctoria* Wall.; *P. webbiana* Wall.

Family: Pinaceae.

English, Hindi and Sanskrit Names: **Eng.** : East Himalayan Silver Fir, Indian Silver Fir; **Hindi** : Talishpatra; **Sans.** : Talishpatra.

Distribution : Himalayas (Kashmir – Assam): 1,600-4,500m; also grown ornamentally.

Parts Used : Leaf.

Flowering & Fruiting : April – May.

Folk Use : Juice of leaves administered as a tonic in Bengal in cases of parturition.

Active Constituents : Leaves yield betuloside, methyl betuloside, n-triacontanol and β-sitosterol (*leaf*).

References : Asolkar *et al.* (1992), Dhiman (2005), Kirtikar & Basu (1984), Sharma (2003), Watt (1972).

4. ***Abroma augusta*** L.F.

Syn.: Ambroma augusta L.F.

Family: Sterculiaceae.

English, Hindi and Sanskrit Names: **Eng.** : Devil's Cotton, Perennial Indian Hemp; **Hindi** : Sanukapashi, Ulatkambal; **Sans.** : Karpasa, Pisacha, Utkamal.

Distribution : Warmer parts of West Bengal, Bihar, Tripura etc.

Parts Used : Root. Root Bark.

Flowering & Fruiting : Flowering: August – September; Fruiting: January – March.

Folk Use : Roots used as uterine tonic. Root bark used as a tonic.

Active Constituents : These are α-amyrin, maslinic acid, protocatechuic, vanillic and caffeic acids in their free, glycosidic and ester forms, rhamnose, arabinose, xylose, mannose, galactose, glucose, galactouronic acid and glucuronic acid (*root bark*).

References : Ambasta (1986), Anonymous (1948), Asolkar *et al.* (1992), Kirtikar & Basu (1984), Prajapati *et al.* (2006), Thakur *et al.* (1989), Watt (1972).

5. ***Abrus precatorius*** L. (Pl. 1B)

Syn.: A. minor Desv.; *A. pauciflorus* Desv.

Family: Fabaceae.

English, Hindi and Sanskrit Names: **Eng.** : Crab's Eye, Indian Liquorice, Jequirity, Rosary Pea; **Hindi :** Chirami, Gunchi, Rati; **Sans.** : Aruna, Bhilabhushana, Durmogha, Gunja, Kanchi, Shitapaki.

Distribution : Common throughout India, ascending upto 1,060m in outer Himalayas; also found in Andaman & Nicobar islands.

Parts Used : Leaves. Root. Seeds.

Flowering & Fruiting : September – November.

Folk Use : Leaves, root and seeds used as a tonic.

Active Constituents : Active properties due to abrine, hypathorine, choline, 5β-cholanic acid, trigonelline & its gallic acid ester precatorine, methylester of N,N – dimethyltryptophanmethocation and several amino acids (*seeds*), glycyrrhizin and pinitol (*leaf*), precol, abrol, glycyrrhizin, abrasine and precasine (*root*).

References : Ambasta (1986), Asolkar *et al.* (1992), Dhiman (2005), Dwivedi (2007), Farooq (2005), Gogoi & Das (2003), Kaushik & Dhiman (2000), Kirtikar & Basu (1984), Kshirsagar *et al.* (2003), Kumar (2002), Parrotta (2001), Prajapati *et al.* (2006), Sharma (2003), Singh & Srivastava (2000), Thakur *et al.* (1989), Trivedi (2006), Watt (1972).

6. *Abutilon indicum* (L.) Sweet

Syn.: A. asiaticum (L.) Sweet; *Sida guineenis* Schumach.; *S. indica* L.

Family: Malvaceae.

English, Hindi and Sanskrit Names: **Eng.** : Country Mallow; **Hindi :** Jhampi, Kanghi, Tepari; **Sans.** : Atibala, Shita, Vatyapushpika.

Distribution : Weed in the Sub-Himalayan tracts & hills upto 1,200m, hotter parts of India & throughout tropics.

Parts Used : Root.

Flowering & Fruiting : July – December.

Folk Use : Roots used as a nervine tonic.

Active Constituents : Roots possess asparagin.

References : Ambasta (1986), Asolkar *et al.* (1992), Bhattacharyya (1996), Dhiman (2005), Kaushik & Dhiman (2000), Trivedi (2006).

7. *Acacia catechu* (L.F.) Willd. (Pl. 1C)*

Syn.: A. catechuoides (Roxb.) Benth.; *A. polyacantha* Willd.; *A. sundra* (Roxb.) Bedd.; *A. wallichiana* DC; *Mimosa catechu* L.F.; *M. catechuoides* Roxb.

Family: Fabaceae.

English, Hindi and Sanskrit Names: **Eng. :** Black Catechu, Cutch Tree, Pegu Catechu; **Hindi :** Katha, Khadir, Khair; **Sans. :** Dantadhavana, Khadirah, Medhya, Vakrakantaka, Yajnika.

Distribution : Sub-Himalayan tracts and semi-arid areas of Central & South India.

Parts Used : Bark. Heart Wood.

Flowering & Fruiting : May – July.

Folk Use : Katha obtained by boiling its heartwood given to women along with gum guggul as tonic after childbirth. Bark also used as a tonic.

Active Constituents : α, β, γ-catechin and 1-epicatechin (*wood*).

References : Deshmukh *et al.* (2000), Kirtikar & Basu (1984), Prajapati *et al.* (2006), Reddi *et al.* (2005), Thakur *et al.* (1989), Watt (1972).

8. ***Acacia nilotica*** (L.) Willd. ex Del.

Syn.: A. arabica auct. non (Lam.) Willd.; *A. vera* Willd.; *Mimosa arabia* Lam.; *M. nilotica* L.

Family: Fabaceae.

English, Hindi and Sanskrit Names: **Eng. :** Black Babool, Indian Gum Arabic Tree; **Hindi :** Babul, Kikar; **Sans. :** Babbula, Sukshmapatra, Yugmakanta.

Distribution : Dry areas of the country.

Parts Used : Bark. Flower.

Flowering & Fruiting : July – September.

Folk Use : Gum obtained from decoction of bark used as a nutritive tonic, particularly in sexual debility and for pregnant ladies. Flowers used as a tonic.

Active Constituents : Bark rich in quercetin, catechol, gallic acid, (+)-catechin, (-) epicatechin, (-) epigallocatechin-5,7-digallate, chlorogenic acid and m-digallic acid.

References : Asolkar *et al.* (1992), Kirtikar & Basu (1984), Mishra (2008), Parrotta (2001), Prajapati *et al.* (2006), Thakur *et al.* (1989), Watt (1972).

9. ***Acacia senegal*** (L.) Willd.

Syn.: A. rupestris Stocks; *A. verek* Guill. & Perrot; *Mimosa senegal* L.; *M. senegalensis* Lam.

Family: Fabaceae.

English, Hindi and Sanskrit Names: **Eng. :** Gum Arabic Tree; **Hindi :** Kumat, Kumatiyo, Kumttha; **Sans. :** Babul, Svetakhadira.

Distribution : Punjab, Rajputana, Sind, Baluchistan; Arabia and tropical Africa.

Parts Used : Stem Bark (Gum).

Flowering & Fruiting : August – December.

Folk Use : Gum Arabic obtained from stem bark used as tonic.

Active Constituents : Sugars, d-galactose, l-arabinose, l-rhamnose and d-glucoronic acid characterized from *gum*.

References : Bhatt (2002), Bhatt & Mitaliya (1999), Bhatt *et al.* (2003).

10. *Acalypha indica* L.

Syn.: A. canescana L.; *A. ciliata* L.; *A. spicata* L.

Family: Euphorbiaceae.

English, Hindi and Sanskrit Names: **Eng. :** Indian Acalypha; **Hindi :** Khokli, Kuppi; **Sans. :** Arittamunjariye, Harita-manjari.

Distribution : Throughout the plains of India.

Parts Used : Whole Plant.

Flowering & Fruiting : June – December.

Folk Use : Plant has hypnotic properties.

Active Constituents : These are kaempferol, sitosterol, triacetonamine (*plant*), acalyphine, acalypus, acalyphamide, quinine, sterols and cyanogenic glycoside (*leaf, twigs*).

References : Ambasta (1986), Prajapati *et al.* (2006).

11. *Acampe praemorsa* Blatter & McCann

Syn.: A. wightiana Lindl.; *Saccolabium praemorsum* Hook. f.; *S. wightianum* Hook. f.; *Vanda wightiana* Lindl. *ex* Wight.

Family: Orchidaceae.

English and Sanskrit Names: **Eng. :** Acampe Orchid, Brittle Orchid; **Sans. :** Gandhanakuli, Rasna.

Distribution : W. Bengal, lower Himalayan region, Assam, Karnataka and Maharashtra.

Parts Used : Whole Plant.

Flowering & Fruiting : March.

Folk Use : Plant used as a bitter tonic.

References : Ambasta (1986), Kirtikar & Basu (1984).

12. *Acanthopanax trifoliatus* L.

Syn.: A. aculeatum Seem.

Family: Araliaceae.

Distribution : Wild in mountaineous regions & forests of eastern Asia, Malaysia.

Parts Used : Root Bark. Stem Bark.

Flowering & Fruiting : September – November.

Folk Use : Drug prepared from root bark and stem bark consid-ered as CNS stimulant and tonic for improving memory.

References : Khatoon & Goel (2010), Prajapati *et al.* (2006).

13. *Achillea millefolium* L.

Syn.: A. lanulosa Nutt.

Family: Asteraceae.

English and Hindi Names: **Eng. :** Milfoil, Yarrow; **Hindi :** Biranjasif, Gandana, Rojmari.

Distribution : Western Himalayas (Kashmir – Kumaon) : 1,066-3,657m, especially around Shimla.

Parts Used : Whole Plant. Flower. Leaf.

Flowering & Fruiting : August – December.

Folk Use : Powdered plant, leaves and flower heads used as a tonic. Plant and decoction of leaves considered stimulant. Flowers used as brain tonic (Yunani).

Active Constituents : Properties due to rutin, apigenin, cosmosin, luteolin and its glucoside (*leaf*).

References : Ambasta 1986), Anonymous (1948), Asolkar *et al.* (1992), Bhattacharjee (2001), Dhiman (2005), Kala (2003), Kaul (1997), Kirtikar & Basu (1984), Prajapati *et al.* (2006), Shah (2010), Sharma (2003), Thakur *et al.* (1989), Watt (1972).

14. *Achyranthes aspera* L.

Syn.: Aerva aspera Spreng.; *Digera muricata* Mart.

Family: Amaranthaceae.

English, Hindi and Sanskrit Names: **Eng. :** Rough Chaff Flower; **Hindi :** Apang, Chirchira, Latjira; **Sans. :** Aghamargava, Malakanta, Mayurata, Sthalamangari.

Distribution : Tropical & sub-tropical regions of India upto 1,000m.

Parts Used : Seeds. Stem.

Flowering & Fruiting : May – October.

Folk Use : Seeds used as brain tonic; also seeds boiled with milk used as tonic for growth. Fresh stem twigs used for retentive memory by local inhabitants of Gujarat.

Active Constituents : Known to possess saponin A, saponin B and saponin β-methyl ester, hentriacontane, 10-octacosanone, 10-triacosanone and 4-tritriacontanone (*seeds*).

References : Bhattacharjee (2001), Khatoon & Goel (2010), Singh & Pandey (1998).

15. *Aconitum bisma* (Ham.) Rapaics

Syn.: A. palmatum D. Don.; *Caltha bisma* Hamilt.; *Nirbisia bisma* G. Don.

Family: Ranunculaceae.

English, Hindi and Sanskrit Names: **Eng.** : Aconite, Crowfoot; **Hindi** : Bikhma; **Sans.** : Prativisha.

Distribution : Alpine Himalayas of Sikkim and the adjoining part between 3,048-3,487m.

Parts Used : Root.

Flowering & Fruiting : July – August.

Folk Use : Roots possess tonic properties.

Active Constituents : Five diterpene alkaloids – vakognavine, palmatisine, vakatisine, vakatisinine and vakatidine (*root*).

References : Ambasta (1986), Dhiman (2005), Kirtikar & Basu (1984).

16. *Aconitum chasmanthum* Stapf. ex Holmes

Syn.: A. dissectum Watt.; *A. hians* Watt.

Family: Ranunculaceae.

English, Hindi and Sanskrit Names: **Eng.** : Aconite, Indian Napellus; **Hindi** : Bachhnaag, Banbalnag; **Sans.** : Vatsanagaka, Visa, Vajranaga.

Distribution : Sub-alpine and alpine Himalayas: 2,133-3,657m.

Parts Used : Tuberous Roots.

Flowering & Fruiting : Flowering: July – August; Fruiting: September – October.

Folk Use : Tuberous roots used as sedative.

Active Constituents : Attributed properties due to chasmanine, indacontine and new alkaloids – base A, base B, chasmaconitine and chasmanthinine (*root*).

Reference : Dhiman (2005).

17. *Aconitum deinorrhizum* Stapf.

Syn.: A. atrox Goris; *A. ferox* var. *laciniatum* Watt.

Family: Ranunculaceae.

English, Hindi and Sanskrit Names: **Eng.** : Indian Aconite; **Hindi** : Safed Bish; **Sans.** : Vatsnabh.

Distribution : Common in the sub-alpine and alpine zones of the Himalayas up to 4,500m.

Parts Used : Leaf. Root.

Flowering & Fruiting : July – September.

Folk Use : Leaves and roots used as a tonic in diarrhoea and cough.

Active Constituent : Roots rich in psuedacinitine.

References : Prajapati *et al.* (2006), Sharma (2003).

18. *Aconitum ferox* Wall. ex Ser.

Syn.: A. dissectum Stein.; *A. virosum* D. Don.

Family: Ranunculaceae.

English, Hindi and Sanskrit Names: **Eng.** : Indian Aconite; **Hindi** : Bacchang, Bish, Mahoor, Mitha Zahar; **Sans.** : Vatsanava, Visha.

Distribution : Wild in alpine Himalayas and in Kashmir upto 3,600m.

Parts Used : Root.

Flowering & Fruiting : Flowering: July – August; Fruiting: October – December.

Folk Use : Alkaloids extracted from roots used as sedative.

Active Constituents : Active properties due to picro-aconine, aconine, benzyl aconine, homo-napalline, bikhaconitine, chasmaconitine, indaconitine, ψ aconitine, veratroyl-ψ-aconine and di-Ac-ψ-aconitine (*root*).

References : Ambasta (1986), Anonymous (1948), Dhiman (2005), Prajapati *et al.* (2006), Watt (1972).

19. *Aconitum heterophyllum* Wall. ex Royle

Syn.: A. atees Royle; *A. cordatum* Royle; *A. ovatum* Lindl.

Family: Ranunculaceae.

English, Hindi and Sanskrit Names: **Eng.** : Atis Root, Indian Atis; **Hindi** : Atish, Patis; **Sans.** : Ativisha, Bhangura, Kashia, Pravisha, Shringi, Vira, Virupa, Visha.

Distribution : Common in alpine and sub-alpine belts of Himalayan north-western regions between 1,800-4,500m.

Parts Used : Root.

Flowering & Fruiting : Flowering: July – August; Fruiting: October – December.

Folk Use : Roots used as a bitter tonic.

Active Constituents : Known to contain atisine, atidine, histisine, hetisine, hetidine, heterophyllisine, heterophylline, isoatisine, dehydroatisine, hetisinone and benzoyl heteratisine (*root*).

References : Ambasta (1986), Anonymous (1948), Dhiman (2005), Farooq (2005), Hemlata *et al.* (2007), Kala (2003), Kapur & Singh (1996), Kaul (1997), Kirtikar & Basu (1984), Kumar (2002), Nadeem *et al.* (2001), Nautiyal *et al.* (2003), Prajapati *et al.* (2006), Rawat *et al.* (2007), Thakur *et al.* (1989), Uniyal & Malhotra (1989), Watt (1972).

20. ***Aconitum koreanum*** R. Rkaymond

Family: Ranunculaceae.

English Name: **Eng.** : Korean Monk's Hood.

Distribution : Grassy areas in mountain valleys or on slopes.

Parts Used : Rhizome.

Flowering & Fruiting : July – August.

Folk Use : Rhizome used as a cardiotonic.

Active Constituents : Attributed properties due to hypaconitines A,B,C,D,E, aconitine, mesaconitine, atisine, higenamine, coryneine, isoatisine, and 19-epi-isoatisine (*rhizome*).

Reference : Prajapati *et al.* (2006).

21. ***Aconitum napellus*** L.

Family: Ranunculaceae.

English, Hindi and Sanskrit Names: **Eng.** : Helmet Flower, Indian Aconite, Monk's Hood; **Hindi** : Bachnag, Mithavis; **Sans.** : Vatsanabhah, Visha.

Distribution : Alpine and sub-alpine belt of the Himalayas.

Parts Used : Whole Plant. Root.

Flowering & Fruiting : June – August.

Folk Use : Plant and roots used as narcotic and nervine tonic. Roots also used as sedative.

Active Constituents : Constituents isolated are aconitine, mesaconitine, neoline, songoramine, deoxyaconitine, hypaconitine and napelline (*root*).

References : Dhiman (2005), Prajapati *et al.* (2006), Reddi *et al.* (2005).

22. ***Aconitum violaceum*** Stapf.

Syn.: A. multifidum Royle; *A. nepellus* var. *dissectum* Duthie.

Family: Ranunculaceae.

English, Hindi and Sanskrit Names: **Eng.** : Patis; **Hindi** : Dudhiabis, Mitha Telia, Telia Kachnag, Tilla; **Sans.** : Prativishaa, Shyaamkandaa.

Distribution : Open slopes of sub-alpine and alpine meadows of cold arid region from Kashmir – Uttaranchal Himalayas between 3,400-4,500m.

Parts Used : Root.

Flowering & Fruiting : August – November.

Folk Use : Roots used as tonic.

Active Constituents : Active properties due to indaconitine and bikhaconitine (*root*).

References : Ambasta (1986), Kaul (1997).

23. *Acorus calamus* L.

Family: Araceae.

English, Hindi and Sanskrit Names: **Eng.** : Calamus, Flag Root, Myrtle Root, Sea Sedge, Sweet Cinnamon, Sweet Flag, Sweet Root, Sweet Rush; **Hindi** : Bach, Bhoja, Gorbacc; **Sans.** : Bhadra, Jalaja, Ugragandha, Vacha.

Distribution : Marshy places almost throughout India, ascending Himalayas upto 1,828m in Sikkim.

Parts Used : Rhizome.

Flowering & Fruiting : Flowering: May – June; Fruiting: July – August.

Folk Use : Rhizome used as bitter tonic, nervine tonic and tonic; also eaten alongwith sugar candy to improve memory.

Active Constituents : Rhizome yields α-asarone, β-asarone [1-3], calamol, calamine [4], luteolin-6,7-diglucoside.

References : Ambasta (1986), Anonymous (1948), Asolkar *et al.* (1992), Bhattacharjee (2001), Dhiman (2005), Dwivedi (2007), Farooq (2005), Kaul (1997), Kirtikar & Basu (1984), Kumar (2002), Nadeem *et al.* (2001), Parrotta (2001), Prajapati *et al.* (2006), Rajan *et al.* (2003), Rana *et al.* (1996), Sharma (2004), Thakur *et al.* (1989), Upadhyay (2003), Watt (1972).

24. *Acorus gramineus* Solander

Family: Araceae.

English, Hindi and Sanskrit Names: **Eng.** : Gross-leaved Sweet Rush, Japanese Sweet Flag; **Hindi** : Bajai, Vasa Bach; **Sans.** : Vacha.

Distribution : Grows wild in Sikkim Himalayas up to 2,000m and Khasia hills.

Parts Used : Rhizome.

Flowering & Fruiting : May – August.

Folk Use : Rhizome possesses sedative properties.

Active Constituents : Asaron, as aryl aldehyde and a bitter glucoside, acorin have been isolated (*rhizome*).

References : Kirtikar & Basu (1984), Prajapati *et al.* (2006).

25. *Actaea spicata* L.

Syn.: A. acuminata Wall. *ex* Royle; *A. racemosa* Gilib.

Family: Ranunculaceae.

English Names: **Eng.** : Grapewort, Herb Christopher.

Distribution : Native to Britain; occurs in temperate Uttaranchal Himalayas between 2,500-3,500m.

Parts Used : Root.

Flowering & Fruiting : Flowering: May – June; Fruiting: July – August.

Folk Use : Root possesses nerve sedative properties.

References : Anonymous (1948), Sharma (2003).

26. *Actinopteris australis* (L.F.) Link

Syn.: A. dichotoma Kuhn.; *A. radiata* Link.

Family: Actinopteridaceae.

English, Hindi and Sanskrit Names: **Eng. :** Peacock's Tail; **Hindi :** Morpankhi, Morpath; **Sans. :** Mayurshikha.

Distribution : Deccan Peninsula below 1,300m, Nilgiris up to 600m, Kumaon hills.

Parts Used : Rhizome.

Flowering & Fruiting : June – December.

Folk Use : Rhizome used as tonic.

Reference : Singh & Viswanathan (1996).

27. *Adiantum capillus-veneris* L. (Pl. 1D)

Family: Adiantaceae.

English, Hindi and Sanskrit Names: **Eng. :** Maidenhair Fern, Maria's Fern, Our Lady's Hair; **Hindi :** Hansraj, Mubaraka, Pursha; **Sans. :** Hanspadi.

Distribution : Common in south India, Maharashtra, Punjab & Bihar; also occurs mainly in western Himalayas ascending upto 2,400m and extending to Manipur.

Parts Used : Whole Plant. Fronds.

Flowering & Fruiting : Sori formed throughout the greater part of the year.

Folk Use : Fern and its fronds used as tonic.

Active Constituents : Isolated components are 3α, 4α-epoxyfilicane, 21-hydroxyadiantone, adiantone, adiantoxide, astragalin, isoquercitrin, nicotiflorin, kaempferol-3-glucuronide, rutin and querciturone (*plant*), β-sitosterol, stigmasterol and campesterol (*fronds*).

References : Ambasta (1986), Dhiman (2005), Sharma (2004), Singh & Viswanathan (1996).

28. *Adiantum incisum* Forsk

Syn.: A. caudatum L.

Family: Adiantaceae.

English, Hindi and Sanskrit Names: **Eng. :** Tailed Maiden Hair; **Hindi :** Marshikha; **Sans. :** Nilkantha Shikha.

Distribution : Known to occur in lower slopes of hills, old walls and shady places of Maharashtra, Rajasthan, Punjab, Tamil Nadu and West Bengal.

Parts Used : Fronds.

Flowering & Fruiting : June – December.

Folk Use : Fronds used as a tonic.

References : Kaushik & Dhiman (2000), Singh & Viswanathan (1996), Subramaniam (2000) Watt (1972).

29. *Adiantum pedatum* L.

Family: Adiantaceae.

English Names: **Eng. :** Canadian Maidenhair, Five-fingered Fern, Northern Maidenhair Fern.

Distribution : N.W. Himalaya (Garhwal – Sikkim) : 2,000-3,000m; Japan, N. America.

Parts Used : Rhizome.

Folk Use : Rhizome used as stimulant and tonic.

References : Singh (2000b), Singh & Viswanathan (1996).

30. *Adiantum vensutum* G. Don

Family: Adiantaceae.

English, Hindi and Sanskrit Names: **Eng. :** Black Maidenhair Fern, Venus Hair; **Hindi :** Hansraj, Kalijhanp; **Sans. :** Hansapadi.

Distribution : N.E. Himalayas, Kashmir and in Shimla : 1,350-3,150m.

Parts Used : Whole Plant. Fronds.

Flowering & Fruiting : Sori formed in winters and spring.

Folk Use : Herb and fronds used as tonic.

Active Constituents : Plant yields adiantone, traces of 3-filicane, new ketol-21-hydroxy-30-norhopan-22-one (I), triterpenoid keto-alcohol, α-carotene monoepoxide, leucopelargonidin, kaempferol and quercitin glucosides.

References : Farooq (2005), Kaushik & Dhiman (2000), Kirtikar & Basu (1984), Sharma (2003), Singh & Viswanathan (1996), Watt (1972).

31. *Adonis aestivalis* L.

Family: Ranunculaceae.

English Names: **Eng. :** Pheasant's eye, Summer Adonis, Summer Pheasant's Eye.

Distribution : W. Himalaya; temperate Europe and Asia.

Parts Used : Whole Plant.

Flowering & Fruiting : April – July.

Folk Use : Plant used as cardiotonic.

Active Constituents : Aerial parts possess cymarin, two new cardenolides – strophanthidin-3-O-β-D-digitixose (4→1) diginose (4→1)-glucoside along with 3-epiperiplogenin and helveti-coside.

References : Ambasta (1986), Kala (2003), Kirtikar & Basu (1984).

32. ***Aegle marmelos*** Correa ex Roxb.*

Syn.: Belou marmelos (L.) Lyons; *Bilacus marmelos* (L.) Kuntze; *Crateava marmelos* L.; *C. reliogosa* Ainstie.; *Feronia pellucida* Roth.

Family: Rutaceae.

English, Hindi and Sanskrit Names: **Eng. :** Bael Tree, Bengal Quince, Holy Fruit Tree; **Hindi :** Bel, Bilwa, Sirphal; **Sans. :** Sivadrumah, Sriphalah, Vilvah.

Distribution : Dry hilly areas throughout India, gardens and road sides; also cultivated in various places in India.

Parts Used : Fruit. Root Bark.

Flowering & Fruiting : Flowering: February – March; Fruiting: September – December.

Folk Use : Fruits used as tonic and its pulp considered as cardio tonic. Root bark forms an ingredient of 'Dashmool' and used as restorative tonic.

Active Constituents : These are marmelosin and abscisic acid (*fruit*), γ-fagarine, marmesin and umbelliferone (*root bark*).

References : Ambasta (1986), Dhiman (2005), Kirtikar & Basu (1984), Prajapati *et al.* (2006), Sood *et al.* (2009), Thakur *et al.* (1989), Trivedi (2006), Varghese (1996), Watt (1972).

33. ***Aerva lanata*** L. Juss. ex Schultes

Syn.: Achyranthus lanata L.; *Aerva elegans* Moq.; *Illecebrum lanatum* (L.) Muer.

Family: Amaranthaceae.

English, Hindi and Sanskrit Names: **Eng. :** Polpala; **Hindi :** Chaya, Gorakhganja, Kapurijadi; **Sans. :** Astmabayda, Bhadra.

Distribution : Wastelands throughout India.

Parts Used : Whole Plant. Root.

Flowering & Fruiting : September – March.

Folk Use : Plant and roots used as a tonic.

Active Constituents : Therapeutic properties due to α-amyrin, campesterol, β-sitosterol, β-sitosteryl palmitate, chrysin, four flavonoid glycosides, hentriacontanr, daucosterol, syringic acid, vanillic acid, feruloylhomovanillylamine, four new alkaloids – aervine, methylaervine, aervoside and aervola-nine, canthin-6-one and β-carboline-propionic acid (*plant*).

References : Parrotta (2001), Prajapati *et al.* (2006).

34. *Aesculus hippocastanum* L.

Family: Hippocastanaceae.

English and Hindi Names: **Eng. :** Conker Tree, Horse - chestnut; **Hindi :** Pu.

Distribution : N. India, occasionally cultivated in other places as an ornamental plant.

Parts Used : Bark. Flower.

Flowering & Fruiting : Flowering: April – May; Fruiting: September – October.

Folk Use : Bark regarded as tonic and narcotic. Flowers said to be tonic.

References : Kaul (1997), Sharma (2003).

35. *Ageratum conyzoides* L.

Syn.: A. cordifolium Roxb.

Family: Asteraceae.

English, Hindi and Sanskrit Names: **Eng. :** Appa Grass, Bastard Agrimony, Conyzoid Floss-flowers, Goat Weed, White Weed, **Hindi :** Nil Phool; **Sans. :** Visamustih.

Distribution : Native of America; now got naturalized throughout India, ascending upto 1,650m in the Himalayas.

Parts Used : Whole Plant. Leaf.

Flowering & Fruiting : During cold season.

Folk Use : Plant used as a nervine tonic. Leaves employed as stimulant as well as tonic in diarrhoea.

Active Constituents : Major constituents are friedelin, β-sitosterol, hydrocarbons, 2 unidentified esters, stigmasterol, α-spinas-terol, coumarin, flavones, antigonadotropic hormones and precocenes (*plant*).

References : Asolkar *et al.* (1992), Dhiman (2005), Farooq (2005), Gogoi & Das (2003), Kaushik & Dhiman (2000), Kirtikar & Basu (1984), Parrotta (2001), Prajapati *et al.* (2006).

36. *Ailanthus excelsa* Roxb.

Syn.: Pongelion wightii Tiegh.

Family: Simaroubaceae.

English, Hindi and Sanskrit Names: **Eng. :** Tree of Heaven; **Hindi :** Arua, Limbado, Maharukha; **Sans. :** Aralu, Araluka, Araluvrksa, Atarusha, Mahanimba, Pisasha.

Distribution : Indigenous to India and more common in Bihar, Gujarat, Madhya Pradesh, Orissa and South India; also cultivated in roadsides and gardens.

Parts Used : Bark. Leaf. Root Bark.

Flowering & Fruiting : Flowering: February – March; Fruiting: April – May.

Folk Use : Bark, root bark and leaf juice used as a tonic.

Active Constituents : Properties mainly due to glaucarubin, glaucarubinone, glaurubol, 15-isovelurate, 18-glaucarubol, excelsin, 13,18-dihydro excelsin, ailanthione, ailantic acid, malanthin, 2,6-dimethoxy benzquinone, β-sitosterol, canthure-6-one, 1-methoxy-canthine-6-one, 5-methoxycan-thine-one, 5-methoxycanthine-6-one, 8-hydroxycanthine-6-one, tricontane and hexatricontane (*bark*); ailanic acid, β-sitosterol and vilexin (*leaf*).

References : Anonymous (1948), Dwivedi (2007), Farooq (2005), Kirtikar & Basu (1984), Kothari & Londhe (2000), Kumar & Lal (2003), Reddi *et al.* (2005), Sharma (2003), Singh & Pandey (1998), Watt (1972).

37. *Ailanthus triphysa* (Dennst.) Alston.

Syn.: A. malabarica DC.

Family: Simaroubaceae.

English, Hindi and Sanskrit Names: **Eng. :** White Bean, White Palle, White Siris; **Hindi :** Guggul Dhup, Matti, Muddhe Dhup, Perumaram, Pongilyam; **Sans. :** Atisarahita, Mahanimbu.

Distribution : Konkan, Deccan, W. Ghats from N. Kanara and Mysore to Travancore up to 1,000m, Sri Lanka, eastern slopes of the Pegu Yoma in Myanmar; often planted in S. India.

Parts Used : Bark.

Flowering & Fruiting : Flowering: February – March; Fruiting: April – May.

Folk Use : Bark used as a tonic.

Active Constituents : Bark rich in eight β-carbaline alkaloids.

References : Ambasta (1986), Kirtikar & Basu (1984).

38. *Ajuga bracteosa* Wall. ex Benth.

Family: Lamiaceae.

English, Hindi and Sanskrit Names: **Eng. :** Bungle; **Hindi :** Kauributi; **Sans. :** Nilkanthi.

Distribution : As a weed in field borders, roadsides, open slopes and even on rock crevices upto 1,500m.

Parts Used : Whole Plant. Leaf.

Flowering & Fruiting : June – September.

Folk Use : Aromatic leaves regarded as stimulant and tonic. Plant considered as a bitter tonic and tonic.

Active Constituents : Plant known to contain ceryl alcohol, β-sitosterol, γ-sitosterol and another compound from nonsaponificable fraction, cerotic, palmitic, oleic and linoleic acids from saponifiable fraction, glucose, arabinose, phenolic, acidic and neutral bitter compounds present in alcoholic extract.

References : Anonymous (1948), Bhattacharjee (2001), Farooq (2005), Kala (2003), Kapur *et al.* (1996), Kaul (1997), Kumar & Lal (2003), Prajapati *et al* (2006), Sharma (2003), Singh (2001), Verma & Chauhan (2006), Watt (1972).

39. ***Alangium salviifolium*** (L.F.) Wang.

Syn.: A. decapetalum Lamk.; *A. lamarckii* Thw.; *A. tomentosuum* Lamk.; *Grewia salvifolia* L.F.

Family: Alangiaceae.

English, Hindi and Sanskrit Names: **Eng. :** Sage – leaved Alangium; **Hindi :** Akola, Angol, Dhela, Dhera; **Sans. :** Ankodah, Ankolah, Ankota, Bhushita, Guptasneha, Rochana.

Distribution : Throughout drier parts of India; also grows vigourously in forests of South India.

Parts Used : Bark. Fruit. Seed.

Flowering & Fruiting : Flowering: March – May; Fruiting: June – July.

Folk Use : Decoction of bark used as liver tonic whereas fruits and seeds employed as body tonic.

Active Constituents : These are benzo[α]pyrido[3,4-γ]quinolizidine and bhartamine with unoxygenated ring D derived biogenetically from loganin, emetine, cephaeline, N-Me-cephaeline, psychotrine, minor phenolic alkaloids alangamide, novel quinolizine alkaloids, alangimarine, alamarine, alang-imaridine, deoxytubulosine, betulin, betulinaldehyde, lupeol, betulinic acid and its hydroxyl lactone A, β-sitosterol (*seeds*), alangiside, cephaeline, deoxytubulosine, N-methylcephaeline, β-sitosterol (*fruit*), and lamarckinine (*bark*).

References : Ambasta (1986), Asolkar *et al.* (1992), Dhiman (2005), Hebbar *et al.* (2003), Kaushik & Dhiman (2000), Kirtikar & Basu (1984), Kumar & Goel (2008), Parrotta (2001), Raju (2000), Reddi *et al.* (2005), Watt (1972).

40. *Albizzia lebbeck* (L.) Benth.*

Syn.: Acacia lebbek (L.) Willd.; *A. macrophylla* Bunge; *A. sirisa* Ham.; *A. speciosa* (Jacq.) Willd.; *Albizzia latifolia* Boivin; *Mimosa lebbeck* L.; *M. Sirissa* Roxb.; *M. speciosa* Jacq.

Family: Fabaceae.

English, Hindi and Sanskrit Names: **Eng. :** Fry Wood Tree, Indian Walnut, Parrot Tree, Silk Flower Tree, Siris Tree, Sizzling Tree; **Hindi :** Garso, Kalshish, Lasrin, Mathirshi, Siris, Tantia; **Sans.:** Bhandi, Kapitaka, Krishnasiris, Lamashapushpa, Murdh-pushpa, Shankhiniphala, Sirisha, Vishaghati.

Distribution : Throughout India, ascending upto 1,219m in the Himalayas and in Andaman.

Parts Used : Bark. Flower. Leaf. Seeds.

Flowering & Fruiting : Flowering during rainy season and fruiting in winters.

Folk Use : Bark, flowers, leaves and seeds used as tonic.

Active Constituents : Isolated active constituents are D-catechin, friedelin, β-sitosterol, isomers of leucocyanidin, melacacidin and a new leucoantho-cyanidin debbecacidin (*bark*), vicenin-2, rcynoutrin, rutin, myricitin, robibine and albegenin (*leaves*), benzyl acetate, benzyl benzoate and crocetin (*flower*).

References : Farooq (2005), Kaushik & Dhiman (2000), Prajapati *et al.* (2006), Ranjan (2000), Sharma (2003), Singh *et al.* (2007), Siwakoti & Varma (1996).

41. *Alisma plantago-aquatica* L.

Syn.: Alisma plantago L.

Family: Alismataceae.

English Names: **Eng. :** Mad-dog Weed, Water-plantain.

Distribution : Bengal, Bihar, lower Himalayas from Kashmir – Manipur and Myanmar.

Parts Used : Root.

Flowering & Fruiting : May – July.

Folk Use : Roots used as a tonic.

References : Asolkar *et al.* (1992), Ambasta (1986).

42. *Allium cepa* L.*

Syn.: A. cepa Roxb.; *A. cepa* Wall.; *A. comaria* Herb.

Family: Liliaceae.

English, Hindi and Sanskrit Names: **Eng. :** Onion; **Hindi :** Pyaj; **Sans. :** Dirghrapatra, Mahakanda, Palanduh, Rajastha, Raktakanda.

Distribution : Cultivated throughout India.

Parts Used : Bulbs.

Flowering & Fruiting : Plant sown in cold season and harvested in April – May.

Folk Use : Bulbs used as a tonic and stimulant.

Active Constituents : Reported earlier to possess oleanilic acid, a new amino acid – cycloallin, two trisaccharides – 1-β-fructo-sylsucrose and 6-β-fructosylsucrose, arabinose, rhamnose, xylose, ribose, (+) S-methyl-L-cysteine sulfoxide and (+) S-propyl-L-cysteine sulfoxide (*bulb*).

References : Anonymous (1948), Asolkar *et al.* (1992), Kaushik & Dhiman (2000), Kirtikar & Basu (1984), Pal & Paul (2003), Prajapati *et al.* (2006), Reddi *et al.* (2005), Sharma (2004), Siwakoti & Varma (1996).

43. *Allium fistulosum* L.

Syn.: A. bouddhae O. Debeaux; *A. bakeri* Hoop. non Regel; *Porrum fistulosum* Schur.

Family: Liliaceae.

English and Hindi Names: **Eng. :** Bunching Onion, Greeen Onion, Japanese Leek, Salad Onion, Scallion, Spring Onion, Stone Leek, Welsh Onion. **Hindi :** Vilayati Lahsun.

Distribution : Extensively cultivated as a culinary herb throughout the country.

Parts Used : Bulbs.

Flowering & Fruiting : Flowering during July.

Folk Use : Bulbs used as neurostimulant.

Active Constituents : Propyl disulfide, diallyl disulfide and sulfur compounds characterized from *bulb*.

Reference : Prajapati *et al.* (2006).

44. *Allium sativum* (L.) Cav. (Pl. 1E)*

Syn.: A. sativum Roxb.

Family: Liliaceae.

English, Hindi and Sanskrit Names: **Eng. :** Churl's Treacle, Garlic, Poor Man's Treacle; **Hindi :** Lahsun, Lasan; **Sans. :** Arishtha, Bhutabhna, Grinjana, Lasuna, Ugragandha.

Distribution : Cultivated throughout India.

Parts Used : Bulbs.

Flowering & Fruiting : May – June.

Folk Use : Bulbs used as heart tonic and stimulant.

Active Constituents : Known to yield 0.1-0.3% of volatile oil, allyl disulphide, diallyl disulphide, allin, allicin, mucilage and albumin (*bulb*).

References : Dixit & Kumar (2003), Farooq (2005), Kirtikar & Basu (1984), Siwakoti & Varma (1996), Watt (1972).

45. ***Allophylus serratus*** (Roxb.) Kurz

Family: Sapindaceae.

English, Hindi and Sanskrit Names: **Eng.** : Wild Berry; **Hindi** : Tippani; **Sans.** : Triputah.

Distribution : Throughout South India and Assam hills.

Parts Used : Fruit.

Flowering & Fruiting : Flowering: July – August; Fruiting: September – October.

Folk Use : Fruits used as a nourishing tonic.

Reference : Prajapati *et al.* (2006).

46. ***Aloe barbadensis*** Mill.*

Syn.: A. elongata Murray; *A. indica* Royle; *A. littoralis* Koening; *A. maculata* Forssk.; *A. variegata* Forssk.; *A. vera* Tourn. ex L.; *A. vulgaris* (Bauhin) Lam.

Family: Liliaceae.

English, Hindi and Sanskrit Names: **Eng.** : Barbodos Aloe, Curacao Aloe, Indian Aloe, Jafarabad Aloe; **Hindi** : Ghi Kanwar, Gvarapatha, Kuvarapatha; **Sans.:** Ajara, Bahupatri, Brahmagni, Ghrita Kumari, Vipulasrava.

Distribution : Semi-wild state in all parts from dry westward valleys of the Himalayas; also planted in gardens.

Parts Used : Whole Plant.

Flowering & Fruiting : During late winter.

Folk Use : Plant used as a tonic after pregnancy.

Active Constituents : Chief ingredients are barbaloin aloesin, aloesone β-sitosterol (*leaf*), and four partially acetylated glucomannans, β-barbaloin and aloe-anodin (*jelly*).

References : Asolkar *et al.* (1992), Kaushik & Dhiman (2000), Kirtikar & Basu (1984), Parrotta (2001).

47. ***Aloe perryi*** Baker

Family: Liliaceae.

English Names: **Eng.** : Sucotrine Aloe, Zanzibar Aloe.

Distribution : Indigenous to Socotra island, introduced and naturalized into India.

Parts Used : Leaf.

Flowering & Fruiting : Summers.

Folk Use : Juice of leaves used as tonic.

Reference : Ambasta (1986).

48. *Alpinia calcarata* Rosc.

Family: Zingiberaceae.

English, Hindi and Sanskrit Names: **Eng.** : Cardamom Ginger, Indian ginger, Java Galangal, Snap Ginger; **Hindi** : Kulanjan; **Sans.** : Kulanjana, Rasna, Sugandhamoola.

Distribution : Native of India, cultivated in Madhya Pradesh, Maharashtra, Tamil Nadu, Tripura and Central Deccan Plateau.

Parts Used : Whole Plant.

Flowering & Fruiting : Flowering during May.

Folk Use : Plant used as a tonic.

Reference : Behura & Rout (2003).

49. *Alpinia galanga* L. Willd.

Syn.: Amomum galanga (L.) Lour.

Family: Zingiberaceae.

English, Hindi and Sanskrit Names: **Eng.** : Blue Ginger, Greater Galangal, Languas Galangal, Langkwas, Siamease Ginger; **Hindi** : Barakalijan, Kulanjana, Rasna; **Sans.** : Aruna, Dhuwal, Elaparni, Kulanjan, Ugragandha.

Distribution : Sub-Himalayan region of West Bengal, Bihar, Assam; cultivated throughout India.

Parts Used : Rhizome.

Flowering & Fruiting : May – November.

Folk Use : Dried rhizomes considered stimulant and tonic.

Active Constituents : Active ingredients are kaempferide, galangin, alpinin, di(p-hydroxy-cis-styryl)methane and p-hydroxy cinnamaldehyde (*rhizome*).

References : Ambasta (1986), Behura & Rout (2003), Bhattacharjee (2001), Farooq (2005), Kirtikar & Basu (1984), Kumar (2002), Prajapati *et al.* (2006), Reddi *et al.* (2005), Sharma (2004), Tripathi & Goel (2001).

50. *Alpinia officinarum* Hance

Family: Zingiberaceae.

English and Hindi Names: **Eng.** : Lesser Galangal, Small Galangal; **Hindi** : Shitta-rattai.

Distribution : Grassland slopes, wood or thickets, cultivated in North India.

Parts Used : Rhizomes.

Flowering & Fruiting : Flowering: April – September; Fruiting: May – November.

Folk Use : Rhizomes used as an aromatic stimulant and nervine tonic.

Active Constituents : Chief components identified are 7-(4"-hydroxyphenyl)-1-phenyl-4-hepten-3-one, 5-methoxy-7-(4"-hydroxyphenyl)-1-phenyl-3-heptanone, 5-methoxy-1,7-diphenyl-3-heptanone and (3R, 5R) 1-(4"-hydroxyphenyl)-7-phenylheptane-3,5 diol (I) (*rhizome*).

Reference : Ambasta (1986).

51. ***Alstonia scholaris*** (L.) R. Br.

Syn.: Echites pala Ham. *ex* Spreng.; *E. scholaris* L.

Family: Apocynaceae.

English, Hindi and Sanskrit Names: **Eng. :** Devil's Tree, Dita Bark, The Satan Tree; **Hindi :** Chatian, Chhativan, Saittan ka Jat; **Sans. :** Ayugmaparna, Devavriksha, Saptaparna, Vishalatvaka.

Distribution : Wild in sub-Himalayan tract, West Bengal, Bihar, Peninsular India, Andaman; also cultivated.

Parts Used : Bark. Flower.

Flowering & Fruiting : Flowering: December – March; Fruiting: May – June.

Folk Use : Flowers act as a depressant on CNS. Bark of the plant used as a bitter tonic, cardio tonic, sedative and tonic.

Active Constituents : Active properties due to picrinine, srictamine, tetrahydroalstonine, n-hexacosane, lupeol, β-amyrin, palmitic and ursolic acid (*flower*), echitamine chloride, a-amyrin, its OAc, lupeol OAc, stigmasterol, β-sitosterol and campesterol (*bark*).

References : Ambasta (1986), Anonymous (1948), Asolkar *et al.* (1992), Bhattacharjee (2001), Dhiman (2005), Dwivedi (2007), Farooq (2005), Kaushik & Dhiman (2000), Kirtikar & Basu (1984), Kumar (2002), Prajapati *et al.* (2006), Reddi *et al.* (2005), Sharma (2004), Thakur *et al.* (1989), Tigga *et al.* (2003), Trivedi (2006), Watt (1972).

52. ***Alysicarpus pubescens*** Law

Family: Fabaceae.

English Name: **Eng. :** Bicolor Alyce Clover.

Distribution : Found in Andhra Pradesh, Bihar, Goa, Gujarat, Karnataka, Kerala, Madhya Pradesh, Orissa, Maharashtra and West Bengal.

Parts Used : Whole Plant.

Flowering & Fruiting : September – October.

Folk Use : Plant used as a tonic.

Reference : Bhalla & Bhalla (2002).

53. *Amaranthus spinosus* L.

Family: Amaranthaceae.

English, Hindi and Sanskrit Names: **Eng. :** Prickly Amaranth, Spiny Pigweed, Thorny Amaranth; **Hindi :** Cauleyi, Kateli; **Sans. :** Ghanasvana, Sushaka, Tenduliyah, Vishaghna.

Distribution : Wasteplaces throughout India.

Parts Used : Whole Plant. Leaf. Root.

Flowering & Fruiting : August – November.

Folk Use : Plant used as a tonic. Leaves especially used as iron tonic. Juice of crushed roots mixed with equal quantity of milk used as tonic.

Active Constituents : Roots rich in a saponin mixture composed of oleanolic acid, D-glucose, D-glucuronic acid, β-sitosterol, stigmasterol, campesterol, cholesterol, stearic, oleic, linoleic acids, quercetin, rutin, α-spinasterol octacosanoate and a new saponin – β-D-glucopyranosyl (1→4)-β-D-glucopyranosyl (1→4)-β-D-glucuronopyranosyl (1→3) oleanolic acid (I).

References : Jamir *et al.* (2008); Mishra (2008), Prajapati *et al.* (2006), Sarma *et al.* (2002), Srivastava *et al.* (2003a).

54. *Amberboa divaricata* Kuntze

Syn.: A. moschata (L.) Less.; *A. ramosa* (Roxb.) Jeffrey; *Centaurea moschata* L.; *Volutorella divaricata* Benth. & Hook. f.; *V. ramosa* Roxb.

Family: Asteraceae.

Distribution : Central and Southern India, from Bihar and the upper Gangetic plains to Lahore; and from Sind to Mysore and the Deccan; ascending to 1,000m in the N.W. Himalaya.

Parts Used : Whole Plant.

Flowering & Fruiting : January – March.

Folk Use : Herb used as a tonic.

Active Constituents : Isolated constituents are two new triter-penes-22 (R) cycloart-25-ene-3β-22-diol and cycloart-20-en-3β,25-diol along with cycloartanol and cycloart-23-ene-3β,25-diol (*plant*).

References : Ambasta (1986), Anonymous (1948), Bhalla *et al.* (1996), Kumar & Lal (2003).

55. *Ammi majus* L.

Family: Apiaceae.

English and Sanskrit Names: **Eng.** : Bishop's Weed, Bullwort, Greater Ammi, Lady's Lace, Laceflower. **Sans.** : Yavdni.

Distribution : Native to the Mediterranean region and as far east as Iran; introduced into India, grown at Jammu and Delhi.

Parts Used : Fruit.

Flowering & Fruiting : April – June.

Folk Use : Fruits used as a tonic.

Active Constituents : These are new furocoumarin-isopimpinellin, xanthotoxin, bergapten, marmesin, marmesinin, ammajine and isoimperatorin (*fruit*).

References : Ambasta (1986), Sharma & Sood (1997), Trivedi (2006).

56. *Amomum subulatum* Roxb.

Family: Zingiberaceae.

English, Hindi and Sanskrit Names: **Eng.** : Greater Cardamom, Large Cardamom, Nepal Cardamom; **Hindi** : Bari Elachi, Bari Llayaca; **Sans.** : Bhadraila, Ela, Nishkuti, Prithvi, Sthulaila, Triputa.

Distribution : Grows wild in eastern Himalayas from Nepal to Sikkim; cultivated in north west Bengal, Sikkim and Assam between 500-800m on slopes under chequered shades.

Parts Used : Seeds.

Flowering & Fruiting : Flowering: April – May; Fruiting: June – October.

Folk Use : Seeds used as stimulant, cardiac tonic and liver tonic.

Active Constituents : Seeds possess 1,8-cineole(74%) & limonene (10.3%), subulin, petunidin-3,5-diglucoside, leucocyanidin-3-O-β-D-glucopyranoside, cardamonin, alpinetin[4-6] and chalcone.

References : Kirtikar & Basu (1984), Prajapati *et al.* (2006), Rajan *et al.* (2003), Reddi *et al.* (2005), Thakur, *et al.* (1989), Tripathi & Goel (2001), Watt (1972).

57. *Amomum xanthioides* Wall. ex Baker

Syn.: A. villosum Lour. var. *xanthioides.*

Family: Zingiberaceae.

English and Hindi Names: **Eng.** : Bastard Cardamom, False Cardamom, Malabar Cardamom, Tavoy Cardamom; **Hindi** : Ilayechi.

Distribution : S.W. India (Malabar Coast), Myanmar, Thailand; cultivated in India and China.

Parts Used : Seeds.

Folk Use : Seeds regarded as stimulant.

References : Ambasta (1986), Kirtikar & Basu (1984), Watt (1972).

58. *Amorphophallus companulatus* Blume ex Decne.

Syn.: Amorphophallus paeoniifolius Dennst.

Family: Araceae.

English, Hindi and Sanskrit Names: **Eng.** : Elephant Foot Yam; **Hindi** : Suran, Jamikand; **Sans.** : Bahukanda, Durnamari, Suranah.

Distribution : Cultivated throughout India.

Parts Used : Corms.

Flowering & Fruiting : May – June.

Folk Use : Corms used as a liver tonic.

Active Constituents : Corms yield betulinic acid, β-sitosterol, stigmasterol, triacontane, β-sitosterol palmitate, glucose, galactose, rhamnose and exylose.

References : Farooq (2005), Prajapati *et al.* (2006), Siwakoti & Varma (1996), Watt (1972).

59. *Anacyclus pyrethrum* DC.

Syn.: Anacyclus depressus Maire; *A. freynii* Porta & Rigo; *A. officinarum* Hayne; *Anthemis pyrethrum* Hayne.

Family: Asteraceae.

English, Hindi and Sanskrit Names: **Eng.** : Bertram, Roman Pellitory, Spanish Pellitory; **Hindi** : Akarkara, Akorkaro; **Sans.** : Agragrahi, Akallaka, Akarakalabha, Akarakarava.

Distribution : Native to North Africa, globally distributed in the Mediterranean region; grown on an experimental scale at 900m at Katra (Jammu & Kashmir).

Parts Used : Root.

Flowering & Fruiting : April – July.

Folk Use : Roots used as a stimulant and tonic.

Active Constituents : Anacyclin, pellitorine, enetriyne alcohol, hydrocarolin, inulin, seasamin, deca-dodeca & tetradeca-trans-2,trans-4-dienamide have been isolated from *roots*.

References : Ambasta (1986), Prajapati *et al.* (2006), Reddi *et al.* (2005), Watt (1972).

60. *Anagallis arvensis* L.

Family: Myrsinaceae.

English and Hindi Names: **Eng. :** Bird's Eye, Red Chickweed, Scarlet Pimpernel, Shepherd's Delight, Tom Pimpernowl, Weather-glass; **Hindi :** Jangamami, Jonkmari, Krishna-neel.

Distribution : Mountains of Bengal and of the N.W. India and the Himalayas generally, from Nepal westwards, ascending to 2,700m; common in neighborhood of Shimla, on rubbish-heaps and walls around fields; Central India, the Nilgiris and Sri Lanka.

Parts Used : Whole Plant.

Flowering & Fruiting : December – April.

Folk Use : Herb used as a stimulant.

Active Constituents : Plant contains anagalligenone B, anagallin, anagalligenin B, bitter cucurbitacin glycosides, arvenin I & II and cucurbitacin E.

References : Ambasta (1986), Kaul (1997).

61. ***Ananas comosus*** L. Merrill

Syn.: Ananas sativus Schultes f., *Ananassa sativa* Lindl.; *Bromelia ananas* L.; *B. comasa* L.

Family: Bromeliaceae.

English, Hindi and Sanskrit Names: **Eng. :** Ananas, Pineapple; **Hindi:** Ananas; **Sans.:** Anamnasam, Bahunetraphalam, Paravati.

Distribution : Cultivated throughout India.

Parts Used : Fruit.

Flowering & Fruiting : Summers.

Folk Use : Unripe fruits used as cardio and uterine tonic whereas ripe ones used as general tonic.

Active Constituents : Active constituents are vitamin A,B,C, protein, fat, fibre, carbohydrates and minerals calcium, phos-phorus, thiamine, riboflavin, niacin, oxalic acid, citric acid, malic acid, p-coumaric, caffeic, ferulic and sinapic acid (*fruit*).

References : Khatoon & Goel (2010), Prajapati *et al.* (2006), Reddi *et al.* (2005), Sharma (2004).

62. ***Andrographis echioides*** Nees.

Syn.: Justicia echioides L.; *Indoneesiella echioides* L.

Family: Acanthaceae.

English Name: **Eng. :** False Waterwillow.

Distribution : Tropical India in the drier districts, Sri Lanka.

Parts Used : Whole Plant.

Flowering & Fruiting : Flowering: March – June; October – December.

Folk Use : Decoction of the plant used as a tonic.

Active Constituents : Echioidinin characterized as 5, 2'-dihydroxy-7-methoxyflavone and its glucoside, echioidin (*plant*).

Reference : Mishra (2008).

63. ***Andrographis paniculata*** (Burm.f.) Wall. ex Nees

Syn.: Justicia paniculata Roxb.

Family: Acanthaceae.

English, Hindi and Sanskrit Names: **Eng. :** Chiretta, Creat, King of Bitter; **Hindi :** Chirayata, Kalmegh, Kirayat; **Sans. :** Bhumimba, Kirata, Mahatikta.

Distribution : Throughout the plains of India; sometimes cultivated.

Parts Used : Whole Plant. Leaf. Root.

Flowering & Fruiting : October – December.

Folk Use : Leaves and roots used as tonic and liver tonic. Powdered plant (1 tsp) given with a cup of water and 5g of sugar for 30 days as liver tonic.

Active Constituents : Reported to yield andrographin(5-hydroxy-7,7,2'-trimethoxy flavone), panicolin(5,2'-dihydroxy-7,8- dimethoxy flavone), apigenin-4',7-dimethyl ether, mono-O-methyl wightin, 5-hydroxy-7,8,2',3'-tetramethoxy flavone, paniculides, andrographolides, caffeic acid and chlorogenic acid (*root*), β-sitosterol, glucoside and unknown compounds (*leaf*).

References : Ambasta (1986), Anonymous (1948), Banerjee & Ghora (1996), Bhattacharjee (2001), Chopra *et al.* (2007), Dhiman (2005), Dwivedi (2007), Jain *et al.* (2003), Jha & Varma (1996), Kaushik & Dhiman (2000), Kirtikar & Basu (1984), Parrotta (2001), Maya *et al.* (2003), Pal (2000), Raju (2000), Ranjan (2003), Rao (1996), Reddi *et al.* (2005), Sharma (2003), Thakur *et al.* (1989), Watt (1972).

64. ***Aneilema scapiflorum*** Wight.

Syn.: A. formosanum N.E. Br.; *A. tuberosum* Buch.; *Commelina edulis* Stokes; *C. scapiflora* Roxb.; *Murdannia tuberosa* Royle.

Family: Commelinaceae.

English and Hindi Names: **Eng. :** Black Musala; **Hindi :** Siyahmusli.

Distribution : Temperate and tropical Himalaya, Bhutan, W. Peninsula and Sri Lanka.

Parts Used : Root.

Flowering & Fruiting : September – November.

Folk Use : Roots used as a tonic.

References : Kirtikar & Basu (1984), Ranjan (2003), Watt (1972).

65. *Anemone pulsatilla* L.

Family: Ranunculaceae.

English Names: **Eng.** : European Pasque Flower, Pasque Flower, Wind Flower.

Distribution : Introduced in India; grows in W. Himalaya.

Parts Used : Root.

Flowering & Fruiting : March - May

Folk Use : Root acts as a stimulant in minute doses.

Reference : Bhattacharjee (2001).

66. *Anethum graveolens* L.

Syn.: A. graveolens L. var. *sowa* Roxb.; *A. sowa* Roxb. *ex* Flem.; *Peucedanum graveolens* L.; *P. sowa* Roxb.

Family: Apiaceae.

English, Hindi and Sanskrit Names: **Eng.** : Anet, Dill, Garden Dill; **Hindi** : Soya, Sutopsha; **Sans.** : Atichhatra, Madhari, Satapuspa, Vajana.

Distribution : Cultivated throughout India; grown in Gangetic plains and Central India.

Parts Used : Fruit. Seeds.

Flowering & Fruiting : August – October.

Folk Use : Fruits and seeds used as stimulant. Fruits also used as cardio tonic.

Active Constituents : Carvone, dihydrocarvone and limonene characterized from *fruit.*

References : Kaushik & Dhiman (2003), Prajapati *et al.* (2006), Reddi *et al.* (2005), Sharma (2003).

67. *Angelica archangelica* L. var. ***himalaica*** (C.B. Clarke) Kris Badhwar

Syn.: A. archangelica L.; *A. officinalis* Hoffmann; *Archangelica officinalis* Hoffm. var. *himalaica* C.B. Clarke.

Family: Apiaceae.

English and Hindi Names: **Eng.** : Angelica; **Hindi** : Angelica.

Distribution : Found in Kashmir.

Parts Used : Rhizome. Leaf. Flowering Tops. Fruit.

Flowering & Fruiting : June – August.

Folk Use : Dried rhizomes, dried leaves, fruit and flowering tops used as stimulant.

References : Ambasta (1986), Bhattacharjee (2001).

68. *Angelica glauca* Edgew.

Family: Apiaceae.

English and Hindi Names: **Eng.** : Angelica; **Hindi** : Chora, Chura.

Distribution : Distributed in Western Himalayas from Kashmir – Shimla : 2,438-3,048m.

Parts Used : Fruit. Root.

Flowering & Fruiting : July – September.

Folk Use : Roots used as stimulant and tonic. Fruits used as tonic.

Active Constituents : Major components are lactones, sesquiterpenes, d-α-pinene, d-α-phellandrene, selinene, d-α-cadinene, umbelliprenin and terpene alcohols (*root*) and a lactone (*fruit*).

References : Ambasta (1986), Anonymous (1948), Bhattacharjee (2001), Dhiman (2005), Maikhuri *et al* (2001), Negi *et al.* (1999), Prajapati *et al.* (2006), Rawat *et al.* (2007), Sharma (2003, 2009), Sharma & Rana (2000).

69. *Anisomeles indica* O. Kuntze

Syn.: A. ovata R. Br.; *Ballota distichia* L.; *Nepata indica* L.

Family: Lamiaceae.

English, Hindi and Sanskrit Names: **Eng.** : Indian Catmint, Malabar Catmint; **Hindi** : Gobara, Gopali, Kala Bhangra; **Sans.** : Kala Bhangra.

Distribution : Throughout India, ascending up to 1,828m in the Himalayas.

Parts Used : Whole Plant.

Flowering & Fruiting : Flowering: rainy season; Fruiting: cold season.

Folk Use : 2tsp paste of a branch/plant used twice a day for 7 days as tonic.

Active Constituents : Known to possess ovatodiolide and its isoderivatives (*leaf*), stigmasterol, β-sitosterol, paraffins and fatty acids (*root*).

References : Farooq (2005), Kaushik & Dhiman (2000), Kirtikar & Basu (1984), Mahato & Chaudhary (2005), Parrotta (2001), Ranjan (2003), Siwakoti & Varma (1996), Watt (1972).

70. *Annona squamosa* L.

Family: Annonaceae.

English, Hindi and Sanskrit Names: **Eng.** : Custard Apple, Sugar Apple, Sweet Sop; **Hindi** : Atasitaphal, Shariphal, Sitaphal; **Sans.** : Agrimakhya, Bahubijaka, Gandagatra, Sitaphalam.

Distribution : Throughout India, cultivated.

Parts Used : Fruit. Leaf.

Flowering & Fruiting : Flowering: May – June; Fruiting: June – November.

Folk Use : Fruits used as sedative, stimulant and tonic. Leaves used as cardio tonic.

References : Asolkar *et al.* (1992), Kirtikar & Basu (1984), Parrotta (2001), Prajapati *et al.* (2006), Singh & Srivastava (2000).

71. *Anodendron manubrium* Merr.

Family: Apocynaceae.

English Name: **Eng. :** Andamanese Bowstring Plant.

Distribution : Distributed from India and Sri Lanka throughout Bangladesh, Myanmar, Thailand, Cambodia, Laos, Vietnam and Malaysia to Indonesia and Philippines.

Parts Used : Root.

Folk Use : Roots used as a bitter tonic.

Reference : Tigga *et al.* (2003).

72. *Anogeissus latifolia* (Roxb. *ex* DC.) Wall *ex* Guill.

Syn.: Conocarpus latifolia Roxb. *ex* DC.

Family: Combretaceae.

English, Hindi and Sanskrit Names: **Eng. :** Button Tree, Dindiga Tree, Ghatti Tree; **Hindi :** Bankli, Dhawa; **Sans. :** Dhavah, Gaura, Kashaya, Pandura.

Distribution : Deciduous forests of South and Central India ascending up to 1,000m and in hilly places in India.

Parts Used : Gum.

Flowering & Fruiting : Flowering: September – January; Fruiting: November – February.

Folk Use : Gum used as a general tonic by mixing with equal volume of water and sugar by Koya tribe of Andhra Pradesh.

References : Asolkar *et al.* (1992), Dhiman (2005), Hosagoudar & Henry (1996), Upadhyay & Chauhan (2003).

73. *Anthemis nobilis* L.

Syn.: Chamaemelum nobile L.

Family: Asteraceae.

English and Hindi Names: **Eng. :** Ground Apple, Low Chamomile, Roman Chamomile, True Chamomile, Whig Plant; **Hindi :** Babuine.

Distribution : Cultivated in gardens throughout India, temperate regions of the Himalayas; wild at several places.

Parts Used : Flower.

Flowering & Fruiting : June – August.

Folk Use : Dried flower heads used as stimulant and nervine tonic.

Active Constituents : A sesquiterpene lactone – nobilin characterized from *flower*.

References : Ambasta (1986) Anonymous (1948), Watt (1972).

74. *Anthocephalus chinensis* (Lam.) Rich. ex Walp

Syn.: A. cadamba (Roxb.) Miq.; *A. indicus* Rich.; *Cephalanthus chinensis* Lamk.; *Neolamarckia cadamba* (Roxb.) Bosser

Family: Rubiaceae.

English, Hindi and Sanskrit Names: **Eng. :** Wild Chinchona; **Hindi :** Kadamb; **Sans. :** Ashokari, Kadambah, Kutsitangah, Nipa.

Distribution : Throughout Western Ghats at low levels in wet places.

Parts Used : Bark.

Flowering & Fruiting : Flowering and fruiting throughout the year, chiefly during December – July.

Folk Use : Bark used as a tonic.

Active Constituents : Known to yield sapogenins, cadambagenic acid, quinovic acid, β-sitosterol, saponins A,B,C & D, steroids, alkaloids, fats and reducing sugars (*bark*).

References : Kaushik & Dhiman (2000), Kirtikar & Basu (1984), Prajapati *et al.* (2006), Siwakoti & Varma (1996).

75. *Antidesma ghaesembilla* Gaertn.

Syn.: A. paniculatum Roxb.; *A. pubescens* Willd.

Family: Euphorbiaceae.

English, Hindi and Sanskrit Names: **Eng. :** Black Currant Tree; **Hindi :** Umtova; **Sans. :** Janupulisaru.

Distribution : Shimla – Bhutan and southwards to Sri Lanka; Singapore, Perak, Penang

Parts Used : Bark.

Flowering & Fruiting : Flowering: March – September; Fruiting: June – December.

Folk Use : Bark used as a tonic.

Reference : Kumar & Lal (2003).

76. *Apium graveolens* L.

Syn.: A. dulee Mill.; *A. lobatum* Gilib.

Family: Apiaceae.

English, Hindi and Sanskrit Names: **Eng.** : Celery, Marsh Parsley, Smallage; **Hindi** : Ajmod, Bari, Shalari; **Sans.** : Ajmoda, Brahmakoshi, Karavi, Modini, Ugragandhika.

Distribution : Native of Europe; naturalized in foot hills of N.W. Himalaya and outlying hills of Jammu & Kashmir, Himachal Pradesh and Uttar Pradesh.

Parts Used : Fruit. Seeds.

Flowering & Fruiting : May – September.

Folk Use : Fruits used as sedative, stimulant and nervine tonic. Seeds used as tonic, its fatty oil considered nerve stimulant and depressant.

Active Constituents : Choline, myristic acid, myristcic acid, myristoleic acid, stearic acid, apiin, isoquercitrin, apigenin, apigravin, apiumetin, apiumoside, bergapten, celereoside, rutaretin, α-eudesmol, β-eudesmol, santalol, 3-n-butyl phthalide and sedanenolide have been characterized from *fruits*.

References : Ambasta (1986), Anonymous (1948), Asolkar *et al.* (1992), Kayal *et al.* (2003), Kirtikar & Basu (1984), Pal & Paul (2003), Prajapati *et al.* (2006), Rajan *et al.* (2003), Sharma (2003), Singh *et al.* (2007), Thakur *et al.* (1989), Watt (1972).

77. ***Aquilaria agallocha*** Roxb.

Syn.: A. malaccensis Lamk.

Family: Thymelaeaceae.

English, Hindi and Sanskrit Names: **Eng.** : Agar Wood Tree, Aloe – wood, Calambac, Eagle – wood, Lignum Aloes; **Hindi** : Agar; **Sans.** : Agaru, Bhringaja, Jonk, Lohakhya, Yogaja.

Distribution : Eastern parts of India in the forests of Arunachal Pradesh and hilly tracts of Assam, Manipur, Nagaland and Tripura.

Parts Used : Whole Plant. Wood.

Flowering & Fruiting : Flowering: May – June; Fruiting: July – August.

Folk Use : Plant used as tonic and stimulant. Wood used as a tonic and cardio tonic. Agar obtained from plant considered as stimulant.

Active Constituents : Known to possess selinene, dihydroselinene, agoral, a sesquiterpene alcohol, a hydroxyl ketone, 5-isopropyl-7-methyl-5,5,5a,6,7,8-hexahydro-3n-naphtha, liridenine, jinkohol (*heartwood*), two new tri-2- (phenylethyl) chromones-AH19a and AH19b, three new esequiterpenes-(-) guaia-1 (10), 11 dien-15-ol, (-)selina-3,11-dien-9 – one (II) and (+) selina-3,11-dien-9-ol (III) and a new norsesquiterpene ketone-(+)1,5-epoxynor-ketoguaiene (*agarwood*).

References : Ambasta (1986), Anonymous (1948), Dhiman (2005), Kirtikar & Basu (1984), Kumar (2002), Prajapati *et al.* (2006), Saikia & Nath (2003b), Sharma (2004).

78. *Aquilegia vulgaris* L.

Syn.: A. karelini Baker.

Family: Ranunculaceae.

English Names: **Eng. :** Common Columbine, European Columbine, Granny's Nightcap.

Distribution : Western temperate and subalpine Himalayas; temperate Europe and Asia.

Parts Used : Herb. Wood.

Flowering & Fruiting : Flowering during summers and fruiting in October.

Folk Use : Herb and boiled wood pieces have tranquilizing effect.

References : Ambasta (1986), Kapur (1996a), Kapur & Srivastava (1996), Viswanathan (2000).

79. *Arachis hypogaea* L.

Family: Fabaceae.

English, Hindi and Sanskrit Names: **Eng. :** Chinese Almond, Groundnut, Monkeynut, Peanut; **Hindi :** Mungphali, Sheng, Sheng-dana, Vilayeti Mung; **Sans. :** Bhucanakah, Mandapi, Tribija.

Distribution : Throughout India, cultivated.

Parts Used : Seeds.

Flowering & Fruiting : August – October.

Folk Use : Seeds used as a tonic.

Active Constituents : Recorded to possess α-resorcyclic, p-hydroxybenzoic acid, cis- and trans-p-coumaric, phloretic and vanillic acids, protocatechuic, chlorogenic acids, stilbene, phytoalexins (I - III) (*seeds*).

Reference : Prajapati *et al.* (2006).

80. *Ardisia solanacea* Roxb.

Syn.: A. humilis Vahl.; *A. litoralis* Adr.; *A. polycephula* Wight.; *A. umbellata* Roth.

Family: Myrsinaceae.

English and Hindi Names: **Eng. :** Shoe Button Ardesia; **Hindi :** Banjan.

Distribution : Throughout India, up to 1,700m excepting desert regions.

Parts Used : Root Bark.

Flowering & Fruiting : April – June.

Folk Use : Infusion of root bark prescribed as a tonic after delivery.

Active Constituent : Bauerenol (*root*).

Reference : Jain (1996).

81. *Areca catechu* L.

Family: Arecaceae.

English, Hindi and Sanskrit Names: **Eng. :** Areca Nut Palm, Betel Nut Palm, Cashoo Nut Tree, Fasel Nut, Pinang Palm; **Hindi :** Supari, Supyari; **Sans. :** Akota, Chikkana, Ghonta, Puga.

Distribution : Native to Malaya; Cultivated in India in the coastal regions of southern Bombay and Chennai, Mysore, Bengal and Assam.

Parts Used : Whole Plant. Nut.

Flowering & Fruiting : Flowering: August – September; Fruiting: January – May.

Folk Use : Plant used as nervine tonic whereas nuts used as nervine- and cardio tonic.

Active Constituents : Active properties due to two procyanidin tetramers, two trimers and a dimer together with (+) catechin, (-) epicatechin and procynadins A-1, B-1 and B-2 and four bioactive polyphenolic components-NPF-861A, NPF-861B, NPF-8611A and NPF-8611B (*seeds*); Ptholic, lauric, myristic acid, palmitic and stearic acids and bis (2-ethylhexyl) ester (*plant*).

References : Parrotta (2001), Ranjan (2000), Saikia & Nath (2003b), Siwakoti & Varma (1996), Watt (1972).

82. *Argemone mexicana* L.

Family: Papaveraceae.

English, Hindi and Sanskrit Names: **Eng. :** Mexican Poppy, Prickly Poppy, Yellow Mexican Poppy; **Hindi :** Bhatbhamt, Pharamgi Dhattura, Satyanasi; **Sans. :** Brahmadanti, Svamaksiri.

Distribution : Throughout hotter parts of India, in areas up to 1,500m.

Parts Used : Whole Plant. Seeds.

Flowering & Fruiting : November – May.

Folk Use : Plant used as sedative and tonic. Seeds used as narcotic and sedative.

Active Constituents : Major isolated constituents are allocryptopine, berberine, cheilanthifoline, chelerythrine, coptisine, dihydrosanguinarine, norchelerythrine, norsanguinarine, protopine, oxyhydrastinine, sanguinarine, β-scoiulerine, methrohydroxide, α-tetrahydropalmatine methohydroxide, mexicanol, mexicanic acid, 6-acetonyl dihydrosanguinarine, α- and β-stylopine methrohydroxide (*plant*), a flavonoid – isohamnetin – 7 – diglucoside, two flavonal glycosides – isorhamnetin – 3 – glucoside, isorhamnetin – 3, 7 diglucoside, 3 – methoxyquercetin, vanillic acid, cryptopine and β – sitosterol (*flower*), argmexitin (5, 7, 2', 6' - tetrahydroxyflavone) and 5, 7 – dihydroxychromone – 7 – neohesperidoside (I) (*seeds*), isorhamnetin – 3 – glucoside, β – amyrin, cysteine, phenylalanine and protopine (*leaf*), norsanguinarine, berberine, protopine, β – sitosterol and allocryptopine (*root*), tryptophan, methionine, proline, alanine, tyrosine, histidine, glycine, leucine, aspartic acid and glutamic acid (*aerial parts*).

References : Asolkar *et al.* (1992), Kirtikar & Basu (1984), Kumar (2002), Mishra & Das (2003), Prajapati *et al.* (2006), Punjani (2003), Rajendran & Mehrotra (1996), Singh *et al.* (2007), Watt (1972).

83. ***Argyreia nervosa*** (Burm. f.) Boj.

Syn.: Argyreia speciosa Sweet; *Convolvulus speciosus* L.; *Lettsomia nervosa* Roxb.; *Rivea nervosa* (Burm. f.) Hall. f.

Family: Convolvulaceae.

English, Hindi and Sanskrit Names: **Eng. :** Elephant Creeper, Woolly Morning-glory; **Hindi :** Samandar – ka – pat, Samdrasos, Tameshar; **Sans. :** Bastantri, Samudrapalaka, Vrddhadarukah.

Distribution : Throughout India, in areas up to 900m.

Parts Used : Root.

Flowering & Fruiting : August – November.

Folk Use : Root used as tonic, cardio tonic, nervine tonic and brain tonic.

References : Ambasta (1986), Anonymous (1948), Dwivedi (2007), Kirtikar & Basu (1984), Kothari & Londhe (2000), Kumar (2002), Parrotta (2001), Prajapati *et al.* (2006), Watt (1972).

84. ***Aristolochia indica*** L.

Syn.: A. lanceolata Wight.

Family: Aristolochiaceae.

English, Hindi and Sanskrit Names: **Eng. :** The Indian Birthwort; **Hindi :** Isharmul, Ishwar Muli; **Sans. :** Garudi, Isaramula, Ishvari, Jata.

Distribution : Wild throughout the low hills and plains of India.

Parts Used : Root. Rhizome.

Flowering & Fruiting : Flowering: September – October; Fruiting: February – March.

Folk Use : Roots and rhizome used as a bitter tonic.

Active Constituents : Associated properties due to ishwarane, aristolochene, ishwarol, two sesquiterpene alcohols-(125)-7,12-secoishwaran-12-ol and (+)ledol, 2-hydroxyl-1-methoxy-4H-dibenzo [de,g]quinoline-4,5-(6H)-dione(I), aristololide, 5α-stigmastan-3,6-dione(-)cubebin, (-)hinokinin, new naphthoquinone-aristolineliquinone (*root*).

References : Ambasta (1986), Bhattacharjee (2001), Dhiman (2005), Kirtikar & Basu (1984), Mishra (2008), Thakur *et al.* (1989), Watt (1972).

85. *Aristolochia saccata* Wall.

Family: Aristolochiaceae.

English Name: **Eng. :** Birthwort.

Distribution : Eastern and Central Himalaya, N.E. India.

Parts Used : Root.

Flowering & Fruiting : November – January.

Folk Use : Roots used as tonic and stimulant.

Reference : Kumar (2002).

86. *Aristolochia tagala* Cham.

Syn.: A. megalophylla K. Schumann; *A. mindanaesis* Warb.; *A. roxburghiana* Klotzsch.

Family: Aristolochiaceae.

English Name: **Eng. :** Dutchman's Pipe, Indian Birthwort.

Distribution : Bengal, N.E. India and W. Peninsula.

Parts Used : Root.

Flowering & Fruiting : Throughout the year.

Folk Use : Root considered as a tonic.

Active Constituents : Aristolochic acid.

References : Ambasta (1986), Asolkar *et al.* (1992), Kumar (2002).

87. *Armoracia rusticana* Gaertn. *et al.*

Syn.: A. lapathifolia Gilib.; *Cochlearea armoracia* L.

Family: Brassicaceae.

English Name: **Eng. :** Horse-radish.

Distribution : Native to Europe and W. Asia; herb widely cultivated for its root.

Parts Used : Root.

Flowering & Fruiting : Flowering during May – June; fruiting in September – October.

Folk Use : Roots considered as stimulant.

Active Constituents : Glucosilinates, asparagines, resin and vitamin C characterized from roots.

Reference : Ambasta (1986).

88. *Arnica montana* L.

Family: Asteraceae.

English and Hindi Names: **Eng. :** Arnica, Leopard's Bane, Mountain Daisy, Mountain Tobacco, Wolf's Bane; **Hindi :** Arnica.

Distribution : Mountain woods and pastures in Central Europe, the Pyrenees, Siberia and North – Western US & Canada; cultivated sparingly in India.

Parts Used : Flower.

Flowering & Fruiting : Flowering during September.

Folk Use : Dried flower heads used as tonic.

Active Constituents : Active components are synarine and sterol (*flower*), helenalin methacrylate, helenalin acetate,11,13-dihydrohelenalin, 6-O-isobutyryl, 6-O-tigloyl, 6-O-isovaleryl and 6-O-(2-methylbutyryl)-helenalin, 6-O-isobutyryltetrahydrohelenalin (I), 2β-ethoxy-6-O-isobutyryl-2,3-dihydrohelenalin (II), hispidulin-7-O-β-D-glucoside, isorhamnetin-3-O-β-D-glucoside, spinacetin-3-O-β-D-glucoside, 6-methoxykaempferol-3-O-β-D-glucoside, patuletin-3-O-β-D-glucoside, quercetin-3-O-β-D-(6''-O-acetyl)glucoside, 3-O-β-D-glucuronides of quercetin, kaempferol, isorhamnetin, patuletin and 6-methoxykaempferol, tricin and 3,5,7-trihydroxy-6,3',4'-trimethoxy-flavones (*flower*).

Reference : Ambasta (1986).

89. *Artemisia abrotanum* L.

Family: Asteraceae.

English Names: **Eng. :** Appleringie, Boy's Love, Garderobe, Lad's Love, Lover's Plant, Old Man, Oldman Wormwood, Sitherwood, Southernwood.

Distribution : Native of Southern Europe; cultivated in gardens in India.

Parts Used : Aerial Parts.

Folk Use : Aerial parts used as tonic.

Active Constituents : Volatile oil, abrotonin and tannins characterized from aerial parts.

Reference : Prajapati *et al.* (2006).

90. ***Artemisia absinthium*** L.

Syn.: Absinthium officinale Lam.; *A. vulgare* Gaertn.

Family: Asteraceae.

English, Hindi and Sanskrit Names: **Eng. :** Absinthe, Madderwood, Wormwood; **Hindi :** Vilayatiafsanthin; **Sans. :** Aphasantin, Damara, Indhana.

Distribution : Found in Himalayan region.

Parts Used : Whole Plant. Flower. Leaf.

Flowering & Fruiting : September – October.

Folk Use : Plant used as tonic. Leaves and flowers considered stimulant.

Active Constituents : Therapeutic properties due to essential oil-absinth, artemisetin, absinthin, artobasin, myrcene, a pinene, thujyl alcohol and thujyl acetate, α-thujane, α-pinene, camphene, p-cymene, 1,18-cineole, methyleptenone, β-phellandrene, caryophyllene oxide, α-terpineal, geraniol, caryphyllene, α-himachalene, α-cadiene, kaempferol, its methyl ether, quercetagetin-3,6,3'-trimethyl,-3,6, 7-trimethyl, -3,6,7,3'-tetramethyl and -3,6,7,3',4'-pentamethyl ether, homoditerpene peroxides (I and II), artanolide, deacetylglobicin, parishin B and parishin C (*aerial parts*), scopoletin, umbelliferone, caffeoylquinic acid and chlorogenic acid (*plant*), artenolide (*foliage*).

References : Ambasta (1986), Dwivedi (2007), Kayal *et al.* (2003), Kirtikar & Basu (1984), Prajapati *et al.* (2006), Watt (1972).

91. ***Artemisia cina*** Berg.

Family: Asteraceae.

English Names: **Eng. :** Levant Wormseed, Russian Artemisia, Santonica.

Distribution : N.E. India.

Parts Used : Leaf.

Flowering & Fruiting : Mid summers.

Folk Use : Leaf used as a tonic and stimulant.

Reference : Bhattacharjee (2001).

92. ***Artemisia dracunculus*** L.

Family: Asteraceae.

English Names: **Eng. :** Dragon Mugwort, Estragon, Tarragon.

Distribution : Western Tibet (4,650 – 5350m) and in Kinnaur and Lahul; distributed to Aghanistan, Western Asia, south and middle Russia.

Parts Used : Leaf.

Flowering & Fruiting : May – October.

Folk Use : Leaves used as stimulant.

Active Constituents : Isolated constituents are scopoletin and herniarin, β-sitosterol, quercetin, hyperoside, isorhamnetin and artemidinol (*leaf*).

References : Ambasta (1986), Bhattacharjee (2001).

93. *Artemisia indica* Willd.

Syn.: A. vulgaris auct. non L.; *A. nilagirica* (Clarke) Pamp.

Family: Asteraceae.

English, Hindi and Sanskrit Names: **Eng. :** Fleabane, Indian Wormwood; **Hindi :** Dauna, Davana, Nagadouna; **Sans. :** Damanakah, Damanah, Nagadamni.

Distribution : Throughout the mountainous regions of India, ascending up to 3,048m in Western Himalayas and up to 2,438m in Sikkim and Khasia Hills and in Mount Abu and Western Ghats.

Parts Used : Root. Leaf.

Flowering & Fruiting : August – October.

Folk Use : Roots and juice of leaves used as a good tonic.

Active Constituents : Known to possess trideca-1,3,5-trien-7,9,11-triyne, Artemisia ketone and 1-acetoxytetradeca-4, 6-dien-8,10,12-triyne (*root*), threonine (14.28), leucine and isoleucine (12.21), glycine (12.82), glutamic acid (9.8) and arginine present in protein (7.2%) (*leaf*).

References : Chaudhury & Neogi (2000), Dhiman (2005), Farooq (2005), Kumar (2002), Negi *et al.* (1999), Prajapati *et al.* (2006), Tiwari & Tiwari (1996), Watt (1972).

94. *Artemisia maritima* L.

Syn.: A. fragrans Willd.; *A. spicigera* C. Koch.

Family: Asteraceae.

English, Hindi and Sanskrit Names: **Eng. :** Drooping Sea Wormseed, Sea Mugwort, Sea Wormwood, Wormseed; **Hindi :** Ajavayana, Chhuari; **Sans. :** Chauhara, Gandha, Parasi, Yavania.

Distribution : W. Himalaya, (Kashmir – Kumaon) : 2,350-3,500m, coasts of Europe and saline tracts of N. Asia.

Parts Used : Leaf.

Flowering & Fruiting : May – August.

Folk Use : Leaves used as tonic and stimulant.

Active Constituent : Erivanine (*leaf*).

Reference : Bhattacharjee (2001).

95. ***Artemisia pallens*** Wall. ex DC.

Family: Asteraceae.

English, Hindi and Sanskrit Names: **Eng. :** Davana; **Hindi :** Davana; **Sans. :** Davanam.

Distribution : Certain parts of south India especially in the Mysore; also cultivated in Poona.

Parts Used : Whole Plant.

Folk Use : Plant accredited with tonic properties.

Active Constituents : Active properties due to artemone, davana ether (*plant*), 3,4-epoxyisodavanone, cirsimaritin (*aerial parts*), α-and β-dihydrorosefurans, hydroxydihydrorosefuran, its acetyl derivative, trans-hydroxydavanone, camphene, γ-cadiene, sabiene, p-cymene, linalool and its oxide, borneol, nerol, geraniol, davanol, farnesol, nor-davanone, cis- and trans-davanones, cis-hydroxydavanone, cis- and trans-davanafurans, artemone, davana acid and cinnamyl cinnamate (*essential oil*).

Reference : Ambasta (1986).

96. ***Artocarpus communis*** Forst.

Syn.: A. altilis (Park) Fosberg; *A. incisus* (Thunb.) L. f.

Family: Moraceae.

English, Hindi and Sanskrit Names: **Eng. :** Bread Fruits; **Hindi :** Bilatti Kathal, Kathal; **Sans. :** Ksudrapanasah.

Distribution : Indigenous to Malay, cultivated in the tropics; successfully introduced into Bombay, south India and Sri Lanka.

Parts Used : Fruit.

Flowering & Fruiting : March – May.

Folk Use : Fruits used as cardio tonic.

Active Constituents : α-amyrin and β-amyrin acetate, ceryl ester, iodine and fluorine characterized from *fruit*.

Reference : Prajapati *et al.* (2006).

97. ***Artocarpus heterophyllus*** Lam.

Syn.: A. integrifolia auct. non L.; *Polyphema jaca* Lour.; *Sitodium cauliflorum* Gaertn.

Family: Moraceae.

English, Hindi and Sanskrit Names: **Eng.** : Jaca Tree, Jackfruit Tree, Orange Wood; **Hindi** : Cakki, Kathal, Panas; **Sans.** : Apushpa, Mahasarja, Sthula.

Distribution : Widely cultivated throughout India (indigenous in the W. Ghats).

Parts Used : Fruit. Wood.

Flowering & Fruiting : Flowering during winter; fruiting in rainy season.

Folk Use : Fruit used as tonic while wood used as a nervine sedative.

Active Constituents : Reported to contain β-carotene, cycloartenone, cycloartenol, β-sitosterol, butyrospermol, leucine, isoleucine, tyrosine, valine (*fruit*), morin, dihydromorin, cyanomaclurin, artocarpin, isoartocarpin, cycloartocarpin, artocarpesin, oxydihydroarto-carpesin, norartocarpetin, cycloartinone, sucrose and ellagic acid (*wood*).

References : Asolkar *et al.* (1992), Kaushik & Dhiman (2000), Kirtikar & Basu (1984), Parrotta (2001), Prajapati *et al.* (2006), Varghese (1996).

98. *Asarum europaeum* L.

Syn.: A. lucidum Salisb.; *A. officinale* Moench.

Family: Aristolochiaceae.

English, Hindi and Sanskrit Names: **Eng.** : Asarabaca; **Hindi** : Tagar, Taggar; **Sans.** : Upana.

Distribution : Central and South Europe, temperate Asia and Asia Minor.

Parts Used : Whole Plant.

Flowering & Fruiting : May – July.

Folk Use : Plant used as a stimulant.

Active Constituents : Found to possess asarine, asarone, boryl acetate and methyl eugenol, lysine, serine, arginine, aspartic acid, asparagines, tryptophan,, glycine, glutamic acid, glutamione, proline, tyrosine, phenylalanine, methionine, valine and histidine (*leaf, rhizome*), isoquercitrin, trifolin, kaempferol-3-rutinoside and quercetin (*plant*), β-carotene and α-tocopherol (*leaf*), α-asarone (*root*), chologenic acid and is chlorogenic acid (*root, leaf*).

References : Prajapati *et al* (2006), Watt (1972).

99. *Asparagus adscendens* Roxb.

Syn.: Asparagopsis adscendens (Roxb.) Kunth.; *Asparagus volubilis* Wall.; *A. satawur* James A. Murray.

Family: Liliaceae.

English, Hindi and Sanskrit Names: **Eng.** : Asparagus; **Hindi** : Hazarmuli, Musli, Satavar; **Sans.** : Safed Musli.

Distribution : Western Himalayas and Punjab, from Himachal Pradesh – Kumaon, up to 1,500m; Aghanistan.

Parts Used : Root. Root Bark.

Flowering & Fruiting : Flowering during September – October; fruiting in November – December.

Folk Use : Root bark with cow's milk used as tonic. 1 cup of root decoction taken by woman as a tonic thrice a day for 15 days after delivery.

Active Constituents : Roots rich in asparagine and saponin steroidal, asparanins C and D and asparosides C and D, two new glycosides-3β-O-β-D-(2'-O-tetracosyl) xylopyranosyl, stigmasterol and 3β-O-β-D-glucopyranosyl(1→2)-α-L-arabinopyranosylstigmasterol.

References : Aswal (1996), Dhiman (2005), Dwivedi (2007), Kirtikar & Basu (1984), Mahato & Chaudhary (2005), Watt (1972).

100. *Asparagus racemosus* Willd. (Pl. 1F)*

Syn.: Asparagopsis abyssinica Kunth.; *A. acerosa* Kunth.; *A. floribunda* Kunth.; *A. javanica* Kunth.; *Asparagus acerosus* Roxb.; *A. jacquemontii* Baker; *A. saundersiae* Baker; *A. zeylanicus* (Baker) Hook. f.; *Protasparagus racemosus* (Willd.) Oberm.

Family: Liliaceae.

English, Hindi and Sanskrit Names: **Eng.** : Indian Asparagus, Wild Asparagus; **Hindi** : Chatwal, Satavar, Satmuli; **Sans.** : Abhiru, Atirasa, Indivari, Satavari.

Distribution : Common at low altitudes in shade and in tropical climate throughout India.

Parts Used : Root. Leaf. Tuber.

Flowering & Fruiting : Flowering during October – November; fruiting in cold season.

Folk Use : Exudates of dry root, kept in water used as a nutritive and nervine tonic. 50ml of root decoction used 3 times daily for 15 days in delivery as a tonic. Roots also considered stimulant. Leaves, root and tubers used as tonic.

Active Constituents : These are quercetin-3-glucuronide (*leaf*), four glycosides-compound A, shatavarin I, Shatavarin II, and shatavarin IV, sitosterol, 4,6-dihydroxy-2-O-(2'-hydroxyisobutyl) benzaldehyde, undecanyl cetanoate and asparagamine A (*root*).

References : Farooq (2005), Gopal & Chandra (2003), Jain & Goel (2005), Kaushik & Dhiman (2000), Khatoon & Goel (2010), Kirtikar & Basu (1984), Kothari & Rao (2000), Maheshwari *et al.* (1997), Nadeem *et al.* (2001), Pandey & Rout (2006), Pathak &

Pathak (2007), Parrotta (2001), Prajapati *et al.* (2006), Prasad *et al.* (2002, 2003), Punjani (2006), Reddi *et al.* (2005), Sharma (2003, 2004), Sivakumar *et al.* (2003), Thakur *et al.* (1989).

101. *Asparagus scandens* Thunb.

Family: Liliaceae.

English Names: **Eng. :** Asparagus Fern, Climbing Asparagus, Snakefeather.

Distribution : Cultivated in Andamans.

Parts Used : Leaf.

Flowering & Fruiting : Flowering during August – October; fruiting in September – November.

Folk Use : Paste of leaves taken as tonic for vitality.

Reference : Singh & Pandey (1998).

102. *Asperula odorata* L.

Syn.: Gallium odoratum (L) Scop.

Family: Rubiaceae.

English Names: **Eng. :** Master of the Wood, Sweet Woodruff, Wild Baby's Breath, Woodruff.

Distribution : Native of Europe, found in Asia and North Africa; grows in woodlands and shaded places of Western Himalayas.

Parts Used : Aerial Parts.

Flowering & Fruiting : May – July.

Folk Use : Aerial parts considered as tonic.

Active Constituents : Isolated constituents are iridioids, coumarins, tannins, anthraquinons and flavonoids (*aerial parts*).

Reference : Prajapati *et al.* (2006).

103. *Asplenium nidus* L. f.

Syn.: A. ficifolium Goldm.; *Thamnopteris nidus* (L.) Presl.

Family: Aspleniaceae.

English and Hindi Names: **Eng. :** Bird's Nest Fern; **Hindi :** Ruai – Mangma.

Distribution : Found in eastern and north western Himalayas, Tamil Nadu and Karnataka.

Parts Used : Whole Plant.

Flowering & Fruiting : January – March.

Folk Use : Plant used as sedative.

Active Constituents : Kaempferol-3-O-gentiobioside-7,4'-bisglucoside (I), kaempferol-3-O-diglucoside, kaempferol-3,7-diglucoside and kaempferol-3-O-vicianoside have been isolated from *aerial parts.*

References : Asolkar *et al.* (1992), Singh & Viswanathan (1996).

104. *Asplenium trichomanes* L.

Family: Aspleniaceae.

English Names: **Eng. :** Common Spleenwort, Maiden Hair.

Distribution : Nilgiris, Kashmir – Kumaon: 1700-3500m; all over the world.

Parts Used : Whole Plant.

Folk Use : Plant used as a tonic.

Active Constituents : Known to possess kaempferol-3-O-α-L-rhamnoside-7-O-α-L-arabinopyranoside (I), kaempferol-3-O-α-L-arabinopyranoside-7-O-α-L-rhamnopyranoside, kaempferitrin, 3-hydroxyflavone and rutin (*plant*).

Reference : Singh & Viswanathan (1996).

105. *Atropa belladonna* C.B. Clarke

Syn.: A. acuminata Royle.

Family: Solanaceae.

English, Hindi and Sanskrit Names: **Eng. :** Belladona, Deadly Nightshade; **Hindi :** Angurshefa, Luckmuna, Sag-angur; **Sans. :** Suchi.

Distribution : Western Himalayas (Kashmir – Shimla): 2,000-4,000m.

Parts Used : Leaf. Root.

Flowering & Fruiting : Flowering during autumn; fruiting in winters.

Folk Use : Leaves and roots used as sedative and narcotic. Also, leaves used as stimulant.

Active Constituents : Isolated constituents are atropine and hyoscyamine, flavonoglycosiderutin, 3-rhamnogalactoside of kaempferol, quercetin-7-monoglucoside and kaempferol-7-monoglucoside[2], hyoscine, rutin, kaempferol-3-rhamnogalactoside, kaempferol-7-monoglucoside, quercetin-7-monoglucoside (*leaf*), hyoscyamine N – oxide (*root*).

References : Ambasta (1986), Anonymous (1948), Dhiman (2005), Harborne & Baxter (2001), Kapur (1996a), Kapur & Singh (1996), Kapur *et al.* (1996), Kaul (1997), Kirtikar & Basu (1984), Prajapati *et al.* (2006), Reddi *et al.* (2005), Sharma (2003), Thakur *et al.* (1989), Uniyal & Malhotra (1989), Watt (1972).

106. *Atylosia scarabaeoides* (L.) Benth.

Syn.: Cajanus scarabaeoides (L.) Thouars; *Dolichos scarabaeoides* L.

Family: Fabaceae.

English, Hindi and Sanskrit Names: **Eng.** : Wild Kulthi; **Hindi** : Jangli Tur, Kulthi; **Sans.** : Vana-kulathha.

Distribution : Throughout India, up to 1800m on the W. Himalayas; Malaya, China, Mauritius and Madagascar.

Parts Used : Whole Plant.

Flowering & Fruiting : August – November.

Folk Use : Decoction of plant used as tonic after delivery.

Active Constituents : Leaves owe its properties due to hentriacontane, β-sitosterol glucoside, D (+) pinitol, vitexin and hexacosanyl esters of palmitic, stearic, arachidic and oleic acid.

References : Maheshwari *et al.* (1997), Painuli & Maheshwari (1996), Reddi *et al.* (2005), Siwakoti & Varma (1996).

107. *Avena sativa* L. (Pl. 2A)

Family: Poaceae.

English, Hindi and Sanskrit Names: **Eng.** : Groats, Oats; **Hindi** : Javi, Jei, Wilayati-jau; **Sans.** : Yavikaa.

Distribution : Native to Northern Europe, grown worldwide in temperate regions as a cereal crop; cultivated in north India.

Parts Used : Seeds. Straw.

Flowering & Fruiting : November – May.

Folk Use : Straw and seeds used as tonic.

Active Constituents : Sterols, cholesterol, brassicasterol, β-sitosterol, campesterol, stigmasterol, Δ^7-stigmasterol and stigmastdienols have been isolated (*seeds*).

Reference : Prajapati *et al.* (2006).

108. *Averrhoa carambola* L.

Family: Oxalidaceae.

English, Hindi and Sanskrit Names: **Eng.** : Carambola Apple, Coromandel Gooseberry; **Hindi** : Kamarakh, Kamaramga, Karmal; **Sans.** : Brihaddala, Karmarangah, Karukah, Rujakara.

Distribution : Cultivated throughout India.

Parts Used : Fruit.

Flowering & Fruiting : Flowering: June – September; Fruiting: September – October.

Folk Use : Fruits used as a tonic

Active Constituents : Reported to yield proteins, amino acid compounds, oxalic acid, Ca and P, phytofluene, ζ-carotene, β-cryptoflavin, mutatoxanthin, β carotene, β-apo-8'-carotenal, β-cryptoxanthin, β-cryptochrome and lutein (*fruit*).

References : Kirtikar & Basu (1984), Prajapati *et al.* (2006), Reddi *et al.* (2005).

109. *Azadirachta indica* A. Juss. (Pl. 2B)*

Syn.: Antelaea indica (L.) Adelb.; *Melia azadirachta* L.; *M. indica* (A. Juss.) Brandis.

Family: Meliaceae.

English, Hindi and Sanskrit Names: **Eng.** : Indian Lilac, Margosa Tree, Neem Tree; **Hindi** : Nim, Nimb; **Sans.** : Nimbah, Prabhadrah.

Distribution : Everywhere in Indian plains including some parts of arid zone.

Parts Used : Whole Plant. Bark. Flower. Fruit. Gum. Root Bark.

Flowering & Fruiting : March – July.

Folk Use : Plant, flowers, root bark and fruits used as tonic. Bark valued as liver tonic and bitter tonic. Gum used as stimulant and tonic.

Active Constituents : Isolated components are β-sitosterol, quercetin, myricetin, quercetin-3-galactoside, kaempferol-3-glucoside and myricetin-3-arabinoside, tannin, non-tannin and red dye, margosinone, margosinolone, nimbosodione, nimbisonal, demethylnimbionol, margosone, margosolone, isonimbinolide, nimbinonone, nimbionol, nimosone, nimbosone, methyl nimbionone, margolone, margolonone, argolonone, isomargolonone, methyl grevillate, a water soluble polysaccharide, CSP-I composed of galactose, arabinose and glucose in ratio of 4: 2: 1 and having (1→3) linked galactopyranosyl backbone, kulinone, kulactone, kulolactone and methyl kulonate, 6β-hydroxyl-4-stigmasten-3-one (I), 6β-hydroxyl-4-campesten-3-one (*bark*), nimolicinol, azardirachtol, azadirachtin, nimocin, nimbocinol, nimolinone, isonimolicinolide, nimolicinoic acid, nimbochalcin, nimbocetin, 5-hydroxymethylfurfural, 1-cinnamoyl melianolone and 1-cinnamoyl-3,11-dihydroxymeliacarpin (*fruit*), nimocinol, azadirachtanin, nimbocinone, nimbocinolide, nimbaflavone, quercetin, its glucoside, isorhamnetin, quercetin-3-O-β-D-glucoside, -3-O -rutinoside, -3-O-α-L-rhamnoside, kaempferol-3-O-rutinoside, -3-O-β-D-glucoside, myricetin-3- O-rutinoside, 6-deacetylnimbinol, 28-deoxonimbolide, β-sitosterol-β-D-glucoside, n-hexacosanol and β-carotene (*leaf*).

References : Ambasta (1986), Anonymous (1948), Bhattacharjee (2001), Dwivedi (2007), Kaushik & Dhiman (2000), Kumar

(2002), Kirtikar & Basu (1984), Nargas & Trivedi (2000), Parrotta (2001), Prajapati *et al.* (2006), Ranjan (2000, 2003), Reddi *et al.* (2005), Sharma (2003), Siwakoti & Varma (1996), Thakur *et al.* (1989).

110. *Bacopa monnieri* (L.) Penell.

Syn.: Bacopa monnieria (L.) Wettst; *Gratiola monnieria* L.; *Herpestis monnieria* (L.) Kunth.; *Lysimachia monnieria* L.

Family: Scrophulariaceae.

English, Hindi and Sanskrit Names: **Eng.** : Thyme Leaved Gratiola, Water Hyssop; **Hindi** : Brahmi, Jalnim; **Sans.** : Eiendri, Nira Brahmi, Sarasvati.

Distribution : Marshy, damp and wet areas throughout India.

Parts Used : Whole Plant. Leaf.

Flowering & Fruiting : During winter season.

Folk Use : Plant used as brain, nervine and cardio tonic. Leaves taken to improve memory.

Active Constituents : Active properties identified due to brahmine, B1 oxalate, B2 oxalate, B3 chloroplatinate, herpestine, bacoside A & B, monnierin, hersaponin, betulic acid, d-mannitol, stigmasterol and β-sitosterol (*plant*), brahmine, herpestine and mixture of 3 alkaloids (*leaves*).

References : Anonymus (1948), Ansarali & Sivadasan (2009), Arya (2004), Asolkar *et al.*(1992), Banerjee & Ghora (1996), Bhattacharjee (2001), Dhiman (2005), Farooq (2005), Joshi (2004), Kaushik & Dhiman (2000), Khan *et al.* (2003), Khatoon & Goel (2010), Kothari & Londhe (2000), Kumar (2006), Kumar & Lal (2003), Maya *et al.* (2003), Nayak *et al.* (2003), Pal (2000), Parrotta (2001), Prajapati *et al.* (2006), Raju (2000), Reddi *et al.* (2005), Rosakutty *et al.* (2000), Sharma (2003, 2004), Sharma & Sood (1997), Singh (2001), Singh & Pandey (1998), Singh & Kumar (2000b), Siwakoti & Varma (1996).

111. *Baliospermum montanum* (Willd.) Muell. – Arg.

Syn.: B. axillare Bl.; *B. polyandrum* Wight.; *Croton polyandrous* Roxb.

Family: Euphorbiaceae.

English, Hindi and Sanskrit Names: **Eng.** : Red Physic Nut, Wild Castor, Wild Croton; **Hindi** : Danti, Hakun; **Sans.** : Anukheti, Bhadra, Danti, Jayapala, Ruksha, Upachitra.

Distribution : Himalayas, Assam, Khasia Hills, Bihar, Bengal, Madhya Pradesh and Peninsular India, ascending to 1,800m.

Parts Used : Leaf, Root, Seed.

Flowering & Fruiting : January – March.

Folk Use : Roots used as tonic. Seeds and leaves used as stimulant.

Active Constituents : Characterstic properties due to montain, balliospermin, 12-deoxyphorbol and 12-deoxy-16-OH-phorbol as 13-palmitates, 12-deoxy- 5β-OH-phorbol-13 myristate (*root*).

References : Farooq (2005), Prajapati *et al.* (2006), Singh *et al.* (2007).

112. *Bambusa arundinacea* (Retz.) Willd.

Syn.: Arundo bambos L.; *Bambusa bamboos* Druce; *B. orientalis* Nees.

Family: Bambusaceae.

English, Hindi and Sanskrit Names: **Eng. :** Bamboo Tree, Spiny Bamboo, Thorny Bamboo; **Hindi :** Bans, Kantabans, Magarbans; **Sans. :** Vamsah, Vansh, Venuh.

Distribution : Wild throughout the greater parts of India, especially in hill forests of Western and Southern India, ascending up to 914m on the Nilgiris, also in Orissa, Assam and Eastern Bengal, also cultivated in the lower Himalayas and the valleys of Ganga.

Parts Used : Shoot. Root. Culm.

Flowering & Fruiting : Flowering in summer; fruiting during September – October.

Folk Use : Young shoots, culms and leaves considered stimulant and tonic. Roots used as tonic.

Active Constituents : Known to yield cyanogenetic glucoside, benzoic acid, resin and waxes (*root*).

References: Dhiman (2005), Dwivedi (2007), Kaushik & Dhiman (2000), Kirtikar & Basu (1984), Prajapati *et al.* (2006), Raju (2000).

113. *Baptisia leucophaea* Nutt.

Family: Fabaceae.

English Names: **Eng. :** Cream Wild Indigo, False Indigo, Longbract Wild Indigo, Plains Wild Indigo.

Distribution : Native to USA.

Parts Used : Leaf.

Flowering & Fruiting : April – July.

Folk Use : Decoction of leaves used as a stimulant.

Reference : Bhattacharjee (2001).

114. *Barbarea vulgaris* R. Br.

Family: Brassicaceae.

English Names: **Eng. :** Bitter Winter Cress, Hedge Mustard, Land Cress, Scurvy Cress, Winter Rocket, Wound Rocket, Yellow Rocket.

Distribution : Temperate and subalpine Himalayas : 1,800 – 3,750m and W. Tibet; Europe, W. Asia, N. and S. Africa and Australia.

Parts Used : Whole Plant.

Flowering & Fruiting : Flowering during April – July; fruiting in July – August.

Folk Use : Plant used as stimulant.

Reference : Kapur & Srivastava (1996).

115. *Barleria prionitis* L. subsp. ***dicantha*** Blatt. & Hallb.

Syn.: B. prionitis Willd.

Family: Acanthaceae.

English, Hindi and Sanskrit Names: **Eng. :** Thornnail Dye; **Hindi :** Kat-sareya, Vajradanti; **Sans. :** Ananta, Bana, Jhinika, Pura, Udyanapaki, Vira.

Distribution : Throughout India and Sri Lanka, Tropical Asia, Tropical and S. Africa.

Parts Used : Whole Plant.

Flowering & Fruiting : September – December.

Folk Use : Decoction of whole plant taken orally as a health tonic.

Active Constituents : These are scutellarein-7-rhamnosylglucoside (*flower*), barlerin and acetylbarlerin (*leaf, stem*), a new acylated flavonoid-luteolin-7-O-(2''-p-coumaroyl)-β-D-glucopyranoside (*root*).

References : Kirtikar & Basu (1984), Singh & Pandey (1998).

116. *Barringtonia acutangula* (L.) Gaertn. ssp. ***spicata*** (Bl.) Payens

Syn.: B. edaphocarpa Gagnep.; *B. pedicellata* Ridley.; *B. spicata* Bl.; *Stavadium acutangulum* (L.) Miers.

Family: Lecythidaceae.

English, Hindi and Sanskrit Names: **Eng. :** Indian Oak; **Hindi :** Hijjal, Ingar, Samudraphal; **Sans. :** Hijjala, Nishula, Samudra-phalah.

Distribution : Throughout India, particularly in the sub-Himalayan tracts from the Ganges eastwards to Assam and in Madhya Pradesh extending into peninsular India (commonly found along banks of streams, edges of swamps and in moist places).

Parts Used : Bark. Fruit. Leaf. Root.

Flowering & Fruiting : Flowering: April – May; fruiting during September – October.

Folk Use : Fruits used as tonic. Leaves and roots used as bitter tonic. Root and bark valued as stimulant.

Active Constituents : Associated properties due to barringtegnol B, C and D, barrigenic acid, methylbarringtogenate and methylacutagenate (*fruit*), barringtogenin, barringtogentin, 2α, 3β, 19α-trihydroxy-olean-12-en-23, 28-dioic acid and 28-O-β-D-glucopyranoside (*seeds*), β-amyrin, barringtogenic acid, oleanolic acid, acutagulic acid, tangulic acid, β-sitosterol and stigmasterol-3β-O-D-glucoside (*leaf*).

References : Ambasta (1986), Bhattacharjee (2001), Farooq (2005), Kaushik & Dhiman (2000), Kumar (2002), Parrotta (2001), Prajapati *et al.* (2006), Siwakoti & Varma (1996).

117. *Basella alba* (L.) Stewart

Syn.: B. rubra L.

Family: Basellaceae.

English, Hindi and Sanskrit Names: **Eng.** : Indian Spinach; **Hindi :** Bonpoe, Lalbaclu, Poikivel; **Sans.** : Apodike, Madushaka, Potaki, Upodika, Vishala.

Distribution : Wild or cultivated throughout India.

Parts Used : Herb. Leaf. Stem.

Flowering & Fruiting : March – December.

Folk Use : Herb used as tonic and narcotic. Stem used as sedative and tonic. Also, stem and leaves used by Irulas of Nilgiri Biosphere Reserve to increase memory power.

Active Constituents : Herb possesses iodine, fluorine, carotenoids, organic acids and vitamin – K.

References : Balasubramanian & Prasad (1996), Kirtikar & Basu (1984), Prajapati *et al.* (2006).

118. *Bauhinia tomentosa* L.

Family: Fabaceae.

English, Hindi and Sanskrit Names: **Eng.** : St. Thomas Tree, Wild Champak; **Hindi** : Kachnar, Kanchana; **Sans.** : Phalgu, Pita, Ushmadugha.

Distribution : N.W. regions, Circars, Carnatic, in dry forests from the Chilka Lake – Tinnevelly, in other parts of India often cultivated; Sri Lanka, China and tropical Africa.

Parts Used : Seeds.

Flowering & Fruiting : July – December.

Folk Use : Seeds used as a tonic.

References : Ambasta (1986), Asolkar *et al.* (1992).

119. *Bauhinia vahlii* Wight and Arn. (Pl. 2C)*

Syn.: B. racemosa Vahl.; *Phanera vahlii* Benth.

Family: Fabaceae.

English, Hindi and Sanskrit Names: **Eng.** : Camel's Foot Climber; **Hindi** : Jallur, Mahul, Maljhan; **Sans.** : Murva.

Distribution : Distributed in Sub-Himalayan tract up to 915m, Assam, Bihar and Madhya Pradesh.

Parts Used : Seeds.

Flowering & Fruiting : Flowering: April – June; fruiting during winter.

Folk Use : Seed paste taken with milk as tonic.

References : Chhetri (2005), Dhiman (2005), Farooq (2005), Kaushik & Dhiman (2000), Kirtikar & Basu (1984), Kumar (2002), Parrotta (2001), Singh *et al.* (2007), Watt (1972).

120. *Bauhinia variegata* L.

Syn.: B. candida Roxb.; *B. chinensis* (DC.) Vogel; *B. decora* Uribe; *B. purpurea* Wall.; *Phanera variegata* (L.) Benth.

Family: Fabaceae.

English, Hindi and Sanskrit Names: **Eng.** : Buddhist Bauhinia, Mountain Obony, Orchid Tree; **Hindi** : Kachnar, Kandan, Rakta Kanchan; **Sans.** : Kachnar, Kanchanara, Kovidarah.

Distribution : Sub-Himalayan tract from Indus eastwards, dry forests of Central, Eastern and Southern India.

Parts Used : Bark. Gum.

Flowering & Fruiting : Flowering: November – December; Fruiting: January – February.

Folk Use : Bark and gum used as tonic.

Active Constituents : Chief constituents are rutin, kaempferol-3-O-α-galactoside, kaempferol-3-O-α-L-rhamnoglucoside, hentriacontane, octacosan-1-ol, sitosterol, stigmasterol, lupeol, 5,7-dimethoxyflavanone-4'-O-α-L-rhamnopyranosyl-β-D-glucopyranoside and 5,7-dihydroflavanone-4'-O-α-rhamnopyranosyl-β-D-glucopyranoside (*bark*).

References : Ambasta (1986), Anonymous (1948), Dhiman (2005), Farooq (2005), Kaushik & Dhiman (2000), Kirtikar & Basu (1984), Kumar *et al.* (2003), Kumar & Lal (2003), Malla & Chhetri (2009), Parrotta (2001), Thakur *et al.* (1989), Watt (1972).

121. *Benincasa hispida* (Thunb.) Cogn.

Syn.: B. cerifera Savi.

Family: Cucurbitaceae.

English, Hindi and Sanskrit Names: **Eng.** : Ash Gourd, Ash Pumpkin, Tallow Gourd, White Gourd; **Hindi** : Golkaddu,

Kohala, Kondha, Petha, Raksa; **Sans.** : Brihatphala, Kusmandah, Pushpaphala, Suphala, Timisha.

Distribution : Cultivated throughout India in plains and hills.

Parts Used : Fruit.

Flowering & Fruiting : October – November.

Folk Use : Fruits used as tonic.

Active Constituents : n-triacontanol, mannitol, several, amino acids, glucose, rhamnose, lupeol and β-sitosterol (*fruit*).

References : Kirtikar & Basu (1984), Prajapati *et al.* (2006), Siwakoti & Varma (1996), Watt (1972).

122. ***Berberis aristata*** DC. (Pl. 2D)*

Syn.: B. cariaria Royle ex Lindl.

Family: Berberidaceae.

English, Hindi and Sanskrit Names: **Eng.** : Indian Berberry; **Hindi :** Chitra, Kashmal, Rasaut; **Sans.** : Daruharidra, Kata, Suvarnavarna.

Distribution : Himalayas: 1,829m-3,048m and Nilgiri hills.

Parts Used : Stem. Root Bark.

Flowering & Fruiting : Flowering during April – June and fruiting in August – October.

Folk Use : Dried stems and root bark used as a bitter tonic for intermittent fevers.

Active Constituents : Aromoline, oxyberberine, oxyacanthine, berberine and taxilamine have been isolated from *root bark*.

References : Ambasta (1986), Dhiman (2005), Thakur *et al.* (1989).

123. ***Berberis chitra*** Lindley

Family: Berberidaceae.

English, Hindi and Sanskrit Names: **Eng.** : Berberis; **Hindi** : Chitra; **Sans.** : Daruharidra.

Distribution : Himalayas: Kashmir – Nepal (1,500-2,400m).

Parts Used : Root. Stem.

Flowering & Fruiting : March – June.

Folk Use : Root and stem extract used as bitter tonic for intermittent fevers.

Active Constituents : Therapeutic properties due to hentriacontane, triacontane, cetyl alcohol, β-sitosterol, γ-sitosterol, dihydrokaempferol, quercetin and oleic, stearic, palmitic and linoleic acids and dihydropalmatine N – oxide (*root*).

Reference : Raju (2000).

124. *Berberis lycium* Royle

Syn.: B. insignis Hook. f.; *B. parkeriana* C.K. Schn.

Family: Berberidaceae.

English, Hindi and Sanskrit Names: **Eng. :** Raisin Berberry; **Hindi :** Chitra, Kushmal; **Sans. :** Daruharidra.

Distribution : Sub-tropical and temperate regions from Kashmir – Uttaranchal on the outer N.W. Himalayas on clearances and along roadsides between 800-2,500m.

Parts Used : Bark. Root.

Flowering & Fruiting : April – July.

Folk Use : Bark and root extract used as bitter tonic.

Active Constituents : Isolated components are umbellatine, berberine, berbamine, berbericine hydrochloride, berbericinine hydroiodide and oxyberberine (*root bark*).

References : Ballabh & Chaurasia (2006), Dwivedi (2007), Kaul (1997), Sharma (2003).

125. *Berberis ulicina* Hk. f. & Th.

Family: Berberidaceae.

English Name: **Eng. :** Berberry.

Distribution : Western Tibet; stony dry places in Nubra: 4,650-5350m.

Parts Used : Bark. Fruit. Root.

Flowering & Fruiting : June – August.

Folk Use : Bark, fruit and roots used as tonic.

References : Ballabh & Chaurasia (2006), Chaurasia *et al.* (2000), Kala (2003).

126. *Berberis umbellata* Wall.

Syn.: B. garhwalensis C. K. Sch.; *B. jaeschkeana* C. K. Sch.

Family: Berberidaceae.

English Name: **Eng. :** Berberry.

Distribution : Temperate Himalaya: 3,000-3,650m, Kumaon – Bhutan.

Parts Used : Root Bark.

Flowering & Fruiting : Flowering during May – June; fruiting in October – November.

Folk Use : Root bark acts as a tonic if taken in small doses.

Reference : Kapur (1996a).

127. *Bergenia ciliata* (Haw) Sternb. (Pl. 2E)

Syn.: B. ligulata Wall.; *Megasea ciliata* Haw.; *Saxifraga ciliata* (Haw.) Royle.

Family: Saxifragaceae.

English, Hindi and Sanskrit Names: **Eng.** : Rockfoil; **Hindi** : Pakhanabheda; **Sans.** : Ashmaghna, Bhimayojini, Giribhita, Shveta.

Distribution : Afghanistan - S.E. Tibet; Temperate Himalaya from Kashmir – Bhutan, Khasia hills, 1,350m.

Parts Used : Leaf. Root.

Flowering & Fruiting : Flowering during March – July; fruiting in August – November.

Folk Use : Leaves and roots used as tonic.

Active Constituents : (+)-afzelachin, bergenin, a C-glycoside, β-sitosterol, (+)catechin-3-gallate (*root*).

References : Asolkar *et al.* (1992), Kirtikar & Basu (1984), Kumar (2002), Manandhar (2002).

128. *Bergenia stracheyi* (Hook f. & Thoms) Engl.

Syn.: Saxifraga ciliata Lindl. non Royle; *S. stracheyi* Hook. f. & Thoms.

Family: Saxifragaceae.

English, Hindi and Sanskrit Names: **Eng.** : Himalayan Bergenia; **Hindi :** Pasanbhed, Pashanbheda; **Sans. :** Pashanbheda.

Distribution : Commonly found from Kashmir - Uttar Pradesh: 2,400 – 4,800m.

Parts Used : Rhizome.

Flowering & Fruiting : June – September.

Folk Use : Rhizome acts as a tonic.

Active Constituents : Isolated components are gallic acid, tannic acid, mucilage, wax, glucoside, albumens and starch, bergenin, β-sitosterol and (+)-catechin-3-gallate (*rhizome*).

References : Jain & Goel (2005), Sharma (2003).

129. *Beta vulgaris* L.

Syn.: B. maritima L.

Family: Chenopodiaceae.

English, Hindi and Sanskrit Names: **Eng.** : Beetroot, Garden Beet; **Hindi :** Chukkander; **Sans. :** Palakya.

Distribution : Cultivated in many parts of India; grown in northern India in winter and in Bombay and parts of south India throughout the year.

Parts Used : Leaf. Root.

Folk Use : Leaves and roots used as tonic.

References : Kirtikar & Basu (1984), Prajapati *et al.* (2006), Reddi *et al.* (2005).

130. *Betula utilis* D. Don.

Syn.: B. bhojpattra Wall.; *B. castanae* Herb.; *B. jacquemontii* Spach.

Family: Betulaceae.

English, Hindi and Sanskrit Names: **Eng. :** Himalayan Silver Birch, Indian Birch Tree, Indian Paper Birch; **Hindi :** Bhojpatra, Bhujpatar; **Sans. :** Bhurjah, Charmi, Shitri, Vichitraka.

Distribution : Alpine regions of the country between 2,900-3,900m.

Parts Used : Bark.

Flowering & Fruiting : June – September.

Folk Use : Bark used as tonic.

Active Constituents : Betulin, lupeol, oleanolic acid, acetyl oleanolic acid and leucocyanidin, karachic acid have been isolated from *bark.*

References : Kirtikar & Basu (1984), Reddi (2005), Sharma (2003).

131. *Bidens pilosa* auct. (Non L.) Hook. f.

Syn.: B. biternata (Lour.) Merr. & Sherff ; *B. chinensis* Willd.

Family: Asteraceae.

Hindi Name: **Hindi :** Chirchitta.

Distribution : Common throughout India.

Parts Used : Whole Plant.

Flowering & Fruiting : June – September.

Folk Use : Herb used as tonic and stimulant.

References : Ambasta (1986); Parrotta (2001).

132. *Bidens tripartita* L.

Family: Asteraceae.

English Names: **Eng. :** Bur Marigold, Three-lobed Butterbur, Three-lobe Beggarticks, Water Agrimony.

Distribution : Throughout Europe and in other temperate regions including Australia and New Zealand.

Parts Used : Whole Plant.

Flowering & Fruiting : July – November.

Folk Use : Plant used as sedative.

Active Constituents : Associated properties due to flavonoids, xanthophylls, volatile oil, acetylenes, sterols and tannins, essential oil, several flavons, luteolin and its 7-glucoside, butin glucosides, coumarins, umbelliferon, scopoletin and dihydroxycoumarin (*herb*).

References : Ambasta (1986), Asolkar *et al.* (1992).

133. *Biophytum sensitivum* (L.) DC.

Syn.: Oxalis sensitiva L.

Family: Oxalidaceae.

English, Hindi and Sanskrit Names: **Eng. :** Sensitive Wood-sorrel; **Hindi :** Lakchana, Lajjalu, Laksana; **Sans. :** Jalapushpa, Krichhraha, Viparitalajjalu.

Distribution : Throughout the hotter parts of India, weed in moist shady places; Tropical Asia, Africa and America.

Parts Used : Whole Plant.

Flowering & Fruiting : October – November.

Folk Use : Plant used as tonic and a mild stimulant.

Active Constituents : Known to possess insulin (*plant*).

References : Asolkar *et al.* (1992), Farooq (2005), Kirtikar & Basu (1984), Parrotta (2001), Prajapati *et al.* (2006), Singh & Srivastava (2000).

134. *Bistorta affinis* (D. Don.) Greene

Syn.: B. vivipara (L.) S.F. Gray; *Persicaria affinis* (D. Don) Ronse Decr.; *Polygonum affine* D. Don.

Family: Polygonaceae.

English and Hindi Names: **Eng. :** Himalayan Fleece Flower, Knot Weed; **Hindi :** Chookroo.

Distribution : Himalayas (Kashmir to Nepal): 2,400 – 4,800m.

Parts Used : Root.

Flowering & Fruiting : Flowering during August – September.

Folk Use : Roots used as tonic.

Reference : Maity *et al.* (2003).

135. *Bixa orellana* L.

Family: Bixaceae.

English, Hindi and Sanskrit Names: **Eng. :** Arnatto, Chinese Dye Tree, Lipstick Tree, Monkey Tree, Roucou; **Hindi :** Kesari, Latkan, Vatkana; **Sans. :** Raktabija, Sinduri, Trivapushpi.

Distribution : Largely cultivated throughout India; indigenous to tropical America.

Parts Used : Seeds.

Flowering & Fruiting : July – December.

Folk Use : Seeds used as tonic.

Active Constituents : Seeds possess geranylgeraniol, bixin, faresyl-acetone, geranylgeranyl octadecanoate, geranylgeranyl formate, γ-tocotrienol and apocarotenoid.

Reference : Dwivedi (2007).

136. *Blepharis asperrima* Nees

Syn.: Cynarospermum asperrimum (Nees) Vollesen.

Family: Acanthaceae.

English and Hindi Names: **Eng. :** Hill Blepharis; **Hindi :** Akada.

Distribution : W. India, widely distributed from the Nilgiris to the northern end of the W. Ghats.

Parts Used : Leaf. Seeds.

Flowering & Fruiting : November – March.

Folk Use : Leaves and seeds used as tonic.

Reference : Srivastava *et al.* (2003b).

137. *Blepharis linariaefolia* Pers.

Syn.: Acanthium grossum Wight.; *B. sindica* Stocks ex T. Anders.

Family: Acanthaceae.

Hindi and Sanskrit Names: **Hindi :** Bhaugari, Gokhru; **Sans. :** Ushtrakaandi, Utangan.

Distribution : Punjab, Haryana, Rajasthan and Gujarat.

Parts Used : Whole Plant.

Flowering & Fruiting : August – November.

Folk Use : Plant boiled with milk taken as a tonic for sound health.

Active Constituents : Major constituents are betaine, allantoin, oleanolic acid, apigenin, terniflorin, pruning-6''-O-coumarate, blepharin, β-sitosterol, n-hentriacontane (*seeds*).

References : Asolkar *et al.* (1992), Singh & Pandey (1998).

138. *Blepharis persica* (Burm. f.) Kuntze

Syn.: B. edulis Pers.

Family: Acanthaceae.

English, Hindi and Sanskrit Names: **Eng. :** Acanthus; **Hindi :** Chaupatia; **Sans. :** Babhru, Chachu, Kukkuta, Shikhi.

Distribution : Known to occur wild in Punjab and western Rajasthan.

Parts Used : Leaf.

Flowering & Fruiting : August – September.

Folk Use : Leaves used as a tonic.

References : Ambasta (1986), Dhiman (2005).

139. ***Blumea lacera*** (Burm. f.) DC.

Syn.: B. subcapitata DC.; *Conyza lacera* Burm. f.

Family: Asteraceae.

English, Hindi and Sanskrit Names: **Eng. :** Blumea, Malay Blumea; **Hindi :** Jangali Muli, Kukaraundha, Kukronda; **Sans. :** Kukkuradry, Kukundarah, Kukuradru.

Distribution : Throughout India, in areas up to 600m.

Parts Used : Whole Plant. Leaf.

Flowering & Fruiting : Flowering and fruiting during winter season.

Folk Use : Plant used as a bitter tonic and stimulant. Juice of leaves also considered stimulant.

Active Constituents : Known to yield 5-hydroxy-3,6,7,3'4-pentanethoxyflavone, α-amyrin, hentriacontane, hentriancontanol, lupeol and its acetate (*plant*), coniferyl alcohol derivatives, campesterol, flavonoides, essential oil, cineol and d-fenchone (*leaf*).

References : Anonymous (1948), Bhalla *et al.* (1996), Dhiman (2005), Kaushik & Dhiman (2000), Prajapati *et al.* (2006), Ranjan (2003), Sharma (2004), Siwakoti & Varma (1996), Watt (1972).

140. ***Boerhavia diffusa*** L.*

Syn.: B. glabrata Blume.; *B. procumbens* Roxb.; *B. repens* L.

Family: Nyctaginaceae.

English, Hindi and Sanskrit Names: **Eng. :** Hogweed, Pigweed; **Hindi :** Bishkhapra, Gadahpurna, Sant; **Sans. :** Punarnava, Raktakanda, Sophaghni.

Distribution : Throughout India, as a weed in wastelands and road sides.

Parts Used : Whole Plant. Root. Seeds.

Flowering & Fruiting : Major part of the year.

Folk Use : Plant considered as cardio tonic. Seeds and root of the plant crushed with little amount of water taken with milk in the morning as tonic.

Active Constituents : Isolated constituents are triacontanol, hentriacontane, β-sitosterol, ursolic acid, 5,7-dihydroxy-3,4dimethoxy-6,8-dimethyl flavones, glucose, fructose, sucrose and hypoxanthine-9-L-arabinoside, β-ecdysone (*root*).

References : Asolkar *et al.* (1992), Khanna *et al.* (1996), Kirtikar & Basu (1984), Parrotta (2001), Prajapati *et al.* (2006), Sharma (2009).

141. *Bombax ceiba* L.*

Syn.: B. heptaphylla Cav.; *B. malabaricum* DC.; *Gossampinus malabarica* (DC.) Merrill; *G. rubra* Ham.; *Salmalia malabarica* (DC.) Schott & Endl.

Family: Bombacaceae.

English, Hindi and Sanskrit Names: **Eng.** : Silkcotton Tree; **Hindi** : Kaanti-senbal, Pagun, Rakat-senbal, Shembal; **Sans.** : Kantakadruma, Rakta-pushpa, Salmali.

Distribution : Common up to 1,500m throughout India.

Parts Used : Whole Plant. Bark. Gum. Root.

Flowering & Fruiting : February – May.

Folk Use : Plant, bark and gum used as tonic. Root used as tonic and stimulant. Stem bark decoction prescribed as a cooling tonic.

Active Constituents : These are lupeol, β-sitosterol, D-glucoside (*bark*), arabinose, galactose, galacturonic acid, catechu-tannic acid and traces of rhamnose (*gum*), 3', 4' – 5,7-tetrahydroxy-6-methoxyflavan-3-O-β-D-glucopyranosyl-α-D-xylopyranoside (I), β-sitosterol and 1,6-dihydroxy-3-methyl-5-isopropyl-7-methyoxy-8-naphthalenecarboxylic acid (8→) lactone (VI) (*root*).

References : Ambasta (1986), Asolkar *et al.* (1992), Chauhan *et al.* (2007), Farooq (2005), Girach & Aminuddin (1995), Karnick *et al.* (1981), Kaushik & Dhiman (2000), Kirtikar & Basu (1984), Kumar & Lal (2003), Nadeem *et al.* (2001), Parrotta (2001), Prajapati *et al.* (2006), Ranjan (2000), Sharma (2003, 2004), Singh (2001), Singh & Pandey (1998), Siwakoti & Varma (1996), Thakur *et al.* (1989), Watt (1972).

142. *Borassus flabelliffer* L.

Syn.: B. flabelliformis Roxb.

Family: Arecaceae.

English, Hindi and Sanskrit Names: **Eng.** : Brab Tree, Desert Palm, Fan Palm, Palmyra Palm; **Hindi** : Tad, Tal, Tar; **Sans.** : Chirayu Talah, Taladrumah.

Distribution : Dry parts, more or less all over India; common along coastal areas of Peninsula, Bihar and Bengal. Also, grown as ornamental plant.

Parts Used : Root. Stem. Fruit.

Flowering & Fruiting : December – July.

Folk Use : Fermented juice of fruits used as a tonic; fruits also used as sedative. Roots considered stimulant. 'Toddy' – a fermented drink obtained from upper portion of stem given as tonic for weakness.

References : Asolkar *et al.* (1992), Dhiman (2005), Kaushik & Dhiman (2000), Kirtikar & Basu (1984), Kumar & Rao (2001), Prajapati *et al.* (2006), Reddi *et al.* (2005), Sharma (2004), Watt (1972).

143. *Borreria articularis* (L. f.) Willd.

Syn.: B. hispida (L.) K. Schum.; *Spermacoce articularis* L.f.; *S. hispida* L.

Family: Rubiaceae.

English, Hindi and Sanskrit Names: **Eng.** : Shaggy Button Plant; **Hindi** : Madanaghanti, Vasuka; **Sans.** : Madanaghanti, Pasupata.

Distribution : Himalayas from neighbourhood of Shimla eastwards and south to Sri Lanka, Malay Peninsula, Malay Archipelago, Philippines and China.

Parts Used : Seeds.

Flowering & Fruiting : August – November.

Folk Use : Seeds used as stimulant in Darjeeling.

Active Constituents : Known to yield isorhamnetin (*seeds*).

References : Mukherjee *et al.* (1986), Parrotta (2001).

144. *Boswellia serrata* Roxb. ex Colebr.

Syn.: B. glabra Roxb.; *B.thurifera* Roxb.; *Libanus thurifera* Colebr.

Family: Burseraceae.

English, Hindi and Sanskrit Names: **Eng.** : Indian Frankincense, Indian Olibanum Tree; **Hindi :** Kundur, Luban, Salai; **Sans.** : Ashvamutri, Kunduruskah, Sallaki, Yakshadhupa.

Distribution : Dry hills throughout Central and N. India, common at the foot of W. Himalayas, Rajasthan, Gujarat, Maharashtra, Madhya Pradesh, Bihar, Orissa, Andhra Pradesh and further south in the Peninsula.

Parts Used : Bark. Whole Plant.

Flowering & Fruiting : Flowering during February – April; fruiting in May – June.

Folk Use : Bark used as a tonic. Plant valued as cardio tonic.

Active Constituents : Active properties due to tannins, pentosans, lignin, holocellulose and β-sitosterol (*plant*), carbohydrates, glucosides, β-sitosterol (*bark*).

References : Asolkar *et al.* (1992), Kirtikar & Basu (1984), Prajapati *et al.* (2006).

145. *Brassica juncea* Czern. & Coss.

Syn.: Sinapis juncea L.; *S. ramosa* Roxb.; *S. rugosa* Roxb.

Family: Brassicaceae.

English, Hindi and Sanskrit Names: **Eng.** : Brown Mustard, Indian Mustard; **Hindi** : Mohari, Rayi, Sarson; **Sans.** : Rajika, Sarsapah.

Distribution : Cultivated throughout India, westwards to Egypt and eastwards to China.

Parts Used : Seeds.

Flowering & Fruiting : Throughout the year.

Folk Use : Seeds used as tonic.

Active Constituents : Associated properties due to sinigrin, gluconapin, sinapine base and a volatile isothycoynate, 24-methylene-25-methylcholesterol (*seeds*).

Reference : Prajapati *et al.* (2006).

146. *Brassica oleracea* L.

Family: Brassicaceae.

English, Hindi and Sanskrit Names: **Eng.** : Colewort, Sea Cabbage, Wild Cabbage; **Hindi** : Bantgopi, Karamkalla; **Sans.** : Dalamalini, Kembukah, Kemukh, Polini.

Distribution : Cultivated on hill tops of India.

Parts Used : Leaf.

Flowering & Fruiting : Flowering during April.

Folk Use : Leaves used as cardio tonic.

Active Constituents : Known to possess 3-sophoroside-7-glucosides of kaempferol, quercetin, isorhamnetin, kaempferol-3-O-β-D-[2E-p-coumaroyl-β-D-glucosyl(1→2)-glucopyranoside]-7-O-β-D-glucopyranoside, kaempferol-3-O-β-D-[2E-caffeoyl-β-D-glucosyl-(1→2)-glucopyranoside]-7-O-β-D-glucopyranoside, quercetin-3-O-β-D-[2E-feruloyl-β-D-glucosyl (1→2)-glucopyranoside]-7-O-β-D-glucopyranoside and quercetin-3-O-β-D-[2E-caffeoyl-β-D-glucosyl (1→2)-glucopyranoside]-7-O-β-D-glucopyranoside (*leaf*).

References : Kirtikar & Basu (1984), Prajapati *et al.* (2006).

147. *Brassica oleracea* L. var. ***botrytis*** L.*

Family: Brassicaceae.

English and Hindi Names: **Eng.** : Broccoli, Cauliflower; **Hindi** : Phool Gobhi.

Distribution : Native of W. and S. Europe; cultivated in India as vegetable in hills as well as in the plains.

Parts Used : Flower.

Flowering & Fruiting : Flowering during April.

Folk Use : Cauliflower used as cardio tonic.

Active Constituents : S-Me-L-cysteine sulfoxide, vitamin U and folic acid have been isolated from *cauliflower*.

Reference : Kaushik & Dhiman (2000).

148. *Broussonetia papyrifera* Vent. (Pl. 2F)

Family: Moraceae.

English and Hindi Names: **Eng. :** Paper-mulberry, Tapa – cloth Tree; **Hindi :** Jangali Toot.

Distribution : Punjab - Arunachal Pradesh, Meghalaya, W. Bengal and in Peninsular India up to 1,000m.

Parts Used : Seeds.

Flowering & Fruiting : Flowering: March – May; Fruiting: April – August.

Folk Use : Seeds used as tonic.

References : Ambasta (1986), Asolkar *et al.* (1992), Parrotta (2001), Raju (2000).

149. *Bryonopsis laciniosa* (L.) Naud.

Syn.: Bryonia laciniosa L.

Family: Cucurbitaceae.

English, Hindi and Sanskrit Names: **Eng. :** Lollipop Plant; **Hindi :** Gargumaru, Ishwaralingi, Shivalingi; **Sans. :** Apastambhini, Bahupatra, Chandra, Isvari, Lingini, Pandoli, Shivavalli.

Distribution : Throughout India from the Himalaya – Sri Lanka; Mauritius, tropical Africa, Malaya, Philippines and Australia.

Parts Used : Whole Plant.

Flowering & Fruiting : August – December.

Folk Use : Herb used as tonic.

References : Ambasta (1986), Anonymous (1948), Kirtikar & Basu (1984).

150. *Buchanania lanzan* Spreng.

Syn.: B. latifolia Roxb.

Family: Anacardiaceae.

English, Hindi and Sanskrit Names: **Eng. :** Almondi Tree, Cheronjee; **Hindi :** Achar, Char, Ciraumci, Piyar; **Sans. :** Akhatta, Dhanushpatta, Priyalah, Upavata.

Distribution : Hot and drier parts of India from Kumaon through Central India up to Western Peninsula.

Parts Used : Leaf. Seeds. Fruit. Gum.

Flowering & Fruiting : Flowering during January – March; fruiting in March – May.

Folk Use : Fruits, seeds, gum and leaves considered as tonic, nervine tonic and cardio tonic.

Active Constituents : These are amino acids, linoleic, myristic, oleic, palmitic and stearic acids, gallotannins, triterpenoides, saponins and reducing sugars, kaempferol-7-O-glucoside, quercetin-7-O-rhamnoside and quercetin-3-O-rhamnoglucoside (*seeds*), moisture, protein, minerals calcium, phosphorus, iron, oxalic acid, magnesium, sodium, potassium, copper, sulphur, chlorine, thiamine, riboflavin, niacin and vitamin C (*fruit*).

References : Kaushik & Dhiman (2000), Kirtikar & Basu (1984), Painuli & Maheshwari (1996), Prajapati *et al.* (2006).

151. *Bunium persicum* (Boiss.) Fedt.

Syn.: Carum bulbocastanum Clarke

Family: Apiaceae.

English, Hindi and Sanskrit Names: **Eng. :** Black Caraway, Black Cumin; **Hindi :** Himali Jira, Kala-jirah, Kashmiri Jira, Shahi Jira; **Sans. :** Krishna Jiira.

Distribution : Found only in Kashmir and Himachal Pradesh.

Parts Used : Fruit. Seeds.

Flowering & Fruiting : July – August.

Folk Use : Fruits and seeds valued as stimulant.

Active Constituents : p – cymene and cuminaldehyde have been isolated (*seeds*).

References : Chauhan (2001), Sharma (2003).

152. *Bupleurum falcatum* L.

Syn.: B. chinense DC.; *B. scorzonerifolium* Willd.

Family: Apiaceae.

English and Hindi Names: **Eng. :** Sickle Hare's Ear, Thorow-wax; **Hindi :** Jangli Jeera, Singu.

Distribution : Himalayas (Kashmir – Bhutan) 1,000-4,000m; Temperate E. Asia to Japan, W. and Central Asia, S. and Central Europe.

Parts Used : Root.

Flowering & Fruiting : June – October.

Folk Use : About 4cm of root grounded with water taken as a tonic.

Active Constituents : Associated properties due to malonylsaikosaponins a and d, a new chromone-saikochromone A, saikogenin D, saikogenin F, saikogenin G, 24-ethylcholesta-7,22E-dien-3β, 5α,6β-triol, 24ethylcholest-22E-en-3β,5α,6β-triol, nortrachelogenin, saikodiynes A, B and C (*root*).

Reference : Shah & Singh (1990).

153. *Butea monosperma* (Lam.) Taub.*

Syn.: B. frondosa Koenig ex Roxb.; *Erythrina monosperma* Lam.; *Plaso monosperma* Kuntze.

Family: Fabaceae.

English, Hindi and Sanskrit Names: **Eng. :** Bal Kino, Bastard Teak, Flame of The Forest; **Hindi :** Chichra, Dhak, Palas; **Sans. :** Bijasneha, Ksharashreshtha, Palasah, Vatapatha.

Distribution : Deciduous forests throughout India, up to 1,200m and in open areas.

Parts Used : Bark. Flower. Gum. Leaf. Seeds.

Flowering & Fruiting : March – May.

Folk Use : Bark, flowers, gum, leaves and seeds considered as tonic. Also, seed powder mixed with milk given daily to children as a body tonic.

Active Constituents : These are butrin, isobutrin and coreopsin (*flower*), β-sitosterol, leucoanthocynidin, amyrin and palasonin (*root*), tetramers of leucocyanidin (*bark*), α-amyrin, β-sitosterol, its glucoside, sucrose and a new lactone-n-hemicosanoic acid-δ-lactone (*seeds*).

References : Bhandary *et al.* (1996), Dwivedi (2007), Kirtikar & Basu (1984), Kumar (2002), Kumar & Lal (2003), Mishra (2008), Parrotta (2001), Prajapati *et al.* (2006), Sanjay *et al.* (2007), Sharma (2003), Singh & Pandey (1998), Siwakoti & Varma (1996), Watt (1972).

154. *Butea superba* Roxb.

Family: Fabaceae.

English, Hindi and Sanskrit Names: **Eng. :** Red Kwao Kruva; **Hindi :** Bodha, Chihumt, Palas Lata; **Sans. :** Latapalasha.

Distribution : Throughout Central and Southern India.

Parts Used : Root. Seeds.

Flowering & Fruiting : April (flowering); June (fruiting).

Folk Use : Decoction of seeds used as sedative. Roots valued as cardiotonic.

References : Ambasta (1986), Parrotta (2001).

155. *Caccinea glauca* Savi.

Syn.: C. crassifolia O. Kuntze.

Family: Boraginaceae.

Distribution : Fleshy herb of Persia, Baluchistan; sold in Indian market as Gaozaban.

Parts Used : Whole Plant.

Folk Use : Plant accredited with tonic properties.

Active Constituents : Active properties due to caccigenin (2α,3β,21β,23-tetrahydroxyolean-12-en-28-oic acid), caccigenin lactone and 23-deoxycoccigenin (*leaf, stem*), and new pyrrolizidine alkaloid G – 1 (*flower*).

References : Ambasta (1986), Anonymous (1950), Kirtikar & Basu (1984).

156. *Caesalpinia crista* L.

Syn.: Bianacaea scandens Tod.; *Caesalpinia bonduc* (L.) Roxb.; *C. bonducella* (L.) Flem.; *C. ferox* Hassk.; *C. japonica* Sieber & Zucc.; *C. sepiaria* Roxb.; *Guilandina bonduc* L.

Family: Fabaceae.

English, Hindi and Sanskrit Names: **Eng. :** Bonduc Nut, Fever Nut, Physic Nut; **Hindi :** Karanju, Kat-karanja; **Sans. :** Kantakikaranjah, Kuberakshi, Lata-karanj, Putikaranja.

Distribution : Throughout the hotter parts of India particularly along the sea coast and up to 830m, W. Bengal and South India; cultivated as hedge plant in villages.

Parts Used : Fruit. Seeds.

Flowering & Fruiting : July – September (flowering); December January (fruiting).

Folk Use : Fruits and seeds used as tonic and hallucinogenic.

Active Constituents : Known to contain L-γ-ethylidenegluatamic acid and amino acids, α-, β-, γ-, δ-, ξ– and ψ-casesalpins caesalpin-F and 7-hydroxyl-4'-methoxyhomoisoflavone (*seeds*).

References : Ambasta (1984), Anonymous (1950), Dhiman (2005), Kaushik & Dhiman (2000), Kirtikar & Basu (1984), Parrotta (2001), Prajapati *et al.* (2005), Ranjan (2003), Schultes (1993), Watt (1972).

157. *Caesalpinia pulcherrima* (L.) Sw.

Syn.: Poinciana pulcherrima L.

Family: Fabaceae.

English, Hindi and Sanskrit Names: **Eng. :** Barbados Pride, Peacock Flower, Spanish Casrnation; **Hindi :** Gulutora; **Sans. :** Ratnagandhi, Sidhanasha.

Distribution : Universally cultivated throughout India.

Parts Used : Bark. Leaf.

Flowering & Fruiting : Flowering: September – November; Fruiting: March – April.

Folk Use : Bark and leaves used as tonic and stimulant.

Active Constituents : β-sitosterol, sebacic acid, quercimertrin, lecodelphinidin, 2-ellagitannins and caesalpin have been identified (*bark*).

References : Ambasta (1986), Anonymous (1950), Asolkar *et al.* (1992), Deshmukh *et al.* (2000), Farooq (2005), Kirtikar & Basu (1984), Parrotta (2001), Singh & Srivastava (2000).

158. *Calamintha clinopodium* Benth.

Syn.: C.vulgaris (L.) Druce.; *Clinopodium vulgare* L., *Melissa clinopodium* Benth.; *Satureja vulgaris* (L.) Fritsch.

Family: Lamiaceae.

English Name: **Eng. :** Wild Basil.

Distribution : W. Temperate Himalayas (Kashmir – Kumaon) : 1,350-4,000m; Europe, N. Africa, N. and W. Asia and Canada.

Parts Used : Whole Plant. Shoot.

Flowering & Fruiting : July – September.

Folk Use : Plant used as tonic. Shoots used as heart tonic and stimulant.

Active Constituents : Active properties due to mixture of ursolic acid, oleanolic acid and betulin (*plant*).

References : Ambasta (1986), Kala (2003), Kirtikar & Basu (1984).

159. *Calamus rotang* L.

Syn.: C. roxburaghii Griff.

Family: Arecaceae.

English, Hindi and Sanskrit Names: **Eng. :** Chair-bottom Cane, Common Rattan, Rattan; **Hindi :** Bemt, Bet; **Sans. :** Abhrapushpa, Latavamsa, Nichula, Ratha, Vetrah.

Distribution : Dry tracts of Central and South India up to 450m.

Parts Used : Root.

Flowering & Fruiting : Flowering: beginning of rainy season; Fruiting: cold season.

Folk Use : Roots used as tonic.

References : Kirtikar & Basu (1984), Prajapati *et al.* (2006), Reddi *et al.* (2005).

160. *Calendula officinalis* L. (Pl. 3A)

Family: Asteraceae.

English and Hindi Names: **Eng. :** African Marigold, Marigold, Pot Marigold; **Hindi :** Gul-e-asharfi, Sadberg, Zergul.

Distribution : Native to southern Europe, cultivated in temperate regions around the world; found in the fields of the Punjab and Sind.

Parts Used : Whole Plant.

Flowering & Fruiting : March – May.

Folk Use : Plant used as stimulant and tonic.

Active Constituents : Associated properties due to β-carotene, lycopene, violaxanthin, rubixanthin, hentriacontane, ceryl alcohol, stigmasterol, farodiol, calendin, isorhamnetin-3-rutinoside, isorhamnetin-3-glucoside, five glycosides of oleanolic acid: 3-glucuronide (I), 3-galactosyl-glucuronide (II), 3-galactosyl-glucuronide-17-glucoside (III), 3-galactosyl(glucosyl)-glucuronide (IV) and 3-galactosyl(glucosyl)-glucuronide-17-glucoside (V) (*plant*).

References : Ambasta (1986), Anonymous (1950), Bhalla *et al.* (1996), Bhattacharjee (2001), Kayal (2003).

161. *Callicarpa arborea* Roxb.

Family: Verbenaceae.

Distribution : Upper Gangetic plains, lower hills of Kumaon – Sikkim, Assam, Khasia hills, Bengal, Myanmar and Sumatra.

Parts Used : Bark.

Flowering & Fruiting : Flowers at various times throughout the year.

Folk Use : Bark used as tonic.

Active Constituents : Bark yields Me-betulinate, barurenol and β-sitosterol-OAC.

References : Kirtikar & Basu (1984), Tiwari & Tiwari (1996), Watt (1972).

162. *Calotropis gigantia* (L.) R. Br. ex Ait.

Syn.: Ascelepias herbacea Roxb.

Family: Asclepiadaceae.

Hindi and Sanskrit Names: **Hindi :** Aak, Lalak, Lalmadar, Madar, Safedak; **Sans. :** Aditya, Aharbandhava, Arka, Bhanu, Divakar, Ganarupa, Haridashva, Jambhala, Kandaka, Prabhakara, Pratapa, Ravi, Shitapushpaka, Tulaphala.

Distribution : Commonly occurs in wastelands, roadsides and railway embankments, ascending up to 1,000m in the Himalayas from Punjab – Assam.

Parts Used : Flower. Root.

Flowering & Fruiting : Flowering and fruiting is almost throughout the year.

Folk Use : Flowers and roots used as tonic.

Active Constituents : Known to yield calotroposides A and B, five new oxypregnane oligoglycosides-calotroposides C, D, E, F and G (*root*).

References : Dhiman (2005), Dwivedi (2007), Kirtikar & Basu (1984), Parrotta (2001), Subramaniam (2000), Watt (1972).

163. *Calotropis procera* (Ait.) Ait. f. ssp. ***hamiltonii*** (W.) Ali

Syn.: Asclepias procera Willd.; *Calotropis hamiltonii* W.; *C. procera* auct. non (Ait.) (f.); *C. wallichii* Wight.

Family: Asclepiadaceae.

English, Hindi and Sanskrit Names: **Eng. :** Gigantic Swallow Wort, Madar Milk-weed; **Hindi :** Aak, Aakada, Madar; **Sans. :** Alarka, Bimbora, Ganarupi, Raktarka, Shambhu, Tapana, Vasuka, Vedha, Vrittamallika.

Distribution : More or less throughout warm dry places of India from Punjab – Western, Central and Southern India. Abundant in Sub-Himalayan tracts and the adjacent plains in north-east.

Parts Used : Whole Plant. Flower.

Flowering & Fruiting : February – September.

Folk Use : Flowers and dried plant used as a tonic.

Active Constituents : Isolated components are α-and β-amyrins, an alkaline phosphate, cyanidin-3-rhamnoglucoside, cycloart-23-en-3β, 25-diol, cyclosadol, multiflorenol, procesterol, quercetin-3-rutinoside, β-sitosterol, β-sitost-4-en-3-one and stigmasterol (*flower*).

References : Dhiman (2005), Kaushik & Dhiman (2000), Kirtikar & Basu (1984), Prajapati *et al.* (2005), Reddi *et al.* (2005).

164. *Calycopteris floribunda* Lam.

Syn.: Getonia floribunda Roxb.

Family: Combretaceae.

Hindi and Sanskrit Names: **Hindi :** Kokaray, Minnargodi; **Sans. :** Shvetadhataki, Susavi.

Distribution : Distributed in deciduous forests in South India, Orissa, Assam and Chittagong.

Parts Used : Leaf.

Flowering & Fruiting : March – May.

Folk Use : Leaves considered as tonic.

Active Constituents : These are n-octacosanol, sitosterol, calycopterin, ellagic acid and quercetinproanthocyanidin (*leaf*).

References : Ambasta (1986), Dhiman (2005).

165. *Camellia sinensis* (L.) O. Kuntze*

Syn.: C. thea Link.; *C. theifera* Griff.; *Thea assamica* Masters; *T. sinensis* L.

Family: Theaceae.

English, Hindi and Sanskrit Names: **Eng. :** Assam Tea, China Tea, Tea Plant; **Hindi :** Chai; **Sans. :** Caha, Chavika, Syamaparni.

Distribution : Throughout India, cultivated at high altitudes.

Parts Used : Leaf.

Flowering & Fruiting : June – September.

Folk Use : Leaves used as nervine tonic and its beverage considered as stimulant.

Active Constituents : Accredited properties due to nicotinic acid, pantothenic acid, ascorbic acid, malic acid, oxalic acid, carotene, riboflavin, adenine, theophylline, theobromine, xanthine, hypoxanthine, gums, dextrins and inositol (*leaf*).

References : Farooq (2005), Prajapati *et al.* (2006), Reddi *et al.* (2005), Sharma (2004), Sood *et al.* (2009).

166. *Canna edulis* Ker Gawl.

Syn.: C. bifida Roem.; *C. chinensis* Willd.; *C. floribunda* Bouche; *C. indica* L.; *C. occidentalis* Roscoe; *C. orientalis* Rosc.; *C. variabilis* Willd.; *Xyphostylis lutea* (Mill.) Raf.

Family: Cannaceae.

English, Hindi and Sanskrit Names: **Eng. :** Indian Bead, Indian Reed, Indian Shot; **Hindi :** Keli, Sabbajaya, Sarvajjya; **Sans. :** Devakili, Sarvajaya, Silarumba.

Distribution : Native of Central and South America, naturalized in Asia and West Indies; common all over India and Sri Lanka, chiefly in gardens as an ornamental flowering plant.

Parts Used : Whole Plant. Root.

Flowering & Fruiting : Throughout the year.

Folk Use : Plant used as a stimulant. 'Rokshi', a local drink prepared from its roots and corns of *Zea mays* L. given in a small quantity before and after food as a health tonic.

References : Kirtikar & Basu (1984), Kumar & Rao (2001), Saikia & Nath (2003b).

167. *Cannabis sativa* L. (Pl. 3B)*

Syn.: C. indica Lamk.

Family: Cannabinaceae.

English, Hindi and Sanskrit Names: **Eng. :** Marijuana, Soft Hemp, True Hemp; **Hindi :** Bhang, Charas, Ganja, Kinnab, Siddhi; **Sans. :** Bhanga, Ganjka, Indrasana, Shivpriya, Ujaya, Vijaya.

Distribution : Native of Central and Western Asia, naturalized in the Sub-Himalayan tract and abundant in wastelands from Punjab eastwards to Bengal and Bihar extending southwards.

Parts Used : Whole Plant. Flower. Leaf.

Flowering & Fruiting : March – May.

Folk Use : Dried flowering tops of female plant used as sedative and narcotic. Plant and its leaves valued as tonic.

Active Constituents : Plant rich in cannabin, tetrahydrocannabinol, cannabidiol, cannabichromene, cannabicyclol, cannabigerol, cannabielsion, cannabitetrol, cannabinodiol, cannabicumaronone and cannabiglendol.

References : Ambasta (1986), Anonymous (1950), Asolkar *et al.* (1992), Bhattacharjee (2001), Chauhan *et al.* (2007), Chhetri & Shrestha (2004), Dhiman (2005), Dwivedi (2007), Farooq (2005), Kaushik & Dhiman (2000), Kirtikar & Basu (1984), Kumar (2002), Kumar *et al.* (2003), Negi *et al.* (1999), Prajapati *et al.* (2006), Ranjan (2000), Reddi *et al.* (2005), Saikia & Nath (2003b), Sharma (2003, 2004), Singh (2000a, 2001), Singh & Kumar (2000a), Singh & Pandey (1998), Siwakoti & Varma (1996), Trivedi (2002), Watt (1972).

168. *Canscora decussata* (Roxb.) Schult.

Syn.: Pladera decussate Roxb.

Family: Gentianaceae.

English, Hindi and Sanskrit Names: **Eng.** : Canscora; **Hindi** : Kalameg, Samkhaphuli, Shankhini; **Sans.** : Akshapida, Dandotpala, Netramila, Tunduli, Yavi.

Distribution : Throughout India, in moist areas up to 1,500m; S.E. Asia, Australia and Africa.

Parts Used : Whole Plant.

Flowering & Fruiting : August – September.

Folk Use : Plant used as a tonic and nervine tonic.

Active Constituents : Active properties due to loliolide, gluanone, canscoradione, β-amyrin, n-alkanes, n-alkanols, friedelin, friedelan-3β-ol, sitosterol, stigmasterol and campesterol (*plant*), gluanone, canscoradione, friedelin, friedelan-3β-ol, β-amyrin, sitosterol, stigmasterol, campesterol, 1-methoxy-3,5-dihydroxy-dihydroxyxanthone (IV) and its 3-O-rutinosyl derivative (*aerial parts*), 3,6-dihydroxy-1,5,7-trimethoxyxanthone (I), 5-glucosyl-1,6-dihydroxy-3,7-dimethoxyxanthone (II) and 1,3,5-trihydroxy-6,7-dimethoxyxanthone (*flower*).

References : Kirtikar & Basu (1984), Parrotta (2001), Prajapati *et al.* (2006), Reddi *et al.* (2005), Subramaniam (2000), Watt (1972).

169. *Canscora diffusa* (Vahl.) R. Br.

Syn.: Pladera virgata Roxb.

Family: Gentianaceae.

Distribution : Throughout India, Sri Lanka; Malaya, Australia and Tropical Africa.

Parts Used : Whole Plant.

Flowering & Fruiting : August – October.

Folk Use : 2 tsp plant paste taken orally twice a day for 15 days as a nervine tonic.

References : Khatoon & Goel (2010), Kirtikar & Basu (1984), Mahato & Chaudhary (2005), Siwakoti & Varma (1996).

170. *Canthium parviflorum* Lam.

Syn.: Plectronia parviflora Bedd.; *Webera tetrandra* Willd.

Family: Rubiaceae.

English, Hindi and Sanskrit Names: **Eng.** : Carray Cheddie, Wild Jessamine; **Hindi** : Kadbar, Kirma; **Sans.** : Chayatinisah, Gangeruki, Naga-vala.

Distribution : Scrub forests and dry plains throughout India.

Parts Used : Root.

Flowering & Fruiting : Flowering: hot season; Fruiting: September.

Folk Use : Root used as tonic.

Reference : Prajapati *et al.* (2006).

171. *Capparis deciduas* (Forsk.) Edgew.

Syn.: C. aphylla Roth.; *Sodaela decidua* Forsk.

Family: Capparaceae.

English, Hindi and Sanskrit Names: **Eng.** : Berry, Caper; **Hindi** : Karil, Ker, Kurel; **Sans.** : Granthila, Karira, Tikshnasara.

Distribution : Dry areas throughout India.

Parts Used : Whole Plant. Root.

Flowering & Fruiting : Flowering: March – April & September – October; Fruiting: May – June & November.

Folk Use : Plant used as tonic. Roots considered as stimulant and tonic.

Active Constituents : Capparine, capparilline, capparinine, n-pentacosane, n-triacontanol, β-sitosterol and l-stachydrine, arabinose, galactose and polyphenols (*plant*).

References : Kirtikar & Basu (1984), Prajapati *et al.* (2006).

172. *Capparis sepiaria* L.

Syn.: C. incanescens DC.

Family: Capparaceae.

English, Hindi and Sanskrit Names: **Eng.** : Indian Caper; **Hindi** : Hiun, Kanthari; **Sans.** : Dupravesha, Hinsra, Kakadani.

Distribution : Throughout dry regions of India including the Andaman.

Parts Used : Whole Plant.

Flowering & Fruiting : April – August.

Folk Use : Decoction of the plant given as tonic.

Active Constituents : These are α-amyrin, β-amyrin, taraxasterol, erythrodiol, β-sitosterol, betulin and triterpenoid alcohol (*leaf*), betulin-28-acetate, octacosanol, β-sitosterol, its glucoside and α-amyrin (*plant*).

References : Alagesboopathi *et al.* (2000), Ambasta (1986), Farooq (2005), Kirtikar & Basu (1984), Nayak *et al.* (2003), Parrotta (2001), Ranjan (2003).

173. *Capparis spinosa* L.

Syn.: C. leucophylla DC.

Family: Capparaceae.

English, Hindi and Sanskrit Names: **Eng. :** Caper Bush, Caper Plant; **Hindi :** Kabra, Kachra, Karer, Kiari, Koyal Ber; **Sans. :** Himsa.

Distribution : Native to Mediterranean region; occurs in India in the central and northern parts of the Punjab.

Parts Used : Bark. Root. Root Bark.

Flowering & Fruiting : February – May.

Folk Use : Bark, root and root bark accredited with tonic properties.

Active Constituents : Properties due to glucobrassicin, neoglucobrassicin and 4-methoxyglucobrassicin (*root*).

References : Ambasta (1986), Anonymous (1950), Bhattacharjee (2001), Kirtikar & Basu (1984), Raju (2000), Sharma (2003).

174. *Capparis zeylanica* L.

Syn.: C. accuminata Roxb.; *C.brevispina* DC.; *C. horrida* L. f.

Family: Capparaceae.

English, Hindi and Sanskrit Names: **Eng. :** Caper Plant; **Hindi :** Ardanda; **Sans. :** Karambha, Krishangi, Vartala.

Distribution : Throughout the greater parts of India – Java and Philippines.

Parts Used : Root Bark.

Flowering & Fruiting : November – January.

Folk Use : Root bark used as sedative.

References : Ambasta (1986), Anonymous (1950), Farooq (2005).

175. *Capsella bursa-pastoris* (L.) Medik.

Family: Brassicaceae.

English Names: **Eng. :** Blind Weed, Pick Pocket, Shepherd's Purse, Tooth-wort, Witches Pouches.

Distribution : Cultivated places in temperate India, a weed of cultivation.

Parts Used : Seeds.

Flowering & Fruiting : Throughout the year.

Folk Use : Seeds used as stimulant.

References : Jain & Goel (2005), Kirtikar & Basu (1984), Singh *et al.* (1996).

176. *Capsicum annum* L. (Pl. 3C)*

Syn.: C. frutescens Sesu Cl.

Family: Solanaceae.

English, Hindi and Sanskrit Names: **Eng. :** Red Chilli, Red Pepper; **Hindi :** Gachmirch, Lal Mirch, Wangru; **Sans. :** Lanka, Marich.

Distribution : Extensively cultivated throughout India.

Parts Used : Fruit.

Flowering & Fruiting : Flowering and fruiting almost throughout the year.

Folk Use : Fruits used as stimulant.

Active Constituents : Amino acids, ascorbic acid, folic acid, glutathione, citric, glyceric, isochlorogenic, malic, malonic, oxalic, quinic, shikimic and succinic acids, biflavonoids, α-, β- and ζ-carotenes, capsanthin and its mono- and diester, cryptocapsin, capsorubin diester, lutein, neoxanthin, phytoene, phytofluene, violaxanthin, capsaicin, 6,7-dihydroxy capsaicin, bis-homo, tris-homo capsaicin, dihydrocapsaicin, homo capsaisin I & II, vanillyl amides of caprylic, nonylic and decyclic acids, capsaicinoid O-methyl derivative, N-(13-methyltetradecyl) acetamide, capsidiol (sesqui-terpenoidal diol) luteolin-7-O-glucoside, polyphenol oxidase and a tocopherol have been characterized from *fruits*.

References : Kaushik & Dhiman (2000), Rajan *et al.* (2003), Sharma (2003), Watt (1972).

177. *Caralluma attenuata* W.

Syn.: C. fimbriata Wall.

Family: Asclepiadaceae.

English and Sanskrit Names: **Eng. :** Famine Food; **Sans. :** Yugmaphallottama.

Distribution : South Deccan mountains, Nilgiris, Pulneys, Cochin, Sri Lanka; between Kandy and Bandulla.

Parts Used : Stem.

Folk Use : Stem pieces toasted in castor oil (*Ricinus communis*) prescribed along with salt and onion (*Allium cepa*) as a general tonic for all ailments.

Reference : Binu *et al.* (2003).

178. *Cardamine pratensis* L.

Family: Brassicaceae.

English Names: **Eng. :** Apple Pie, Bird's-eye, Cuckoo-flower, May-flower, Lady's Glove.

Distribution : Kashmir, N. & W. Asia, Europe, Abyssinia and N. America.

Parts Used : Whole Plant. Seeds.

Flowering & Fruiting : April – July.

Folk Use : Plant and seeds used as stimulant.

References : Ambasta (1986), Anonymous (1950), Bhattacharjee (2001), Kirtikar & Basu (1984).

179. *Cardiospermum halicacabum* L.

Syn.: C. microcarpum Kunth.

Family: Sapindaceae.

English, Hindi and Sanskrit Names: **Eng. :** Ballon Vine, Blister Creeper, Heart Pea, Winter cherry; **Hindi :** Kanphuti, Kapalphoti; **Sans. :** Indravalli, Jyotir, Sakralata.

Distribution : Throughout Indian plains.

Parts Used : Whole Plant. Foliage. Seeds.

Flowering & Fruiting : September – November.

Folk Use : Plant has sedative action on the CNS. Seeds used as tonic. Juice extracted from foliage added to 'kanji' (a local carminative drink) and given as health tonic to pregnant women.

Active Constituents : These are l-triacontanol, n-pentacosane, n-triacontane, n-triacontanol, n-non-acosane, plargonidin-3-galactoside, glucocappasolin, β-sitosterol and phthalic acid (*plant*), (+)pinitol, 7-O-glucuronides of apigenin, chrysoeriol and luteolin (*leaf*).

References : Kirtikar & Basu (1984), Parrotta (2001), Prajapati *et al.* (2006), Prasad *et al.* (1996), Sharma (2004), Trivedi (2002), Watt (1972).

180. ***Careya arborea*** Roxb.

Family: Myrtaceae.

English, Hindi and Sanskrit Names: **Eng. :** Patana Oak, Wild Guava; **Hindi :** Kamba, Kuba, Kumbhi, Vakamba; **Sans. :** Bhadrendrani, Girikarnika, Vishaghnika.

Distribution : Throughout India, Sri Lanka, Malay Peninsula and Siam.

Parts Used : Flower. Fruit.

Flowering & Fruiting : March – May.

Folk Use : A cup of flower infusion taken as tonic after child birth for seven days. Fruits also consumed as tonic.

References : Kirtikar & Basu (1984), Kumar & Lal (2003), Painuli & Maheshwari (1996), Parrotta (2001), Reddi *et al.* (2005), Upadhye *et al.* (1994), Watt (1972).

181. ***Carica papaya*** L. (Pl. 3D)*

Syn.: Papaya carica Gaertn.

Family: Caricaceae.

English, Hindi and Sanskrit Names: **Eng. :** Melon Tree, Papaya, Paw-paw; **Hindi :** Anda Kharbuja, Papeeta; **Sans. :** Chirbita, Eranda, Nai Kadala.

Distribution : Cultivated mostly throughout India.

Parts Used : Fruit.

Flowering & Fruiting : Greater part of the year.

Folk Use : Fruits used as tonic and cardio tonic.

Active Constituents : Known to possess myristoleic acid, palmitoleic acid, arachidic acid, pectin and free sugars (*fruit*).

References : Anonymous (1950), Kumar *et al.* (2003), Parrotta (2001), Sood *et al.* (2009).

182. ***Carum carvi*** L.

Syn.: Apium carvi L.; *Seseli carvi* L.

Family: Apiaceae.

English, Hindi and Sanskrit Names: **Eng. :** Caraway, Common Caraway; **Hindi :** Shiajira, Zira; **Sans. :** Bahugandha, Jarana, Nilakana, Sushavi.

Distribution : Native to Europe and W. Asia; now cultivated in Bihar, Jharkhand, Orissa, Punjab, Bengal, Andhra Pradesh and in hills of Uttaranchal, Kashmir and Himachal Pradesh. Also, found wild in Northern Himalayan region.

Parts Used : Whole Plant. Fruit.

Flowering & Fruiting : June – August.

Folk Use : Plant used as a tonic. Fruits considered stimulant.

Active Constituents : Fruits rich in umbelliferone, scopoletin, herniarin and lipids.

References : Kirtikar & Basu (1984), Sharma (2003), Sood *et al.* (2001), Watt (1972).

183. *Caryota urens* L.

Family: Arecaceae.

English, Hindi and Sanskrit Names: **Eng. :** Bastard Sago Palm, Fish Tail Palm, Indian Sago-palm, Toddy Palm; **Hindi :** Ban-khajur, Mari; **Sans. :** Dirgha, Madyadruma, Rajju, Vitanaka.

Distribution : Moist forest regions of the West & East coast, Bihar, Orissa and N.E. India up to 2,000m; cultivated elsewhere in country.

Parts Used : Nut.

Flowering & Fruiting : May – August.

Folk Use : Nut used as a tonic in Assam.

References : Asolkar *et al.* (1992), Karnick *et al.* (1981).

184. *Casearia tomentosa* Roxb.

Syn.: C. anavinga Dals. & Gibs.; *C. canziala* Ham.; *C. elliptica* Willd.; *C. ovata* Roxb.

Family: Samydaceae.

English, Hindi and Sanskrit Names: **Eng. :** Toothed Leaf Chilla; **Hindi :** Bheri, Chilla, Kirci; **Sans. :** Chilhaka, Kampillika.

Distribution : Distributed throughout India up to 915m in the Himalayas; commonly occurs in South and Central India.

Parts Used : Root Bark.

Flowering & Fruiting : March – May.

Folk Use : Root bark used as tonic by tribals of Bastar in anaemia.

References : Asolkar *et al.* (1992), Kaushik & Dhiman (2000). Kothari *et al.* (2003), Parrotta (2001).

185. *Cassia absus* L.

Family: Fabaceae.

English, Hindi and Sanskrit Names: **Eng. :** Four-leaved Cassia, Pig's Senna; **Hindi :** Banar, Bankultthi, Chaksu, Chiksi, Chimas; **Sans. :** Arangakulitthika, Chipita, Drikaprasada, Kulatthika, Kulani, Kulmasha, Kumbhakarini, Lochanahita, Vanyakulattha.

Distribution : Throughout India, Sri Lanka; Tropical Asia, Australia and Africa.

Parts Used : Seeds.

Flowering & Fruiting : August – February.

Folk Use : Seeds used as stimulant.

Active Constituents : Major components are β-sitosterol-β-glucoside, a D-galacto-D-mannan composed of galactose (1 mole) and mannose (3 moles), palmitic, gentisic, 5-O-D-glucopyranosylgentisic acids, ethyl-α-D-galactopyranoside, apigenin, luteolin, hydnocarpin and isohydrnocarpin (*seeds*).

Reference : Deshmukh *et al.* (2000).

186. *Cassia fistula* L. (Pl. 3E)*

Syn.: Bactyrilobium fistula Willd.; *Cassia excelsa* Kemth.; *C. rhombifolia* Roxb.; *Cathartocarpus fistula* Pers.

Family: Fabaceae.

English, Hindi and Sanskrit Names: **Eng. :** Golden Shower, Indian Laburnum, Purging Fistula; **Hindi :** Amaltas, Girimala, Kirala, Sinar; **Sans. :** Aragvadha, Krtamalah, Suvarnaka.

Distribution : Wild in many parts of India; also cultivated as an ornamental plant for attractive yellow blossoms in pendant racemes.

Parts Used : Root. Fruit. Seeds.

Flowering & Fruiting : March – July.

Folk Use : Pulp of fruits, root and seeds used as tonic.

Active Constituents : Properties accredited due to fistucacidin, flavonoids and phlobaphenes, Butulinic acid, 7-methylphyscion and β- sitosterol (*root*), rhein, oil, fistulic acid and anthraquinone (*fruit*), 5,7,3',4'-tetrahydroxy-6,8-dimethoxyflavone-3-O-α-L-arabinopyranoside, 5,7,4'-tri-hydroxyl-6,8,3'-trimethoxyflavone-3-O-α-l-rhamnosyl(1→2)-β-D-glucopyranoside and 1,8-dihydroxy-3,7-dimethoxyxanthon-4-O-α-L-rhamnosyl(1→2)-β-D-glucopyranoside (*stem bark*).

References : Bhatt & Jain (2009), Dhiman (2005), Farooq (2005), Kaushik & Dhiman (2000), Parrotta (2001), Prajapati *et al.* (2006), Sharma (2009), Watt (1972).

187. *Cassia occidentalis* L.

Syn.: C. foetida Pers.; *C. sophera* Wall.; *Senna occidentalis* Roxb.

Family: Fabaceae.

English, Hindi and Sanskrit Names: **Eng. :** Coffee Senna, Negro Coffee, Stinking Weed; **Hindi :** Barikasaumdi, Chakunda, Kasondi; **Sans. :** Arimarda, Jarana, Kasmarda.

Distribution : Found scattered from Himalayas to South India.

Parts Used : Whole Plant.

Flowering & Fruiting : Flowering and fruiting throughout the year; chiefly during July – December.

Folk Use : Plant used as a tonic.

Active Constituents : Plant possesses chrysophanol, emodin, glycosides, physcion, crude fibre, metteucinol-7-rhamnoside, jaceidin-7-rhamnoside, ash, calcium, phosphorus, iron, niacin, ascorbic acid, β-sitosterol, emodin, crude protein and 4,4,5,5-tetrahydroxy-2,2-dimethyl-1,1-bianthraquinone.

References : Dhiman (2005), Deshmukh *et al.* (2000), Kaushik & Dhiman (2000), Kirtikar & Basu (1984), Kumar (2002), Parrotta (2001), Prajapati *et al.* (2006).

188. *Cassytha filiformis* L.

Family: Lauraceae.

English, Hindi and Sanskrit Names: **Eng.** : Sita's Yarn; **Hindi** : Amarbeli; **Sans.** : Akasvalli.

Distribution : Throughout coastal India, Sri Lanka. Tropical E. Asia and islands, Polynesia, Australia, Tropical Africa and America.

Parts Used : Whole Plant.

Flowering & Fruiting : Almost throughout the year

Folk Use : Plant used as a tonic.

Active Constituents : 13 alkaloids isolated. Of these, structure of two new oxoaporphine bases established-cassamedine as 1,2,9,10-bis(methylenedioxy)-3-methoxydibenzo[de,g]quinolin-7-one and cassameridine as 1,2,9,10-bis(methylenedioxy)-3-hydroxydibenzo-[de,g]quinolin-7-one, cassythine (cassyfiline) and cassythidine have been characterized.

References : Kapur & Nanda (1996), Kirtikar & Basu (1984), Nayak *et al.* (2003), Watt (1972).

189. *Castanea sativa* Mill.

Syn.: C. vulgaris Lam.

Family: Fagaceae.

English Names: **Eng.** : Spanish chestnut, Sweet Chestnut.

Distribution : Native to the Mediterranean, Asia Minor and the Caucasus; introduced in the Himalayas and grown in various localities and especially in a large number of places in Punjab and hills of the N.W. regions, in Darjeeling and the Khasia hills.

Parts Used : Leaf.

Flowering & Fruiting : July – October.

Folk Use : Leaves used as tonic.

Active Constituents : Active properties due to ursolic acid, waxes, lupeol, betulin, fatty acids, myricetol, quercitol, rutoside, myricitroside, quercitroside, narcissi, helichryoside, genkwanin and a new acylated flavonoid glycoside-kaempferol-3-O-[glucosyl(2''-p-coumaroyl)](6→1)-α-L-rhamnopyranoside (*leaf*).

References : Ambasta (1986), Anonymous (1950), Farooq (2005).

190. *Catha edulis* Forsk.

Family: Celastraceae.

English Names: **Eng. :** Abyssinian Tea, African Tea, Catha, Khat.

Distribution : Native to middle east and the Horn of Africa; it reported to have been successfully introduced in Bombay, Mysore and Sri Lanka.

Parts Used : Leaf.

Flowering & Fruiting : July – December.

Folk Use : Leaves chewed for their stimulating effect.

Active Constituents : Accredited activities due to ephedrine, norpseudoephedrine, cathinine, cathidine, β-sitosterol, its glucoside, frieddin, dihydromyricetin and its 3-O-rhamnoside, d-norisoephedrine (*leaf*).

References : Ambasta (1986), Anonymous (1950), Prajapati *et al.* (2006).

191. *Catharanthus roseus* G. Don (Pl. 3F)*

Syn.: Ammocallis rosea (L.) Small.; *Lochnera rosea* (L.) Reichb.; *Vinca rosea* L.

Family: Apocynaceae.

English, Hindi and Sanskrit Names: **Eng. :** Old Maid, Pink Periwinkle, Red Periwinkle, Rosy-flowered Indian Periwinkle; **Hindi :** Sadabahar, Sadaphul, Sadasuhagen; **Sans. :** Nityakalyani, Rasna, Sadampuspa.

Distribution : Grown as an evergreen ornamental in gardens and houses; cultivated on a large scale in some areas of Tamil Nadu.

Parts Used : Root.

Flowering & Fruiting : Throughout the year, mainly during August – January.

Folk Use : Roots exhibit sedative and tranquillizing properties.

Active Constituents : Known to contain raubasin, ursolic acid, oleanolic acid, alstonine, bornesitol, d-yohimbine, ajmalicine, serpentine and reserpine, venalstonine, vindolinine, perivine, lochnericine, catharanthine, vindoline, leurosine and tetrahydroalstonine (*root*).

References : Ambasta (1986), Kaushik & Dhiman (2000), Parrotta (2001), Raju (2000), Thakur *et al.* (1989), Trivedi (2006).

192. *Cedrus deodara* (Roxb. *ex* D. Don) G. Don (Pl. 4A)

Syn.: Abies deodara Lindl.; *Cedrus libani* var. *deodara* Hook.f.; *Pinus deodara* Roxb.

Family: Pinaceae.

English, Hindi and Sanskrit Names: **Eng. :** Deodar, Himalayan Cedar, True Cedar; **Hindi :** Dedwar, Deodar, Devdarru; **Sans. :** Devadaru, Indravriksha, Kilima, Mastadaru, Shatapadapa.

Distribution : Temperate western Himalayas: 1,500-3,000m.

Parts Used : Wood.

Flowering & Fruiting : June – July.

Folk Use : Oil obtained by distillation of dark coloured wood used as a bitter tonic and cardio tonic in ayurvedic preparations – Kwatha and Arista.

Active Constituents : These are cedeodarin, cedrin, dihydromyricetin, cedrinoside, taxifolin mesosecoisolariciresinol and cedrusinin (*wood*).

References : Prajapati *et al.* (2006), Thakur *et al.* (1989).

193. *Ceiba pentandra* (L.) Gaertn.

Syn.: Bombax pentandrum L.; *Eriodendron pentandrum* Kurz.

Family: Bombacaceae.

English, Hindi and Sanskrit Names: **Eng. :** Sinhaporekapok, True Kopok Tree, White Cotton Tree; **Hindi :** Hattian, Saphed Simal; **Sans. :** Chirayu, Kutasalmali, Moch, Svetasalmali.

Distribution : Planted mainly in south India in areas up to 450m. Tropical America, Asia and Africa.

Parts Used : Bark. Root.

Flowering & Fruiting : October – December.

Folk Use : Bark and roots used as tonic.

Active Constituents : Known to yield a water-soluble polysaccharide consisting of rhamnose, xylose, arabinose, glucose, galactose, glucuronic acid, galacturonic acid and alkali soluble xylans composed of fucose, xylose, arabinose, glucose, galactose, glucuronic acid 4-methoxyglucuronic acid (*bark*).

References : Kirtikar & Basu (1984), Prajapati *et al.* (2006).

194. *Celastrus paniculatus* Willd.

Syn.: Ceanothus paniculatus Roth.; *Celastrus alnifolia* Don; *C. dependens* Wall.; *C. multiflora* Roxb.; *C.nutans* Roxb.; *C. rothiana* Roem.

Family: Celastraceae.

English, Hindi and Sanskrit Names: **Eng.** : Climbing Staff Tree, Intellect Tree; **Hindi** : Malkangani, Malkunki; **Sans.** : Jyotishmati, Malkangni, Pitataila.

Distribution : Tropical and sub-tropical Himalayas ascending up to 1,400m from Punjab – Assam, Bihar, U.P. and South India.

Parts Used : Bark. Leaf. Seeds.

Flowering & Fruiting : April – June (flowering); cold season (fruiting).

Folk Use : Infusion of leaves mixed with juice of *Centella asiatica* (L.) Urban taken as nervine tonic. Bark also used as brain tonic. Seeds and its oil used as tonic, nervine tonic, stimulant and brain tonic.

Active Constituents : Properties attributed to acetic, benzoic, formic, linoleic, linolenic, palmitic and stearic acids, celapagine, celapinigine, celapanine, celastrol, celastrine, paniculatine, malkanguinol and related polyalcohols, malkangunin, paniculatediol, β-amyrin, β-sitosterol and 5-astigmasten-3β-ol (*seeds*).

References : Ambasta (1986), Anonymous (1950), Asolkar *et al.* (1992), Dhiman (2005), Jadeja (1999), Kaushik & Dhiman (2000), Khatoon & Goel (2010), Kirtikar & Basu (1984), Kothari *et al.* (2003), Kothari & Londhe (2000), Kumar & Lal (2003), Parrotta (2001), Prajapati *et al.* (2006), Raju (2000), Reddi *et al.* (2005), Sharma (2003), Singh & Pandey (1998), Siwakoti & Varma (1996), Thakur *et al.* (1989).

195. *Celosia argentea* L.

Family: Amaranthaceae.

English, Hindi and Sanskrit Names: **Eng.** : Cox Comb Plant; **Hindi** : Debkoti, Safed Murgha, Survari; **Sans.** : Pothura, Vitunna.

Distribution : Throughout India, Sri Lanka. Tropical Asia, Africa and America; cultivated or introduced.

Parts Used : Seeds.

Flowering & Fruiting : August – November (flowering); October – January (fruiting).

Folk Use : Seeds used as tonic.

Reference : Punjani (2002).

196. *Centaurea cyanus* L.

Syn.: C. lanata Roxb.; *Leucacantha cyanus* (L.) Nieuwl. & Lunell.

Family: Asteraceae.

English Names: **Eng.** : Bachelor's Button, Bluebottle, Blue Poppy, Boutonniere Flower, Cornflower, Cyani Flower, Hurtsickle.

Distribution : Native to the near east; grows wild in all temperate regions, often in corn fields (N.W. India, Punjab, Kashmir and Mysore).

Parts Used : Florets.

Flowering & Fruiting : June – October.

Folk Use : Florets considered as tonic and stimulant.

Active Constituents : Centarocyanin identified as 3-O-(6"-O-succinyl-β-D-glucosyl) cyaniding and apigenin-4'-O-(6-O-malonyl-β-D-glucoside-7-O-β-D-glucuronide (I) (*flower*).

References : Ambasta (1986), Anonymous (1950), Prajapati *et al.* (2006).

197. *Centella asiatica* (L.) Urban (Pl. 4B)*

Syn.: Hydrocotyle asiatica L.; *H. erecta* L. f.

Family: Apiaceae.

English, Hindi and Sanskrit Names: **Eng. :** Indian Pennywort, Thick-leaved Pennywort; **Hindi :** Bemgsag, Brahmamanduki, Brahmi, Khulakhudi; **Sans. :** Bhekaparni, Brahmi, Divya, Mandukparni.

Distribution : Found throughout marshy places up to 1,828m in India.

Parts Used : Whole Plant. Leaf.

Flowering & Fruiting : July – August (flowering); October – December (fruiting).

Folk Use : Plant used as a tonic. Powdered leaves with cow milk given early in the morning to improve memory.

Active Constituents : These are centoic, centellic, centic acids, brahmoside, brahminoside, two triterpenic acids – brahmic acid and iso-brahmic acid, betulic acid and stigmasterol (*plant*).

References : Anonymous (1950), Arya (2004), Aswal (1996), Bhattacharjee (2001), Bhattacharyya (1996), Chhetri (2005), Dhiman (2005, 2007), Dwivedi (2003, 2007), Farooq (2005), Hemlata *et al.* (2007), Jamir *et al.* (2008), Katewa & Arora (1997), Kaushik & Dhiman (2000), Kayal *et al.* (2003), Khanna (2002), Khatoon & Goel (2010), Kirtikar & Basu (1984), Kshirsagar & Singh (2007), Kumar (2002), Kumar & Lal (2003), Kumar & Pullaiah (1998, 2000), Maya *et al.* (2003), Nadeem *et al.* (2001), Nayak *et al.* (2003), Negi *et al.* (1999), Panda & Das (2000), Prajapati *et al.* (2006), Prusti (2007), Rajendran (2003), Ranjan (2000, 2003), Rao (1996), Reddi *et al.* (2005), Rosakutty *et al.* (2000), Sahu (2004), Sharma (2003, 2004), Sharma *et al.* (2007), Singh (2000a, 2001), Singh & Pandey (1998), Siwakoti & Varma (1996), Sood & Thakur (2004), Subramaniam (2000), Tiwari & Tiwari (1996), Verma *et al.* (1995), Yasodamma *et al.* (2009).

198. *Centratherum anthelminticum* (L.) Kuntze

Syn.: Conyza anthelmintica L.; *Serratula anthelmintica* Roxb.; *Vernonia anthelmintica* (L.) Willd.

Family: Asteraceae.

English, Hindi and Sanskrit Names: **Eng. :** Purple Flebene; **Hindi :** Bakshi, Kali Jiri, Vapchi; **Sans. :** Agnibija, Putiphali, Somraji, Vakushi.

Distribution : Throughout India up to 1,676m in the Himalayas; often cultivated near villages.

Parts Used : Seeds.

Flowering & Fruiting : Flowering: rainy season; Fruiting: spring season.

Folk Use : Infusion of powdered seeds used as tonic. Seeds also used as stimulant.

Active Constituents : Known to contain vernolic acid and its triglycerides, flavonoid glycosides, sitosterol, resin acid, 12-13-epoxy-9-octaecenoic acid and vernodalol (*seeds*).

References : Bhalla *et al.* (1996), Bhattacharjee (2002), Dhiman (2005), Kaushik & Dhiman (2000), Kirtikar & Basu (1984), Parrotta (2001), Sharma (2003), Sharma & Sood (1997).

199. *Cephalenthus occidentalis* L.

Family: Rubiaceae.

English Names: **Eng. :** America Button Wood, Button-bush, Button willow.

Distribution : Cultivated as an ornamental plant in Indian gardens.

Parts Used : Bark.

Flowering & Fruiting : June – September.

Folk Use : Bark used as a tonic.

Reference : Bhattacharjee (2001).

200. *Ceratopteris siliquosa* (L.) Copel

Syn.: C. deltoidea Benedict.; *C. thalictroides* (L.) Brongn.; *Pteris siliquosa* (L.) Beauvois; *P. thalictroides* (L.) Desvaux.

Family: Parkeriaceae.

English Names: **Eng. :** Water-fern, Water Horn Fern, Water Sprite.

Distribution : Found growing in tanks, rice fields and swampy places in India, Sri Lanka and Malay Peninsula.

Parts Used : Fronds.

Folk Use : Fronds used as a tonic.

References : Maya *et al.* (2003), Singh & Viswanathan (1996).

201. *Ceropegia bulbosa* Roxb.

Family: Asclepiadaceae.

English and Hindi Names: **Eng.:** Goglet Flower; **Hindi:** Khapparkadu.

Distribution : Punjab, Upper Gangetic plains, Konkan, S. Kanara, Malabar Deccan and Carnatic.

Parts Used : Root. Tubers.

Flowering & Fruiting : July – December.

Folk Use : Root and tubers used as a tonic.

References : Ambasta (1986), Anonymous (1950), Kirtikar & Basu (1984), Trivedi (2006).

202. *Ceropegia tuberosa* Roxb.

Syn.: C. acuminata Dalz. & Gibs.; *C. candelabrum* subsp. *tuberosa* (Roxb.) H. Huber.

Family: Asclepiadaceae.

English and Hindi Names: **Eng. :** Glabrous Goglet Flower; **Hindi :** Ptala-tumbi.

Distribution : Travancore, Courtallan and Dindygul; Anamallay hills, 850m; Sri Lanka, in the hotter part of the island.

Parts Used : Tuber.

Flowering & Fruiting : September – October.

Folk Use : Tubers considered as tonic.

References : Ambasta (1986), Anonymous (1950), Watt (1972).

203. *Cetraria islandica* (L.) Ach.

Family: Parmeliaceae.

English Names: **Eng. :** Cetraria, Iceland Lichen, Iceland Moss.

Distribution : Native to northern and alpine areas of Europe, flourishes in sub-arctic and mountainous regions on rocks and on bark of trees.

Parts Used : Whole Plant.

Folk Use : Decoction of the lichen used as tonic.

Active Constituents : Usnic acid and polysaccharides have been isolated (*plant*).

References : Ambasta (1986), Anonymous (1950), Prajapati *et al.* (2006).

204. *Cheiranthus cheiri* L.

Family: Brassicaceae.

English and Hindi Names: **Eng. :** Banwort, Bleeding Heart, July Flower, Sweet William, Wall Flower; **Hindi :** Lahurishubu, Todrisurkh.

Distribution : Native to S. Europe; cultivated in gardens of India.

Parts Used : Flower. Seeds.

Flowering & Fruiting : June – November.

Folk Use : Flowers considered as stimulant and seeds valued as tonic.

References : Ambasta (1986), Anonymous (1950), Kirtikar & Basu (1984), Prajapati *et al.* (2006).

205. *Chelone glabra* L.

Syn.: C. alba Moench.; *C. capitata* Rafin.; *C. maculata* Rafin.

Family: Scrophulariaceae.

English Names: **Eng. :** Balmony, Salt-rheum Weed, Shellflower, Snakehead, Turtle Head.

Distribution : Wetlands of India.

Parts Used : Whole Plant. Leaf.

Flowering & Fruiting : July – September (flowering); September – October (fruiting).

Folk Use : Powdered plant or leaves used as tonic.

Active Constituents : Resins and bitters (*plant*).

Reference : Bhattacharjee (2001).

206. *Chenopodium album* L.

Syn.: C. giganteum Don; *C. laciniatium* L.; *C. nepalense* Hort.; *C. purpurascens* Ham.; *C. viride* L.

Family: Chenopodiaceae.

English, Hindi and Sanskrit Names: **Eng. :** Bacon Weed, Hen-fat, Rag Jag, Wild Spinach; **Hindi :** Bathua, Bethuasag, Cheel; **Sans. :** Agaralohita, Chilika, Vastukah, Vasuka.

Distribution : Throughout India, in areas up to 4,200m, wild as well as cultivated.

Parts Used : Whole Plant.

Flowering & Fruiting : May – October.

Folk Use : Plant used as tonic; its extract considered CNS depressant.

Active Constituents : Active principles due to ascorbic acid, caffeic acid, β-carotene, β-sitosterol, catechins, gallocatechin, p-coumaric acid, ferulic acid, campesterol, xanthotoxin, n-tri-acontanol, scopoletin, stigmasterol, 7-stigmasterol, imperatorin and spinasterol (*plant*).

References : Asolkar *et al.*(1992), Jain & Goel (2005), Joshi (2009), Kirtikar & Basu (1984), Prajapati *et al.* (2006).

207. *Chimaphila maculata* Pursh.

Family: Ericaceae.

English Name: **Eng. :** Spotted Winter Green.

Distribution : Sikkim Himalaya: 3,000-3,650m; Bhutan, 2,250m; Khasia mountains, 2,150m.

Parts Used : Fruit. Leaf.

Flowering & Fruiting : June – August.

Folk Use : Fruits and leaves used as tonic.

Reference : Bhattacharjee (2001).

208. *Chloranthus glaber* (Thunb.) Makino

Syn.: C. brachystachys Blume; *Sarcandra glaber* (Thunb.) Nakai.

Family: Chloranthaceae.

English Name: **Eng. :** Chloranthus.

Distribution : Khasia mountains, (1,700m); Travancore, on the Pulney mountains, Penang, Sri Lanka, Central regions, (1,000-1,700m); China, Philippine islands and Japan.

Parts Used : Whole Plant.

Flowering & Fruiting : March – May.

Folk Use : Plant accredited with stimulant properties.

Active Constituents : Plant comprises isofraxidin, succinic acid, linolool acetate, guaiacyl lignin, chloranthalactones A and B, chloranthalactone F, chloranthalactone E, atractylenolide II and (-)4β,7α-dihydroxyaromadendrone.

References : Ambasta (1986), Anonymous (1950).

209. *Chlorophytum arundinaceum* Baker

Family: Liliaceae.

English, Hindi and Sanskrit Names: **Eng. :** India Spider Plant; **Hindi :** Biskandri, Safed Musli; **Sans. :** Musali, Sweta-musli.

Distribution : E. Himalaya, Assam, Bihar and Myanmar.

Parts Used : Tuberous Root.

Flowering & Fruiting : Flowering : March – April.

Folk Use : Tuberous roots used as a tonic and sedative.

Active Constituents : Properties exhibited due to 4-hydroxy-8,11-oxidoheneicosanol and pentacosyl docosanoate along with stigmasterol, its glucoside, nonacosane, tetracosanoic acid, triacotanoic acid and a new bibenzyl glycoside (I) characterized as 4,2',4'-trihydroxy-2-xylopyranosyl-bibenzyl (*root*).

References : Anonymous (1950), Bhattacharjee (2001), Jain *et al.* (2003), Khatoon & Goel (2010), Kirtikar & Basu (1984), Parrotta (2001).

210. *Chlorophytum borivilianum* Sant. & Fernand.

Family: Liliaceae.

English, Hindi and Sanskrit Names: **Eng. :** Muscle; **Hindi :** Dholi Musli, Khiruva, Safed Musli; **Sans. :** Musla, Shwet Musli.

Distribution : Plenty in forest areas of Central India.

Parts Used : Root.

Flowering & Fruiting : August – October.

Folk Use : Root powder mixed with milk or water given orally daily in the morning for a month as a tonic.

References : Chaudhari & Hutke (2002), Punjani (2006).

211. *Chondrodendron tomentosum* Ruiz & Pavon

Family: Menispermaceae.

English and Sanskrit Names: **Eng. :** False Pareira Brava, Pareira; **Sans. :** Ambashthaa, Paatha.

Distribution : Wild in rainforests.

Parts Used : Root. Stem.

Folk Use : Root and stem used as tonic.

Reference : Prajapati *et al.* (2006).

212. *Cibotium barometz* J. Smith

Syn.: C. glaucum Bedd. *Polypodium barometz* L.

Family: Dicksoniaceae.

English Names: **Eng. :** Golden Chick Fern, Golden moss, Woolly Fern.

Distribution : Mishmi, Assam, Tavoy, Malay islands and S. China.

Parts Used : Root.

Flowering & Fruiting : Spore formation occurs during October – January.

Folk Use : Root considered as tonic.

References : Ambasta (1986), Anonymous (1950).

213. *Cicer arietinum* L.

Family: Fabaceae.

English, Hindi and Sanskrit Names: **Eng. :** Bengal Gram, Chick Pea, Common Gram; **Hindi :** But, Chana; **Sans. :** Balabhojya, Canakah, Sakalapriya.

Distribution : Cultivated throughout India.

Parts Used : Seeds.

Flowering & Fruiting : February – March.

Folk Use : Seeds used as tonic.

Active Constituents : Principle activities due to occurance of O-α-D-galactopyranosyl-(1→6)-O-α-D-galactopyranosyl-(1→6)-O-α-D-galactopyranosyl-(1→2)-1D-4-O-methylchiroinositol, pyrimidine nucleotides-uracil-5-β-D-fructofuranosyl-1'-monosulphate and uracil-5-β-D-ribofuranosyl-2',3'-cyclic monosulphate, 24-epiclerosterol, 24α- and 24β-methyl-5α-cholest-8(14)-en-3β-ols, 24α-ethyl-5α-cholest-8(14)-en-3β-ol, 14α, 24α- and 14α, 24β-dimethyl-5α-cholest-9(11)-en-3β-ols, 14α, 24α- and 14α-methyl-24α-ethyl-5α-cholest-9(11)-en-3β-ols,14α-methyl-24β-ethyl-5α-cholesta-9(11), 25-dien-ol and β-sitosterol (*seeds*).

References : Kirtikar & Basu (1984), Manandhar (2003), Parrotta (2001), Prajapati *et al.* (2006).

214. *Cichorium endivia* L.

Family: Asteraceae.

English, Hindi and Sanskrit Names: **Eng. :** Endive, Garden Endive; **Hindi :** Hinduba, Kasini, Kashini; **Sans. :** Kasani.

Distribution : N.W. India, Deccan plateau, Kashmir; Iran and many parts of Europe.

Parts Used : Root.

Flowering & Fruiting : April – July.

Folk Use : Roots used as tonic.

References : Ambasta (1986), Kirtikar & Basu (1984).

215. *Cichorium intybus* L.

Family: Asteraceae.

English, Hindi and Sanskrit Names: **Eng. :** Blue Dandelion, Blue-sailors, Chicory, Coffeeweed, Succory, Wild Endive; **Hindi :** Kasini, Kashini; **Sans. :** Kasni.

Distribution : North India; mainly cultivated in Nadiad, Broach, Amalsad (Saurashtra in Gujarat).

Parts Used : Whole Plant. Root.

Flowering & Fruiting : February – May.

Folk Use : Plant used as liver and cardio tonic. Roots valued as tonic in fever.

Active Constituents : Roots contain lactucin, 8-deoxylactucin, lactuco-picrin, crepidiaside B, sonchusides A and C, cichoriolide A, B and

C, ribosylzeatin, caffeic acid, uridine-5'-diphosphoglucose, chicoric acid, chlorogenic acid and neochlorogenic acid.

References : Ambasta (1986), Anonymous (1950), Dhiman (2005), Kirtikar & Basu (1984), Prajapati *et al.* (2006), Reddi *et al.* (2005), Thakur *et al.* (1989), Trivedi (2006), Watt (1972).

216. *Cimicifuga foetida* L.

Syn.: Actaea cimicifuga L.; *C. frigida* Royle

Family: Ranunculaceae.

English Names: **Eng. :** Bugbane, Skunk Bugbane, Stinking Bugbane.

Distribution : On slopes and in grasslands in India, China and Japan; temperate Himalayas from Kumaon – Bhutan: 2,350-4,000m

Parts Used : Root.

Flowering & Fruiting : July – September.

Folk Use : Roots used as nerve sedative.

References : Ambasta (1986), Anonymous (1950), Gupta (1997).

217. *Cimicifuga racemosa* Nutt.

Syn.: Actaea racemosa L.; *Cimicifuga serpentaria* Pursh.

Family: Ranunculaceae.

English Names: **Eng. :** Black Cohosh, Black Snake Root, Rattle Root, Squaw Root.

Distribution : Temperate Himalayas (Kashmir – Bhutan) up to 2,000m.

Parts Used : Rhizome.

Flowering & Fruiting : Flowering during June – July.

Folk Use : Rhizome used as sedative.

Reference : Prajapati *et al.* (2006).

218. *Cinchona calisaya* Wedd.

Family: Rubiaceae.

English Names: **Eng. :** Bolivian Bark, Cinchona, Yellow Bark.

Distribution : Native of Andes of Ecuador and Peru; found in Sikkim and Nilgiris.

Parts Used : Bark.

Folk Use : Bark used as tonic.

Active Constituents : Active principles are quinine, quinidine, cinchonine and cinchonidine (*bark*).

References : Ambasta (1986), Reddi *et al.* (2005), Thakur *et al.* (1989), Watt (1972).

219. *Cinchona ledgeriana* Moens.

Family: Rubiaceae.

English Name: **Eng.** : Ledger Bark.

Distribution : Cultivated in West Bengal; also grown on the Anamalai hills and in Tinnevelly district in South India and Khasia and Jaintia hills.

Parts Used : Bark.

Folk Use : Bark used as tonic.

Active Constituents : Properties exhibited due to cincholic acid, quinonin, norsolorimic acid and β-sitosterol (*bark*).

References : Ambasta (1986), Thakur *et al.* (1989).

220. *Cinchona officinalis* L.

Syn.: C. condaminea L.

Family: Rubiaceae.

English, Hindi and Sanskrit Names: **Eng.** : Peruvian Bark, Quinine Bark; **Hindi** : Quinine; **Sans.** : Kunayanah, Sinkona.

Distribution : Cultivated in India in Nilgiris: 1,800-2,400m.

Parts Used : Bark.

Folk Use : Quinine obtained from the bark used as a cardio tonic.

Active Constituents : Principle constituents are quinine, chinchonidine, cinchonine, javanine, quinamine and quinidine (*bark*).

Reference : Prajapati *et al.* (2006).

221. *Cinnamomum camphora* (L.) Nees & Eberm.

Syn.: Canphora officinarum Nees ex Wall.; *Laurus camphora* L.

Family: Lauraceae.

English, Hindi and Sanskrit Names: **Eng.** : Camphor Tree, Formosan Wood, Gum Camphor, Tormosa Camphor; **Hindi** : Kapur, Karpur; **Sans.** : Chandra, Karpura.

Distribution : Cultivated as a shade-tree in sub-tropical and tropical region; also cultivated in Indian gardens up to 1,219m.

Parts Used : Whole Plant.

Flowering & Fruiting : Flowering: March – April; Fruiting: July – August.

Folk Use : Plant used as sedative and stimulant.

Active Constituents : Plant contains camphor, d-α-pinene, cineol, safrole, terpineol, caryophyllin, limonene, phellandrene, carvacrol, camphorene, azulene, pinene, camphene, dipentene, candinene and other sesquiterpenoids, cyaniding-3-glucoside, lauralistine, cyaniding-3-rhamnoglucoside and reticuline.

References : Ambasta (1986), Kaushik & Dhiman (2000), Parrotta (2001), Reddi *et al.* (2005), Singh & Srivastava (2000).

222. *Cinnamomum obtusifolium* Roxb. ex Nees

Syn.: C. bejolghota (Buch.-Ham.) Sweet.; *Laurus obtusifolium* Roxb.

Family: Lauraceae.

Distribution : Central and E. Himalayas, up to 2350m, Assam, Tenasserim, Andaman Islands.

Parts Used : Bark.

Folk Use : Decoction of bark used as liver tonic.

Reference : Chhetri (2005).

223. *Cinnamomum tamala* Nees & Eberm.*

Syn.: C. albiflorum Nees.; *C. cassia* Don.; *Laurus albiflora* Wall.; *L. cassia* Roxb.; *L. sailyana* Ham.; *L. somcaurium* Ham.; *L. tamala* Ham.; *L. tazia* Ham.; *Persea tamala* Sperng.

Family: Lauraceae.

English, Hindi and Sanskrit Names: **Eng. :** Cassia Cinnamon, Indian Cassia Lignea; **Hindi :** Kikoa, Silkanti, Tajkalam, Tejpatt; **Sans. :** Ankusha, Dala, Patraka, Roma, Surasa, Tamalapatram, Vasanavhaya.

Distribution : Sub-Himalayan, Khasia and Jaintia hills; cultivated in certain parts of India.

Parts Used : Bark. Leaf.

Flowering & Fruiting : March – September.

Folk Use : Leaves considered stimulant, tonic and tonic to the brain. Bark valued as narcotic.

Active Constituents : Known to possess cinnamic aldehydes, linalool, eugenol, eugenol actetate, β-caryphyllene, benzaldehyde, camphor, cardinene, α-terpineol, α- and β-pinene, p-cymene, limonene, geraniole, ocimone, γ-terpinene, β-phellandrene, benzyl cinnamte and benzyl acetate (*leaf*).

References : Farooq (2005), Kirtikar & Basu (1984), Malla & Chhetri (2009), Prajapati *et al.* (2006), Rajan *et al.* (2003), Rana *et al.* (1996), Rawat *et al.* (2007), Saikia & Nath (2003b), Sharma (2003).

224. *Cinnamomum zeylanicum* Breyn.

Syn.: C. verum Presl.

Family: Lauraceae.

English, Hindi and Sanskrit Names: **Eng. :** Ceylon Cinnamon, Cinnamon; **Hindi :** Dalchini, Daruchini, Qualamidarchini; **Sans. :** Darusita, Gudatvacha, Hridya, Sinhala, Tvaka, Utkata.

Distribution : Grows wild and also cultivated at lower altitudes in tropical areas of South India and East India up to 1,500m.

Parts Used : Bark.

Flowering & Fruiting : March – April.

Folk Use : Bark used as stimulant and tonic.

Active Constituents : Reported to yield cinnamaldehyde, linalool, α-terpineol, α-humulene, β-caryophyllene, cumin aldehyde, cinnamyl acetate and eugenol (*bark*).

References : Ambasta (1986), Dwivedi (2007), Kirtikar & Basu (1984), Parrotta (2001), Prajapati *et al.* (2006), Rajan *et al.* (2003), Thakur *et al.* (1989), Watt (1972).

225. ***Cirsium arvense*** (L.) Scop.

Syn.: Aplotaxis cirsiodes DC.; *Breea arvensis* (L.) Less.; *Carduus lanatus* Roxb.; *Cnicus arvensis* Hoffm.; *C. candicans* Wall.; *Sarratula arvensis* L.

Family: Asteraceae.

English and Hindi Names: **Eng. :** Canada Thistle, Corn Thistle, Creeping Thistle, Field Thistle, Green Thistle, Hard Thistle, Prickly Thistle; **Hindi :** Kandai.

Distribution : Bengal and Gangetic Plains, from the Sundarbans north-westwards to the Punjab; Western Himalaya: Kashmir – Kumaon; Western Tibet: 3,650-4,350m; N. Asia and westwards to Atlantic.

Parts Used : Whole Plant.

Flowering & Fruiting : May – June.

Folk Use : Plant considered as a tonic.

Active Constituents : Plant contained glucuronides of apigenin and acacetin, taraxasterol, its –OAC and stigmasterol, alkane mixture, β-amyrin, taraxasterol acetate, tetracosanol, hexacosanol and octacosanol.

References : Ambasta (1986), Anonymous (1950).

226. ***Cirsium verutum*** (D. Don) Spreng.

Syn.: C. argyracanthum DC.; *Cnicus argyracanthum* DC.

Family: Asteraceae.

English Name: **Eng. :** Common Thistle.

Distribution : Temperate Himalayas.

Parts Used : Root.

Flowering & Fruiting : July – September.

Folk Use : Root used as tonic.

Reference : Kala (2003).

227. *Cissampelos pareira* L.

Syn.: C. caapeba L.; *C. delicatula* Miers.; *C. diversa* Miers.; *C. elata* Miers.; *C. eriantha* Miers.; *C. grallatoria* Miers.; *C. heradiflora* Wall.; *C. nepalensis* Rhode; *C. obectata* Wall.; *C. orbiculata* Buch. DC.; *Cocculus orbiculatus* DC.; *C. villons* Wall.; *Menispermum orbiculatum* L.

Family: Menispermaceae.

English, Hindi and Sanskrit Names: **Eng. :** False Pareira Brava, Velvet Leaf; **Hindi :** Akanadi, Pari, Patha; **Sans. :** Laghupatha, Patha.

Distribution : Distributed throughout the tropical and sub-tropical India.

Parts Used : Root. Stem.

Flowering & Fruiting : Rainy and autumn season (flowering); winter (fruiting).

Folk Use : Extract of root and stem used as CNS depressant. 1tsp of root infusion given twice a day for one month as health tonic.

Active Constituents : Properties exhibited due to alkaloids, isoquinoline, pelosine, berberine, saponin, respine, cissampeline, hyatin, hyatinin, a quercitel and a sterol, L-curine and d-isochondrodendrine (*root*).

References : Anonymous (1986), Dhiman (2005), Jha & Varma (1996), Kaushik & Dhiman (2000), Kirtikar & Basu (1984), Reddi *et al.* (2005), Sharma (2003), Watt (1972).

228. *Citrullus vulgaris* Schrad.

Syn.: Citrullus lanatus (Thunb.) Matsumara & Nakai; *Colocynthis citrullus* (L.) Kuntze; *Cucurbita citrullus* L.; *Momordica lanata* Thunb.

Family: Cucurbitaceae.

English, Hindi and Sanskrit Names: **Eng. :** Watermelon; **Hindi :** Halinda, Samanka, Tarbuj; **Sans. :** Chayaphula, Godumba, Kalinda, Tarambuja.

Distribution : Native to tropical Africa; cultivated especially in Upper and North India.

Parts Used : Fruit. Seeds.

Flowering & Fruiting : May – June.

Folk Use : Fruits and seeds considered as tonic. Seeds also used as tonic to the brain (Yunani).

Active Constituents : These are ferulic, caffeic and chlorogenic acids, cucurbitacins and folic acids (*fruit*), β-(1-pyrazolyl) alanine, codisterol, 25(27)-dehydroporiferasterol, cleosterol, isofucosterol, stigmasterol, campesterol, sitosterol, 25(27)-dehydrofungisterol,

25(27)-dehydrochondrillasterol, 24β-ethyl-25(27)-dehydrolathosterol, avenasterol, spinasterol, 24ξ-methyllathosterol and 22-dihydrospinasterol (*seeds*).

References : Ambasta (1986), Kaushik & Dhiman (2000), Kirtikar & Basu (1984), Singh *et al.* (2007), Singh & Srivastava (2000).

229. *Citrus aurantifolia* (Christm.) Swingle

Syn.: C. medica L. var. *acida* Watt.; *L. aurantifolia* Christm.

Family: Rutaceae.

English, Hindi and Sanskrit Names: **Eng. :** Lemon; **Hindi :** Kagzi Nimbu; **Sans. :** Jambira.

Distribution : Tropical parts of India including valleys of the Himalayas and Khasia hills; also cultivated in West Bengal, Punjab, Madhya Pradesh, Maharashtra, Karnatka, Andhra Pradesh, Tamil Nadu and Uttar Pradesh.

Parts Used : Aerial Parts.

Flowering & Fruiting : March – April (flowering); May – June (fruiting).

Folk Use : Ethyl alcohol (50%) extract of the aerial parts CNS depressant.

References : Asolkar *et al.* (1992), Kaushik & Dhiman (2000).

230. *Citrus aurantium* L. (Pl. 4C)

Syn.: C.aurantium L. var. *bigaradia* Hook. f.

Family: Rutaceae.

English, Hindi and Sanskrit Names: **Eng. :** Loose Skinned Orange, Mandarin Orange, Portugal Orange, Seville Orange, Sour Orange, Tangerine Orange; **Hindi :** Khatta, Naurangi, Santara; **Sans. :** Chakradhivasi, Naranga, Nimbuphala, Tvagagandha.

Distribution : Widely cultivated in sub-tropical regions of India, particularly in Assam, Meghalaya, West Bengal, Sikkim, U.P., Punjab, Kashmir, M.P., Maharashtra and Karnataka.

Parts Used : Flower. Fruit.

Flowering & Fruiting : January – February (flowering); September – October (fruiting).

Folk Use : Flowers considered stimulating whereas fruits used as tonic.

Active Constituents : Known to contain hesperidin, neohesperidin, limonene, naringin, poncirin and umbelliferone (*plant*), β-sitosterol, ergosterol and desmosterol (*flower*).

References : Farooq (2005), Kirtikar & Basu (1984).

231. ***Citrus limon*** (L.) Burm. f.*

Syn.: C. limonum Risso.; *C. medica* var. *limon* L.; *C. medica* var. *lomonum* L. (Risso) Hook. f.

Family: Rutaceae.

English, Hindi and Sanskrit Names: **Eng.** : Galgal, Hill Lemons, Lemon, Lime; **Hindi** : Bara Nimbu, Jamiri Nimbu, Paharikaghzi, Paharinembu; **Sans.** : Jambirah, Mahanimbu, Nimbuphala.

Distribution : Throughout India, cultivated in plains and hills up to 1,200m.

Parts Used : Fruit.

Flowering & Fruiting : November – February.

Folk Use : Fruits used as nervine tonic, stimulant and for improving memory.

Active Constituents : Properties exhibited due to apigenin, luteolin, chrysoseriole, quercitin, isorhamnetin, new flavonols-limocitrin and limocitrol-sinapic acid, p-coumaric acid, scopoletin, umbelliferone, hesperidin, citrusins A-D, conferin, syringing and dehydro-diconeferyl alcohol-4β-glucoside (*fruit*).

References : Prajapati *et al.* (2006), Reddi *et al.* (2005), Viswanathan & Singh (1996).

232. ***Citrus medica*** L.

Syn.: C. medica var. *medica* Watt.

Family: Rutaceae.

English, Hindi and Sanskrit Names: **Eng.** : Adams Apple, Citron, Melon Lime; **Hindi :** Bijaura, Baranimbu, Nimbu, Turanj; **Sans.** : Beejapura, Begapura, Matulungah, Rochanaphala.

Distribution : Throughout India, on forests along streams, in areas up to 1,200m.

Parts Used : Whole Plant. Bud. Flower. Fruit.

Flowering & Fruiting : April-June.

Folk Use : Plant used as cardio tonic. Buds, flowers and fruits considered stimulant. Fruits valued as tonic.

Active Constituents : Principle activities due to ascorbic acid, alanin, niacin, aspartic acid, glutamic acid, glycine, lysine, proline, serine and threonine (*fruit*).

References : Farooq (2005), Prajapati *et al.* (2006), Saikia & Nath (2003b), Watt (1972).

233. *Cleome viscosa* L.

Syn.: Cleome icosandra L.; *Polanisia icosandra* Wight. & Arn.; *P. viscosa* DC.

Family: Capparaceae.

English, Hindi and Sanskrit Names: **Eng.** : Dog Mustard, Sticky Cleome, Wild Mustard; **Hindi** : Jangli Harrar, Hulhul, Hurhur; **Sans.** : Adityabhakta, Bhaskareshta, Pasugandha.

Distribution : Throughout tropical and warm India, and the rest of the world.

Parts Used : Root. Seeds.

Flowering & Fruiting : July – October.

Folk Use : Roots and seeds considered stimulant.

Active Constituents : Kaempferide-3-glucuronide, dehydrokaempferide-3-glucuronide, docosoanoic acid, naringenin-4'-galactoside, dihydrokaepferol-4'-xyloside, 3',4',5-trihydroxyflavone-7-o-α-L-rhamnopyranoside, β-amyrin and lupeol have been isolated (*root*).

References : Kirtikar & Basu (1984), Prajapati *et al.* (2006), Ranjan (2000), Sharma (2004).

234. *Clerodendrum indicum* (L.) O. Kuntze

Syn.: C. siphonanthus R. Br.; *Siphonanthus indica* L.

Family: Verbenaceae.

English, Hindi and Sanskrit Names: **Eng.** : Tube – flower; **Hindi** : Bharangi; **Sans.** : Bharangi.

Distribution : Wasteplaces, hedges and roadsides throughout the warmer parts of India.

Parts Used : Leaf.

Flowering & Fruiting : Flowering: May – July; Fruiting: cold season.

Folk Use : Leaves used as a bitter tonic.

Active Constituents : Known to contain (24s)-ethycholesta-5,24,25-trien-3β-ol hispidulin, 7-O-glucoronides of hispidulin and scuteuarein (*leaf*).

References : Anonymous (1950), Kaushik & Dhiman (2000).

235. *Clerodendrum infortunatum* auct.

Syn.: C. viscosum Vent.

Family: Verbenaceae.

English, Hindi and Sanskrit Names: **Eng.** : Glory tree; **Hindi** : Bhant, Tinabharmi, Titabhamt; **Sans.** : Barhichuda, Bhandira, Bhantaka, Harita, Kiravarnika, Kukkura, Shirnaroma, Shukabarha.

Distribution : Wasteplaces and roadsides throughout India.

Parts Used : Leaf.

Flowering & Fruiting : Flowering: January – March; Fruiting: March – April.

Folk Use : Fresh leaf juice used as a health tonic.

Active Constituents : Leaf contained hispidulin-7-O-glucuronide and scutellarin.

References : Ambasta (1986), Anonymous (1950), Dhiman (2005), Kaushik & Dhiman (2000), Mudgal *et al.* (1999), Prajapati *et al.* (2006), Raju (2000), Watt (1972).

236. *Clerodendrum multiflorum* (Burm. f.) O. Kuntze

Syn.: C. phlomidis L. f.; *Volkameria multiflora* Burm. f.

Family: Verbenaceae.

English, Hindi and Sanskrit Names: **Eng.** : Wind Killer; **Hindi** : Arni, Piran, Urni; **Sans.** : Agnimantha, Agnimanthah, Vataghani.

Distribution : Throughout India, chiefly in U.P., Bihar, West Bengal and Deccan Peninsula.

Parts Used : Leaf. Root.

Flowering & Fruiting : April – May (flowering); May – June (fruiting).

Folk Use : Leaves and root decoction used as bitter tonic.

Active Constituents : Active ingredients are pectolinarigenin and secutellarein together with (24s) ethylcholesta-5 and clerosterol (*leaf*), clerodin, clerosterol, clerodendrin A, cerolic acid, ceryl alcohol and raffinose (*root*).

References : Asolkar *et al.* (1992), Dhiman (2005), Farooq (2005), Kaushik & Dhiman (2000), Parrotta (2001), Watt (1972).

237. *Clerodendrum serratum* (L.) Moon

Syn.: C. fortunatum Burm. *ex* Maldenke.; *C. serrulatum* Spreng. *ex* Razi.; *Volkameria serrata* L.

Family: Verbenaceae.

English, Hindi and Sanskrit Names: **Eng.** : Beetle Killer, Blue Fountain Bush; **Hindi** : Bharangi; **Sans.** : Bharangi, Kharasakah.

Distribution : More or less throughout India.

Parts Used : Leaf. Root.

Flowering & Fruiting : May – August (flowering); September – November (fruiting).

Folk Use : Leaves used as tonic and bitter tonic. Roots considered stimulant.

Active Constituents : Saponin, mannitol and stigmasterol have been characterized (*root*).

References : Kaushik & Dhiman (2000), Parrotta (2001), Sharma (2003, 2004).

238. *Clitoria ternatea* L.

Family: Fabaceae.

English, Hindi and Sanskrit Names: **Eng. :** Butterfly-pea, Candiflower, Clitoria; **Hindi :** Aparajita, Kajina, Kowa; **Sans. :** Aparajita, Bhumilagna, Girikarnika.

Distribution : Common garden plant throughout India; wild in Assam and South India.

Parts Used : Root. Seeds.

Flowering & Fruiting : May – June (flowering); winters (fruiting).

Folk Use : Root used as tonic. Seed powder prescribed to increase memory.

Active Constituents : Attributed properties due to γ-sitosterol, hexacosanol, β-sitosterol, an anthoxanthin (*seeds*).

References : Kaushik & Dhiman (2000), Parrotta (2001), Prajapati *et al.* (2006), Reddi *et al.* (2005), Thakur *et al.* (1989).

239. *Cnicus benedictus* L.

Syn.: Carbenia benedicta L.; *Carduus benedictus* L.

Family: Asteraceae.

English Names: **Eng. :** Bitter Thistle, Blessed Thistle, Holy Thistle, Lady's Thistle, Spotted Thistle, St. Benedict's Thistle.

Distribution : Native to Mediterranean region; flourishes on dry stony grounds and wastelands.

Parts Used : Whole Plant.

Flowering & Fruiting : May – October.

Folk Use : Plant used as a tonic.

Active Constituents : Isolated constituents are lignans, vsesquiterpene lactones, volatile oil, polyacetylenes, flavonoids, triterpenes, tannins and phytosterols (*plant*).

Reference : Prajapati *et al.* (2006).

240. *Coccinia grandis* (L.) Voigt.

Syn.: Bryonia grandis L.; *Cephalandra indica* (Wight. & Arn.) Naud.; *Coccina cordifolia* Cogn.

Family: Cucurbitaceae.

English, Hindi and Sanskrit Names: **Eng. :** Ivy Gourd; **Hindi :** Bimb, Kanduri; **Sans. :** Bimbi, Bimbika, Govhi, Ostopamaphala.

Distribution : Throughout India, cultivated as well as found growing wild.

Parts Used : Fruit.

Flowering & Fruiting : March – December.

Folk Use : Fruits used as liver tonic.

Active Constituents : Attributed properties due to taraxerone, sitosterol, taraxerol and 24-ethylcholest-5-en-3β-ol, β-amyrin, lupeol, a bitter glycoside containing cucuritoacin B, β-sitosterol, β-carotene, lycopene, cryptoxanthin, apo-6'-lycopenol (*fruit*).

Reference : Prajapati *et al.* (2006).

241. *Cocculus hirsutus* (L.) Diels

Syn.: C. aristolochiae DC.; *C. sepium* Colebr.; *C. villosus* Lam.; *Menispermum hirsutum* L.; *M. villosum* Lam.

Family: Menispermaceae.

English, Hindi and Sanskrit Names: **Eng. :** Broom Creeper, Ink Berry; **Hindi :** Jaljamni, Patalagarudi; **Sans. :** Chilihindah, Patalagarudah, Vanatiktika.

Distribution : Dry localities throughout India.

Parts Used : Root. Stem.

Flowering & Fruiting : September – December.

Folk Use : Roots and stem used as tonic and sedative.

Active Constituents : Isolated constituents are D-trilobine and DL-coclaurine (*root*), trilobine, isotrilobine, coclaurine and magnoflorine (*stem, root*).

References : Asolkar *et al.* (1992), Kirtikar & Basu (1984), Prajapati *et al.* (2006), Reddi *et al.* (2005).

242. *Cocculus pendulus* (Forsk.) Diels

Syn.: C. glabra Wight. & Arn.; *C. laurifolius* DC.; *C. leaeba* DC.

Family: Menispermaceae.

Distribution : Outer Himalayas up to 1,700m.

Parts Used : Root.

Flowering & Fruiting : March – August.

Folk Use : Roots used as tonic.

Reference : Ambasta (1986).

243. *Cochlospermum religiosum* (L.) Alston

Syn.: Bombax gossypium L.; *C. gossypium* DC.

Family: Cochlospermaceae.

English, Hindi and Sanskrit Names: **Eng. :** Golden Silk Cotton Tree, Yellow-flowered Cotton Tree; **Hindi :** Galgal, Ganiar, Kumbi; **Sans. :** Girisalmalika, Silakarpasika.

Distribution : Stony hills in dry forests throughout India.

Parts Used : Leaf. Flower. Gum.

Flowering & Fruiting : January – March (flowering); April – September (fruiting).

Folk Use : Dried leaves and flowers used as stimulant. Gum valued as tonic and sedative.

Active Constituent : Naringenin (*flower*).

References : Anonymous (1950), Bhattacharjee (2001), Kirtikar & Basu (1984), Kumar & Lal (2003), Parrotta (2001), Prajapati *et al.* (2006), Raju (2000), Reddi *et al.* (2005), Sharma (2003), Watt (1972).

244. *Cocos nucifera* L.

Family: Arecaceae.

English, Hindi and Sanskrit Names: **Eng. :** Cocoa Nut, Coconut Palm, Coconut Tree; **Hindi :** Nariel, Nariyal, Nariyalkaper; **Sans. :** Dakshinatriya, Garikera, Jataphala, Langoli, Mahaphala, Nilataru, Payodhara, Rasaphala Shiraphala, Varaphala .

Distribution : Mainly on sea coasts throughout India.

Parts Used : Coconut Milk. Fruit.

Flowering & Fruiting : Throughout the year.

Folk Use : Tender coconut milk and fruits used as tonic.

References : Kirtikar & Basu (1984), Rosakutty *et al.* (2000).

245. *Codonopsis clematidea* (Schrenk) C.B. Clarke

Family: Campanulaceae.

English Name: **Eng. :** Asian Bellflower.

Distribution : Western Himalayas.

Parts Used : Whole Plant.

Flowering & Fruiting : June – July.

Folk Use : Decoction of the plant considered as tonic and stimulant.

References : Ballabh & Chaurasia (2006), Singh *et al.* (1996).

246. *Codonopsis ovata* Benth.

Family: Campanulaceae.

English Names: **Eng. :** Bonnet Bellflower, Kashmir Bellflower.

Distribution : Native to W. Himalayas; occurring between 2,600-3,800m of Himachal Pradesh and from Kashmir – Garhwal up to 4,000m.

Parts Used : Root.

Flowering & Fruiting : June – September.

Folk Use : Roots considered as a stimulant and a good physical and sexual tonic.

References : Bhattacharjee (2001), Sharma (2003).

247. *Codonopsis rotundifolia* Benth.

Syn.: C. lurida Lindl.; *Glosocomia lurida* (Lindl.) Lindl.; *G. rotundifolia* (Benth.) Rupr.; *Wahlenbergia lurida* (Lindl.) G. Manneti; *W. rotundifolia* (Benth.) A. DC.

Family: Campanulaceae.

Distribution : Native to W. Himalayas, sporadically occurring along with alpine ferns between 3,500-3,800m of Himachal Pradesh.

Parts Used : Root.

Flowering & Fruiting : July – October.

Folk Use : Roots considered as stimulant.

Reference : Bhattacharjee (2001).

248. *Coffea arabica* L.

Family: Rubiaceae.

English, Hindi and Sanskrit Names: **Eng. :** Arabian Coffee, Coffee; **Hindi :** Caphi, Pilu; **Sans. :** Rajapiluh.

Distribution : Native of Abyssinia; introduced to India through Arabia; cultivated in South India.

Parts Used : Seeds.

Flowering & Fruiting : January – April.

Folk Use : Dried ripe seeds used as stimulant.

Active Constituents : Caffeine, 2-methyl-3-furanthiol, 2-furfuryl thiol, acetoin, prenyl mercaptan, 3-mercapatn, 2- heptanol, α-terpineol, limonene, methional, 3-isopropyl-2-methoxypyrazine, β-damascenone, 2-ethyl-3,5-dimethyl pyrazine, 3-hydroxy-4,5-dimethyl-2-furanone, 4-ethyl guaincol, 5-ethyl-3-hydroxy-4-methyl-2-furanone and 4-vinyl guaincol (*seeds*).

References : Ambasta (1986), Asolkar *et al.* (1992), Parrotta (2001), Prajapati *et al.* (2006), Sharma (2003), Watt (1972).

249. *Coffea benghalensis* Hyne ex Roem. & Schult.

Syn.: C. horsfieldiana Miq.

Family: Rubiaceae.

English and Hindi Names: **Eng. :** Bengal Coffee; **Hindi :** Coffee.

Distribution : Indigenous to Ethiopia and Sudan; cultivated in India.

Parts Used : Seeds.

Flowering & Fruiting : December – April.

Folk Use : Seeds valued as stimulant and nervine tonic.

Reference : Kayal *et al.* (2003).

250. ***Coix lacryma-jobi*** L.

Syn.: C. agrestis Lour.; *C. lacrymal* L.

Family: Poaceae.

English, Hindi and Sanskrit Names: **Eng. :** Christ's Tears, Job's Tears; **Hindi :** Baru, Dabhir, Gurlu, Samkru; **Sans. :** Gavedhukah, Jargadi, Kshudra, Kunta.

Distribution : Throughout India, in areas up to 2,100m.

Parts Used : Seeds.

Flowering & Fruiting : Autumn.

Folk Use : Seeds used as a tonic.

Active Constituents : Known to yield trans-ferulyl stigmastanol and trans-ferulyl campestanol (*seeds*).

References : Asolkar *et al.* (1992), Kirtikar & Basu (1984), Parrotta (2001), Prajapati *et al.* (2006), Reddi *et al.* (2005), Siwakoti & Varma (1984), Srivastava *et al.* (2003b).

251. ***Cola acuminata*** (Beauv.) Schott & Endl.

Syn.: Cola macrocarpa (G. Don) Schott & Endl.; *Sterculia macrocarpa* G. Don; *S. acuminata* Beauv.

Family: Sterculiaceae.

English Names: **Eng. :** Abata Kola, Cola Nut, Cola Tree, Goora Nut, Kola Nut.

Distribution : Native to West Africa, cultivated widely in the tropics especially in Nigeria, Brazil and the West Indies; introduced in India.

Parts Used : Seeds.

Flowering & Fruiting : Flowering almost throughout the year mainly at the start of rainy season.

Folk Use : Kola nut stimulates the CNS.

Active Constituents : Active properties due to caffeine, theobromine, tannins, phlobaphene and an anthocyanin, cycloartenol, 24-methylenecycloartenol (*nut*).

References : Asolkar *et al.* (1992), Prajapati *et al.* (2006).

252. *Cola nitida* (Vent.) Schott & Endl.

Syn.: C. vera K. Schum.; *Sterculia nitida* Vent.

Family: Sterculiaceae.

English Names: **Eng. :** Bitter Cola, Sudan Cola Nut Tree.

Distribution : Introduced in India.

Parts Used : Seeds.

Flowering & Fruiting : May – July (flowering); October – December (fruiting).

Folk Use : Dried cotyledons used as stimulant, nervine and tonic.

Active Constituent : Caffeine (*seeds*).

References : Ambasta (1986), Anonymous (1950).

253. *Colebrookia oppositifolia* Smith (Pl. 4D)*

Syn.: C. ternifolia Roxb.; *Elsholtzia oppositifolia* Poir.

Family: Lamiaceae.

English and Hindi Names: **Eng. :** Indian squirrel tail; **Hindi :** Binda, Bindu, Pansra.

Distribution : Tropical and sub-tropical parts of India, ascending to 1,350m.

Parts Used : Leaf.

Flowering & Fruiting : November – February.

Folk Use : 2 tsp decoction of leaves along with extract of *Plantago major* (3:1) prescribed twice a day for 10 days as a tonic.

Active Constituents : 5,6,7-tri- and 5,6,7,4"-tetra-MeO-flavones (*leaf*).

Reference : Sood & Thakur (2004).

254. *Coleus amboinicus* Lour.

Syn.: C. aromaticus Benth.; *Plectranthus amboinicus* (Lour.) Sperng.

Family: Lamiaceae.

English, Hindi and Sanskrit Names: **Eng. :** Countryborage, Indian Borage; **Hindi :** Amroda, Patharcur, Patta Ajavayin; **Sans. :** Karpuravalli, Sugandhavalakam.

Distribution : Cultivated in gardens throughout India.

Parts Used : Leaf.

Flowering & Fruiting : Flowering during May – June.

Folk Use : Leaves used as liver tonic.

Active Constituents : Attributed properties due to cirsimaritin, β-sitosterol-β-D-glucoside, oleanolic, 2α,3α-dihydroxyolean-12-en-28-oic, crategolic, pomolic euscaphic, tormentic, ursolic ans 2α,3α,19α,23-tetrahydroxyurs-12-en-28-oic acids, salvigenin, 6-

methoxy-genkwain, quercetin, chysoeriol, luteolin, apigenin, eridictyol, eridictyol and taxifolin (*leaf*).

Reference : Prajapati *et al.* (2006).

255. *Colocasia esculenta* (L.) Schott (Pl. 4E)

Syn.: Alocasia dussii Dammer; *A. illustris* W. Bull; *Aron colocasium* (L.) St.-Lag.; *Arum esculentum* L.; *Caladium colocasia* (L.) W. Wight.; *Colocasia antiquorum* Schott.; *Leucocasia esculenta* Nakai.; *Steudnera virosa* (Roxb.) Prain.

Family: Araceae.

English, Hindi and Sanskrit Names: **Eng. :** Cocoyam, Eddo, Elephant's Ear, Taro; **Hindi :** Arvi, Gagli, Kachalu, Kachu; **Sans. :** Alukam, Alupam, Kachu.

Distribution : Cultivated throughout the hotter parts of India.

Parts Used : Leaf.

Flowering & Fruiting : Summer crop sown in February – March and rainy season crop in June – July.

Folk Use : Leaf juice used as a stimulant.

References : Kaushik & Dhiman (2000), Prajapati *et al.* (2006), Reddi *et al.* (2005).

256. *Commiphora mukul* (Hook. *ex* Stocks) Engl.

Syn.: Balsamodendron mukul Hook. *ex* Stocks; *Commiphora wightii* (Arn.) Bhandari

Family: Burseraceae.

English, Hindi and Sanskrit Names: **Eng. :** Bolellium, Indian Bdellium Tree, Mukul Myrrh Tree; **Hindi :** Gugal, Guggul, Mukul, Ranghanturb; **Sans. :** Bhutahara, Gugguluh, Mahisaksah, Marudishta, Puta, Uddipta, Usha.

Distribution : Foot hills of Aravalli range, in arid, semi-arid land including desert areas, M. P., Karnataka.

Parts Used : Resinous Gum.

Flowering & Fruiting : Flowering during March – April; Fruiting later.

Folk Use : Gum used as a nervine tonic, stimulant, liver tonic and general tonic.

Active Constituents : Known to contain guggulsterone-E,Z, guggulsterone-I-IV, cholesterol, seasamin, camphorene and cambrane A (*gum*).

References : Kirtikar & Basu (1984), Prajapati *et al.* (2006), Reddi *et al.* (2005).

257. *Commiphora myrrha* (Nees) Engl.

Syn.: Balsamodendron myrrha T. Nees., *C. cispidata* Choiv.; *C. coriacea* Engl.; *C. molmol* (Engl.) Engl.

Family: Burseraceae.

English, Hindi and Sanskrit Names: **Eng. :** Myrrh; **Hindi :** Bol, Hirabol; **Sans. :** Bolah.

Distribution : Indigenous to N.E. Africa especially Somalia Island; sold in Indian bazaars.

Parts Used : Gum.

Flowering & Fruiting : Flowering during late spring.

Folk Use : Gum used as stimulant and tonic.

Reference : Prajapati *et al.* (2006).

258. *Conium maculatum* L.

Family: Apiaceae.

English Names: **Eng. :** Poison Hemlock, Spotted Hemlock.

Distribution : Commonly found in Europe, also grows in temperate regions of Asia and N. America; introduced in India.

Parts Used : Fruit.

Flowering & Fruiting : May – September.

Folk Use : Fully grown unripe fruits used as sedative.

Active Constituents : Coniine, N-Me-conine and γ-coniceine (*fruit*).

References : Ambasta (1986), Asolkar *et al.* (1992), Kayal *et al.* (2003), Prajapati *et al.* (2006), Raju (2000).

259. *Convallaria majalis* L.

Family: Convallariaceae.

English Names: **Eng. :** Lily of the Valley, May-lily.

Distribution : Grown as ornamental in gardens.

Parts Used : Whole Plant. Flower.

Flowering & Fruiting : May – July.

Folk Use : Plant used as cardiac tonic. Crushed flowers given with sugar to strengthen memory.

References : Harborne & Baxter (2001), Khatoon & Goel (2010).

260. *Convolvulus arvensis* L.

Syn.: C. chinensis Ker.; *C. divaricatus* Wall.; *C. malcolmi* Roxb.; *Strophocaulos arvensis* (L.) Small.

Family: Convolvulaceae.

English, Hindi and Sanskrit Names: **Eng. :** Deer's Foot; **Hindi :** Harinpadi, Hiranpaddi; **Sans. :** Bhadrabala, Rajbala.

Distribution : Kashmir – Deccan, ascending to 3,000m in the Himalayas.

Parts Used : Flower.

Flowering & Fruiting : September – March.

Folk Use : Flowers used as brain tonic.

Reference : Dwivedi (2003).

261. *Convolvulus microphyllus* Sieb. ex Spreng.

Syn.: C. parviflorus Spreng.; *C. pluricaulis* Choisy.; *C. prostratus* Forssk.; *Evolvulus pilosus* Roxb.

Family: Convolvulaceae.

English, Hindi and Sanskrit Names: **Eng. :** Bindweed; **Hindi :** Kshirapushpi, Shanka ahavaa, Shankahuki, Shankhpushpi; **Sans. :** Shankhpushpi.

Distribution : A common weed in cultivated fields throughout most of India, ascending to 2,000m in the Himalayas.

Parts Used : Whole Plant. Flower.

Flowering & Fruiting : August – December.

Folk Use : Dried plant and open flowers collected early morning grounded with 'mishri' (crystalline sugar) in 3:1 ratio given with water as brain tonic for improving memory.

Active Constituents : Principle activities due to 6-methoxy-7-hydroxycoumarin, glucose, maltose, β-sitosterol, kaempferol, kaempferol-3-glucoside, 3,4-dihydroxycinnamic acid, rhamnose, sucrose, starch, n-hexacosanol, n-octacosanol, n-triacosanol, n-dotriacontanol, β- and ε-sitosterols (*plant*).

References : Dwivedi (2000), Khatoon & Goel (2010), Kumar & Chauhan (2005), Parrotta (2001), Shukla *et al.* (2003), Singh & Pandey (1998).

262. *Coptis teeta* Wall.

Family: Ranunculaceae.

English, Hindi and Sanskrit Names: **Eng. :** Coptis, Gold Thread, Mouth Root, Vegetable Gold; **Hindi :** Mamira, Mamiran; **Sans. :** Mamiram-Chini, Mishmitikta, Pita, Supita.

Distribution : Temperate regions of Mishmi hills at the Northern frontier of Assam.

Parts Used : Rhizome. Root.

Flowering & Fruiting : Flowering during Septmber – October and fruiting later.

Folk Use : Rhizome and roots used as tonic.

Active Constituents : Coptin, coptisine, berberine, palmatine and jatrorrhizine (*rhizome*).

References : Ambasta (1986), Anonymous (1950), Bhattacharjee (2001), Dhiman (2005), Kirtikar & Basu (1984), Prajapati *et al.* (2006), Sharma (2003, 2004), Thakur *et al.* (1989), Tiwari & Tiwari (1996).

263. *Corchorus capsularis* L.

Family: Tiliaceae.

English, Hindi and Sanskrit Names: **Eng. :** White Jute; **Hindi :** Narcha, Pat, Titapat; **Sans. :** Kalasaka.

Distribution : Throughout hotter parts of India; may have been introduced from China.

Parts Used : Leaf.

Flowering & Fruiting : August – October.

Folk Use : Leaves considered stimulant and also consumed along with diet as a tonic.

Active Constituents : Leaves possess corchoral, capsularol, capsularone, β-sitosterol and its glucoside, capsin, capsugenin-30-O-glucopyranoside and capsugenin-25,30-diglucoside.

References : Ambasta (1986), Kirtikar & Basu (1984), Singh & Srivastava (2000), Srivastava *et al.* (2003a), Watt (1972).

264. *Corchorus depressus* (L.) C. Chr.

Syn.: C. antichorus Raeusch.; *C. humilis* Munro & Stocks; *C. microphyllus* Fres.

Family: Tiliaceae.

English, Hindi and Sanskrit Names: **Eng. :** Shrubby Jute; **Hindi :** Baphuli; **Sans. :** Bhedani, Katuka, Tvakasara.

Distribution : Punjab, Sind, Baluchistan, Cutch, Gujarat, Deccan. Afganistan, Arabia, N. Africa, Cape Verde Islands and tropical Africa.

Parts Used : Whole Plant. Seeds. Leaf.

Flowering & Fruiting : Almost throughout the year, but profusely during April – June.

Folk Use : Plant, leaf and seeds used as tonic.

Active Constituents : Associated properties due to three new α-amyrin derivatives-cordepressic acid, cordepressenic acid and cordepressin, sitosterol, apigenin, luteolin and 2α,3β,20β-Urs-12-en-23β, 28-dioic acid, 2,3-diacetate (*plant*), quercetin and kaempferol (*leaf, flower*).

References : Ambasta (1986), Asolkar *et al.* (1992), Kirtikar & Basu (1984), Trivedi (2006).

265. *Corchorus fascicularis* Lam.

Family: Tiliaceae.

English, Hindi and Sanskrit Names: **Eng. :** Hoof; **Hindi :** Bankosta, Khetapat; **Sans. :** Chanchu, Kalabhi, Sushaka, Vijala.

Distribution : Hotter parts of India; tropical Africa and Australia.

Parts Used : Leaf. Fruit.

Flowering & Fruiting : October – November.

Folk Use : Decoction of fruits used orally as a tonic for health. Leaves used as stimulant and tonic.

References : Kirtikar & Basu (1984), Singh & Pandey (1998).

266. *Corchorus olitorius* L.

Family: Tiliaceae.

English, Hindi and Sanskrit Names: **Eng. :** Jew's Mallow, Tossa Jute; **Hindi :** Koshta Pata, Singinjanascha; **Sans. :** Brihatchanchu, Dirghapatri, Nadika, Vishari.

Distribution : Generally distributed in all tropical countries; cultivated in Western and Southern India.

Parts Used : Whole Plant. Leaf.

Flowering & Fruiting : July – November.

Folk Use : Leaves and decoction of plant taken orally as a health tonic.

References : Kirtikar & Basu (1984), Parrotta (2001), Singh & Pandey (1998), Singh & Srivastava (2000), Watt (1972).

267. *Cordia dichotoma* Forst. f.

Syn.: C. indica Lamk.; *C. latifolia* Roxb.; *C. lowriana* Brandis; *C. myxa* Thwaites; *C. obliqua* Willd.

Family: Boraginaceae.

English, Hindi and Sanskrit Names: **Eng. :** Large Sebestan, Sebesten Plum; **Hindi :** Chhotalasura, Lasura; **Sans. :** Bahuvarah, Kshudrashle, Shmataka, Slesmatak.

Distribution : Warmer regions throughout India; wild and cultivated up to 650m.

Parts Used : Bark. Fruit.

Flowering & Fruiting : April – July.

Folk Use : Bark and unripe fruits used as a health tonic.

Active Constituents : Major constituents isolated are tannins, lactose, D-ribose, L-rhamnose, alanine, aspartic acid, cystine, glucoronic acid, glutamic acid, glycine, leucine, proline, threonine, arachiolic, behenic, linoleic, oleic, palmitic, stearic acids and lysine (*bark*), d-arabinose, L-fructose, D-glucose, D-xylose, galactouronic acid, β-glucoronic acid, arabinoglucan and polysaccharide composed of (1-6) linked D-glucopyranosyl and arabinofuranosyl residues (*fruit*).

References : Kirtikar & Basu (1984), Sharma (2004).

268. Cordia macleodii (Griff.) Hook. f. & Thoms.

Syn.: Hemigymnia macleodii Griff.

Family: Boraginaceae.

Hindi Names: **Hindi :** Dahipalas, Dhagan, Dhaiman, Dhen-gan.

Distribution : Central India, Concan and Belgaum, Chota Nagpur, North Kanara and Deccan.

Parts Used : Seeds.

Flowering & Fruiting : Flowering during March – April.

Folk Use : Seeds used as brain tonic for mental disorders.

Reference : Kothari *et al.* (2003).

269. *Coriandrum sativum* L. (Pl. 4F)*

Family: Apiaceae.

English, Hindi and Sanskrit Names: **Eng. :** Coriander; **Hindi :** Dhania, Dhanya; **Sans. :** Dhanya, Dhanaka, Kustumburi.

Distribution : Cultivated throughout the warmer parts of India as cold season crop.

Parts Used : Fruits.

Flowering & Fruiting : Flowering and fruiting during cold season.

Folk Use : Half a spoonful of coriander fruits with a little sugar taken to strengthen memory and as stimulant.

Active Constituents : Principle constituents are umbelliferone, scopoletin, rutin, coumarins, quercetin, kaempferol and aflatoxins, β-sitosterol, D-mannitol, flavonoid glycosides, gnaphaloside A, gnaphaloside B, isrhamnetin and luteolin (*fruit*).

References : Ambasta (1986), Anonymous (1950), Bhattacharjee (2001), Kaushik & Dhiman (2000), Khatoon & Goel (2010), Kirtikar & Basu (1984), Pal & Paul (2003), Prajapati *et al.* (2006), Rajan *et al.* (2003), Reddi *et al.* (2005), Sharma (2003, 2004), Singh & Srivastava (2000), Singh *et al.* (1996), Siwakoti & Varma (1996), Watt (1972).

270. *Corydalis flabellata* Edgew.

Family: Papaveraceae.

English Names: **Eng. :** Bosc's Hawthorn, Fanleaf Hawthorn, New England Hawthorn.

Distribution : Himalayas (Kashmir – Nepal): 1,000-1,350m.

Parts Used : Whole Plant.

Flowering & Fruiting : July – September.

Folk Use : Plant used as a tonic.

Active Constituents : These are severzinine and (+) sibiricine (*aerial parts*), and chelidimerine (*root*).

Reference : Singh *et al.* (1996).

271. *Corydalis govaniana* Wall.

Family: Papaveraceae.

English, Hindi and Sanskrit Names: **Eng. :** Govan's Corydalis; **Hindi :** Bhutkesi, Bhutkis; **Sans. :** Bhutakesi.

Distribution : W. Himalayas (Kashmir – Kumaon): 2,700-4,000m.

Parts Used : Root.

Flowering & Fruiting : Flowering during March – May.

Folk Use : Root used as tonic and liver tonic.

References : Ambasta (1986), Anonymous (1950), Chhetri (2005), Kapur & Nanda (1996), Kirtikar & Basu (1984), Negi *et al.* (1999), Sharma (2003), Watt (1972).

272. *Corylus colurna* L.

Syn.: C. lacera Wall.

Family: Betulaceae.

English and Hindi Names: **Eng. :** Turkish Hazel; **Hindi :** Bhutea Badam.

Distribution : W. temperate Himalayas (Kashmir – Kumaon): 1,850-3,500m, Asia Minor, Thrace and Hungary.

Parts Used : Nut.

Flowering & Fruiting : April – August.

Folk Use : Nuts used as tonic.

References : Kapur & Nanda (1996), Watt (1972).

273. *Coscinium fenestratum* Colebr.

Syn.: Menispermum fenestratum Gaertn.

Family: Menispermaceae.

English, Hindi and Sanskrit Names: **Eng. :** Ceylon Calumba Root, False Calumba, Tree Turmeric; **Hindi :** Jhari-huldi; **Sans. :** Daru Haridra, Pitadru Darvi.

Distribution : Indigenous to South Vietnam; occurring in south India particularly in W. Ghats, Nilgiris, Travancore and Sri Lanka.

Parts Used : Root.

Folk Use : Root considered as tonic and bitter tonic.

Active Constituents : Berberine, jatrarrhizine, berbarrubine and N,N-di-Me-lindacarpine and thalifendine and palmitine (*root*).

References : Ambasta (1986), Asolkar *et al.* (1992), Kirtikar & Basu (1984), Watt (1972).

274. *Costus speciosus* (Koenig) Sm.

Syn.: Banksia speciosa Koenig.; *Costus speciosus* var. *nepalensis* (Rosc.) Baker; *Hellenia grandiflora* Retz.

Family: Costaceae.

English, Hindi and Sanskrit Names: **Eng.** : Spiral Ginger, Wild Ginger; **Hindi :** Keu, Kust; **Sans. :** Bramhatirtha, Canda, Kemuka, Subandhu, Vira.

Distribution : Known to occur more or less throughout India up to 1,320m in the Himalayas; also cultivated as an ornamental plant.

Parts Used : Rhizome. Root.

Flowering & Fruiting : July – September.

Folk Use : Rhizome and roots accredited with tonic and stimulant properties.

Active Constituents : Properties attributed due to diosgenin, tigogenin, an essential oil, saponins, genins and β-sitosterol, five new compounds-tetradecyl 13-methylpentadecanoate, tetradecyl 11-methyltridecanoate, 14-oxotricosanoic acid, 14-oxoheptacasanoic acid and 15-oxooctacosanic acid and methyl 3-(4-hydroxyphenyl)-2E-propenoate (*rhizome*), 31-norcycloartanone, cycloartanol, cycloartenol, cycloaudenol, 8-hydroxy-triacontane-25-one and methyltritriacontanoate (*root*).

References : Ambasta (1986), Anonymous (1950), Asolkar *et al.* (1992), Bhattacharjee (2001), Chaudhari & Hutke (2002), Dwivedi (2007), Farooq (2005), Gogoi & Das (2003), Kaushik & Dhiman (2000), Kothari & Londhe (2000), Parrotta (2001), Sharma & Sood (1997), Siwakoti & Varma (1996), Watt (1972).

275. *Crataegus oxyacantha* L.

Syn.: C. laevigata DC.; *Mespilus oxycantha* Crantz.

Family: Rosaceae.

English Name: **Eng. :** English Hawthorn.

Distribution : N. W. Himalayas : 1,800-3,000m.

Parts Used : Fruit.

Folk Use : Fruit juice used as a cardiac tonic for valvular insufficiency and as nervine tonic.

Active Constituents : Cynogenetic glycosides, crategin, amines viz. O-methoxyphenethylamine, tyramine, rutin, hyperoside, vitex glycosides, procyanidins, citric, tartaric and crataegus acid (*fruit*).

References : Ambasta (1986), Anonymous (1950), Kaul (1997).

276. *Crataeva nurvala* Buch.-Ham.

Syn.: C. magna (Lour.) DC. var. *magna*; *C. religiosa* Hook. f. & Th.

Family: Capparaceae.

English, Hindi and Sanskrit Names: **Eng.** : Capertra, Thorn Wing, Three Leaved Coper; **Hindi** : Barna, Bilasi, Varvunna; **Sans.** : Ajapa, Marutapaha, Shvetadru, Urumana.

Distribution : Native to India; found throughout the greater parts of India.

Parts Used : Bark.

Flowering & Fruiting : February – June.

Folk Use : Bark used as sedative and tonic.

Active Constituents : Betrulinic acid, ceryl alcohol, disogenin, friedelin, lupeol, cadabicine and its diacetate have been isolated (*bark*).

References : Anonymous (1950), Jain & Borthakur (1980), Kirtikar & Basu (1984), Sharma (2003), Watt (1972).

277. *Cressa cretica* L.

Family: Convolvulaceae.

English, Hindi and Sanskrit Names: **Eng.** : Littoral Bind Weed; **Hindi** : Lana, Rudravanti; **Sans.** : Palitaka, Rudanti, Rudantika, Sanjivani.

Distribution : Along sandy shores throughout India.

Parts Used : Whole Plant.

Flowering & Fruiting : November – March.

Folk Use : Herb decoction used as tonic.

Active Constituents : Known to possess n-octa-cosanol, scopoletin, umbelliferone, isopimpinellin, β-sitosterol and its D-glucose and quercetin (*plant*).

References : Ambasta (1986), Anonymous (1950), Kirtikar & Basu (1984), Kothari & Rao (2000), Prajapati *et al.* (2006), Reddi *et al.* (2005), Singh & Pandey (1998), Watt (1972).

278. *Crinum asiaticum* L.

Syn.: C. toxicarium Roxb.

Family: Amaryllidaceae.

English, Hindi and Sanskrit Names: **Eng.** : Poison Bulb; **Hindi** : Chindar, Kanmu, Kanwal, Pindar; **Sans.** : Bala, Dusaha, Jambati, Jambu, Mota, Nagadamani, Vritta.

Distribution : Throughout tropical India, Sri Lanka, wild or cultivated.

Parts Used : Bulbs.

Flowering & Fruiting : Throughout the year excepting winters.

Folk Use : Bulbs used as tonic.

Active Constituents : Known to yield two phenanthridone alkaloids-crinasiatine and crinasiadine (*bulbs*).

References : Anonymous (1950), Kirtikar & Basu (1984), Subramaniam (2000).

279. *Crocus sativus* L.

Family: Iridaceae.

English, Hindi and Sanskrit Names: **Eng. :** Saffron, Saffron Crocus, Spanish Saffron; **Hindi :** Kesar, Zafran; **Sans. :** Agnishekhara, Aruna, Asra, Dhira, Dipaka, Harichandana, Jaguda, Keshara, Vara.

Distribution : Cultivated mainly in Kashmir valley.

Parts Used : Flower.

Flowering & Fruiting : October – December.

Folk Use : Dried stigmas and tops of styles considered as stimulant, sedative, psychoactive and tonic.

Active Constituents : Safranol, isophorene, glucoside, corcin, crocetin, picro-crocin, β and R carotene and lycopene (*flower*).

References : Ambasta (1986), Anonymous (1950), Asolkar *et al.* (1992), Chauhan (2001), Kaul (1997), Kirtikar & Basu (1984), Prajapati *et al.* (2006), Rajan *et al.* (2003), Reddi *et al.* (2005), Sharma (2003), Sharma & Sood (1997), Thakur *et al.* (1989), Watt (1972).

280. *Crotalaria albida* Heyne ex Roth.

Syn.: C. montana Roxb.

Family: Fabaceae.

English Name: **Eng. :** Hill Rattle Wort.

Distribution : Hotter parts of India, Sri Lanka, Myanmar, Malay Peninsula and Archipelago to China and the Philippines.

Parts Used : Leaf.

Flowering & Fruiting : September – April.

Folk Use : 5g of leaf paste mixed with water consumed once a day for 5 days as tonic.

Reference : Viswanathan *et al.* (2001).

281. *Cryptolepis buchananii* Schult.

Syn.: C. reticulata Wall.; *Echites cuspidata* Heyne.; *E. reticulata* Roth.; *Nerium reticulum* Roxb.

Family: Asclepiadaceae.

English, Hindi and Sanskrit Names: **Eng. :** Indian Sarsaparilla, Wax Leaved Climber; **Hindi :** Dudhi, Karanta; **Sans. :** Krishnasariva.

Distribution : Throughout India, ascending to 1,350m.

Parts Used : Leaf.

Flowering & Fruiting : August – October.

Folk Use : Leaves used as a tonic.

Active Constituents : Known to possess cardenolidecryptosin, crytanoside B, cryptanoside A (buchanin), cryptonoside B and cryptosin (*leaf*).

References : Parrotta (2001), Srivastava *et al.* (2003), Subramaniam (2000).

282. *Cucumis melo* var. ***utilissimus*** (Roxb.) Duthie & Fuller

Syn.: C. utilissimum Roxb.

Family: Cucurbitaceae.

English, Hindi and Sanskrit Names: **Eng. :** Musk Melon, Sweet Melon; **Hindi :** Kakri; **Sans. :** Bahukanda, Chirbhati, Ervaru, Pinasa.

Distribution : Cultivated throughout North India.

Parts Used : Fruit.

Flowering & Fruiting : May – October.

Folk Use : Fruits used as tonic.

Active Constituents : Ferulic, caffeic and chlorogenic acids have been isolated (*fruit*).

References : Kirtikar & Basu (1984), Prajapati *et al.* (2006).

283. *Cucumis sativus* L.

Syn.: C. hardwickii Royle.; *C. muricatus* Wall.

Family: Cucurbitaceae.

English, Hindi and Sanskrit Names: **Eng. :** Common Cucumber, Musk Melon, Sweet Melon; **Hindi :** Khira; **Sans. :** Bahpushpa, Kandalu, Pitapushpa, Sudhavasa, Trapusah, Tundilaphala.

Distribution : Throughout India, cultivated.

Parts Used : Fruit. Seeds.

Flowering & Fruiting : May – July.

Folk Use : Fruits and seeds used as a tonic.

Active Constituents : Constituents isolated from different parts are ascorbic acid, an enzyme erepsin, proteolytic enzymes, oxidase, succinic and malic dehydrogenases, palmitic, stearic and oleic acids, their glycerides, sterols and squalene (*fruit*), sitosterol, stigmasta-7,22,25-trien-3β-ol, 24β-ethyl-31-norlanosta-8,25(27)-dien-3β-ol (I), 24β-ethyl -25(27)-dehydrolophenol (II), multiflorenol, isomultiflorenol, α- and β-amyrins, taraxerol, lupeol, cycloartenol, 24-methylenecycloartanol, 24-methyl-25(27)-dehydrocycloartanol, 24-methylene-24-dihydrolanosterol, 24-methylene-24-dihydroparkeol, euphol, tirucallol, 24-ethyl-5-α-cholesta-7-trans-22-dien-3β-ol, 24-ethylcholest-5-en-3β-ol, 24-ethyl-5α-cholest-7-en-3β-ol, codisterol, 25(27)-dehydroporiferosterol, clerosterol, avenasterol, isofucosterol, stigmasterol, campesterol, sitosterol, 25(27)-dehydrochondrillasterol, 24β-ethyl-25(27)-dehydrolathosterol, 24ξ-methyllathosterol, spinasterol and 22-dihydrospinasterol (*seeds*).

References : Anonymous (1950), Chauhan *et al.* (2007), Kaushik & Dhiman (2000), Prajapati *et al.* (2006), Sood *et al.* (2009).

284. *Cucurbita maxima* Duchesne ex Lam.

Family: Cucurbitaceae.

English, Hindi and Sanskrit Names: **Eng. :** Pumpkin, Red Gourd, Sweet Gourd; **Hindi :** Kaddu, Kadimah, Lal Kumra, Mitha Kumra, Sitaphal; **Sans. :** Dangari, Gramya, Pitakusmandah.

Distribution : Cultivated throughout India.

Parts Used : Fruit. Seeds.

Flowering & Fruiting : Late winters.

Folk Use : So far known as sedative (fruits), tonic (seeds, fruit) and nervine tonic (seed oil).

Active Constituents : Attributed characterstical properties due to stigmasta-7,24(28)-dien-3β-ol (I), stigmasta-7,25-dienol acetate, cholesterol, ergosterol, campesterol, β-sitosterol, 24β-ethyl-5α-cholesta-7,22,25(27)trien-3β-ol, 24β-ethyl-5α-cholesta-7,25(27)-dien-3β-ol, avenasterol, spinasterol, 24-dihydrospinasterol, 24ξ-methyllathosterol, 25(27)-dehydrofungisterol,25(27)-dehydroporiferasterol, clerosterol, isofucosterol, stigmasterol, sitosterol, campesterol and codisterol (*seeds*).

References : Ambasta (1986), Anonymous (1950), Kirtikar & Basu (1984), Prajapati *et al.* (2006), Watt (1972).

285. *Curculigo orchioides* Gaertn.

Syn.: C. malabarica Wight.

Family: Hypoxidaceae.

English, Hindi and Sanskrit Names: **Eng. :** Common Curculigo, Golden Eye-grass; **Hindi :** Kali Musli, Muslikand, Siyah Musli; **Sans. :** Bhutali, Hema Pushpi, Kharjuri, Talemnlika.

Distribution : Sub-tropical areas of Central Himalaya and Western Ghats.

Parts Used : Rhizome. Root.

Flowering & Fruiting : June – August.

Folk Use : Rhizome (decoction) and roots used as tonic.

Active Constituents : Active properties due to phenolic glycoside curculigoside, curculigine A, orcinol glycoside corchioside A, sitosterol, stigmasterol, hentriacontanol, cycloartenol, lycorine, glycosides of cyclo-aratane sapogenin-curculigosaponins G-M, curculigenin B & C and glycoside yuccagenin (*rhizome*), and methoxy ketone (*root*).

References : Anonymous (1950), Chaudhari & Hutke (2002), Dhiman (2005), Dwivedi (2007), Farooq (2005), Jain *et al.* (2003), Kirtikar & Basu (1984), Kumar & Lal (2003), Panthi & Chaudhary (2003), Parrotta (2001), Pathak & Pathak (2007), Patil (2007), Prajapati *et al.* (2006), Reddi *et al.* (2005), Siwakoti &Varma (1996), Thakur *et al.* (1989).

286. *Curcuma angustifolia* Roxb.

Family: Zingiberaceae.

English, Hindi and Sanskrit Names: **Eng. :** East Indian Arrow Root, Narrow-leaved Turmeric, Wild Arrow Root; **Hindi :** Tavakhira, Tikhur; **Sans. :** Gavayodbhava, Pishtika, Talakshira, Yavaja.

Distribution : Moist deciduous Sal and mixed forests of M.P., Chhatisgarh and Jharkhand and Bengal, Bombay, Chennai and some of lower Himalyan ranges.

Parts Used : Rhizome.

Flowering & Fruiting : Flowering during July – November.

Folk Use : Rhizome used as a tonic.

References : Kirtikar & Basu (1984), Saikia & Nath (2003a), Sharma (2003).

287. *Curcuma aromatica* Salisb.

Family: Zingiberaceae.

English, Hindi and Sanskrit Names: **Eng. :** Cochin Turmeric, Wild Turmeric, Yellow Zedoary; **Hindi :** Ban-haridra; **Sans. :** Aranyaharidra, Sholi, Vanaharidra.

Distribution : Bengal, W. Peninsula; sometimes cultivated.

Parts Used : Rhizome.

Flowering & Fruiting : Flowering during July – November.

Folk Use : Rhizome used as a tonic.

Active Constituents : Camphene, a and b-pinenes, 1,8-cineole, borneol, isoborneol, camphor and germacrone (*rhizome*).

References : Farooq (2005), Kirtikar & Basu (1984), Lalramnghinglova (2003), Reddi *et al.* (2005), Saikia & Nath (2003a), Watt (1972).

288. *Curcuma caesia* Roxb.

Family: Zingiberaceae.

English, Hindi and Sanskrit Names: **Eng. :** Black Zedoary; **Hindi :** Kali Haldi, Narkachura; **Sans. :** Krishna Haridra, Nar-kachura.

Distribution : Bengal.

Parts Used : Rhizome.

Flowering & Fruiting : Flowering during Autumn.

Folk Use : Rhizome used as stimulant and tonic.

References : Dwivedi (2007), Rahman & Yusuf (1996), Saikia & Nath (2003a), Sharma (2004), Tripathi & Goel (2001).

289. *Curcuma domestica* Valeton (Pl. 5A)*

Syn.: C. longa L.

Family: Zingiberaceae.

English, Hindi and Sanskrit Names: **Eng. :** Indian Saffron, Long Turmeric, Turmeric; **Hindi :** Halad, Haldi; **Sans. :** Aneshta, Gauri, Haridra, Hemaragi, Jvarantika, Kshapa, Lakshmi, Uma, Varangi.

Distribution : Cultivated in humid climatic regions of the plains of India.

Parts Used : Rhizome.

Flowering & Fruiting : Flowering during July – November.

Folk Use : Rhizome used as stimulant and tonic.

Active Constituents : These are α-phellandrene, d-sabinene, cineole, borneol, zingiberene, sesquiterpene, turmerone, ar-termerine and sabinone, cucurcumin, monodesmetoxy curcumin and bidismethoxy curcumin, 2-7% essential oil comprising mainly bisabolane, guaine and germacrane, termerone, arteromerone, curlone (*rhizome*).

References : Ambasta (1986), Anonymous (1950), Bhattacharjee (2001), Kirtikar & Basu (1984), Pal & Paul (2003), Parrotta (2001), Rahman & Yusuf (1996), Saikia & Nath (2003a, b), Sarma *et al.* (2002), Sharma (2003), Sharma & Sood (1997), Thakur *et al.* (1989), Watt (1972).

290. *Curcuma zedoaria* (Christm.) Rosc.

Syn.: C. zerumbet Roxb.

Family: Zingiberaceae.

English, Hindi and Sanskrit Names: **Eng. :** Cochin Turmeric, Round Zedoary, Zedoary; **Hindi :** Kachura, Kalihaladi; **Sans. :** Durlabha, Karchura, Mukhya, Sati.

Distribution : Wild in the Eastern Himalayas and moist deciduous forests of coastal tract of Kanara.

Parts Used : Rhizome.

Flowering & Fruiting : Flowering during summer season.

Folk Use : Rhizome used as stimulant and tonic to brain and heart.

Active Constituents : Principle activities due to α-pinene, camphor, curzerenone, epicurzerenone, iso-furanogermerene [2], curcumadiol, curcumol, curcumenol, iso-curcumenol, procurcumenol [3,4], dehydrocurdione [5], germacrone-4,5-epoxide, germacrone, germacronefuranodienone, curcumenol, iso-curcumenol and cucumenone [6] (*rhizome*).

References : Ambasta (1986), Anonymous (1950), Behura & Rout (2003), Dwivedi (2007), Gogoi *et al.* (2003), Kirtikar & Basu (1984), Rahman & Yusuf (1996), Saikia & Nath (2003a), Sharma (2004), Thakur *et al.* (1989), Watt (1972).

291. *Cuscuta chinensis* Lam.

Family: Convolvulaceae.

English Names: **Eng. :** Common Dodder, Dodder.

Distribution : Greater part of India, Sri Lanka, Persia and Australia.

Parts Used : Seeds.

Flowering & Fruiting : July – September.

Folk Use : Seeds used as a tonic.

Active Constituents : Quercetin, quercetin-3-O-β-galactosido-7-O-β-glucoside, astragalin and hyperin (*seeds*).

References : Kirtikar & Basu (1984), Subramaniam (2000).

292. *Cuscuta reflexa* Roxb.

Syn.: C. elatior Choisy.; *C. grandiflora* Wall.; *C. hookeri* Sweet.; *C. macrantha* G. Don; *C. verrucosa* Sweet.

Family: Convolvulaceae.

English, Hindi and Sanskrit Names: **Eng. :** Dodder; **Hindi :** Akashbel, Amarabela; **Sans. :** Amarvela, Akasavalli, Asparsa, Nilatar, Vyomavallika.

Distribution : Parasitic climber throughout the plains of India, ascending hills up to 2,438m.

Parts Used : Whole Plant. Seeds.

Flowering & Fruiting : Flowering: October – December; December – February (fruiting).

Folk Use : Plant and seeds regarded as tonic.

Active Constituents : Activities characterized due to amarbelin, cuscutin, cerotic, linolenic, linolic, oleic, stearic and palmitic acids and phytosterol (*seeds*), dulcitol, luteolin and its glycoside, quercetin and cuscutin (*stem*), amino acids, cusculatin and cuscutin (*plant*).

References : Ambasta (1986), Anonymous (1950), Dhiman (2005), Farooq (2005), Gogoi & Das (2003), Kaushik & Dhiman (2000), Kirtikar & Basu (1984), Maity *et al.* (2003), Parrotta (2001), Prajapati *et al.* (2006), Raju (2000), Varghese (1996).

293. *Cyamopsis tetragonoloba* (L.) Taub.

Syn.: C. psoralioides (L.) DC.

Family: Fabaceae.

English, Hindi and Sanskrit Names: **Eng. :** Cluster Bean; **Hindi :** Govar; **Sans. :** Bakuchi, Garak, Ksudrasimbi.

Distribution : Throughout India, cultivated.

Parts Used : Pods.

Flowering & Fruiting : September – November.

Folk Use : Pods used as a tonic.

References : Prajapati *et al.* (2006).

294. *Cycas revoluta* Thunb.

Family: Cycadaceae.

English and Hindi Names: **Eng. :** China Sago Palm, Japan Sago Palm, Cycas; **Hindi :** Cycas.

Distribution : Commonly cultivated in Indian gardens.

Parts Used : Whole Plant. Tuber.

Flowering & Fruiting : February – May.

Folk Use : Plant and its tubers used as tonic.

References : Kaushik & Dhiman (2000), Kirtikar & Basu (1984).

295. *Cyclea arnottii* Miers

Syn.: C. burmanni Hook. f. & Thoms.; *C. peltata* Hook. f. & Thoms., non Diels

Family: Menispermaceae.

English, Hindi and Sanskrit Names: **Eng. :** Pata Root; **Hindi :** Parvel, Path, Pathi; **Sans. :** Patha.

Distribution : Throughout India, up to 900m.

Parts Used : Tuber.

Flowering & Fruiting : Throughout the year.

Folk Use : Tubers used as tonic.

References : Ambasta (1986), Prajapati *et al.* (2006).

296. *Cydonia oblonga* Mill.

Syn.: C. vulgaris Pers.

Family: Rosaceae.

English, Hindi and Sanskrit Names: **Eng. :** Quince; **Hindi :** Bihi; **Sans. :** Patala.

Distribution : Kashmir, H.P. and Nilgiris at 1,700m.

Parts Used : Fruit.

Flowering & Fruiting : March – May.

Folk Use : Fruits used as a tonic.

Reference : Chatterjee & Pakrashi (1997).

297. *Cymbopogon citratus* DC. Stapf.

Syn.: Andropogon citratus DC.; *A.citriodendrum* Desf.; *A.roxburghii* Nees.

Family: Poaceae.

English, Hindi and Sanskrit Names: **Eng. :** Citronella Grass, Ginger Grass, Lemon Grass, Sere Grass; **Hindi :** Agin Ghas, Gandhatrina; **Sans. :** Bhustrina, Chhatra, Guchhala.

Distribution : Known in the cultivated state, probably of Indian origin; now widely distributed over the tropics of both hemispheres.

Parts Used : Leaf.

Flowering & Fruiting : August – March.

Folk Use : Leaves used as stimulant.

Active Constituents : Known to possess luteolin, its 7-O-glucoside and 7-O-neohesperoside, homoorientin and its 2''-O-rhamnosyl derivative, chlorogenic, caffeic and p-coumaric acids, fructose, sucrose, octacosanol, triacontanol and dotriacontanol (*leaf*).

References : Parrotta (2001), Srivastava & Singh (2003).

298. *Cymbopogon jwarancusa* (Junes) Schult.

Syn.: Andropogon jwarancusa Jones; *A. laniger* Duthie, non Desf.

Family: Poaceae.

English, Hindi and Sanskrit Names: **Eng. :** Iwarancusa Grass, Russ Grass; **Hindi :** Lamjak; **Sans. :** Lamajjaka.

Distribution : Wild in Himalayas from Kashmir – Assam up to 3,290m and in the north western plains reaching down to Mumbai.

Parts Used : Leaf.

Flowering & Fruiting : July – December.

Folk Use : Infusion of leaves considered stimulant.

References : Dhiman (2005), Watt (1972).

299. ***Cymbopogon martinii*** (Roxb.) Wats

Syn.: Andropogon martinii Roxb.; *A. pachnaodes* Trin.; *A. schoenanthus* L.

Family: Poaceae.

English, Hindi and Sanskrit Names: **Eng. :** Palmarosa, Rosha Grass, Rusa Grass; **Hindi :** Gandhabel, Rusa Ghas; **Sans. :** Dhyamakah, Rohisah.

Distribution : Cultivated in dry localities throughout India.

Parts Used : Whole Plant.

Flowering & Fruiting : August – December.

Folk Use : Infusion of plant valued as stimulant and cardio tonic.

Active Constituents : Geraniol, geranyl acetate, linalool, geranyl octanoate, geranyl butyrate, dihemiacetal, bismonoterpenoid and cymbodiacetal (*plant*).

References : Asolkar *et al.* (1992), Parrotta (2001), Prajapati *et al.* (2006).

300. ***Cynodon dactylon*** (L.) Pers. (Pl. 5B)*

Syn.: C. maritimus H. Bock.; *C. occidentalis* Willd.; *C. radiatus* Roth ex Roem. & Schult.; *Dactylon officinale* Vill.; *Digitaria dactylon* Scop.; *Fibichia umbellata* Koel.; *Panicum dactylon* L.; *Paspalum dactylon* DC.

Family: Poaceae.

English, Hindi and Sanskrit Names: **Eng. :** Bahama Grass, Barmuda Grass, Conch Grass, Dhub Grass; **Hindi :** Dub, Dubra, Kabbar, Mangala, Nanda, Romghas, Shatmula, Vamini; **Sans. :** Amari, Bahuvirya, Durva, Niladurva.

Distribution : Wild, throughout India; also cultivated for decorative purposes.

Parts Used : Whole Plant. Leaf.

Flowering & Fruiting : August – November.

Folk Use : Plant used as a tonic. Leaf paste with 'Sal' flower or fruit and rice also taken as tonic.

Active Constituents : Known to yield β-ionone, 2-propionic, 4-hydroxybenzoic, 3-methoxy-4-hydroxybenzoic acid, phytol, β-sitosterol-D-glucoside, stigmasterol acetate, phytone, glycosides, saponins, tannins, flavonoids and carbohydrates (*plant*).

References : Banerjee (2000a), Nayak *et al.* (2003), Prajapati *et al.* (2006), Rosakutty *et al.* (2003).

301. *Cyperus articulatus* L.

Family: Cyperaceae.

English Name: **Eng. :** Guinea Rush.

Distribution : Warm regions extending from Bengal – Sri Lanka.

Parts Used : Tuber.

Folk Use : Tubers used as tonic and stimulant.

References : Ambasta (1986), Anonymous (1950).

302. *Cyperus esculentus* L.

Syn.: C. tuberosus Rottb.

Family: Cyperaceae.

English, Hindi and Sanskrit Names: **Eng. :** Chufa, Rush Nut, Tiger Nut; **Hindi :** Cicada, Cicoda; **Sans. :** Bhadramusta, Nagaramustakah

Distribution : Punjab – Nilgiri hills, scattered; S. Europe, Africa and America.

Parts Used : Tuber.

Flowering & Fruiting : Summer.

Folk Use : Tubers accredited with stimulant and tonic properties.

Active Constituents : Properties due to isoleucine, leucine, lysine, methionine, phenylalanine, threonine, tryptophan, valine (*tuber*).

References : Ambasta (1986), Kirtikar & Basu (1984), Prajapati *et al.* (2006).

303. *Cyperus iria* L.

Syn.: C. microiria Steud.; *C. microlepis* Baker; *C. panicoides* Lam.; *C. parviflorus* Nees; *C. santonici* Rottb.; *C. umbellatus* Roxb.

Family: Cyperaceae.

English Names: **Eng. :** Rice Flat Sedge, Umbrella Sedge.

Distribution : Rice fields throughout India; Sri Lanka, Indo-China, Australia and Mediterranean.

Parts Used : Whole Plant.

Flowering & Fruiting : August – December.

Folk Use : Plant considered as stimulant and tonic.

References : Ambasta (1986), Anonymous (1950), Kirtikar & Basu (1984), Parrotta (2001), Srivastava *et al.* (2003a).

304. *Cyperus kyllingia* Endl.

Syn.: C. monocephalus (Rottb.) F.V.M.; *Kyllinga monocephala* Rottb.

Family: Cyperaceae.

Hindi and Sanskrit Names: **Hindi :** Nirbishi, Swetgothubi; **Sans. :** Musta, Nirvisha.

Distribution : Distributed throughout India.

Parts Used : Rhizome.

Flowering & Fruiting : July – December.

Folk Use : Decoction of rhizome used as a tonic.

References : Ambasta (1986), Gogoi & Das (2003), Kaushik & Dhiman (2000), Subramaniam (2000).

305. *Cyperus longus* L.

Family: Cyperaceae.

English Names: **Eng. :** Cypress, Cypress-root, Galanga, Galingale, Sweet Galingale.

Distribution : Quetta, Mt. Abu, and westwards to the Atlantic.

Parts Used : Rhizome.

Folk Use : Rhizome used as tonic.

References : Ambasta (1986), Anonymous (1950).

306. *Cyperus rotundus* L.

Syn.: C. albidus Herb.; *C. hexastachyus* Rottb.; *C. tenuiflovus* Royle.

Family: Cyperaceae.

English, Hindi and Sanskrit Names: **Eng. :** Coco Grass, Nut Grass; **Hindi :** Motha, Nagarmotha; **Sans. :** Abda, Bhadrakshi, Hima, Musta, Mustaka.

Distribution : Commonly in wastegrounds, gardens and along roadsides throughout India up to 1,828m.

Parts Used : Tuber.

Flowering & Fruiting : Throughout the year.

Folk Use : Tubers used as stimulant, tonic and nervine tonic.

Active Constituents : Activities attributed to cyperotundone, cyperolone, oleanolic acid-3-O-neohesperidoside, rhamnetin-3-O-rhamnosyl(1→4)rhamnopyranoside (*tuber*).

References : Dhiman (2005), Farooq (2005), Kapur & Singh (1996), Kaushik & Dhiman (2000), Kirtikar & Basu (1984), Nayak *et al.* (2003), Parrotta (2001), Prajapati *et al.* (2006), Prakash (2001), Ranjan (2000), Sharma (2003), Siwakoti & Varma (1996).

307. *Cyperus scariosus* R. Br.

Family: Cyperaceae.

English, Hindi and Sanskrit Names: **Eng. :** Umbrella's Edge; **Hindi :** Nagarmotha; **Sans. :** Chakranksha, Nadeyi, Shishira.

Distribution : Damp places in Bengal, U.P., eastern and southern parts of India.

Parts Used : Tuber.

Folk Use : Decoction of tuber given as tonic after fever.

References : Kumar & Goel (1998), Singh & Kumar (2000b).

308. *Cyperus serotinus* Rottb. var. ***inundates*** (Roxb.) Kukenthal

Syn.: Juncellus inundatus C.B. Clarke.

Family: Cyperaceae.

Hindi Name: **Hindi :** Pati.

Distribution : Kashmir: 1,700 – 1,900m.

Parts Used : Tuber.

Folk Use : Tubers used as stimulant and tonic.

Reference : Ambasta (1986).

309. *Cyperus stoloniferus* Retz.

Syn.: C. juncifolius Klien; *C. littoralis* R. Br.; *C. tuberosus* Baker.

Family: Cyperaceae.

Distribution : Sea – shores of India.

Parts Used : Tuber.

Folk Use : Tubers used as stimulant and cardiac tonic.

References : Ambasta (1986), Asolkar *et al.* (1992).

310. *Cytisus scoparius* L.

Syn.: Sarothamnus scoparius Koch.

Family: Fabaceae.

English Names: **Eng. :** Broom, Scotch Broom, Yellow Broom.

Distribution : Native of West Europe; found growing wild as a garden escape between 2,350-2650m.

Parts Used : Twigs.

Flowering & Fruiting : May – June (flowering); August – September (fruiting).

Folk Use : Green twigs used as cardiac tonic.

Active Constituent : Sparteine (*twigs*).

References : Ambasta (1986), Anonymous (1950).

311. ***Dactylorhiza hatagirea*** (D. Don) Soo.

Syn.: Orchis latifolia auct.

Family: Orchidaceae.

English, Hindi and Sanskrit Names: **Eng.** : Heart Orchid, Marsh Orchid, Salep; **Hindi** : Salam, Salampanja, Salep; **Sans.** : Munjatakah, Salampamisri.

Distribution : Western temperate Himalayas (Kashmir – Sikkim) up to 3,657m.

Parts Used : Root. Tuber.

Flowering & Fruiting : January – July.

Folk Use : Roots valued as nervine tonic. Extract of tuber considered as a tonic.

References : Asolkar *et al.* (1992), Dhiman (2005), Hemlata *et al.* (2007), Kaul (1997), Maikhuri *et al.* (2001), Nadeem *et al.* (2001), Nautiyal *et al.* (2003), Samant *et al.* (2001), Sharma (2003).

312. ***Dalbergia sissoo*** Roxb.

Syn.: D. pendula Tenore.

Family: Fabaceae.

English, Hindi and Sanskrit Names: **Eng.** : Sissoo, South Indian Red Wood; **Hindi** : Shisham, Sissai, Sissu; **Sans.** : Aguru, Dhira, Dhumrika, Kapila, Pichhila, Shinshapa, Simsapa, Vira.

Distribution : Western Himalayas up to 1,500m, Sikkim – upper Assam; extensively planted along roadsides in different parts of India.

Parts Used : Leaf.

Flowering & Fruiting : February – May.

Folk Use : Leaves considered as stimulant.

Active Constituents : Active properties due to isoflavone and sissotrin (*leaf*), dalbergin, methyl dalbergin, 5,5-dihydroxy-4-methoxy isoflavone, 3,5-dihydroxystilbene, 7-methoxy-6-hydroxy-4-phenylchrom-3-one, nordalbergin, isodalbergin, dalbergiphenol, tectorigenin, isotectorigenin, isocavindin, caviunim-7-O-glucoside and 7-O-gentiobioside (*plant*).

References: Anonymous (1952), Chauhan *et al.* (2007), Dhiman (2005), Farooq (2005), Kaushik & Dhiman (2000), Parrotta (2001), Prajapati *et al.* (2006), Shah (2005), Subramaniam (2000).

313. ***Daphne cannabina*** Wall.

Syn.: D. bholua Don.; *D. odora* Don.; *D. papyracea* Wall. *ex* Steud.; *D. papyrifera* Ham.

Family: Thymelaeaceae.

English and Hindi Names: **Eng.** : Nepal Paper Plant; **Hindi** : Satpura, Setburusa, Setburwa.

Distribution : Temperate Himalayas and Khasia hills up to 2,100m.

Parts Used : Leaf.

Flowering & Fruiting : April – October.

Folk Use : Leaves prescribed as bitter tonic.

Active Constituent : Taraxerone (*leaf*).

Reference : Sood *et al.* (2009).

314. *Datura innoxia* Mill.*

Syn.: D. metel Auct.

Family: Solanaceae.

English, Hindi and Sanskrit Names: **Eng.** : Thorn Apple; **Hindi** : Safed Datura; **Sans.** : Dhattura.

Distribution : Found in Bangalore, Ahmadabad, Pune, Lucknow, Oilani and Jammu.

Parts Used : Leaf. Seeds.

Flowering & Fruiting : June – August.

Folk Use : Seeds used as depressant and sedative. Leaves valued as narcotic.

Active Constituents : Known to yield hyoscyamine, scopolamine, meteloidine (*leaf*), quercetin-7-glucoside-3-sophoroside, quercetin-7-glucoside-3-glucogalactoside and their esters, caffeic and p-coumaric acids, kaempferol-7-glucoside-3-sophoroside and kaempferol-7-glucoside-3-glucogalactoside and their esters with caffeic acids, 6β-propionyloxy-3α-tigloyloxytropane (I), daturadiol, daturaolone and α- and β-scopodonnines (*seeds*).

References : Ambasta (1986), Anonymous (1950), Kumar *et al.* (2003), Sharma (2003).

315. *Datura metel* L. (Pl. 5C)*

Syn.: D. alba Nees.; *D. fastuosa* L.; *D. fastuosa* L. var. *alba* (Nees) Cl.

Family: Solanaceae.

English, Hindi and Sanskrit Names: **Eng.** : Downy Thornapple, Thorn-apple, White Datura; **Hindi** : Kanak, Sadahdhatura, Safed Dhatura; **Sans.** : Dhastura, Kanak, Khanakonmatta, Shivashekwa.

Distribution : Native of Mexico, naturalized and very common in wasteplaces, along road sides, railway lines and in scrub jungles throughout the tropical parts of India.

Parts Used : Leaf. Root. Seeds.

Flowering & Fruiting : April – June.

Folk Use : Leaves considered narcotic. Seeds, leaves and roots used in insanity. Seeds valued as narcotic and depressant.

Active Constituents : Properties exhibited due to norhyoscyamine, datumetine, datumetelin, datumelin, daturilinol, niacin, withametelin, scopolamine, hyoscyamine, vitamin C & 1-oxo-21,24S-epoxy-(20S,22S)-witha-2,5,25-trienoline (*leaf*), hyosine, hyoscyamine, atropine, fastudine, fastunine, fastusic acid, lophenol, daturanolene, citrostadienol, cycloeucalenol, obtusifoliol, 4α-methylcholest-8-enol, 31-norcycloartenol, 31-norlanosterol, 31-norlanost-8-enol, 31-norlanost-9 (11)-enol, allantoin, caproic, α- and β-linolic, linoleic and oleic acids, scopolamine and factusine (*seeds*).

References : Ambasta (1986), Bhattacharjee (2001), Dhiman (2005), Farooq (2005), Gogoi & Das (2003), Kaushik & Dhiman (2000), Nayak *et al.* (2003), Parrotta (2001), Prajapati *et al.* (2006), Punjani (2003), Sharma (2003), Sood *et al.* (2009), Subramaniam (2000), Thakur *et al.* (1989), Trivedi (2002).

316. *Datura sanguinea* Ruiz. & Pav.

Family: Solanaceae.

English Names: **Eng.** **:** Red Angel's Trumpet, Scarlet Angel's Trumpet, Scarlet Datura.

Distribution : Native of S. America; grown in gardens on tho hills.

Parts Used : Seeds.

Folk Use : Seeds valued as narcotic.

References : Ambasta (1986), Anonymous (1950).

317. *Datura stramonium* L.*

Syn.: D. ferox Nees.; *D. stramonium* var. *tatula* (L.) Cl.; *D. tatula* L.; *D. wallichii* Dunal; *Stramonium vulgatum* Gaertn.

Family: Solanaceae.

English, Hindi and Sanskrit Names: **Eng.** **:** Jimson Weed, Mad Apple, Stink Weed, Stramonium, Thorn Apple; **Hindi :** Dhatura; **Sans.** **:** Devika, Dhattura, Ghantika, Kanaka, Kitara, Mahamohi, Matta, Shivapriya, Unmattaka.

Distribution : Himalayas from Kashmir – Sikkim up to 2,743m.

Parts Used : Flower. Fruit. Leaf. Seeds.

Flowering & Fruiting : July – December.

Folk Use : Flowering tops, leaves and seeds considered narcotic. Fruits used as sedative.

Active Constituents : Therapeutic properties due to hyoscamine, hyoscine, scopolamine, triterpene alcohols-lanost-8-en-3β-ol (2.0), lanosterol (tr) and 24-methylenelanost-8-en-3β-ol (tr) in addition to cycloartanol (tr), cycloartenol (34.0), 24-methylene-cycloartanol (7.0), lupeol (2.0), β-amyrin (53.0%), daturaolone (tr), daturadiol (tr), two new β-carboline derivatives-fluorodaturatine and homofluorodaturatine and 1,2-dehydrofluorodaturine (*seeds*), β-sitosterol, stigmasterol, campesterol, withanolide, stramolide, withstramonolide, a new vitanolide-vitastramonolide along with daturalactone and its keto derivative, chrysin, quercetin and kaempferol (*leaf*).

References : Ambasta (1986), Anonymous (1950), Bhattacharjee (2001), Dhiman (2005, 2007), Dwivedi (2007), Gogoi & Das (2003), Hemlata *et al.* (2007), Kaul (1997), Kaushik & Dhiman (2000), Kirtikar & Basu (1984), Ranjan (2000), Sharma (2003), Singh (2001), Singh & Srivastava (2000), Subramaniam (2000).

318. *Daucus carota* L.

Syn.: D. carota L. var. *sativa* DC.

Family: Apiaceae.

English, Hindi and Sanskrit Names: **Eng.** : Bee's Nest, Bird's Nest, Queen Anne's Lace, Wild Carrot; **Hindi** : Gajar, Gajor, Shikhamulam; **Sans.** : Gajara, Garjara, Granthimula, Narangavarnaka, Svadumula.

Distribution : Cultivated all over India.

Parts Used : Root. Seeds.

Flowering & Fruiting : February – March (flowering); April – May (fruiting).

Folk Use : Infusion of roots considered as stimulant whereas seeds valued both as stimulant and nervine tonic.

Active Constituents : These are n-alkanes, β-sitosterol, glucose, lipid glycerides and amino acids, apigenin-7-O-β-D-galactopyranosyl(1→4)-O-β-D-mannopyranoside, apigenin-4'-O-β-D-glucoside, kaempferol-3-O-β-D-glucoside (*seeds*), β-carotene, L(+)-lactic acid, iodine, pyruvate kinase, steryl glucoside, acyl transferase, daucic acid, alkane, n-hexacosane, cholesterol, β-sitosterol, glucose, lipid glycerides and amino acids, umbelliferone (α-D-galactopyranosyl-2-α-D-gluco-pyranosyl-2β-D-fructofuranoside), n-alkanes (C13-58), n-nonacosane, n-heptacosane, n-hentriacontane, β-sitosterol, leucine, valine, tryptophan, proline, alanine, glycine, glucose, fructose and a carotenoid protein complex containing α- and β-carotenes and lutein as major carotenoids (*root*).

References : Ambasta (1986), Anonymous (1952), Asolkar *et al.* (1992), Kaushik & Dhiman (2000), Kirtikar & Basu (1984), Ranjan (2000), Sharma (2004), Singh *et al.* (1996, 2007), Watt (1972).

319. *Delphinium denudatum* Wall.

Syn.: D. pauciflorum Royle.

Family: Ranunculaceae.

English, Hindi and Sanskrit Names: **Eng.** : Delphinium, Kylinga Triceps; **Hindi** : Jadwar, Nirbisi; **Sans.** : Apavisha, Nirvisha, Vishalakarani.

Distribution : Commonly found in the western Himalayas, from Kumaon – Kashmir between 2,440-3,600m.

Parts Used : Root. Seeds.

Flowering & Fruiting : Flowering during May – July and fruiting later.

Folk Use : Considered earlier as stimulant and tonic (roots, seeds) and cardio tonic (roots).

Active Constituents : Fatty acid comprises of capric (2.3), lauric (1.6), myristic (3.8), palmitic (16.0), palmitoeic (2.5), stearic (1.3), oleic (23.0) and linoleic acids (30.2) (*root oil*), β-carotene, L(+)-lactic acid, iodine, pyruvate kinase, steryl glucoside, acyl transferase, daucic acid, alkane, n-hexacosane, cholesterol, β-sitosterol, glucose, lipid glycerides and amino acids (*carrot*).

References : Ambasta (1986), Anonymous (1952), Dhiman (2005), Kapur & Singh (1996), Kaul (1997), Prajapati *et al.* (2006), Sharma (2003), Singh (2001), Verma & Chauhan (2006).

320. *Dendrobium alpestre* Royle

Family: Orchidaceae.

English Name: **Eng.** : Wild Orchid.

Distribution : Garhwal Himalaya up to 2,100m.

Parts Used : Whole Plant.

Folk Use : Plant used as a tonic.

References : Bhattacharjee (2001), Sharma (2003).

321. *Dendrobium denudans* D. Don

Syn.: Callista denudans (D. Don) Kuntze.

Family: Orchidaceae.

English Name: **Eng.** : The Bare Dendrobium.

Distribution : Sub-tropical and temperate Himalayas (1,350-2,350m), from Kumaon eastwards to Sikkim.

Parts Used : Whole Plant.

Flowering & Fruiting : September – November.

Folk Use : Infusion of the plant has narcotic or hallucinogenic effect.

Reference : Lalramnghinglova (2003).

322. *Dendrobium normale* Fale.

Family: Orchidaceae.

English Name: **Eng. :** Blue Orchid.

Distribution : Uttaranchal Himalayas up to 2,200m.

Parts Used : Whole Plant.

Folk Use : Plant used as a tonic.

References : Bhattacharjee (2001), Sharma (2003).

323. *Dendrocalamus strictus* (Roxb.) Nees

Syn.: Bambus stricta Roxb.

Family: Poaceae.

English, Hindi and Sanskrit Names: **Eng. :** Male Bamboo, Solid Bamboo; **Hindi :** Bans, Kantwasi, Kopar, Lathi Bans, Narbans; **Sans. :** Vansha, Venu, Yavaphala.

Distribution : Found throughout the greater parts of India.

Parts Used : Culms. Siliceous Matter.

Folk Use : Siliceous matter and culms used as tonic.

References : Dwivedi (2007), Parrotta (2001), Sharma (2003), Watt (1972).

324. *Dendrophthoe falcata* (L. f.) Ettingshausen

Syn.: Loranthus amplexifolius Desr.; *L. bicolor* Roxb.; *L. falcatus* L. f.; *L. longiflorus* Desr.

Family: Loranthaceae.

English, Hindi and Sanskrit Names: **Eng. :** Mistletoe; **Hindi :** Banda; **Sans. :** Vanda, Vrikshabhaksha, Vriksharuha.

Distribution : Parasite on trees throughout India.

Parts Used : Whole Plant. Bark.

Flowering & Fruiting : December – January (flowering); February – April (fruiting).

Folk Use : Plant and bark considered narcotic. Also, plant growing on *Calotropis gigantean* valued for increasing brain power.

Active Constituents : Known to possess casein hydrolysate, quercetin, kaempferol, quercetagetin, quercitin, myricetin, hyperoside, myricitrin, meratin, rutin (*plant*), (+)catechin and leucocyanidin (*bark*).

References : Ambasta (1986), Anonymous (1952), Farooq (2005), Kaushik & Dhiman (2000), Prajapati *et al.* (2006), Raju (2000), Siwakoti & Varma (1996), Subramaniam (2000), Trivedi (2006).

325. *Derris trifoliata* Lour.

Syn.: D. uliginosa Benth.; *Galedupa uliginosa* (Willd.) Roxb.; *Robnia uliginosa* Willd.

Family: Fabaceae.

English, Hindi and Sanskrit Names: **Eng. :** Common Derris; **Hindi :** Panlata; **Sans. :** Angaar Valli.

Distribution : Known to occur wild along the sea-coasts in Maharashtra, Tamil Nadu, Orissa, Sunderbans and Andaman.

Parts Used : Whole Plant.

Flowering & Fruiting : Rainy season (flowering) & winters (fruiting).

Folk Use : Plant used as a stimulant.

Active Constituents : Active properties due to rhamnetin-3-O-neohesperidoside and quercetin-3-O-β-neohesperidoside (*leaf*), dehydrorotenone, lupeol and a straight chain ketone (*root*).

References : Ambasta (1986), Anonymous (1952), Dhiman (2005).

326. *Descurainea sophia* (L.) Webb.

Syn.: Sisymbrium sophia L.; *Sophia chirurgorum* Lobel.

Family: Brassicaceae.

English and Hindi Names: **Eng. :** Flixweed, Fluxweed; **Hindi :** Khubkallana.

Distribution : Kashmir – Kumaon up to 4,650m, E. Himalayas, Salt Range, Peshawar, Baluchistan, Afghanistan, Persia, Central arctic, Subarctic, Mesopotamia and Mediterranean; Europe, China, Japan, Central Asia; wild in S. Africa, N. & S. America and New Zealand.

Parts Used : Seeds.

Flowering & Fruiting : July – August.

Folk Use : Seeds used as tonic and cardiotonic.

Active Constituents : Known to possess strophanthidine, evomonoside, helvitocoside, evobioside and erysimoside (*seeds*).

References : Anonymous (1952), Asolkar *et al.* (1992), Kapur *et al.* (1996), Kirtikar & Basu (1984).

327. *Desmodium gangeticum* (L.) DC.

Syn.: D. collinum Roxb.; *D. muculatum* (L.) DC.; *Hedysarum gangeticum* L.; *H. muculatum* L.

Family: Fabaceae.

English, Hindi and Sanskrit Names: **Eng. :** Tick Tree; **Hindi :** Salpan, Salparni, Salvan, Salwan; **Sans. :** Amsumati, Prsniparni, Prthakparni, Shaliparni.

Distribution : Wild in the forests of India up to 1,500m.

Parts Used : Whole Plant. Root.

Flowering & Fruiting : Flowering during June – August; fruiting in cold season.

Folk Use : Plant considered as a tonic and bitter tonic. Roots used as tonic and nervine tonic.

Active Constituents : Isolated constituents identified as N,N-dimethyltryptamine, 5-methoxy-N,N-dimethyltryptamine, β-phenylthylamine[1,2], N(b)-methyltetrahydroharman, β-carboline, 6-methoxy-2'-methyl β-carbolinium cations and tryptamine (*plant*), desmodin, a pterocarpan and pterocarpanoids (*root*).

References : Anonymous (1952), Deshmukh *et al.* (2000), Kaushik & Dhiman (2000), Kirtikar & Basu (1984), Kumar & Lal (2003), Parrotta (2001), Prajapati *et al.* (2006), Ranjan (2003), Sharma (2003), Singh (2001), Thakur *et al.* (1989).

328. *Desmodium heterophyllum* DC.

Syn.: *Hedysarum heterophyllum* Willd.; *Meibomia heterophyllum* (Willd.) Kuntze.

Family: Fabaceae.

English Name: **Eng. :** Spanish Clover.

Distribution : Eastern Himalayas, Prome, Ava on the banks of Irrawaddi, Western Peninsula and Java.

Parts Used : Root.

Flowering & Fruiting : May – October.

Folk Use : Roots used as a tonic.

Reference : Ambasta (1986).

329. *Desmodium tiliaefolium* G. Don.

Syn.: *D. argenteum* Wall.; *D. nutans* Wall.; *Hedysarum tiliaifolium* Don.

Family: Fabaceae.

Hindi Names: **Hindi :** Chamkat, Gurshagal, Laber, Motha, Saubar, Shamru.

Distribution : Along the Himalayas from the upper Punjab to Tavoy, temperate and tropical regions up to 3,000m.

Parts Used : Root.

Folk Use : Roots used as tonic and brain tonic.

References : Kapur & Nanda (1996), Kapur & Srivastava (1996), Kirtikar & Basu (1984), Watt (1972).

330. *Desmodium triflorum* (L.) DC.

Syn.: Hedysarum stipulaceum Burm. f.; *H. triflorum* (L.) DC.; *Meibomia triflora* (L.) Kuntze.

Family: Fabaceae.

English, Hindi and Sanskrit Names: **Eng. :** Creeping Tickfoil; **Hindi :** Kundaliya; **Sans. :** Hamsapadi, Tripadi.

Distribution : Weed of open habitats in all plains districts, up to 900m in hills of peninsular India and to 2,000m in the Himalayas.

Parts Used : Root.

Flowering & Fruiting : August – November.

Folk Use : Roots used as a tonic.

Active Constituents : Stachydrine, trigonelline, choline, betaine, four phenethylamines and five indole-3-alkylamines (*root*).

Reference : Parrotta (2001).

331. *Desmotrichum fimbriatum* Blume

Syn.: Dendrobium macraei Lindl.

Family: Orchidaceae.

Hindi and Sanskrit Names: **Hindi :** Jiban, Jivanti; **Sans. :** Jivanti, Jivya, Kshudrajiva, Yashasya.

Distribution : W. Ghats of Bombay and Chennai Presidencies, Sri Lanka, Sikkim, Khasia hills, Myanmar, Malay Peninsula; Malay Archipelago to the Philippines.

Parts Used : Whole Plant.

Folk Use : Plant used as stimulant and tonic.

References : Ambasta (1986), Anonymous (1952), Kirtikar & Basu (1984).

332. *Dicentra canadensis* Walp.

Family: Papaveraceae.

English Name: **Eng. :** Squirrel Corn.

Distribution : Cultivated in Indian Gardens.

Parts Used : Tuber.

Folk Use : Tubers valued as a tonic.

Active Constituents : Protopone and corydaline (*tubers*).

References : Ambasta (1986), Anonymous (1952), Asolkar *et al.* (1992).

333. *Dicliptera roxburghiana* Nees

Syn.: D. burmanni Nees; *D. chinensis* (L.) Juss.; *Peristrophe japonica* (Thunb.) Bremk.

Family: Acanthaceae.

English Name: **Eng. :** Chinese Foldwing.

Distribution : Plains of Punjab and Bengal, Assam, Bhutan.

Parts Used : Whole Plant.

Folk Use : Plant used as a tonic.

Active Constituents : Principle constituents are betulin, β-amyrin, β-sitosterol, cetyl alcohol and palmitic, stearic, oleic, oxalic and tartaric acids, glucose, fructose, arabinose and rhamnose (*leaf*), 3,5-pyridinedicarboxamide, lupeol and allantoin, stigmasterol and β-sitosterol along with their respective glucosides (*aerial parts*).

References : Anonymous (1952), Kapur & Nanda (1996), Kirtikar & Basu (1984).

334. *Digitalis lanata* Ehrh.

Family: Scrophulariaceae.

English Names: **Eng. :** Grecian Foxglove, Woolly Foxglove.

Distribution : Plant grown in Kashmir, Chamba, Kinnaur and Kullu.

Parts Used : Leaf.

Flowering & Fruiting : May – August.

Folk Use : Leaves used as tonic, heart tonic and stimulant.

Active Constituents : Isolated principles are lanatoside A, lanatoside B, lanatoside C, tigonin and lanadigin, scutellarein, luteolin, dinatin, digitoxigenin, glucoside-6-deoxy-glucoside, digoxin, gitoxin, digitoxigenin, gitaloxigenin, acetyldigitoxin, lanadoxin, acetylgitaloxin, acetylgitoxin, tigogenin, digitoxigenin, 3-epidigitoxigenin, digifolein, digoxoside, neodigoxoside, acetyl-γ-digoxin, neo-odorobioside G, neo-digitalinum verum, digitoxigenin-O-glucosyl-6-deoxyglucoside, digitoxigenin-β-D-glucoside, 1-methoxy-2-methylanthraquinone, 3-methoxy-2-methylanthraquinone, digitolutein and 3-methylalizarin (*leaf*).

References : Aman & Dar (2003), Kapur *et al.* (1996), Prakash (2001), Sharma (2003), Thakur *et al.* (1989).

335. *Digitalis purpurea* L.

Family: Scrophulariaceae.

English and Hindi Names: **Eng. :** Common Foxglove, Purple Foxglove; **Hindi :** Tilpushipi.

Distribution : Plant cultivated as an ornamental plant between 1,500-1,800m in Kashmir and Himachal Pradesh.

Parts Used : Leaf.

Flowering & Fruiting : April – August.

Folk Use : Leaves used as heart tonic. Infusion of leaves considered sedative and stimulant.

Active Constituents : Active principles are digitoxin, gitoxin and gitalin, gitaloxin, lanatocide C, digipronin, purpnin, purpronin, saponins-degalactotigonin and F-gitonin, 1-methoxy- and 3-methoxy-2-methylanthraquinones, digitolutein, 3-methylalizarin, digitolutein, digitopurpone, phomorin, isochrysophanol, apigenin, dinatin, chrysoeriol, hepetin, desrhamnosyl acteoside, forsthiaside, purpureasides A and B and 3,4-dihydroxyphenethyl alcohol-6-O-caffeoyl-β-D-glucoside (*leaf*).

References : Aman & Dar (2003), Ambasta (1986), Bhattacharjee (2001), Kapur *et al.* (1996), Prajapati *et al.* (2006), Prakash (2001), Raju (2000), Reddi *et al.* (2005), Sharma (2003), Thakur *et al.* (1989).

336. *Dillenia indica* L.

Syn.: D. elliptica Thunb.; *D. speciosa* Thunb.

Family: Dilleniaceae.

English, Hindi and Sanskrit Names: **Eng. :** Elephant Apple; **Hindi :** Chalta, Girnar; **Sans. :** Bharija, Bhavya.

Distribution : Occurs in the moist and evergreen forests of the sub-Himalayan tract, from Kumaon and Garhwal eastwards to Assam and Bengal and southwards to central and southern India.

Parts Used : Fruit.

Flowering & Fruiting : May – July.

Folk Use : Fruits used as tonic.

Active Constituents : Arabinose, galactose and mannose (*fruit*).

References : Ambasta (1986), Anonymous (1952), Hajra & Baishya (1981).

337. *Dimocarpus longan* Lour.

Syn.: Euphoria longan (Lour.) Steud.; *Nephelium longana* (Lam.) Camb.

Family: Sapindaceae.

English Name: **Eng. :** Longan.

Distribution : Occasionally cultivated, roadsides, gardens; also found wild in tropical forests of India.

Parts Used : Fruit.

Folk Use : Fruits used as tonic.

References : Farooq (2005), Khatoon & Goel (2010).

338. *Dioclea javanica* Benth.

Syn.: D. reflexa Hook. f.

Family: Fabaceae.

English Name: **Eng. :** Sea Purse.

Distribution : Forests of Silhet; Penang and Singapore, Tenasserim, Sri Lanka; Malay isles, tropical Africa and tropical America; W. Bengal and Assam.

Parts Used : Seeds.

Folk Use : Seeds used as stimulant and tonic.

References : Ambasta (1986), Anonymous (1952), Asolkar *et al.* (1992).

339. *Dioscorea alata* L.

Syn.: D. atropurpurea Roxb.; *D. globose* Roxb.; *D. purpurea* Roxb.; *D. rubella* Roxb.

Family: Dioscoreaceae.

English, Hindi and Sanskrit Names: **Eng. :** Asiatic Yam, Greater Yam, Water Yam, White Yam; **Hindi :** Chupri Alu, Khamalu; **Sans. :** Alukam, Pindalu.

Distribution : Throughout India, cultivated.

Parts Used : Tuber.

Flowering & Fruiting : October – December.

Folk Use : Tubers used as a tonic.

Active Constituents : Sucrose, D-fructose, D-glucose, maltose, β-amylase, peroxidase, pyruvate kinase, phosphatase and O-dephonalase, cyaniding-3-glucoside, (+)catechin and procyanidins B1 and B3 have been isolated (*tuber*).

Reference : Prajapati *et al.* (2006).

340. *Dioscorea bulbifera* L. (Pl. 5D)*

Syn.: D. pulchella Roxb.; *D. sativa* auct. non L.; *D. tamnifolia* Salisb.; *Helmia bulbifera* Kunth.

Family: Dioscoreaceae.

English, Hindi and Sanskrit Names: **Eng. :** Bulb Bearing Yam; **Hindi :** Bhirvolikanda, Genthi, Karawakand, Karukanda, Zaminkand; **Sans. :** Amrita, Brahmaputri, Kanya, Krodi, Kumari, Magadhi, Trinetra, Vanavasi, Varahi, Vrishvaksenapriya.

Distribution : Throughout India up to 1,850m in Himalayas.

Parts Used : Fruit. Root.

Flowering & Fruiting : June – August.

Folk Use : Fruits and root tubers used as tonic.

Active Constituents : Isolated constituents are three new furanoids nortiterpenes – diosbulbin A, diosbulbin B and diosbulbin C (*root tubers*).

References : Kirtikar & Basu (1984), Mishra (2008), Parrotta (2001), Sharma (2003), Singh (2001).

341. *Dioscorea pentaphylla* L.

Syn.: D. jacquemontii Hook. f.; *D. triphylla* L.

Family: Dioscoreaceae.

English, Hindi and Sanskrit Names: **Eng. :** Fiji Yam, Kawan Yam; **Hindi :** Bhusa, Gazaria, Kanta Alu; **Sans. :** Kantakalu.

Distribution : Native of tropical Asia extending eastwards to the remotest islands of the Pacific; occurs throughout India ascending to 1,850m in the Himalayas and in Andaman Islands.

Parts Used : Tubers.

Flowering & Fruiting : August – September.

Folk Use : Tubers considered as tonic.

Active Constituent : Dioscorine (*tuber*).

References : Ambasta (1986), Anonymous (1952), Kirtikar & Basu (1984).

342. *Diospyros lotus* L.

Family: Ebenaceae.

English and Hindi Names: **Eng. :** Dateplum, Persimmon; **Hindi :** Amlok, Malok.

Distribution : N.W. Himalayas particularly in the forests of Jammu & Kashmir.

Parts Used : Seeds.

Flowering & Fruiting : April – August.

Folk Use : Seeds used as sedative.

Reference : Ambasta (1986).

343. *Diplocyclos palmatus* (L.) Jeffrey

Syn.: Bryonia laciniosa L.; *B. palmatus* L.; *Bryonopsis laciniosa* Sensu. Naudin

Family: Cucurbitaceae.

Hindi and Sanskrit Names: **Hindi :** Isvaralingi, Sivalingi; **Sans. :** Lingini.

Distribution : Throughout India, on hedges and bushes up to 1,200m.

Parts Used : Whole Plant.

Flowering & Fruiting : September – October.

Folk Use : Plant used as a tonic and bitter tonic.

References : Asolkar *et al.* (1992), Farooq (2005), Kothari & Londhe (2000), Parrotta (2001), Prajapati *et al.* (2006), Subramaniam (2000).

344. *Dipsacus japonicus* Miq.

Family: Dipsacaceae.

English Names: **Eng. :** Chinese Teasel, Japanese Teasel.

Distribution : Common weed of Savannas at 1,000m and above.

Parts Used : Root.

Flowering & Fruiting : Flowering during August – September.

Folk Use : Root possesses tonic properties.

Active Constituents : Alkaloids, tannins and sugars (*root*).

Reference : Prajapati *et al.* (2006).

345. *Dolichos biflorus* L.

Syn.: D. uniflorus Lank.; *Macrotyloma uniflorum* (Lam.) Verdc.

Family: Fabaceae.

English, Hindi and Sanskrit Names: **Eng. :** Horsegram, Kool Tee, Madras Gram; **Hindi :** Gahat, Kulthi; **Sans. :** Kulaththa, Svetabija, Tamabija.

Distribution : Native to most parts of India, and up to 1,000m.

Parts Used : Seeds.

Flowering & Fruiting : June – October.

Folk Use : Seeds used as tonic.

Active Constituents : 5-hydroxy-7,3',4'-trimethoxy-8-methylisoflavone-5-neohesperidoside (*seeds*).

References : Ambasta (1986), Anonymous (1952), Bhattacharjee (2001), Prajapati *et al.* (2006), Sharma (2003).

346. *Dorema ammoniacum* D. Don

Family: Apiaceae.

English Name: **Eng. :** Ammoniacum.

Distribution : Native to central Asia, Iran and northern Russia.

Parts Used : Whole Plant.

Folk Use : Plant valued for its stimulant properties.

Active Constituents : Known to yield ammodoremin, chromadiones separated as their acetates and doremone (*resin*).

References : Ambasta (1986), Anonymous (1952), Prajapati *et al.* (2006).

347. *Doronicum falconeri* C.B. Clarke

Family: Asteraceae.

English Name: **Eng. :** Leopard's Bane.

Distribution : Kashmir, 4,350m; Western Tibet; Karakoram, 4,650m.

Parts Used : Root.

Flowering & Fruiting : July – September.

Folk Use : Root used as tonic and good against nervous depression.

References : Anonymous (1952), Jain & Goel (2005), Kala (2003).

348. *Doronicum hookeri* Hook. f.

Syn.: D. scorpioides Clarke.

Family: Asteraceae.

English, Hindi and Sanskrit Names: **Eng. :** Leopard's Bane; **Hindi :** Tarang; **Sans. :** Vrishichka.

Distribution : Sikkim Himalayas: 4,000-4,650m.

Parts Used : Root.

Folk Use : Roots used as aromatic tonic.

References : Ambasta (1986), Anonymous (1952), Kirtikar & Basu (1984).

349. *Doronicum pardalianches* L.

Family: Asteraceae.

English Name: **Eng. :** Great Leopard's Bane.

Distribution : Introduced into India.

Parts Used : Root.

Folk Use : Roots imported into India for use in cardiac and nervine tonic preparations.

Active Constituents : Pardalianchol and its acetophenone derivative, otosenine (*root*).

References : Ambasta (1986), Anonymous (1952).

350. *Doronicum roylei* DC.

Syn.: Fullartonia kumaonensis DC.

Family: Asteraceae.

Distribution : W. Himalayas, from Kashmir – Garhwal up to 300m.

Parts Used : Root.

Folk Use : Roots used as aromatic tonic and cardio tonic.

References : Ambasta (1986), Kirtikar & Basu (1984), Watt (1972).

351. *Drosera peltata* J.E. Sm. ex Willd.

Syn.: D. lunata Buch. Ham.

Family: Droseraceae.

English and Hindi Names: **Eng. :** Sundew; **Hindi :** Chitra, Mukhajali.

Distribution : Along open grassy slopes and moss-clad dripping rocks of higher altitudes throughout India.

Parts Used : Leaf.

Flowering & Fruiting : July – November.

Folk Use : Leaves used as tonic.

Active Constituents : Known to possess nephthaquinones, plumbagin, droserone, hydroxyserone, the flavonoids, quercetin, gossypetin, gosspin and isogossypitin (*leaf*).

References : Anonymous (1952), Prajapati *et al.* (2006), Sharma (2003), Subramaniam (2000).

352. *Drymaria cordata* (L.) Roem. *et* Schult.

Family: Caryophyllaceae.

English Names: **Eng.** **:** Drymaria, Heartleaf Drymary, Tropical Chickweed, West Indian Chickweed.

Distribution : Tropical and subtropical India and Sri Lanka, extending westwards to the Punjab and ascending the Himalaya to 2,350m in Sikkim; Tropical Asia, Africa and America.

Parts Used : Leaf.

Flowering & Fruiting : October – November.

Folk Use : Leaf extract used as heart tonic.

Reference : Chhetri (2005).

353. *Drynaria quercifolia* (L.) J. Smith.

Syn.: Polypodium quercifolium L.

Family: Polypodiaceae.

English, Hindi and Sanskrit Names: **Eng. :** Wood Fern; **Hindi :** Asvakatri, Kankali, Katikajpan; **Sans. :** Ashvakatri.

Distribution : Throughout India, in the plains or very low down in the mountains, on trees, on rocks.

Parts Used : Whole Plant. Rhizome. Root.

Flowering & Fruiting : May – July.

Folk Use : Plant, rhizome and roots used as tonic.

References : Asolkar *et al.* (1992), Islam (1983), Kirtikar & Basu (1984), Singh & Viswanathan (1996).

354. *Drypetes roxburghii* (Wall.) Hurusawa

Syn.: Putranjiva roxburghii Wall.

Family: Euphorbiaceae.

Hindi and Sanskrit Names: **Hindi :** Putranjiva, Putranjvak; **Sans. :** Putranjeeva.

Distribution : Wild or cultivated throughout tropical parts of India.

Parts Used : Fruit. Seeds.

Flowering & Fruiting : Flowering during April – May; fruiting in cold season.

Folk Use : Fruits and seeds used as stimulant and tonic.

References : Dhiman (2005), Kaushik & Dhiman (2000).

355. *Duboisea myoporoides* R. Br.

Family: Solanaceae.

English Names: **Eng. :** Australian Mgmea, Cork Wood Tree.

Distribution : Native to Australia; cultivated in Karnataka as commercial crop.

Parts Used : Leaf.

Folk Use : Leaves valued as sedative and hypnotic.

Active Constituents : Tropane, alkaloids like hyoscine, hyoscyamine, norhyoscyamine, valeroidine, isopelletierine, tigloidine, poroidine, isoporoidine and anabasine have been isolated (*leaf*).

References : Bhattacharjee (2001), Prajapati *et al.* (2006), Sharma (2003).

356. *Durio zibethinus* L.

Family: Bombacaceae.

English Names: **Eng. :** Civet Fruit, Durian.

Distribution : Cultivated in Malaya and East Indies; found in lower elevation of Nilgiris and some parts of the west coast in India.

Parts Used : Fruit.

Flowering & Fruiting : March – April (flowering); July – September (fruiting).

Folk Use : Fruits used as tonic.

Reference : Ambasta (1986).

357. *Echinops echinatus* Roxb.

Family: Asteraceae.

English, Hindi and Sanskrit Names: **Eng. :** Indian Globe Thistle; **Hindi :** Gokhru, Utakatira; **Sans. :** Kantalu, Utakantak, Utati.

Distribution : More or less throughout India, ascending up to 1,524m in the hills.

Parts Used : Whole Plant. Root. Seeds.

Flowering & Fruiting : July – December.

Folk Use : Plant used as nervine tonic. Seeds and decoction of root used as tonic.

Active Constituents : These are hentriacontane, hentriacontanol, β-amyrin and lupeol, apigenin-7-O-β-D-(4"-*cis*-p-coumaroyl) glucoside, apigenin-7-O-glucoside, echinacin, echinaticin, echinozolinone, echinopsine, echinopsidine, taraxasterol, its acetate, lupeol, β-amyrin, β-sitosterol, its glucoside, apigenin and its 7-O-glucoside (*plant*).

References : Anonymous (1952), Bhalla *et al.* (1996), Bhatt *et al.* (2002), Dhiman (2005), Kaushik & Dhiman (2000), Kirtikar & Basu (1984), Kothari & Londhe (2000), Parrotta (2001).

358. *Eclipta prostrata* L.

Syn.: E. alba (L.) Hassk.; *Verbesina prostrata* L.

Family: Asteraceae.

English, Hindi and Sanskrit Names: **Eng. :** Trailing Eclipta; **Hindi :** Babri, Bhangra, Mocakand, Safed Bhannra; **Sans. :** Ajagara, Bhringa, Bhringraj, Kayyoni, Kesaraja, Markava, Tekarajah.

Distribution : Found distributed throughout India up to 1,980m in the Himalayas.

Parts Used : Whole Plant. Leaf. Root.

Flowering & Fruiting : Flowering and fruiting during greater part of the year.

Folk Use : Juice of the plant in combination with 'ajwain' seeds used as a tonic. Leaves and roots also used as tonic.

Active Constituents : Properties associated to coumestin derivatives, wedelolactone, tiopene derivatives like ecliptal, saponins viz. eclalbosaponina demethylwedelolactone, common sterols and triterpenoids viz. pentriacontanol, 14-heptaconosal, flavonoids like luteolin 7-O-glucoside, alkaloid and polypeptide (*plant*), β-sitosterol, 2-an-geloyloxy methylene-5'-dithiophene, 5'-isovaleryloxy methylene-2-dithiophene, stigmasterol (*aerial parts*), 2,5-thiopene, 5'-tigloy-loxymethylene-2-dithiopene, 5'-seneocioyl oxymethylene-2-dithiopene (*root*).

References : Bhalla *et al.* (1996), Das *et al.* (1996), Dhiman (2005, 2007), Dwivedi (2007), Kaushik & Dhiman (2000), Prajapati *et al.* (2006), Ranjan (2003), Reddi *et al.* (2005), Sharma (2003), Subramaniam (2000), Watt (1972).

359. *Elaeagnus infundibularis* Momiyama

Syn.: E. arborea auct. non Roxb.; *E. latifolia* auct. non L.

Family: Elaeagnaceae.

English Names: **Eng. :** Bastard Oleaster, Wild Olive.

Distribution : Throughout the hilly parts of India; on the Himalayas it occurs westwards of Jaunsar, up to 3,000m; also in Myanmar, Penang, South India and Sri Lanka.

Parts Used : Flower. Fruit.

Flowering & Fruiting : Flowering in December; fruiting during March – April.

Folk Use : Flowers and fruits used as cardiac tonic.

Reference : Siwakoti & Varma (1996).

360. *Elaeagnus umbellata* Thunb.

Family: Elaeagnaceae.

English Names: **Eng. :** Autumn Elaeagnus, Autumn Olive, Japanese Silverberry, Spreading Oleaster, Umbellate Oleaster.

Distribution : Temperate Himalayas (Kashmir–Nepal): 1,000-3,350m; Afghanistan, N. China and Japan.

Parts Used : Flower. Seeds.

Folk Use : Flowers and seeds used as stimulant.

References : Ambasta (1986), Kapur & Singh (1996), Kirtikar & Basu (1984).

361. *Elaeocarpus sphaericus* (Gaertn.) K. Schum.

Syn.: E. ganitrus Roxb.; *Ganitrus sphaericus* Gaertn.

Family: Elaeocarpaceae.

English, Hindi and Sanskrit Names: **Eng. :** Utrasum-bead Tree; **Hindi :** Rudraks, Rudraki; **Sans. :** Bhutanashana, Pavana, Rudraksah, Shivaksha, Shivapriya, Trinameru.

Distribution : Bihar, Bengal, Assam, Maharashtra and Madhya Pradesh.

Parts Used : Fruit. Kernel (Seed).

Flowering & Fruiting : Flowering: May – June; Fruiting: November – December.

Folk Use : Fruits considered sedative. Seed kernel accredited with sedative and liver tonic properties.

References : Asolkar *et al.* (1992), Kaushik & Dhiman (2000), Prajapati *et al.* (2006).

362. *Elephantopus scaber* L.

Family: Asteraceae.

English, Hindi and Sanskrit Names: **Eng. :** Prickly Leaved Elephant's Foot; **Hindi :** Gobhi, Gojivha; **Sans. :** Darvika, Gojihva, Hastipadi, Satamulika.

Distribution : Dry localities throughout India.

Parts Used : Whole Plant. Leaf.

Flowering & Fruiting : September – March.

Folk Use : Plant used as a tonic and cardiac tonic. Juice of leaves considered memory enhancer.

Active Constituents : Epiriedelinol, lupeol, stigmasterol, a mixture of triacontan-1-ol and dotriacontan-1-ol, isodeoxyelephantopin, deacylcynaropicrin, glucozaluzanin C, crepiside E and stigmasteryl-3β-glucoside (*plant*).

References : Anonymous (1952), Bhalla *et al.* (1996), Chaudhury & Neogi (2000), Kirtikar & Basu (1984), Maya *et al.* (2003), Prajapati *et al.* (2006), Ranjan (2003).

363. *Elettaria cardamomum* Maton

Syn.: Alpinia cardamomum Roxb.

Family: Zingiberaceae.

English, Hindi and Sanskrit Names: **Eng. :** Cardamom, Lesser Cardamom; **Hindi :** Choti-elachi; **Sans. :** Ela, Ela-suksmalika-ksudraila, Gaurangi, Nishkuti, Putika, Upakunchika.

Distribution : Cultivated in South India.

Parts Used : Fruit. Seeds.

Folk Use : Fruits and seeds used as stimulant and tonic to heart.

References : Ambasta (1986), Anonymous (1952), Bhattacharjee (2001), Dwivedi (2007), Farooq (2005), Kirtikar & Basu (1984), Pal & Paul (2003).

364. *Eleusine coracana* (L.) Gaertn.

Syn.: Cynosurus coracanus L.

Family: Poaceae.

English, Hindi and Sanskrit Names: **Eng. :** African Millet, Finger Millet, Ragi; **Hindi :** Kado, Mandva, Maruva; **Sans. :** Bahupatrica, Guchha, Narttakah, Nrtyakundalah.

Distribution : Cultivated throughout South India in areas up to 1,500m.

Parts Used : Grains.

Flowering & Fruiting : June – September.

Folk Use : Grains used as a tonic.

References : Kirtikar & Basu (1984), Prajapati *et al.* (2006), Srivastava & Singh (2003).

365. *Embelia ribes* Burm. f.

Syn.: E. glandulifera Wight.; *Samara ribes* Kurz.

Family: Myrsinaceae.

English, Hindi and Sanskrit Names: **Eng. :** Embelia, Embelia Fruit; **Hindi :** Baberang, Bidang, Vidang, Wawrung; **Sans. :** Amogha,

Bidanga, Chitra, Jantughna, Krimiripu, Vellah, Vidanga, Vrishanasana.

Distribution : Throughout India up to 1,750m.

Parts Used : Fruit.

Flowering & Fruiting : March – April.

Folk Use : Fruits used as stimulant, tonic and brain tonic.

Active Constituents : Major activities due to embelin, homoembelin, rapanone, chistembine, homoraponone, vilangine and quercitol (*fruit*).

References : Ambasta (1986), Anonymous (1952), Kirtikar & Basu (1984), Prajapati *et al.* (2006), Sharma (2003), Thakur *et al.* (1989).

366. *Emblica officinalis* Gaertn.*

Syn.: Cicca emblica Kurz.; *Dichelastina nodicaulis* Hance.; *Phyllanthus emblica* L.; *P. glomeratus* Roxb.; *P. taxifolius* Don.

Family: Euphorbiaceae.

English, Hindi and Sanskrit Names: **Eng. :** Emblic Myrobalan, Indian Goseberry; **Hindi :** Amala, Anola, Anvula, Aungra, Aura; **Sans. :** Adiphala, 'Akara, Amlika, Amri, Dhatri, Jatiphala, Kamu, Rochni, Shanta, Shiva, Tishya, Triphala, Umrita.

Distribution : Wild or cultivated throughout tropical parts of India.

Parts Used : Fruit.

Flowering & Fruiting : Flowering: February – May; Fruiting: October – March.

Folk Use : Fruits used as tonic, liver tonic, brain tonic and stimulant.

Active Constituents : Trigalloylglucose, terchebin, corilagin, ellagic acid, indole acetic acid and four other auxins-a1, a3, a4 & a5, two growth inhibitors R1 and R2, tannins and vitamin C have been characterized from *fruit*.

References : Asolkar *et al.* (1992), Bhalla *et al.* (1982), Bhatt (2002), Bhatt *et al.* (2003), Bhattacharjee (2001), Dhiman (2005), Kaushik & Dhiman (2000), Khanna (2002), Kshirsagar & Singh (2007), Mishra & Sahu (1984), Panthi & Chaudhary (2003), Parrotta (2001), Rajwar (1983), Sharma (2003, 2004), Singh (2001).

367. *Enicostemma littorale* Blume

Syn.: Adenema nyssopifolia G. Don.; *Cicendia nyssopifolia* Wight & Arn.; *E. axillare* (Lam.) Raynal; *Gentiana axillare* Lam.; *G. verticillata* L.

Family: Gentianaceae.

English, Hindi and Sanskrit Names: **Eng.** : Indian Gentian; **Hindi** : Chota Chirayata, Katvinayi, Nai; **Sans.** : Mamajjakah, Nagajihva, Nahi.

Distribution : Throughout Indian areas up to 450m elevation, common in coastal areas.

Parts Used : Whole Plant.

Flowering & Fruiting : July – September.

Folk Use : Plant used as a bitter tonic and liver tonic. Infusion of plant taken as growth tonic.

Active Constituents : Plant rich in lactonic enamino ketone, gentiocrucine, swertisioside, apigenin, yenkwanin, isovitexin, swertisin, saponarin and 5-O-glucoside derivatives of sylswertism and isoswertisin.

References : Ambasta (1986), Anonymous (1952), Parrotta (2001), Prajapati *et al.* (2006), Singh & Pandey (1998), Watt (1972).

368. *Entada phaseoloides* Merrill

Syn.: *E. pursaentha* DC.; *E. scandens* (L.) Benth.; *E. schefferi* Ridl.

Family: Fabaceae.

English, Hindi and Sanskrit Names: **Eng.** : Climbing Entada, Elephant Creeper, Garbee Bean, Mackay Bean, Match Box Bean, Nicker Bean; **Hindi** : Chian, Garbi, Gila; **Sans.** : Gilla.

Distribution : Central and E. Himalayas, Nepal, W. Peninsula, Sri Lanka. Tropics generally.

Parts Used : Seeds.

Flowering & Fruiting : April – June.

Folk Use : Seeds considered as tonic.

Active Constituents : Associated property due to dopamine, dopamine glucoside and tryosin glucoside, O-β-D-quercitin, gallic acid, cyanidin chloride, prosapogenin A, β-sitosterol, lupeol, entadamide A and phaseoloidin (*seeds*).

References : Ambasta (1986), Anonymous (1952), Parrotta (2001).

369. *Entada rheedii* Spreng.

Family: Fabaceae.

English and Hindi Names: **Eng.** : African Dream Herb, Sea Heart, Snuff Box Sea Bean; **Hindi** : Barabi, Chian, Gila.

Distribution : Central and Eastern Himalayas, Nipai, ascending to 1,350m in Sikkim, Western Peninsula, Sri Lanka, Malacca; Cosmopolitan in the tropics.

Parts Used : Seeds.

Folk Use : Seeds used as tonic.

Reference : Kothari & Rao (2000).

370. *Ephedra gerardiana* Wall. ex Stapf.

Syn.: E. vulgaris Hook. A. Rich.

Family: Ephedraceae.

English, Hindi and Sanskrit Names: **Eng.** : Ephedra; **Hindi** : Asmania, Batsur, Tut Gundha; **Sans.** : Soma.

Distribution : Drier parts of temperate and alpine Himalayas (Kashmir – Sikkim): 2,133-4,876m.

Parts Used : Whole Plant. Root.

Folk Use : Plant and its roots used as tonic. Drug ephedrine from the plant valued as nerve stimulant.

References : Ambasta (1986), Reddi *et al.* (2005), Singh *et al.* (1996).

371. *Eragrostis cynosuroides* Beauv.

Syn.: Briza bipinnata L.; *Desmostachya bipinnata* Stapf.; *Poa cynosuroides* Retz.

Family: Poaceae.

Hindi and Sanskrit Names: **Hindi** : Dab, Davoli, Kusha; **Sans.** : Barhi, Darbha, Hrasva, Suchyagra.

Distribution : Throughout India in hot and dry parts.

Parts Used : Culms.

Flowering & Fruiting : August – November (flowering); cold season (fruiting).

Folk Use : Culms used as stimulant.

Active Constituents : Pectose, lignin, cellulose and ash (*culms*).

References : Kaushik & Dhiman (2000), Kirtikar & Basu (1984), Srivastava *et al.* (2003b), Watt (1972).

372. *Erigeron asteroides* Roxb.

Family: Asteraceae.

Distribution : Bengal and the western Peninsula up to 1,350m on the eastern Himalayas.

Parts Used : Whole Plant.

Folk Use : Plant considered stimulant.

Reference : Kirtikar & Basu (1984).

373. *Erigeron canadensis* L.

Syn.: Conyza canadensis (L.) Cronq.

Family: Asteraceae.

English and Sanskrit Names: **Eng.** : Canada Fleabane; **Sans.** : Jarayupriya, Makshikavisha, Patila.

Distribution : Native to N. America; Punjab, upper Gangetic Plain, Western Himalayas up to 1,000m, Assam and Western Ghats and Nilgiris up to 2,350m in India.

Parts Used : Whole Plant.

Folk Use : Plant used as a stimulant.

References : Ambasta (1986), Anonymous (1952), Kapur *et al.* (1996), Kirtikar & Basu (1984).

374. *Eriobotrya japonica* (Thunb.) Lindl.

Syn.: Crataegus bibas Lour.; *Mespilus japonica* Banks; *Photinia japonica* (Thunb.) Franch. & Sav.

Family: Rosaceae.

English and Hindi Names: **Eng. :** Japanese Medlar, Japanese Plum, Loquat; **Hindi :** Lokat.

Distribution : Native of China and Japan; naturalized in India and grown nearly throughout the country up to 1,700m.

Parts Used : Fruit.

Flowering & Fruiting : July – September.

Folk Use : Fruits considered as sedative.

Active Constituents : Levulose, sucrose, malic acid, citric acid, tartaric acid, succinic acid, cryptoxanthin, β-carotene and neo-β-carotene (*fruit*).

References : Ambasta (1986), Kaushik & Dhiman (2000), Kirtikar & Basu (1984), Parrotta (2001).

375. *Eruca sativa* Mill.

Syn.: Brassica eruca L.; *B. erucoides* Roxb.; *E. sativa* Lamk.

Family: Brassicaceae.

English, Hindi and Sanskrit Names: **Eng. :** Rocket-salad, Roquette, Tara-mira; **Hindi :** Seoha, Taramira; **Sans. :** Bhutaghna, Bimbala, Daradharsha, Rakshitaphala, Siddartha.

Distribution : Western Himalayas up to 3,300m.

Parts Used : Leaf. Seeds.

Flowering & Fruiting : February – April.

Folk Use : Leaves considered stimulant. Half teaspoon of dried seed powder prescribed with warm water for a month as tonic.

Active Constituents : Associated property due to isorhamnetin and its 3-glucoside (*leaf*) and glucoerucin (4-methylthiobutyl glucosinolate) (*seeds*).

References : Ambasta (1986), Anonymous (1952), Kirtikar & Basu (1984), Parrotta (2001), Sood & Thakur (2004).

376. *Eryngium billardieri* Delar.

Family: Apiaceae.

English Name: **Eng. :** Billardiere's Eryngo.

Distribution : Kashmir : 1,700-2,000m, Ladakh, Scinde; Western Asia.

Parts Used : Root.

Folk Use : Roots used as tonic.

Reference : Kapur (1996a).

377. *Eryngium caeruleum* Bieb.

Syn.: E. planum L.

Family: Apiaceae.

Hindi Name: **Hindi :** Dhudholi.

Distribution : Kashmir and western Himalayas: 1,700-2,000m; Afghanistan, Persia, Turkestan.

Parts Used : Root.

Folk Use : Root considered as tonic and nervine tonic.

Active Constituents : Saponin, mixture, aglycone, barringenol R1 and barringenol A1 have been isolated (*root*).

References : Ambasta (1986), Anonymous (1952), Kirtikar & Basu (1984), Watt (1972).

378. *Eryngium foetidum* L.

Family: Apiaceae.

English Names: **Eng. :** Duck-tongue Herb, Fitweed, Mexican Coriander, Saw-leaf Herb, Sawtooth, Spiritweed, Wild Coriander.

Distribution : Occurs in Assam up to 1,700m.

Parts Used : Whole Plant.

Folk Use : Plant used as a tonic for general weakness.

Reference : Singh *et al.* (2003).

379. *Erythrina stricta* Roxb.

Family: Fabaceae.

English, Hindi and Sanskrit Names: **Eng. :** Coral Tree; **Hindi :** Mura; **Sans. :** Mura.

Distribution : Commonly met with in the deciduous forests of Konkan, North Kanara, Malabar and Travancore; also found in Assam and Manipur, extending westwards to Nepal.

Parts Used : Flower.

Flowering & Fruiting : August – September.

Folk Use : Flowers used as a tonic in Assam.

References : Asolkar *et al.* (1992), Jain & Borthakur (1980).

380. *Erythrina variegata* L.

Syn.: E. variegata L. var. *orientalis* L.; *E. indica* Lam.

Family: Fabaceae.

English, Hindi and Sanskrit Names: **Eng. :** East Indies Coral Tree, India Bean, India Coral-bean, Mochi Wood Tree; **Hindi :** Mandara, Pamkara, Pangra, Pharhad; **Sans. :** Mandar, Nimbataru, Parijata.

Distribution : Grows wild in deciduous forests, cultivated in gardens as ornamental plant.

Parts Used : Leaf.

Flowering & Fruiting : February – July.

Folk Use : Leaves used as sedative.

Active Constituents : Erythrinaline, erythaline-HC1, erystorine, erythrinine and de-N-Me-orientaline (*leaf*).

Reference : Prajapati *et al.* (2006).

381. *Erythroxylum coca* Lam.

Family: Erythroxylaceae.

English Names: **Eng. :** Coca, Cocaine Plant.

Distribution : Native to Peru and Bolivia; found as an ornamental plant in Indian gardens.

Parts Used : Leaf.

Folk Use : Leaves used as tonic and cerebral stimulant for producing some pleasant excitement.

Active Constituents : Known to possess cocaine, cinnamylescaine, β-truxilline, tropacocaine, valerine, hygroline and cuscohygrine (*leaf*).

References : Ambasta (1986), Anonymous (1952), Asolkar *et al.* (1992), Bhattacharjee (2001), Harborne & Baxter (2001), Prajapati *et al.* (2006), Watt (1972).

382. *Erythroxylum monogynum* Roxb.

Syn.: E. indicum (DC.) Bedd.; *Seethia indica* DC.

Family: Erythroxylaceae.

English and Sanskrit Names: **Eng. :** Bastard Sandal, Deccany Deodar, Red Cedar; **Sans. :** Devataru, Kattuchandanam.

Distribution : Dry evergreen forests of Peninsular India up to 925m.

Parts Used : Stem Bark.

Flowering & Fruiting : June – November.

Folk Use : Stem bark pounded with water given as a tonic for curing stomachache.

References : Kumar & Pullaiah (1998), Watt (1972).

383. *Eucalyptus globules* Labill (Pl. 5E)*

Syn.: E. gigantea Dehn.; *E. glauca* DC.; *E. perfoliata* Desf.; *E. pulverulenta* Link.

Family: Myrtaceae.

English, Hindi and Sanskrit Names: **Eng. :** Blue Gum, Eucalypt; **Hindi :** Yukeliptas; **Sans. :** Nilaniryasa, Tailaparnah.

Distribution : Cultivated in various parts of India.

Parts Used : Oil.

Flowering & Fruiting : January – August.

Folk Use : Oil used as cardio tonic and stimulant.

Active Constituents : Associated properties due to eucalyptin, 8-desmethyleucalyptin, sideroxylin, quercetin, quercitrin, quercetol, chrysin, rutin hyperoside, phloroglucin derivatives, n-triacontan-16,18-dione, 4-hydroxytritriaconate-16,18-dione, 16-hydroxy-18-tritriacontanone, 11,12-dehydroursolic lactone acetate, caffeic, ferulic, gallic, gentisic, protocatechuic, maslinic and oleanolic acids (*oil*), α-pinene, an unsaturated α-ketone, gallic, caffeic, ferulic and protocatechnic acids, quercitol, quercitrin, rutin, hyperoside and quercitol-3-glucoside (*leaf*).

References : Parrotta (2001), Prajapati *et al.* (2006), Reddi *et al.* (2005), Watt (1972).

384. *Eulophia campestris* Wall.

Syn.: Bletia dabia D. Don; *Eulophia dabia* (D. Don) Hochr.

Family: Orchidaceae.

English, Hindi and Sanskrit Names: **Eng. :** Witton Root; **Hindi :** Salibmisri; **Sans. :** Amrita, Pranada, Sudhamuli.

Distribution : Throughout the greater parts of Indian plains.

Parts Used : Rhizome.

Flowering & Fruiting : March – May (flowering); July – August (fruiting).

Folk Use : Rhizomes esteemed as tonic.

References : Ambasta (1986), Anonymous (1952), Bhattacharjee (2001), Dhiman (2005), Kirtikar & Basu (1984), Watt (1972).

385. *Euodia officinalis* Dode.

Family: Rutaceae.

English Name: **Eng. :** Euodia.

Distribution : Mostly cultivated in Korea and China.

Parts Used : Fruit.

Folk Use : Fruits used as cardio tonic.

Active Constituents : Evodiamine, rutaecarpine, higenamine, dehydroevodiamine (*fruit*).

Reference : Prajapati *et al.* (2006).

386. *Eupatorium cannabinum* L.

Syn.: E. nodiflorum Wall.; *E. punduanum* Wall.

Family: Asteraceae.

English Names: **Eng. :** Hemp Agrimony, Hemp Eupatorium, Water Hemp.

Distribution : Native to Europe; occurs in temperate Himalayas in India: 1,000-3,500m.

Parts Used : Whole Plant.

Flowering & Fruiting : July – August.

Folk Use : Plant used as a tonic.

Active Constituents : Reported to possess volatile oil, sesquiterpene lactones, flavonoids, pyrrolizidine alkaloids and polysaccharides (*plant*), caffeic, chlorogenic and isochlorogenic acid, rutin and hyperin (*herb*), hispidulin, eupafolin, astragalin, kaempferol-3-rutinoside, hyperoside, isoquercitrin and rutin (*aerial parts*), alantolactone, isoalantolactone and dihydroalantolactone (*aerial & underground parts*).

Reference : Prajapati *et al.* (2006).

387. *Eupatorium perfoliatum* L.

Family: Asteraceae.

English Names: **Eng. :** Agueweed, Boneset, Feverwort, Sweating-plant.

Distribution : Native to eastern North America; meadows and marshlands in India.

Parts Used : Whole Plant.

Flowering & Fruiting : July – October.

Folk Use : Plant used as a tonic.

Active Constituents : Known to possess sesquiterpene lactones, polysaccharides, flavonoides, diterpenes, sterols and volatile oil (*aerial parts*).

Reference : Prajapati *et al.* (2006).

388. *Eupatorium purpureum* L.

Syn.: Eutrochium purpureum (L.) E.E. Lamont

Family: Asteraceae.

English Names: **Eng. :** Gravel Root, Sweet Joe-pye-weed.

Distribution : Native to eastern and central N. America.

Parts Used : Root.

Folk Use : Roots used as stimulant and tonic.

Reference : Bhattacharjee (2001).

389. *Eupatorium triplinerve* Vahl.

Syn.: E. ayapana Vent.

Family: Asteraceae.

English, Hindi and Sanskrit Names: **Eng. :** Ayapana Tea; **Hindi :** Ayapana; **Sans. :** Ajaparnah, Ayaparnah.

Distribution : Native of Brazil, in marshes throughout India, also as a weed in plantations.

Parts Used : Whole Plant.

Folk Use : Plant used as stimulant, tonic and cardio tonic.

Active Constituents : These are daphnetin dimethyl ether, hydrangetin, daphnetin-7-methyl ether, ayapin, umbelliferone daphenetin and herniarin (*plant*).

References : Ambasta (1986), Prajapati *et al.* (2006).

390. *Euphorbia neriifolia* L.

Syn.: E. ligularia Roxb.

Family: Euphorbiaceae.

English, Hindi and Sanskrit Names: **Eng. :** Common Milk Hedge; **Hindi :** Pattonkisend, Sehund, Sij, Thohar, Thuhar; **Sans. :** Patrasnuhi, Snuhi, Snuk, Svarasana, Vajra.

Distribution : Rocky sites throughout W. Peninsular India and Orissa; Malay islands.

Parts Used : Leaf.

Flowering & Fruiting : February – April.

Folk Use : Decoction of leaves taken internally as tonic.

Active Constituents : Glut-5(10)-en-1-one, friedelan-3α- and 3β-ol and taraxerol (*leaf*).

Reference : Sarma *et al.* (2002).

391. *Euphorbia pilosa* L.

Syn.: E. procera Bieb.

Family: Euphorbiaceae.

English Names: **Eng. :** Saatala, Saptalaa.

Distribution : Western Himalayas, from Garhwal westwards to Murree; Soongaria, eastern Siberia and westwards to Atlantic.

Parts Used : Leaf.

Flowering & Fruiting : April – July.

Folk Use : 2 tsp decoction of leaves given twice a day for 15 days prescribed as a tonic for children.

Reference : Sood & Thakur (2004).

392. *Euphorbia thymifolia* L.

Syn.: E. prostrata Ait.; *Chyamaesyce thymifolia* (L.) Mill.

Family: Euphorbiaceae.

English, Hindi and Sanskrit Names: **Eng. :** Chicken Weed, Thyme-leaved Spurge; **Hindi :** Chotadudhi, Choti-dudhi; **Sans. :** Dugdhika, Laghududhika, Nagarjuni, Raktabindu.

Distribution : Pasture lands, railway embankments and in gardens; weed up to 2,000m throughout India.

Parts Used : Leaf. Seeds.

Flowering & Fruiting : December – March.

Folk Use : Leaves and seeds used as stimulant and sedative.

References : Ambasta (1986), Anonymous (1952), Kapur & Singh (1996), Kaushik & Dhiman (2000), Kirtikar & Basu (1984), Parrotta (2001), Ranjan (2000), Siwakoti & Varma (1996), Watt (1972).

393. *Euphrasia officinalis* L.

Syn.: E. latifolia Pursh.; *E. maculata* Gilib.; *E. minima* Jacq.; *E. pumila* Kern. A.; *E. rostkoviana* Hayna.

Family: Scrophulariaceae.

English Names: **Eng. :** Eye Bright, Eye-bright Euphrasia.

Distribution : Temperate Himalayas (Kashmir – Kumaon): 1,350-4,350m, and in Sikkim at 3,350-4,000m.

Parts Used : Whole Plant.

Folk Use : Herb used as tonic.

Active Constituents : Known to yield phenol carboxylic acid, flavones and Me-flavones, derivatives of cinnamic acid, diosmetin, kaempferol, quercetin and quercetin diglucoside (*plant*).

References : Ambasta (1986), Anonymous (1952), Kaul (1997).

394. *Euryale ferox* Salisb.

Syn.: Anneslea spinosa Roxb.; *Euryale india* Planch.

Family: Nymphaeaceae.

English, Hindi and Sanskrit Names: **Eng. :** Fox Nut, Gorgon Fruit; **Hindi :** Machana, Makhana; **Sans. :** Mukkhuana, Padma, Padmini.

Distribution : Water bodies and fresh tanks of Kashmir, Rajasthan, U.P., Bihar, West Bengal, Assam, Tripura and Manipur.

Parts Used : Leaf. Seeds.

Flowering & Fruiting : April – June.

Folk Use : Leaves and seeds used as tonic.

Active Constituents : α-, β-, γ- and δ-tocopherol (*seeds*).

References : Ambasta (1986), Anonymous (1952), Baruah & Sarma (1987), Das *et al.* (1996), Gogoi & Das (2003), Kirtikar & Basu (1984), Raju (2000), Siwakoti & Varma (1996).

395. *Evodia lunu-ankenda* Merrill.

Syn.: E. roxburghiana Benth.; *Fagara triphylla* Roxb.; *Zanthoxylum triphyllum* Juss.

Family: Rutaceae.

Sanskrit Name: **Sans. :** Vanashampaga.

Distribution : Throughout the hills of Deccan Peninsula and in eastern Himalayas, Assam and Andaman.

Parts Used : Whole Plant. Flower. Leaf.

Folk Use : Plant used as tonic. Infusion of flowers and leaves used as tonic.

References : Ambasta (1986), Anonymous (1952).

396. *Evodia rutaecarpa* Hook. f. & Thoms.

Family: Rutaceae.

English Name: **Eng. :** Medicinal Evodia.

Distribution : Sikkim Himalayas and Assam: 1,700-3,500m.

Parts Used : Fruit.

Folk Use : Dried fruits used as a stimulant.

References : Ambasta (1986), Anonymous (1952).

397. *Evolvulus alsinoides* L.

Syn.: Convolvulus linifolius L.; *C. valerianoides* Blanco.; *Evolvulus angustifolius* Roxb.; *E. hirsutus* Lam.; *E. sericeus* Wall.

Family: Convolvulaceae.

English, Hindi and Sanskrit Names: **Eng. :** Dwarf Morning Glory; **Hindi :** Shankpushpi, Shyamakranta; **Sans. :** Laghuvishnukranta, Nilapushpi, Shankpushpi, Tandikodebaha, Vishnukranta.

Distribution : Common throughout the drier parts of India, ascending up to 1,828m.

Parts Used : Whole Plant. Flower.

Flowering & Fruiting : July – December.

Folk Use : Tribals of western Maharashtra take 1/2tsp powder of the herb as a brain tonic for a month. Fresh flowers with sugar also prescribed as brain tonic.

Active Constituents : Evolvine, betaine, β-sitosterol, pentatriacontane and triacontane (*plant*).

References : Anonymous (1952), Asolkar *et al.*(1992), Dhiman (2005), Kirtikar & Basu (1984), Parrotta (2001), Ranjan (2000), Singh (2001), Singh & Pandey (1998), Sood & Thakur (2004), Subramaniam (2000), Trivedi (2002), Upadhye *et al.* (1994), Watt (1972).

398. *Exacum tetragonum* Roxb.

Family: Gentianaceae.

English and Hindi Names: **Eng. :** Bicolor Persian Violet; **Hindi :** Avachiretta, Titakhana.

Distribution : Himalayas (Shimla – Bhutan) up to 1,700m, upper Gangetic Plains, from Bengal and Chota Nagpur to C. India, N. Circars, Khasia hills to Myanmar, Malay Peninsula, China, Borneo and Philippines.

Parts Used : Whole Plant.

Flowering & Fruiting : Flowering during September – December.

Folk Use : Plant used as a tonic.

Active Constituents : Two seco-iridoids-gentiopicroside and methyl grandifloroside methyl ester have been characterised from *aerial parts*.

References : Ambasta (1986), Anonymous (1952), Kirtikar & Basu (1984), Watt (1972).

399. *Fagonia cretica* L.

Syn.: F. arabica L.; *F. bruguieri* DC.; *F. indica* Burm. f.; *F. mysorensis* Roth.; *F. schweinfurthii* Hadidi.

Family: Zygophyllaceae.

Hindi and Sanskrit Names: **Hindi :** Damahan, Hinguna, Ustarkhar, Usturgar; **Sans. :** Ajabhakshya, Dusparsha, Idankarya, Rodini, Vikanta.

Distribution : Deccan, W. Khandesh, Kutch, Sind, Baluchistan, Waziristan, W. Rajputana, upper Gangetic Plain, Punjab, westwards to Afghanistan; Persia, Arabia and Mediterranean.

Parts Used : Whole Plant.

Folk Use : Plant used as stimulant and tonic.

Active Constituents : Attributed properties due to ceryl alcohol, β-sitosterol, n-triacontanol, harmine, three triterpenoids of oleanane group, alanine, arginine, glycine, isoleucine, leucine, lysine, phenylalanine, proline, tyrosine and valine (*plant*).

References : Ambasta (1986), Anonymous (1956) Kirtikar & Basu (1984), Singh & Pandey (1998).

400. *Fagonia indica* Burm. f.

Syn.: F. parviflora Boiss.

Family: Zygophyllaceae.

English Name: **Eng. :** Fagonia.

Distribution : Indo-Pakistan subcontinent, westwards to north and east tropical Africa in arid and semi-arid regions.

Parts Used : Whole Plant.

Folk Use : Decoction of the plant is taken orally as a tonic.

References : Singh & Pandey (1998).

401. *Fagraea racemosa* Jack.

Syn.: F. morindaefolia Blume.

Family: Loganiaceae.

English Name: **Eng. :** False Coffee Tree.

Distribution : Evergreen forests of Andaman & Nicobar Islands.

Parts Used : Root.

Folk Use : Roots used as a tonic.

References : Ambasta (1986), Asolkar *et al.* (1992).

402. *Feronia limonia* (L.) Swingle

Syn.: F. elephantum Correa; *Limonia acidissima* L.

Family: Rutaceae.

English, Hindi and Sanskrit Names: **Eng. :** Elephant Apple, Monkey Fruit, Wood Apple; **Hindi :** Bilin, Kait, Kavitha; **Sans. :** Bhukapittha, Chirpaki, Kapittha.

Distribution : Native to southern India; occurs wild or cultivated up to 1,500m in western Himalayas.

Parts Used : Leaf. Fruit.

Flowering & Fruiting : February – April.

Folk Use : Used as tonic (fruits, leaves) and cardio tonic (fruits).

Active Constituents : Active principles due to psoralen, bergapten, orientin, vitexin, saponarin (*leaf*), umbelliferone, dictamnine, xanthotaxol, xanthotoxin, scoparone, isopimpinellin, isoimperatorin and marmin (*pericarp*).

References : Ambasta (1986), Anonymous (1956), Kirtikar & Basu (1984), Parrotta (2001), Raju (2000), Ranjan (2000), Watt (1972).

403. *Ferula asafoetida* L.

Syn.: F. foetida Regel.

Family: Apiaceae.

English, Hindi and Sanskrit Names: **Eng.** : Asafoetida; **Hindi** : Hing; **Sans.** : Bahleeka, Hingu, Shulanashana.

Distribution : Grows wild in W. Himalayas, Punjab, Kashmir and W. Baluchistan (found in Chamba, Lahul Spiti and Kinnaur in Himachal Pradesh).

Parts Used : Exudates (Root).

Flowering & Fruiting : March – April.

Folk Use : Resinous exudates (oleoresin) of roots used as tonic and sedative.

Active Constituents : Isolated constituents are allyl persulphide, essential oil, ferulic acid and organic sulphur compounds, dimethyl trisulphide, 2-butylmethyl disulphide, 2-butylmethyl trisulphide, di-2-butyl disulphide, di-2-butyltrisulphide, di-2-butyltetrasulphide, assafoetidin and ferocalicin (*oleoresin*).

References : Asolkar *et al.* (1992), Prajapati *et al.* (2006), Rajan *et al.* (2005), Watt (1972).

404. *Ferula galbaniflua* Boiss.

Syn.: F. gummosa

Family: Apiaceae.

English Name: **Eng.** : Galbanum.

Distribution : Native to Central Asia.

Parts Used : Whole Plant. Root.

Folk Use : Plant used as stimulant. Decoction of roots good for improving memory.

Active Constituents : Volatile oil, resin gums, and a coumarin and a new sesquiterpenoid – 10-epijunenol have been identified (*plant*).

References : Ambasta (1986), Khatoon & Goel (2010).

405. *Ferula narthex* Boiss.

Syn.: Narthex asafoetida Falr.

Family: Apiaceae.

English, Hindi and Sanskrit Names: **Eng.** : Asafoetida, Devil's Dung, Persian Fennel-gaint; **Hindi** : Hing, Hingra; **Sans.** : Agudagandha, Balhika, Bhutari, Dipta, Grihini, Hinghka, Jatuka, Kesara.

Distribution : N.W. Himalayas, also occurs in Kashmir.

Parts Used : Rhizome. Stem.

Folk Use : Resinous gum obtained by incision in the living rhizome of the plant considered stimulant. Stem used as tonic to the brain (Yunani).

References : Bhattacharjee (2001), Kirtikar & Basu (1984), Sharma (2003).

406. *Ferula suaveolens* Aitch. & Hemsel.

Family: Apiaceae.

English Name: **Eng. :** Musk-root, Sumbul Root.

Distribution : Central Asia, Khorasan, on the hills to the south of Bezol.

Parts Used : Rhizome.

Folk Use : Sliced and dried rhizomes used as sedative.

References : Ambasta (1986), Anonymous (1956).

407. *Ficus benghalensis* L. (Pl. 5F)*

Syn.: F. banyan Oken; *F. cotonaefolia* Vahl.; *F. indica* L.; *Urostigma bengalense* Gasp.

Family: Moraceae.

English, Hindi and Sanskrit Names: **Eng. :** Banyan Tree; **Hindi :** Bar, Bargad, Bor; **Sans. :** Bahupada, Bhringi, Chira, Dhruva, Mana, Nyagrdhah, Rohina, Vatah, Yakshaves, Yamapriya.

Distribution : Throughout the plains and forest tracts of India, also planted in avenues for shade.

Parts Used : Bark. Leaf. Aerial Roots. Fruit. Seeds.

Flowering & Fruiting : Flowering during summers and fruiting in rainy season.

Folk Use : Fruits, seeds, infusion of bark and extract of aerial roots used as a tonic. Prop roots (10g) pounded in curd (20ml) taken daily to enhance memory.

Active Constituents : These are bengalenoside, flavonoid, glycoside, leucocyanidin 3-O-β-D-galactosyl cellobioside 5,3'-dimethylether, leucopelargonidin-3-O-α-L-rhamnoside,5,7-dimethylether, leucopelagonidin glycoside and leucocyanidin (*plant*), delphinidin-3-O-α-L-rhamnoside (I), pelargonidin-3-O-α-L-rhamnoside (II), leucocyanidin-3-O-β-D-galactosylcellobioside, 20-tetratriaconten-2-one, pentatriacontan-5-one and 6-heptatriaconten-10-one (*bark*), a polysaccharide which on hydrolysis afforded xylose, arabinose and small amount of glucose and galactose (*fruit*).

References : Ambasta (1986), Anonymus (1956), Dhiman (2005), Dwivedi (2007), Farooq (2005), Kaushik & Dhiman (2000), Kirtikar & Basu (1984), Parrotta (2001), Prajapati *et al.* (2006),

Ranjan (2003), Satapathy & Brahmam (1996), Sharma (2004), Singh & Pandey (1998), Subramaniam (2000), Watt (1972).

408. *Ficus hispida* L. f. (Pl. 6A)*

Syn.: Covellia assamica Miq.; *C. daemonum* Miq.; *C. dasycarpa* Miq.; *C. setulosa* Miq.; *C. wightiana* Miq.; *Ficus daemonum* Koem.; *F. mollis* Willd.; *F. oppositifolia* Willd.; *F. prominens* Wall.; *F. scabra* Jacq.; *Sycomorpha roxburghii* Miq.

Family: Moraceae.

English, Hindi and Sanskrit Names: **Eng.** : Fig; **Hindi** : Daduri, Dagurin, Gobla, Kagsha, Katgularia; **Sans.** : Ajaji, Kakodumbara, Kushthaghni, Phalguni, Rajika.

Distribution : Throughout the outer Himalayas from Chenab eastwards to Bengal, central and south India and Andaman Islands.

Parts Used : Fruit.

Flowering & Fruiting : June – August.

Folk Use : Fruits used as a tonic.

Active Constituents : n-triacontanyl, β-amyrin and gluanol (*bark*).

References : Ambasta (1986), Anonymous (1956), Asolkar *et al.* (1992), Kirtikar & Basu (1984), Parrotta (2001), Sharma (2004), Watt (1972).

409. *Ficus racemosa* L.

Syn.: Covellia glomerata Miq.; *C. mollis* Miq.; *Ficus chittagonga* Miq. *F. gloomerata* Roxb.; *F. goolereea* Roxb.; *F. mollis* Miq.

Family: Moraceae.

English, Hindi and Sanskrit Names: **Eng.** : Cluster Fig, Country Fig, Gular Fig; **Hindi** : Gular, Lelka, Paroa, Umar; **Sans.** : Brahmavriksha, Haritaksha, Jantuphala, Udumbarah, Yojanaphala.

Distribution : Outer Himalayas, plains and low hills of India, from Rajputana and the Salt Range to the Khasia mountains, Myanmar, Deccan Peninsula and Sri Lanka.

Parts Used : Bark. Root. Fruit.

Flowering & Fruiting : April – July.

Folk Use : Fruits given as a tonic to pregnant women. 1 tsp of root and bark decoction given orally twice a day as a tonic.

Active Constituents : Constituents isolated are hentriacontane, β-sitosterol, gluanol acetate and glucose (*fruit*), lupeol, friedelin, β-sitosterol, stigmasterol, ceryl behenate lupeol acetate and α-amyrin acetate (*bark*), leucopelarogonidin-3-O-α-rhamnopyranoside, β-sitosterol and lupeol (*root bark*).

References : Bhatt & Jain (2009), Dhiman (2007), Kirtikar & Basu (1984), Sivakumar *et al.* (2003), Watt (1972).

410. *Ficus religiosa* L. (Pl. 6B)*

Syn.: F. affinior Griff.; *Urostigma affine* Miq.; *U. religiosum* Gasp.

Family: Moraceae.

English, Hindi and Sanskrit Names: **Eng. :** Bo Tree, Peepal Tree, Sacred Fig Tree; **Hindi :** Pipal, Pipar, Pipli; **Sans. :** Asvatthah, Bodhidruma, Chaityadru, Devatma, Gajapatra, Kapitana, Mangalya, Nagabandhu, Pavitraka, Vriksharaja, Yajnika.

Distribution : Wild in the sub-Himalayan forests, Bengal and in C. India, planted elsewhere.

Parts Used : Bud. Leaf.

Flowering & Fruiting : March – October.

Folk Use : 1g powdered dried young buds and leaves given twice a day with water as tonic.

Reference : Kalpesh *et al.* (2004).

411. *Ficus virens* Ait.

Syn.: F. infectoria Sensu Roxb.; *F. lacor* Buch.-Ham.; *F. lucescens* Blume.

Family: Moraceae.

English, Hindi and Sanskrit Names: **Eng. :** White Fig; **Hindi :** Khabar, Pakar, Pilkhan, Ramanjir; **Sans. :** Plaksha.

Distribution : Northern India, Madhya Pradesh and Peninsular India; frequently planted as an avenue and ornamental tree.

Parts Used : Fruit.

Flowering & Fruiting : Flowering during rainy season; fruiting in February – May.

Folk Use : Juice of fruits used as cardio tonic.

Reference : Kaushik & Dhiman (2000).

412. *Flacourtia indica* (Burm. f.) Merr.

Syn.: F. latifolia Cooke.; *F. occidentalis* Blatter; *F. ramontchi* L. Herit.; *F. sapida* Roxb.; *F. sepiara* Roxb.; *Gmelina indica* Burm. f.

Family: Flacourtiaceae.

English, Hindi and Sanskrit Names: **Eng. :** Governor's Plum, Madagascar Plum; **Hindi :** Baichi, Bilangru, Kakein, Katar; **Sans. :** Aghori, Gopakanta, Svadukantaka, Wadukantaka.

Distribution : Throughout India up to 1,000m.

Parts Used : Bark.

Flowering & Fruiting : March – April.

Folk Use : Pounded bark with water given as a tonic after delivery.

Active Constituents : Bark afforded a tannin, a flacourtin and a phenolic glucoside ester.

References : Parrotta (2001), Sikarwar & Kaushik (1992).

413. *Foeniculum vulgare* Mill. (Pl. 6C)*

Syn.: Anethum foeniculum L.; *A. panmorium* Roxb.; *F. capillaceum* Gilib.; *F. dulce* Bauh.; *F. officinale* All.; *F. panmorium* DC.; *Ozodia foeniculacea* Wight. & Arn.

Family: Apiaceae.

English, Hindi and Sanskrit Names: **Eng. :** Fennel, Finkel, Indian Sweet Fennel; **Hindi :** Barisounf, Saunf, Sonp, Sont; **Sans. :** Ahichhatra, Avakapshpi, Bhuripushpa, Gandhadhika, Karavi, Madhurica, Methica, Potika, Shaleya.

Distribution : Native of S. Europe and Asia; cultivated mostly as a garden or homeyard crop throughout India up to 2,000m.

Parts Used : Fruit.

Flowering & Fruiting : August – September.

Folk Use : Fruits used as stimulant and tonic.

Active Constituents : Constituents characterized are columbianetin, 7-hydroxy- and 6,7-dihydroxy-coumarins, marmesin, osthenol, psoralen, scoparone, seselin, xanthotoxin, α- and β-amyrins, oleanalic acid, Δ^5-avenasterol, campesterol, Δ^7-stigmasterol, stigmastadienol, linoleic, oleic, palmitic and petroselinic acids (*fruit*).

References : Ambasta (1986), Anonymous (1956), Bhattacharjee (2001), Kapur & Singh (1996), Kaul (1997), Kirtikar & Basu (1984), Pal & Paul (2003), Rajan *et al.* (2003), Reddi *et al.* (2005), Sharma & Sood (1997), Singh *et al.* (2007), Thakur *et al.* (1989), Watt (1972).

414. *Fraxinus excelsior* auct.

Syn.: F. hookeri Wenzig; *F. moorcroftiana* Wall.; *Ornus moorcroftiana* G. Don.

Family: Oleaceae.

English Names: **Eng. :** Ash, Common Ash.

Distribution : N.W. Himalayas: 1,350-3,500m.

Parts Used : Bark.

Folk Use : Bark used as a tonic.

Active Constituents : Known to possess coumarins, flavonoids, tannins, sugars and a volatile oil (*bark*).

References : Kapur & Nanda (1996), Kirtikar & Basu (1984), Prajapati *et al.* (2006).

415. *Fritillaria roylei* Hk. f.

Family: Liliaceae.

English, Hindi and Sanskrit Names: **Eng. :** Fritillary; **Hindi :** Kakoli; **Sans. :** Kakoli.

Distribution : W. temperate Himalayas: 2,700-4,350m (Kashmir – Kumaon).

Parts Used : Bulb. Root.

Folk Use : Bulb used as a tonic. Liquid obtained from crushed roots considered as a source of a strong stimulant.

Active Constituent : Kashmirine (*bulb*).

References : Gaur *et al.* (1983), Kala (2003), Negi *et al.* (1999), Reddi *et al.* (2005), Uniyal (1968).

416. *Fumaria officinalis* L.

Family: Fumariaceae.

English, Hindi and Sanskrit Names: **Eng. :** Fumitory; **Hindi :** Pitapapra; **Sans. :** Parpata.

Distribution : Many parts of India, Nepal and in the Nilgiri mountains.

Parts Used : Whole Plant.

Flowering & Fruiting : May – August.

Folk Use : Plant used as a tonic.

Active Constituents : Principle constituents are corydoline, protopine, bulbocapnine, dicentrine, scoulerine, fumaricine, fumaritine, fumariline, rutoside, quercetin-3-glucoside, chlorogenic acid and caffeic acids, isoquercitrin, quercetin-3,7-diglucoside, quercetin-3-arabinoglucoside, rutin, kaempferol arabinoside, l-hexacosanol, N-methyl-sinactine, fumarofine, (+) fumariline, protopine, cryptopine, (-)fumaricine and fumaritine N-oxide (*plant*).

References : Ambasta (1986), Asolkar *et al.* (1992), Bhattacharjee (2001).

417. *Galipea officinalis* Hancock

Syn.: G. cusparia Dc Cand.

Family: Rutaceae.

English Name: **Eng. :** Angostura.

Distribution : Native to some Caribbean island and to tropical South America.

Parts Used : Bark.

Folk Use : Bark accredited with tonic properties.

Active Constituents : Alkaloids including cusparine and 1-2% volatile oil have been isolated (*bark*).

Reference : Prajapati *et al.* (2006).

418. *Garcinia indica* (Thouars) Choisy

Syn.: Brindonia indica Dupetit-Thouars; *Garcinia purpurea* Roxb.

Family: Clusiaceae.

English, Hindi and Sanskrit Names: **Eng. :** Brindonia Tallow Tree, Kokam Butter Tree, Wild Mangosteen; **Hindi :** Kokam; **Sans. :** Amlavetasa, Chukra, Chukramala, Tintidika, Vrksamla.

Distribution : Forests of Andhra Pradesh, Maharashtra and Chennai.

Parts Used : Fruit.

Flowering & Fruiting : November – February (flowering); April – May (fruiting).

Folk Use : Fruits eaten as cardio tonic.

Active Constituents : Fruits afford garcinol, isogarcinol, hydroxycitric acid, camboginol, cyaniding-3-glucoside, cyaniding-3-sambubioside and benzophenone derivatives.

References : Dhiman (2005), Kirtikar & Basu (1984), Parrotta (2001), Prajapati *et al.* (2006), Rajan *et al.* (2003).

419. *Gardenia jasminoides* Elllis

Syn.: G. augusta Merrill; *G. florida* L.

Family: Rubiaceae.

English, Hindi and Sanskrit Names: **Eng. :** Cape Jasmine; **Hindi :** Gandharaj; **Sans. :** Gandharaj.

Distribution : Grows wild in wet places; cultivated as an ornamental plant.

Parts Used : Fruit.

Flowering & Fruiting : July – October.

Folk Use : Fruits used as stimulant. 1-3g powdered fruits prescribed with cow's milk, twice a day for 8-10 days as nervine tonic.

Active Constituents : Nonacosane, β-sitosterol, δ-mannitol, iridoid glucoside, shanzhiside, gardenoside, genipin-1-β-gentiobioside, geniposide, Me-Ac-asperuloside and picrocrocinic acid (*fruit*).

References : Ambasta (1986), Sood *et al.* (2009).

420. *Gaultheria fragrantissima* Wall.

Family: Ericaceae.

English, Hindi and Sanskrit Names: **Eng. :** Fragrant Wintergreen, Indian Wintergreen; **Hindi :** Gandupura; **Sans. :** Hemant-harit.

Distribution : Distributed from Nepal – Bhutan : 1,828-2,438m; and in Assam and Niligiris.

Parts Used : Whole Plant. Leaf.

Flowering & Fruiting : Flowering during June – July, and fruiting later.

Folk Use : Plant and oil of 'Gaultheria' (the genuine winter green oil obtained from its leaves), used as a stimulant.

Active Constituents : Active properties due to paraffin, aldehyde, ketone, ester, secondary alcohol, β-sitosterol, quercetin-3-galactoside and ursolic acid (*leaf*).

References : Ambasta (1986), Anonymous (1959), Bhattacharjee (2001), Dhiman (2005), Kirtikar & Basu (1984), Sharma (2003).

421. *Gaultheria trichophylla* Royle

Family: Ericaceae.

English Names: **Eng. :** Creeping Snowberry, Himalayan Snowberry, Snowberry.

Distribution : Alpine Himalayas: 3,350-4,350m (Lahul – Sikkim).

Parts Used : Leaf. Fruit.

Flowering & Fruiting : November – February.

Folk Use : Fruits and oil obtained from leaves regarded as stimulant.

Reference : Negi *et al.* (1999).

422. *Gelsemium sempervirens* (L.) Ait. f.

Family: Loganiaceae.

English Names: **Eng. :** Caroline Yellow Jasmine, False Jasmine.

Distribution : Eastern United States.

Parts Used : Rhizome. Root.

Folk Use : Roots and rhizome used as a depressant of CNS.

Reference : Bhattacharjee (2001).

423. *Gentiana algida* Pallas

Family: Gentianaceae.

English Names: **Eng. :** Arctic Gentian, Whitish Gentian.

Distribution : Alpine Himalayas, Kashmir – Sikkim.

Parts Used : Root.

Flowering & Fruiting : July – September.

Folk Use : Decoction of roots used as tonic to promote appetite and digestion.

Reference : Singh *et al.* (1996).

424. *Gentiana kurroo* Royle

Syn.: Pneumananthe kuroo D. Don.

Family: Gentianaceae.

English, Hindi and Sanskrit Names: **Eng. :** Himalayan Gentian, Indian Gentian; **Hindi :** Karu, Kutki, Trayaman; **Sans. :** Anuja, Girija, Palani, Palanika, Trayanti, Tryamana.

Distribution : Kashmir and N.W. Himalayas: 1,524-3,352m.

Parts Used : Whole Plant. Rhizome. Root.

Flowering & Fruiting : August – September (flowering); fruiting later.

Folk Use : Plant and roots used as bitter tonic. Rhizome and roots valued as a tonic.

Active Constituents : These are gentiopicrin, gentianic acid, a pectin and a sugar (*root*), catapol, its 6'-cinnamoyl, 6-O-cinnamoyl, 6-O-vanilloyl and 6-O-feruloyl derivatives and aucubin (*root, rhizome*).

References : Ambasta (1986), Bhattacharjee (2001), Dhiman (2005), Farooq (2005), Gupta (1997), Kaul (1997), Kirtikar & Basu (1984), Sharma (2003), Uniyal & Malhotra (1989), Watt (1972).

425. *Gentiana lutea* L.

Family: Gentianaceae.

English Name: **Eng. :** Yellow Gentian.

Distribution : Native of Alpine and sub-alpine regions of S. Europe; dried roots of the plant imported to India.

Parts Used : Whole Plant. Root.

Folk Use : Plant and roots used as bitter tonic.

References : Ambasta (1986), Bhattacharjee (2001), Sharma (2003).

426. *Gentiana prostrata* Haenke

Syn.: Chondrophylla nutans (Bunge) W. A. Weber; *C. prostrata* (Haenke) J. P. Anderson; *Gentiana nutans* Bunge.

Family: Gentianaceae.

English Name: **Eng. :** Pygmy Gentian.

Distribution : N. W. Himalayas.

Parts Used : Whole Plant.

Flowering & Fruiting : July – September.

Folk Use : Plant used as a tonic.

Reference : Kaul (1997).

427. *Gentiana stracheyi* (Cl.) Kitam.

Family: Gentianaceae.

Distribution : Kashmir, W. Tibet and Himachal Pradesh.

Parts Used : Root.

Flowering & Fruiting : July – August.

Folk Use : Roots used as a tonic.

Reference : Kala (2003).

428. *Gentiana villosa* L.

Family: Gentianaceae.

English Name: **Eng. :** Sanpson's Snake Root.

Distribution : Hilly regions of Indian Himalayas.

Parts Used : Rhizome. Root.

Folk Use : Rhizome and roots used as tonic for nervous distress.

References : Bhattacharjee (2001), Sharma (2003).

429. *Geum elatum* Wall.

Syn.: Acomastylis elata (Wall. ex G. Don) F. Balle; *Sieversia elata* (Wall.) Royle.

Family: Rosaceae.

English Name: **Eng. :** High Avens.

Distribution : Sub-alpine and Alpine Himalayas, from Kashmir (3,000-4,000m) – Sikkim (4,000-5,000m).

Parts Used : Whole Plant.

Folk Use : Plant used as a tonic.

Active Constituents : Hentriacontanol, hentriacontane, β-sitosterol, ellagic acid and its tetra-O-methyl derivative and isoquercitrin have been characterized from *aerial parts.*

Reference : Kapur *et al.* (1996).

430. *Gillenia trifoliata* L.

Family: Rosaceae.

English Names: **Eng. :** Moench Indian Psysic.

Distribution : Eastern United States.

Parts Used : Root.

Folk Use : Roots regarded as tonic.

Reference : Bhattacharjee (2001).

431. *Ginkgo biloba* L. (Pl. 6D)

Family: Ginkgoaceae.

English Names: **Eng. :** Maidenhair Tree, Glory Lily.

Distribution : Widely planted in Korea and China; cultivated in Indian gardens.

Parts Used : Whole Plant. Seeds.

Folk Use : Plant used in traditional medicine to enhance memory and as brain tonic. Seeds valued as sedative.

Active Constituents : Reported to yield 4-methoxypyridoxine, novel phenolic lipids-6-(pentadec-8-enyl) resorcyclic (I) (97) and 6-tridecylresorcylic (3%) acid (*seeds*), β-sitosterol, stigmasterol, kaempferol-3-rhamnoglucoside, rutin, quercetin, octaacetyl isoquercitrin, (-) epicatechin acetate, bilobolide A, quercetin-3-O-α-(6'''-p-coumarylglucosyl-(1→4)rhamnoside), ginkgolides B and C, sciadopitysin, bilobetin, amentoflavone, ginkgetin, isoginkgetin and sciadopitysin (*leaf*).

References : Ambasta (1986), Harborne & Baxter (2001), Khatoon & Goel (2010), Sharma (2003).

432. *Glechoma hederacea* L.

Family: Lamiaceae.

English Name: **Eng. :** Ground Ivy.

Distribution : Native to Europe and western Asia, now naturalized in temperate regions, including N. America.

Parts Used : Aerial Parts.

Folk Use : Used as a tonic.

Active Constituents : Sesquiterpenes, flavonoids, saponin, resin and tannins (*aerial parts*).

References : Bhattacharjee (2001), Prajapati *et al.* (2006).

433. *Glinus oppositifolius* (L.)A. DC.

Syn.: Mollugo spergula L.

Family: Aizoaceae.

English, Hindi and Sanskrit Names: **Eng. :** Bitter Cumin; **Hindi :** Gima; **Sans. :** Grishma-sundaraka.

Distribution : Greater part of India, especially in Assam, W. Bengal and Deccan Peninsula.

Parts Used : Whole Plant.

Flowering & Fruiting : Rainy season.

Folk Use : 10-20g decoction of plant used as a bitter tonic for liver disorders.

References : Chatterjee & Pakrashi (1997).

434. *Gloriosa superba* L.

Syn.: G. simplex Don.; *Methonica superba* Lam.

Family: Liliaceae.

English, Hindi and Sanskrit Names: **Eng. :** Climbing Lily, Malabar Glory Lily, Supper Lily, Tiger Lily; **Hindi :** Karihari, Kuthari, Languli; **Sans. :** Agnimukhi, Agnisikha, Ailni, Garbhaghatini, Kalikari, Langli, Vishalya.

Distribution : Throughout the plains and low hills of India.

Parts Used : Root. Tuber.

Flowering & Fruiting : June – October.

Folk Use : Tubers used as tonic for stomachache and as stimulant. Roots used as tonic.

Active Constituents : Reported to possess β-sitosterol, its glucoside, 2-hydroxy-6-methoxybenzoic acid, a long chain fatty acid, β- and γ-lumicolchicines (*tuber*).

References : Ambasta (1986), Anonymous (1959), Bhattacharjee (2001), Kaushik & Dhiman (2000), Nadeem *et al.* (2001), Raju (2000), Singh (2000c), Singh *et al.* (2003), Subramaniam (2000), Thakur *et al.* (1989), Watt (1972).

435. *Glycyrrhiza glabra* L.

Family: Fabaceae.

English, Hindi and Sanskrit Names: **Eng. :** Licorice, Liquorice; **Hindi :** Jethi-madh, Mulhatti; **Sans. :** Jalayashti, Klitaka, Madhuka, Yashti-madhu.

Distribution : Jammu & Kashmir, Punjab, Sub-Himalayan tracts and in some parts of South India.

Parts Used : Rhizome. Root.

Folk Use : Rhizome and roots used as tonic.

Active Constituents : Attributed properties due to glabrolide, deoxoglabrolide, isoglabrolide, glycyrrhetol, 18α-hydroxyglycyrrhetic acid, liquiritic acid, olean-12-en-3β-30-oic acid, olean-11,13(18)-dien-3β-ol-30-oic acid, glabridin, glabrol, licoisoflavone B, licoisoflavnone, isoliquiritigenin, formononetin, 4',7-dihydroxyflavone, echinotin, glybrol, echinatin, naringenin, licoricone, glycyrin, isoliquiritigenin, four new isoprenoid-subsitituted phenolic constituents-semilicoisoflavone B, 1-methoxyficifolinol, isoangustone A and licoriphenone (*root*).

References : Ambasta (1986), Farooq (2005), Thakur *et al.* (1989).

436. *Gmelina arborea* L.

Syn.: G. indica Burm. f.; *G. rheedii* Hook.; *G. tomentosa* Wall.; *Premna arborea* Roth.

Family: Verbenaceae.

English, Hindi and Sanskrit Names: **Eng. :** Candahar Tree, Cashmere Tree, Coomb Teak, White Teak; **Hindi :** Gambhari,

Kambar, Sewan, Shiwan; **Sans.** : Ashveta, Bhadraparni, Ghambhari, Kasmari, Suphala, Vidari.

Distribution : Roadsides and in forests throughout India.

Parts Used : Bark. Root. Fruit.

Flowering & Fruiting : February – April (flowering); April – July (fruiting).

Folk Use : Known to be used earlier as tonic (roots, fruits), and bitter tonic (roots, bark).

Active Constituents : β-sitosterol, umbelliferone, benzoic acid, octacosanol, ceryl alcohol, gmelofuran, gmelinol, hentriacontanol-1, n-octacosonal and sesquiterpene (*root*).

References : Anonymous (1959), Chauhan *et al.* (2007), Dhiman (2005), Dwivedi (2007), Kaushik & Dhiman (2000), Kirtikar & Basu (1984), Kumar & Lal (2003), Lalramnghinglova (1996), Parrotta (2001), Sharma (2004), Trivedi *et al.* (1982), Watt (1972).

437. *Gossypium arboreum* L. var. ***obtusifolium*** (Roxb.) Roberty

Syn.: G. herbaceum var. *obtusifolium* (Roxb.) Masters; *G. obtusifolium* Roxb.

Family: Malvaceae.

English, Hindi and Sanskrit Names: **Eng.** : Cotton Plant; **Hindi** : Deokapai, Kapas, Nurma; **Sans.** : Karpasamu, Karpashi.

Distribution : Wild or cultivated throughout India.

Parts Used : Flower. Seeds.

Flowering & Fruiting : Flowering during winters and fruiting in summers.

Folk Use : Flowers used as a tonic. Seeds valued as nervine tonic.

References : Kaushik & Dhiman (2000), Ranjan (2000), Reddi *et al.* (2005), Singh *et al.* (2007).

438. *Grevillea robusta* A. Cunn. *ex* R. Br. (Pl. 6E)

Family: Proteaceae.

English and Hindi Names: **Eng.** : Silky Oak, Silver Oak Tree; **Hindi** : Savukkumaram.

Distribution : Native of Queensland and New South Wales; now cultivated throughout India : 700-2,000m.

Parts Used : Aerial Parts.

Flowering & Fruiting : March – April (flowering); July – August (fruiting).

Folk Use : Aerial parts considered CNS depressant.

Active Constituents : Known to possess phenol, robustol, rutin, cinnamic acid derivatives, arbutin derivatives and flavonoid, 2,5-dihydroxycinnamic acid, 6-hydroxycoumarin, quercetin, arbutin, hydroquinone (*leaf*), grevillol and mono and bisnorstriatol (*wood*), epilyoniresinol, its 2-O-glucoside, rhamnocitrin-3-O-rutinoside and kaempferol-3-O-rutinoside (*wood, leaf*).

References : Asolkar *et al.* (1992), Kaushik & Dhiman (2000).

439. *Grewia microcos* L.

Syn.: G. affinis Lindl.; *G. ulmifolia* Roxb.; *Microcos paniculata* Roxb.; *M. syauntoniana* G. Don.

Family: Tiliaceae.

Distribution : E. Bengal, Assam, Myanmar, W. Peninsula, Mysore, Sri Lanka; China and Java.

Parts Used : Fruit.

Folk Use : Ripe fruits used as a cooling tonic.

Reference : Prakash & Mehrotra (1991).

440. *Grewia tiliifolia* Vahl.

Syn.: G. arborea Roxb. *ex* Rotter; *G. subinaequalis* Wall.; *G. variabilis* Wall.

Family: Tiliaceae.

English, Hindi and Sanskrit Names: **Eng.** : Dhaman; **Hindi** : Dhaman, Dhamni, Pharsa; **Sans.** : Dhamni, Dhanurvrksah, Dhanvangah.

Distribution : Deciduous and semi-evergreen forests throughout India up to 1,200m.

Parts Used : Bark. Fruit.

Flowering & Fruiting : April – May (flowering); fruiting June onwards.

Folk Use : Bark and fruits used as tonic.

Active Constituents : Attributed properties due to taraxasterol, β-sitosterol, erythrodiol, β-amyrin and betulin (*bark*), pelargonidin-3,5-diglucoside, quercetin, its 3-O-glucoside, naringenin-7-O-β-D-glucoside, proline, lysine, glutaric acid, phenylalanine, delphinidin-3-glucoside and cyanidin-3-glucoside (*fruit*).

References : Hebbar *et al.* (2003), Prajapati *et al.* (2006), Reddi *et al.* (2005).

441. *Gymnema sylvestre* R. Br. ex Schult

Syn.: G. melicida Edgew.; *Periploca sylvestris* Willd.

Family: Asclepiadaceae.

English, Hindi and Sanskrit Names: **Eng. :** Periploca of the Woods, Small Indian Ipecacuanha, Vine; **Hindi :** Chhotadudhilata, Gurmar, Merasimgi; **Sans. :** Ajaballi, Ajashringi, Bahalchakshu, Chakrashreni, Grihadruma, Karnika, Maurvi, Medhasingi, Nandivruksha, Putrashringi, Vatika, Vishani, Vrikshikali.

Distribution : Commonly occurs in the hills of Bihar, Orissa, Madhya Pradesh and Deccan Peninsula.

Parts Used : Whole Plant. Leaf.

Flowering & Fruiting : July – August (flowering); October – December (fruiting).

Folk Use : Plant used as tonic, cardio and liver tonic. Leaves in powder form used as stimulant.

Active Constituents : Active principles are alanine, γ-aminobutyric acid, isoleucine, valine, adenine, choline, betaine, gymnamine, hentriacontane, pentatriacontane, conduritol A, inositol, d-quercitol, α- and β-chlorophylla, butyric, formic and tartaric acids, β-amyrin, lupeol, stigmasterol and gymnamosaponins (I-IV) (*plant*).

References : Anonymous (1959), Asolkar *et al.* (1992), Dhiman (2005), Farooq (2005), Kirtikar & Basu (1984), Parrotta (2001), Prajapati *et al.* (2006).

442. *Habenaria acuminata* L.

Family: Orchidaceae.

Distribution : Native to Annamalai hills, Kerala and Uttaranchal Himalayas.

Parts Used : Tuber.

Folk Use : Tubers used as a tonic.

References : Bhattacharjee (2001), Sharma (2003).

443. *Habenaria edgeworthii* Hook. f. ex Callett.

Family: Orchidaceae.

English, Hindi and Sanskrit Names: **Eng. :** Edgeworth's Habenaria; **Hindi :** Rddhi; **Sans. :** Chetaniya, Jeevadari, Mangalaya, Pranada, Pushtida, Siddhi.

Distribution : Shimla, Garhwal and U.P. at altitude of 2,500m.

Parts Used : Tuber.

Flowering & Fruiting : July – September.

Folk Use : Tubers used as a tonic.

Reference : Prajapati *et al.* (2006).

444. *Habenaria goodyeroides* Don

Syn.: H. hayataeana Schltr.; *Peristylus plantagineus* (Lindl.) Lindl.; *P. warburgianus* Kraenzl.

Family: Orchidaceae.

Distribution : Found in Jharkhand, Bihar, Orissa, Sikkim, Manipur, Uttaranchal and Khasia hills at 1,000m.

Parts Used : Tuber.

Folk Use : Tubers used as a tonic.

References : Bhattacharjee (2001), Sharma (2003).

445. *Habenaria griffithi* Hook. f.

Syn.: Diphylax griffithii (Hook. f.) Kranz.

Family: Orchidaceae.

English Name: **Eng. :** Griffith's Habenaria.

Distribution : Distributed in Garhwal: 1,500-2,100m.

Parts Used : Tuber.

Flowering & Fruiting : Flowering during April – May.

Folk Use : Tubers used as a tonic.

References : Bhattacharjee (2001), Sharma (2003).

446. *Habenaria intermedia* D. Don

Syn.: H. arietina H. f.

Family: Orchidaceae.

English and Sanskrit Names: **Eng. :** Wild Orchid; **Sans. :** Mangala, Riddhi, Risisrista, Saravajanpriya, Vasu, Yuga.

Distribution : Native to Kashmir – Kumaon (Uttaranchal); Sikkim and Khasia hills: 700-2,300m.

Parts Used : Tuber.

Folk Use : Tubers used as a general tonic.

References : Asolkar *et al.* (1992), Bhattacharjee (2001), Sharma (2003), Singh (2001).

447. *Haematoxylon campechianum* L.

Syn.: Haematoxylum campechianum L.

Family: Fabaceae.

English and Hindi Names: **Eng. :** Campeachy Wood, Logwood Tree; **Hindi :** Patang.

Distribution : Native to tropical America; introduced in Indian gardens for its delicate foliage and fragrant flowers.

Parts Used : Heart Wood.

Folk Use : Heart wood used as a tonic.

Active Constituents : Haematoxylin, tannin, resin, quercetin, traces of volatile oil, oxalic acid, acetic acid, hematoxylol A and hematoxylol B (*heart wood*).

References : Anonymous (1959), Bhattacharjee (2001), Sharma (2003), Watt (1972).

448. *Haldina cordifolia* (Roxb.) Ridsdale

Syn.: Adina cordifolia (Roxb.) Hook. f. *ex* Brandis.; *Nauclea cordifolia* Roxb.

Family: Rubiaceae.

English, Hindi and Sanskrit Names: **Eng. :** Haldu; **Hindi :** Haldu; **Sans. :** Dharakadambah, Girikadambah.

Distribution : Deciduous and semi-evergreen forests throughout India, especially in the Eastern Ghats, Karnataka and Kerala.

Parts Used : Bark.

Flowering & Fruiting : June – August (flowering); October – December (fruiting).

Folk Use : Bark used as a tonic.

Active Constituents : Reported to possess lignin, octasonol, pentosan, cadambine, α-amyrin, naringenin-7-methylether-4-O-α-L-rhamnopyranoside, stigmasta-5,22-dien-3β-O-α-L-rhamnopyranosul (1→4)-β-D-xylopyranosul and tannin (*bark*).

Reference : Prajapati *et al.* (2006).

449. *Hedera nepalensis* Koch. (Pl. 6F)

Syn.: H. himalaica Tobler

Family: Araliaceae.

English and Hindi Names: **Eng. :** Barren Ivy, Ground Ivy, Nepal Ivy, Wood-bind; **Hindi :** Lablab.

Distribution : Throughout the Himalayas: 2,000-3,500m & Khasia hills: 1,350-2,000m, grows in some places from W. Europe – Japan.

Parts Used : Berry. Leaf.

Folk Use : Berries and leaves regarded as stimulant.

Active Constituents : These are saponins and molluscacidal (*berry*), sesquiterpenes, β-elemene, elixene, germacrene B and hederin (*leaf*).

References: Ambasta (1986), Anonymous (1959), Kirtikar & Basu (1984).

450. *Hedychium coronarium* Koenig

Syn.: Amomum filiforme Hunter ex Ridl.; *Gandasulium linglulatum* (Hassk.) Kuntze; *Hedychium chrysoleucum* Hook.; *H. flavum* Roxb.; *H. maximum* Roscoe.; *Kaempferia hedychium* Lam.

Family: Zingiberaceae.

English and Sanskrit Names: **Eng.** : Common Ginger Lily, Garland Flower; **Sans.** : Shati.

Distribution : Throughout India, ascending up to 2,000m.

Parts Used : Rhizome.

Flowering & Fruiting : July – October.

Folk Use : Rhizomes used as tonic and stimulant.

Active Constituents : Properties due to coronarin, A,B,C,D and labda-8,12-diene-15,16-dial, methyl and ethyl ethers of coronarin-D, isocoronarin-D and 14,15,16-trinorlabda-8,11-dien-13-al (*rhizome*).

References : Ambasta (1986), Anonymous (1959), Behura & Rout (2003), Parrotta (2001), Prajapati *et al.* (2006), Sharma (2004).

451. *Hedychium spicatum* Buch.-Ham. *ex* J. F. Smith

Syn.: H. acuminatum Rosc.

Family: Zingiberaceae.

English, Hindi and Sanskrit Names: **Eng.** : Spiked Ginger Lily; **Hindi** : Gandhapalashi, Kapurakachari; **Sans.** : Amlanisha, Gandhashati, Karpurakachali, Sati.

Distribution : Sub-tropical Himalayas and Kumaon between 1,524-2,133m.

Parts Used : Rhizome. Root.

Flowering & Fruiting : August – October.

Folk Use : Rhizomes used as tonic in dyspepsia; also its decoction used as brain tonic by Rawal community of Haat Kali sacred grove of Gangolihat. Roots also valued as tonic.

Active Constituents : Hedychenone, 6-oxo-labda-7,11,14-triene-16 oic acid lactone, β-sitosterol, cryptomeridiol, α- and β-pinene, limonene, γ-terpinene, 1,8-cineole, β-phellandrene, β-terpineol, p-cymene and linalool (*rhizome*).

References : Ambasta (1986), Anonymous (1959), Dhiman (2005), Dwivedi (2007), Farooq (2005), Gupta (1997), Khatoon & Goel (2010), Kirtikar & Basu (1984), Prajapati *et al.* (2006), Samant *et al.* (2001), Sharma (2003), Sharma & Sood (1997), Singh (2001), Thakur *et al.* (1989).

452. *Hedyotis biflora* (L.) Wight & Arn.

Syn.: Oldenlandia biflora L.; *O. paniculata* Burm.

Family: Rubiaceae.

Hindi Name: **Hindi :** Daman-papra.

Distribution : N.E. India and Karnataka.

Parts Used : Whole Plant.

Folk Use : Plant used in nervous depression.

Active Constituents : Known to yield a fraction containing thirteen amino acids, proteins, K^+, Na^+ and Cl^- (*plant*).

Reference : Ambasta (1986).

453. *Hedyotis corymbosa* (L.) Lam.

Syn.: H. pruinosa Wight. & Arn. *Oldenlandia corymbosa* L.

Family: Rubiaeae.

English, Hindi and Sanskrit Names: **Eng. :** Daman Pappar, Old World Diamond Flower; **Hindi :** Daman-papar, Pitpapra; **Sans. :** Kshetraparpata, Parpatakah, Parpatah.

Distribution : As a weed up to 2,000m throughout India, both on dry and wetlands.

Parts Used : Whole Plant.

Flowering & Fruiting : July – September.

Folk Use : Plant decoction used in nervous depression and as liver tonic.

Active Constituents : Known to yield biflorin, biflorone (*air dried plant*), ebanolic acid, sitosterol and urosolic acid (*stored plant*).

References : Ambasta (1986), Prajapati *et al.* (2006), Sharma (2004).

454. *Hedyotis herbacea* L.

Syn.: Oldenlandia herbacea (L.) Roxb.

Family: Rubiaceae.

English, Hindi and Sanskrit Names: **Eng. :** Slender Oldenlandia; **Hindi :** Paper-bhed; **Sans. :** Chayaparpatika, Parpataka.

Distribution : Open wasteplaces up to 1,650m throughout India.

Parts Used : Whole Plant.

Flowering & Fruiting : February – August.

Folk Use : Plant used as a tonic and bitter tonic.

References : Anonymous (1959), Prajapati *et al.* (2006).

455. *Heimia salicifolia* Link.

Syn.: Nesaea salicifolia H. B. & K.

Family: Lythraceae.

English Names: **Eng.** **:** Elixir of the Sun, Shrubby Yellowcrest, Willow-leaf Heimia.

Distribution : Native from Mexico – Argentina and grown in Lloyd Botanic Garden, Darjeeling.

Parts Used : Leaf.

Flowering & Fruiting : Flowering during August – September.

Folk Use : Leaves used as tonic.

Active Constituent : Nessin (*leaf*).

References : Ambasta (1986), Anonymous (1966), Asolkar *et al.* (1992).

456. *Helminthostachys zeylanica* (L.) Hook.

Syn.: H. dulcis Kaulf.; *Osmunda zeylanica* L.

Family: Helminthostachyaceae.

English Name: **Eng. :** Ophioglossum.

Distribution : W. Bengal, Western Ghats of S. India up to 1,000m and Andaman.

Parts Used : Rhizome.

Flowering & Fruiting : Fertile spikes appear at the beginning of winters.

Folk Use : Rhizome used as a tonic; its powder used for improving memory.

Active Constituents : Ugonin A, B, C and D (*rhizome*).

References : Anonymous (1959), Prajapati *et al.* (2006), Singh & Maheshwari (1992), Singh & Viswanathan (1996).

457. *Hemidesmus indicus* (L.) R. Br.

Family: Asclepiadaceae.

English, Hindi and Sanskrit Names: **Eng. :** Country Sarasaparilla, Indian Sarsaprilla; **Hindi :** Anantmul, Magrabu, Salsa, Sariva, Shewat; **Sans. :** Anantamulah, Sarba, Sariva.

Distribution : Upper Gangetic plains, eastwards to West Bengal and from Madhya Pradesh to south India.

Parts Used : Leaf. Root.

Flowering & Fruiting : August – December.

Folk Use : Root considered as a bitter tonic; also its paste mixed with milk taken as a tonic for improving health by Malekudiyas. Infusion of leaves given along with sugar also valued as tonic.

Active Constituents : Attributed properties due to rutin (*leaf*), hentriacontane, lupeol, lupeol octacosanoate, β-amyrin acetate, β-sitosterol, lupeol acetate, α- and β-amyrin and sitosterol [1] (*root*).

References : Anonymous (1959), Dhiman (2005), Farooq (2005), Khanna (2000), Kirtikar & Basu (1984), Kshirsagar & Singh (2007), Kumar & Lal (2003), Maya *et al.* (2003), Parrotta (2001), Prajapati *et al.* (2006), Prameela & Menon (2003), Ranjan (2000, 2003), Reddi *et al.* (2005), Thakur *et al.* (1989), Upadhyay & Singh (2005), Watt (1972).

458. *Hesperethusa crenulata* (Roxb.) M. Roem.

Syn.: Limonia acidissima auct., non L.; *L. crenulata* Roxb.; *Naringi crenulata* (Roxb.) Nicholson

Family: Rutaceae.

English, Hindi and Sanskrit Names: **Eng. :** Elephant Apple, Monkey Fruit; **Hindi :** Beli; **Sans. :** Bilvaparni, Surasi, Vilvaparni.

Distribution : Dry hills in various parts of India, N.W. Himalayas in Shimla and Kumaon, ascending to 1,350m; Monghyr hills in Bihar; Assam, W. Peninsula, from the Bombay Ghats and Coromandal southwards.

Parts Used : Fruit.

Flowering & Fruiting : May – June (flowering); November – December (fruiting).

Folk Use : Fruits used as tonic and useful in malignant and persistent fevers.

Active Constituents : Fruits yield severine palmitate, acidissiminol, acidissiminin epoxide and N-benzoyltyramine.

References : Ambasta (1986), Anonymous (1959), Parrotta (2001).

459. *Heteropogon contortus* (L.) Beauv. *ex* Roem. & Schult.

Syn.: Andropogon contortus L.

Family: Poaceae.

English and Hindi Names: **Eng. :** Bellary Grass, Spear Grass; **Hindi :** Kher, Kumeria, Parwa, Sura.

Distribution : Dry places throughout India ascending up to 2,350m in the Himalayas; occurs in Punjab, U.P., Bihar, Bengal, Assam, Orissa, M.P., Bombay and Chennai.

Parts Used : Root.

Flowering & Fruiting : August – December (flowering); October – January (fruiting).

Folk Use : Roots used as stimulant.

References : Ambasta (1986), Kirtikar & Basu (1984), Parrotta (2001), Srivastava *et al.* (2003a), Srivastava & Singh (2003).

460. *Heynea trijuga* Roxb.

Syn.: Walsura trijuga Kurz.; *Trichilia connaroides* (Wt. & Arn.) Bent.

Family: Meliaceae.

English Name: **Eng.** : Indian Heynea.

Distribution : Sub-Himalayan tract from Kumaon eastwards, Sikkim up to 1,350m, Khasia hills, Manipur, E. Ghats in the forests of Godavari and Vizagapatnam up to 1,500m, W. Ghats from Poona southwards through the Nilgiris and Anamalais – Travancore, up to 2,000m; Myanmar, Cambodia, Malay Peninsula (cult.) and Sumatra.

Parts Used : Bark. Leaf.

Flowering & Fruiting : Flowering during April – June.

Folk Use : Bark and leaves used as a bitter tonic.

Active Constituents : Trijugins A and B and trijugin B acetate (*leaf*).

References : Ambasta (1986), Anonymous (1959), Kirtikar & Basu (1984).

461. *Hibiscus rosa-sinensis* L. (Pl. 7A)*

Family: Malvaceae.

English, Hindi and Sanskrit Names: **Eng.** : Bengal Rose, China Rose, Chinese Hibiscus, Shoe Flower; **Hindi** : Gulhar, Jasum, Jasun, Odhul; **Sans.** : Arkapriya, Aruna, Harivallabha, Japa, Odhrapushpa, Pratika, Pushpi, Roga, Trisandhya.

Distribution : Native to China; cultivated in Indian gardens.

Parts Used : Flower.

Flowering & Fruiting : February – October.

Folk Use : Flowers used as brain tonic.

Active Constituents : Isolated constituents are quercetin-3-diglucoside, 3,7-diglucoside, cyanidin-3,5-diglucoside and cyaniding-3-sophoroside-5-glucoside (*deep yellow flower*), kaempferol-3-xylosylglucoside (*ivory white flower*).

References : Pandey (2003), Sharma (2004).

462. *Hibiscus sabdariffa* L.

Syn.: H.sanguineus Griff.

Family: Malvaceae.

English and Hindi Names: **Eng.** : Indian Sorrel, Jamaica Sorrel, Natal Sorrel, Red Sorrel, Rosella; **Hindi** : Lalambari, Patwa.

Distribution : Cultivated in hotter parts of India and Sri Lanka; tropics of the old world.

Parts Used : Bud. Calyx.

Flowering & Fruiting : December – March.

Folk Use : Infusion of buds and calyces taken as tonic.

Active Constituents : Accredited properties due to hiviscin, delphinidin-3-O-β-D-xylopyranosyl(1→2)-β-D-glucopyranoside (*calyx*).

References : Banerjee (1996), Watt (1972).

463. *Hibiscus sabdariffa* L. var. ***sabdariffa***

Family: Malvaceae.

English, Hindi and Sanskrit Names: **Eng. :** Red Sorrel; **Hindi :** Lalambari; **Sans. :** Ambasthaki.

Distribution : Cultivated in hotter parts of India and in Sri Lanka.

Parts Used : Leaf.

Folk Use : Leaves used as sedative.

Reference : Sharma (2004).

464. *Hippophae rhamnoides* L. subsp. ***turkestanica*** Rousi.

Syn.: H. rhamnoides auct.

Family: Elaeagnaceae.

English Names: **Eng. :** Sallow-thorn, Sea Buck Thorn.

Distribution : N.W. Himalayas, 2,350-5,000m; Afghanistan and westwards to N. and middle Europe, N. and Central Asia.

Parts Used : Fruit.

Folk Use : Dried pulp of fruits used as tonic for weak children.

Active Constituents : Carotene, β-carotene, cryptoxanthin, zeaxanthin and its dipalmitic ester physalien, lycopene, γ-carotene, a flavonol, iso-rhamnetin, malic acid, citric acid, tartaric acid, traces of succinic acid, reducing sugars, pectins, tannins and mannitol (*fruit*).

Reference : Singh *et al.* (2003).

465. *Hiptage benghalensis* (L.) Kurz.

Syn.: Banisteria benghalensis L.; *Gaertnera racemosa* Roxb.; *Hiptage madablota* Gaertn.; *Molina racemosa* Lamk.; *Teria bengalensis* L.

Family: Malpighiaceae.

English, Hindi and Sanskrit Names: **Eng. :** Clustered Hiptage; **Hindi :** Aneta, Hutimukta, Kampti, Madhavilata, Madhumalati; **Sans. :** Atimukha, Bhadralata, Bhushana, Chandravalli, Kamuka, Madhavi, Pundrakalata, Sugandha, Vasanti, Vasantaduti.

Distribution : Throughout India, in forests up to 1,800m.

Parts Used : Bark. Leaf. Flower.

Flowering & Fruiting : Flowering: February – April; Fruiting: April – May.

Folk Use : Bark, flowers and leaves used as cardio tonic.

References : Asolkar *et al.* (1992), Prajapati *et al.* (2006), Reddi *et al.* (2005) Sharma (2004).

466. *Holarrhena antidysenterica* (L.) Wall.

Syn.: Echites antidysentrica Roxb.; *E. pubescens* Buch.-Ham. *Holarrhena codaga* G. Don.; *H. malaccensis* Wight.; *H. pubescens* Wall. *ex* G. Don.; *Wrightia antidysentrica* Graham.

Family: Apocynaceae.

English, Hindi and Sanskrit Names: **Eng. :** Bitter Oleander, Conessi Bark, Easter Tree, Ivory Tree, Kurchi Bark, Tellicherry Bark; **Hindi :** Indrajau, Karchi, Karra, Kuar, Kurchi, Kureya; **Sans. :** Girimallika, Indrayava, Kalinga, Kutaja, Mallikapushpa.

Distribution : Tropical parts of India, abundant in the sub-Himalayan tracts; also planted on roadsides.

Parts Used : Bark. Leaf. Seeds.

Flowering & Fruiting : Flowering: April – July; Fruiting: August – October.

Folk Use : Bark, leaves and seeds used as a tonic.

Active Constituents : Active constituents are holacetine, Kurcholessine, 3α-aminoconan-5-ene, kurcholessine, L-quebrachitol, dihydroisoconessimine and 7-α-oh-conessine and holonomine (*bark*).

References : Ambasta (1986), Anonymous (1959), Bhattacharjee (2001), Dhiman (2005), Kirtikar & Basu (1984), Parrotta (2001), Prajapati *et al.* (2006), Reddi *et al.* (2005), Singh *et al.* (2007), Watt (1972).

467. *Holostemma ada-kodien* Schultes

Syn.: H. annulare (Roxb.) K. Schumn.; *H. rheedei* Wall.; *H. rheedianum* auct. non Spreng.

Family: Asclepiadaceae.

English, Hindi and Sanskrit Names: **Eng. :** Holostemma Creeper; **Hindi :** Charivel, Chirvel; **Sans. :** Arkapushpi, Jalakamuka, Jivanti, Kutumbini, Payasya, Shitala, Sitaparni, Vakrashalya.

Distribution : Hedges and open forests throughout India.

Parts Used : Root.

Flowering & Fruiting : Flowering during April – September; fruiting in January.

Folk Use : Roots used as stimulant; also its juice given as a health tonic once daily.

Active Constituents : Reported to contain α-amyrin, lupeol, alanine, aspartic acid, glycine, serine, threonine, valine, moisture (10.8%), protein (4.07%), sugar (24.0%), starch (32.54%), fibre (12.2%), ash (3.07%), ethanol and β-sitosterol (*root*).

References : Jha & Varma (1996), Prajapati *et al.* (2006), Reddi *et al.* (2005).

468. *Homalonema aromatica* Schott.

Syn.: Colla aromatica Roxb.

Family: Araceae.

Hindi Name: **Hindi :** Sugandh Mantra.

Distribution : N.E. India.

Parts Used : Rhizome.

Folk Use : Rhizome used as stimulant.

Active Constituent : Linalool (*rhizome*).

References : Ambasta (1986), Anonymous (1959), Lalramnghinglova (2003).

469. *Hordeum vulgare* L.*

Syn.: H. sativum Jessen.

Family: Poaceae.

English, Hindi and Sanskrit Names: **Eng. :** Barley; **Hindi :** Jau, Jav, Suj; **Sans. :** Akshata, Divya, Hayapriya, Hayeshta, Medhya, Shaktu, Sitashuka, Tikshnashuka, Yavah.

Distribution : Throughout N. India, cultivated.

Parts Used : Grains.

Flowering & Fruiting : February – April.

Folk Use : Grains used as a tonic.

Active Constituents : Active principles due to arginine, histidine, lysine, tyrosine, tryptophan, phenylalanine, cystine, methionine, threonine, leucine, isoleucine, valine and glycine (*grains*).

Reference : Prajapati *et al.* (2006).

470. *Houttuynia cordata* Thunb.

Family: Saururaceae.

English Names: **Eng. :** Chameleon Plant, Fishwort, Heart Leaf, Lizard Tail, Tsi.

Distribution : Tropical Himalayas (Garhwal – Sikkim): 300-1,700m, Assam and Khasia mountains; Siam, China and Japan.

Parts Used : Whole Plant. Leaf.

Folk Use : Plant and leaves taken internally as tonic.

Active Constituents : These are afzelin, hyperin, rutin, β-sitosterol, chlorogenic, palmitic, linoleic, oleic and stearic acids (*aerial parts*), 3,5-didecanoyl-4-nonyl-1,4-dihydropyridine, 3-decanoyl-4-nonyl-5-dodecanoyl-1,4-dihydropyriline, 3-decanonyl-6-nonylpyridine, 3,5-didecanoylpyridine and 1,3,5-tridecanoylbenzene (*plant*).

References : Asolkar *et al.* (1992), Sarma *et al.* (2002).

471. *Humulus lupulus* L.

Family: Urticaceae.

English Name: **Eng. :** Hops.

Distribution : Native of N. America; cultivated on commercial scale in hills of Himachal Pradesh.

Parts Used : Flower. Strobiles.

Flowering & Fruiting : July – September.

Folk Use : Strobiles used as sedative, hypnotic and tonic. Flowers also used as tonic.

Active Constituents : Reported to possess humulene, flavonoids, polyphenolic tannins, oestrogenic substances and asparagin (*strobiles*).

References : Anonymous (1959), Bhattacharjee (2001), Harborne & Baxter (2001), Kala (2003), Kirtikar & Basu (1984), Prajapati *et al.* (2006), Sharma (2003), Singh *et al.* (1996).

472. *Hybanthus enneaspermus* (L.) F. Muell.

Syn.: Ionidium enneaspermum (L.) Vent.; *I. suffriticosum* (L.) R. & S.; *Viola enneaspermum* L.; *V. suffriticosa* L.

Family: Violaceae.

Hindi and Sanskrit Names: **Hindi :** Ratanpurus; **Sans. :** Amburuha, Charati, Padmacarini.

Distribution : Throughout the warmer parts of India from U.P. southwards to the Deccan Peninsula.

Parts Used : Whole Plant.

Flowering & Fruiting : October – December.

Folk Use : Herb used as a tonic.

Active Constituents : Aurantiamide-OAC, β-sitosterol and isoarborinol (*plant*).

References : Ambasta (1986), Anonymus (1959), Dhiman (2005), Parrotta (2001), Ranjan (2000).

473. *Hydnocarpus laurifolia* (Dennst.) Sleumer

Syn.: H. pentandra Oken; *H. wightiana* Bl.

Family: Flacourtiaceae.

English, Hindi and Sanskrit Names: **Eng.** : Chaulmugra, Marothi Tree; **Hindi** : Calmogara, Kava; **Sans.** : Garudaphala, Katukapitta, Kushthavaeri, Tuvarakah.

Distribution : Forests of the Western Ghats, up to 600m.

Parts Used : Seeds.

Flowering & Fruiting : January – May.

Folk Use : Seeds used as a tonic.

Active Constituents : Known to contain two new cyclopentenoid cyanhydrin glucoside-(1S, 4R) and (1R-4S)-[6-O (α-L-rhamnopyranosyl)-β-D-glucopyranosyl]-4-hydroxy-2-cyclopentane-1-carbonitrile (*seeds*).

References : Prajapati *et al.* (2006), Watt (1972).

474. *Hydrangea arborescens* L.

Family: Hydrangeaceae.

English Name: **Eng.** : Hydrangea.

Distribution : Himalayas, from Ravi eastwards to Assam.

Parts Used : Root.

Folk Use : Roots used as tonic.

References : Asolkar *et al.* (1992), Bhattacharjee (2001).

475. *Hydrastis canadensis* L.

Syn.: Warnera canadensis Mill.; *W. tinctoria* Rofin.

Family: Ranunculaceae.

English Names: **Eng.** : Golden Seal, Yellow Puccoon.

Distribution : Temperate Himalayas.

Parts Used : Root.

Folk Use : Tea prepared from roots valued as tonic.

Active Constituents : Roots rich in hydrastine, berberastine, berberine, canadine and reticuline alkaloids.

Reference : Bhattacharjee (2001).

476. *Hydrocotyle javanica* Thunb.

Syn.: H. polycephala Wight. & Arn.

Family: Apiaceae.

English and Sanskrit Names: **Eng.** : Java Pennywort; **Sans.** : Mahagotukala.

Distribution : Himalayas (Kashmir – Bhutan): 700-2,700m, Khasia mountains, 700-2,000m, mountains of Malabar and Sri Lanka; Myanmar, Malaya – Australia.

Parts Used : Whole Plant.

Flowering & Fruiting : March.

Folk Use : Plant used as tonic and stimulant.

Active Constituents : α-amyrin-OAC, lignoceric acid, stigmasterol and its glucoside (*plant*).

References : Ambasta (1986), Tiwari & Tiwari (1996), Lalramnghinglova (1996).

477. *Hydrocotyle sibthorpioides* Lam.

Syn.: H. rotundifolia Roxb.

Family: Apiaceae.

English, Hindi and Sanskrit Names: **Eng. :** Lawn Pennywort; **Hindi :** Mandukaparni; **Sans. :** Khulkhuri.

Distribution : N.W. Himalayas: 1,350-2,350m; Bengal Plains, Sikkim, Khasia hills, Malabar and Sri Lanka and extending throughout south-eastern Asia and Japan.

Parts Used : Whole Plant.

Flowering & Fruiting : May – August.

Folk Use : Juice of plant extracted by pounding with *Portulaca oleracea* in equal parts given to growing children for improving memory.

Active Constituents : Known to possess stigmasterol, quercetin-3 O-β-D-(6''-caffeoylgalactoside), quercetin, its 3-galactoside and isorhamnetin (*plant*).

Reference : Nath *et al.* (2007).

478. *Hygrophila auriculata* (Sch.) Heine.

Syn.: Asteracantha longifolia (L.) Nees.; *Barleria auriculata* Sch.; *Hygrophila augustifolia* auct.; *H. spinosa* T. Ander.; *Ruellia longifolia* Roxb.

Family: Acanthaceae.

English, Hindi and Sanskrit Names: **Eng. :** Asteracantha, Long-leaved Barleria; **Hindi :** Gokshura, Juliakanta, Kamtakaliya, Talamkhana, Untkataru; **Sans. :** Atichhatra, Ikshugandha, Iksurah, Kakilakshya, Pichhila, Vajra.

Distribution : Moist places throughout India, chiefly in Darbhanga (Bihar).

Parts Used : Whole Plant. Leaf. Root. Seeds.

Flowering & Fruiting : March – May.

Folk Use : Plant, roots and seeds used as tonic. Leaves used as hypnotic.

Active Constituents : Lupeol, betulin and stigmasterol (*plant*).

References : Asolkar *et al.* (1992), Dhiman (2005), Kirtikar & Basu (1984), Kshirsagar & Singh (2007), Parrotta (2001), Prajapati *et al.* (2006), Siwakoti & Varma (1996), Srivastava *et al.* (2003a), Subramaniam (2000), Watt (1972).

479. *Hygroryza aristata* (Retz.) Nees *ex* Wight. & Arn.

Syn.: Leersia aristata (Retz.) Roxb.; *Pharus aristatus* Retz.; *Zizania aristata* (Retz.) Arn. Kunth.

Family: Poaceae.

English, Hindi and Sanskrit Names: **Eng.** : Bengal Wild Rice; **Hindi** : Jungali-dal, Tili, Tini; **Sans.** : Aranyadhanya, Aranyajali, Munidhanya, Nivarah, Prasadhik, Vanavrihi.

Distribution : In tanks, ponds and wet grounds throughout India.

Parts Used : Root.

Flowering & Fruiting : Flowering: August – September; Fruiting: October – November.

Folk Use : Roots used as tonic.

Reference : Prajapati *et al.* (2006).

480. *Hyoscyamus muticus* L.

Family: Solanaceae.

English Name: **Eng.** : Egyptian Henbane.

Distribution : West Punjab, Sind; Kabul, westwards to Egypt.

Parts Used : Whole Plant. Leaf.

Folk Use : Plant used as sedative. Leaves valued both as narcotic and sedative.

Active Constituents : Known to contain hyoscyamine and trophane alkaloids, hentriacontanyl, dotriacontanoate, tritriacontanoic acid, 15-hydroxyheneicosanoic acid, 3-hydroxytetratriacontan-30-one, tritriacontan-3-one and pentatriacontan-3-one (*leaf*), atropine or hyoscyamine and scopolamine (*plant*).

References : Ambasta (1986), Sharma (2003).

481. *Hyoscyamus niger* L.

Family: Solanaceae.

English, Hindi and Sanskrit Names: **Eng.** : Belene, Black Henbane, Chenile, Henbane, Hogsbean; **Hindi** : Khurasaniajvayan, Khurosaniyamani; **Sans.** : Dipya, Medaka, Parasikaya, Shyama, Turushka.

Distribution : Western Himalayas: Kashmir – Kumaon (1,500-3,000m); cultivated in U.P. and M.P.

Parts Used : Leaf. Flower. Seeds.

Flowering & Fruiting : September – March.

Folk Use : Leaves and flowering tops used as sedative. Seeds possess narcotic properties.

Active Constituents : Attributed properties due to hyoscyamine and scopolamine or hyosine, skimmianine, apohyoscine, tropine, α- and β-belladonine, cuscohygrine, apoatropine, choline, hypocyprin and daturamine (*leaf*), alkenes, linoleic, myristic, oleic, palmitic and steric acids (*seeds*).

References : Ambasta (1986), Anonymous (1959), Bhattacharjee (2001), Kapur *et al.* (1996), Kaul (1997), Kirtikar & Basu (1984), Parrotta (2001), Prajapati *et al.* (2006), Ranjan (2000), Sharma (2003), Thakur *et al.* (1989), Uniyal & Malhotra (1989), Watt (1972).

482. *Hypecoum leptocarpum* Hook. f. Thoms.

Family: Papaveraceae.

Distribution : Sandy soil in West Tibet: 3,000-4,000m, interior of Sikkim Himalayas : 4,000-4,650m.

Parts Used : Whole Plant.

Folk Use : Plant used as narcotic and sedative.

Active Constituents : Plant possesses protopine, sanguinorine, chelirubine, coptisine, allocrytopine and chelerythrine, cryptopine, oxyhydrastinine, procumbine, corydine, isocorydine, magnoflorine and (-) trans-N-methylstylopinium hydroxide.

References : Ambasta (1986), Anonymous (1959), Bhattacharjee (2001).

483. *Hypericum cernuum* Roxb.

Syn.: H. govanianum Wall.; *H. oblongifolium* Choisy.; *H. speciosum* Wall.

Family: Hypericaceae.

English Name: **Eng. :** St. John's Wort.

Distribution : W. Temperate Himalayas (Kumaon – Sirmour): 1,700-2,350m.

Parts Used : Seeds.

Flowering & Fruiting : April – October.

Folk Use : Seeds used as stimulant.

Reference : Kapur (1996a).

484. *Hypericum patulum* Thunb. (Pl. 7B)

Syn.: H. oblongifolium Wall.; *H. uralum* Buch.-Ham.

Family: Hypericaceae.

English Name: **Eng. :** St. John's Wort.

Distribution : Khasia hills: 1,700-2,000m, throughout the temperate Himalayas (except Sikkim), 1,000-2,350m, from Bhutan to the Ravi. China, Formosa and Japan.

Parts Used : Seeds.

Flowering & Fruiting : February – May.

Folk Use : Seeds used as stimulant.

References : Ambasta (1986), Anonymous (1959), Kirtikar & Basu (1984), Watt (1972).

485. *Hypericum perforatum* L.

Family: Hypericaceae.

English and Hindi Names: **Eng. :** Amber, Cammock, Devil's Scourge, Grace of God, Hundred Holes, St. John's Wort, Terrestrial Sun, Touch Heal; **Hindi :** Bassant, Dendhu.

Distribution : Shady waysides in forest clearings, grazing grounds, damp forests of western Himalayas: 1,000-2,600m.

Parts Used : Whole Plant.

Flowering & Fruiting : May – October.

Folk Use : Herb possesses sedative properties.

Active Constituents : Active properties due to pseudohypericin, quercitrin, quercetin, hyperoside, rutin, umbelliferone, scopoletin, lysine, ornithine, proline, threonine, glutamine, cysteine, leucine, mixture of alkanols composed of 1-tetracosanol, 1-hexacosanol, 1-octacosanol and 1-triacontanol, 3,8"-biapigenin (*plant*).

References : Harborne & Baxter (2001), Sharma (2003).

486. *Hypochoeris glabra* L.

Family: Asteraceae.

English Name: **Eng. :** Smooth Cat's Ear.

Distribution : Nilgiris, Pulney and Travancore hills: 1,350-2,700m; also recorded from Alka hills in Assam.

Parts Used : Root.

Folk Use : Roots used as tonic.

Active Constituents : Two quaianolides esterified with an unusual sesquiterpenic acid-lactucin-8-o-hypoglabrate (I) and 8α-hypoglabroyloxy-jacquinelin (II) have been isolated (*root*).

References : Ambasta (1986), Anonymous (1959).

487. *Hypoxis aurea* Lour.

Family: Amaryllidaceae.

English, Hindi and Sanskrit Names: **Eng. :** Golden Star Grass; **Hindi :** Bhui Khajur, Golkya.

Distribution : Found in hilly parts throughout India.

Parts Used : Root.

Flowering & Fruiting : May – July.

Folk Use : Roots used as tonic.

References : Ambasta (1986), Anonymous (1959), Kapur (1996a).

488. *Hyptis capitata* Jacq.

Syn.: H. rhomboidea Mart. & Gal.

Family: Lamiaceae.

English, Hindi and Sanskrit Names: **Eng. :** Buttonweed, False Ironwort, Knobweed.

Distribution : Found in Bengal.

Parts Used : Whole Plant.

Folk Use : Plant used as tonic and stimulant.

Active Constituents : Properties mainly due to ferruginol, sugiol, 6-dehydroroyleanone, kaempferol, its 3-O-rutinoside, quercetin-3-O-rutinoside, β-sitosterol, its glucoside, stigmasterol, its glucoside and betulic, ursolic, hyptadienic, 2α-hydroxyursolic and 2α,3α,24-trihydroxyolean-12-en-28oic acids (*plant*)

References : Ambasta (1986), Anonymous (1959).

489. *Hyptis suaveolens* (L.) Poit.

Syn.: Ballota suaveolens L.; *Bysteropogo graveolens* Blume.

Family: Lamiaceae.

English, Hindi and Sanskrit Names: **Eng. :** Hyptis; **Hindi :** Nili Tulsi, Wilayati Tulsi; **Sans. :** Bhustrna.

Distribution : Native of America, naturalized in northern India, Deccan Peninsula and Andaman.

Parts Used : Whole Plant.

Flowering & Fruiting : April – December.

Folk Use : Plant used as stimulant.

Active Constituents : Isolated constituents are essential oil containing menthol (*plant*), campesterol and fucosterol (*leaf, flower*), β-sitosterol, oleanolic and a-peltoboykinolic acids (*root*).

References : Ambasta (1986), Anonymous (1959), Kaushik & Dhiman (2000), Nayak *et al.* (2003), Parrotta (2001).

490. *Hyssopus officinalis* L.

Family: Lamiaceae.

English and Hindi Names: **Eng.** : Hyssop; **Hindi** : Zufah Yabis, Zupha.

Distribution : Wild in western Himalayas: Kashmir – Kumaon (2,438-3,352m).

Parts Used : Whole Plant. Leaf.

Flowering & Fruiting : August – October.

Folk Use : Plant and leaves used as stimulant.

Active Constituents : Plant possesses pinocamphene, a- and b-pinenes, camphene, a-terpinene and pinocampheol.

References : Ambasta (1986), Anonymous (1959), Bhattacharjee (2001), Dhiman (2005), Kirtikar & Basu (1984), Rajan *et al.* (2003), Sharma (2003).

491. *Ichnocarpus frutescens* R. Br.

Syn.: Apocynum frutescens L.; *Echites frutescens* Roxb.; *Ichnocarpus dascalyx* Miq.; *I. radicans* Wall.

Family: Apocynaceae.

English, Hindi and Sanskrit Names: **Eng.** : Black Creeper; **Hindi** : Dhudhi, Kalidudhi, Siamalata; **Sans.** : Palindi, Paravalli, Saria, Syamalata, Ulpalasariba.

Distribution : Plains and lower hills up to 1,200m throughout India.

Parts Used : Root.

Flowering & Fruiting : Flowering: August – December; Fruiting: February – April.

Folk Use : Roots used as a tonic.

Active Constituents : 2 – hydroxyl – 4 – methoxybenzaldehyde (*root*).

References : Ambasta (1986), Anonymous (1959), Farooq (2005), Kaushik & Dhiman (2000), Kirtikar & Basu (1984), Kumar & Lal (2003), Parrotta (2001), Prajapati *et al.* (1986), Ranjan (2003), Sahu (2004), Siwakoti & Varma (1996), Watt (1972).

492. *Ilex paraguariensis* A. St. Hil.

Syn.: I. domestica Reiss.; *I. mate* St.-Hil.; *I. paraguaiensis* Lamb.

Family: Aquifoliaceae.

English Names: **Eng.** : Mate, Paraguay Tea, Yerba Mate.

Distribution : Wild in Northern Argentina, Paraguay, Uruguay and Southern Brazil, cultivated in Argentina, Spain and Portugal; cultivated in some gardens in Northern India.

Parts Used : Leaves.

Folk Use : Leaves stimulate the nervous system.

Active Constituents : Caffeine, theobromine, theophylline and tannins, neochlorogenic acid, 5 – O – caffeoylquinic acid, matesaponin I, 3

– caffeoylquinic acid and 3, 5 – dicaffeoylquinic acid, caffeine and theobromine (*leaf*).

References : Anonymous (1959), Prajapati *et al.* (2006).

493. *Illicium verum* Hook. f.

Family: Magnoliaceae.

English, Hindi and Sanskrit Names: **Eng. :** Chinese Anise, True Star-anise; **Hindi :** Anasphal, Badayan; **Sans. :** Takkolakam.

Distribution : Cultivated in mountainous regions especially in Lang Son province; does not occur in India, but fruits imported from China.

Parts Used : Fruit. Oil.

Flowering & Fruiting : Flowering: June; Fruiting: October – November.

Folk Use : Fruits and oil used as stimulant.

Active Constituents : Active properties due to kaempferol and its 3 – rutinoside and 3 – galactose, quercetin and its 3 – rhamnoside, 3 – rutinoside and 3 – xyloside (*fruit*), myristic, palmitic, stearic, oleic and linoleic acids (*oil*).

References : Ambasta (1986), Bhattacharjee (2001), Rajan *et al.* (2003).

494. *Impatiens balsamina* L.

Family: Balsaminaceae.

English, Hindi and Sanskrit Names: **Eng. :** Balsam, Jewel Weed, The Garden Balsam, Touch-me-not; **Hindi :** Gulmendhi; **Sans. :** Dushpatrijati.

Distribution : Cultivated in gardens throughout the tropical and sub-tropical parts of India.

Parts Used : Flower.

Flowering & Fruiting : August – November.

Folk Use : Flowers used as tonic.

Active Constituents : Cyanidin, leucocyanidin, myricetin, quercetin, kaempferol, 2 – methoxy – 1, 4 – napththaquinone (*flower*).

References : Farooq (2005), Kirtikar & Basu (1984), Ranjan (2003).

495. *Imperata cylindrica* Beauv.

Syn.: Imperata arundinacea Cyr.; *Lagarus cylindrica* L.

Family: Poaceae.

English and Sanskrit Names: **Eng. :** Thatch Grass; **Sans. :** Yagnika.

Distribution : Hotter parts of India from the Punjab southwards and eastwards to Malacca and Sri Lanka and also in all warm countries.

Parts Used : Underground stem.

Flowering & Fruiting : April – May.

Folk Use : Underground stem used as tonic.

References : Gogoi & Das (2003), Saikia & Nath (2003b).

496. *Indigofera articulata* Gouan

Syn.: I. argentea L. var. *caerulea* Baker; *I. caerulea* Roxb.

Family: Fabaceae.

English, Hindi and Sanskrit Names: **Eng. :** Arabian Indigo, Egyptian Indigo, Surat Indigo, Wild Indigo; **Hindi :** Surmainil; **Sans. :** Kalaklitaka.

Distribution : W. India, parts of Bihar, Sindh, Deccan; Arabia, Egypt, Abyssinia.

Parts Used : Leaf. Root.

Flowering & Fruiting : June – January.

Folk Use : Roots and leaves used as bitter tonic.

References : Ambasta (1986), Anonymous (1959), Kirtikar & Basu (1984).

497. *Indigofera gerardiana* Wall. ex Baker

Syn.: I. dosua Haneilt.; *I. hetarantha* Wall. *ex* Brandis.

Family: Fabaceae.

English Name: **Eng. :** Himalayan Indigo.

Distribution : Temperate and subtropical Western Himalayas (salt range – Kumaon): 700-3,500m, Afghanistan.

Parts Used : Flower.

Flowering & Fruiting : May – August.

Folk Use : Flowers boiled in milk for 4-5 minutes consumed as a tonic drink.

Reference : Kumar & Rao (2001).

498. *Indigofera glabra* L.

Syn.: I. pentaphylla L.

Family: Fabaceae.

English Names: **Eng. :** Five Leaved Indigo, Smooth Indigo.

Distribution : All over India, tropical Africa.

Parts Used : Leaf.

Flowering & Fruiting : Flowering: August – September; Fruiting: January – March.

Folk Use : Leaves used as a bitter tonic.

References : Ambasta (1986), Kirtikar & Basu (1984).

499. *Indigofera glandulosa* Roxb. ex Willd.

Family: Fabaceae.

English Name: **Eng. :** Befri.

Distribution : Common weed on dry and sandy soils in U.P., Bihar, Gujarat and the plains of Central India southwards to northern Karnataka and A.P.

Parts Used : Seeds.

Flowering & Fruiting : September – February.

Folk Use : Seeds used as tonic.

References : Bhalla & Bhalla (2002), Deshmukh *et al.* (2000), Kirtikar & Basu (1984), Parrotta (2001), Watt (1972).

500. *Indigofera tinctoria* L.

Syn.: I. sumatrana Gaertn.

Family: Fabaceae.

English, Hindi and Sanskrit Names: **Eng. :** Common Indigo, Indian Indigo; **Hindi :** Gouli, Lil, Nil; **Sans. :** Aklika, Dronika, Gramya, Nilini, Rangapatri, Vijaya.

Distribution : Throughout India, mainly as an escape from cultivation.

Parts Used : Leaf. Root. Stem.

Flowering & Fruiting : August – September (flowering); October – May (fruiting).

Folk Use : Root, stem and leaves used as a tonic.

Active Constituents : Known to possess apigenin, kaempferol, luteolin and quercetin (*root*), apigenin, kaempferol, luteolin and quercetin (*root, stem, leaf*).

References : Deshmukh *et al.* (2000), Prajapati *et al.* (2006).

501. *Indigofera trita* L. f.

Family: Fabaceae.

English Names: **Eng. :** Asian Indigo, Three Leaved Indigo.

Distribution : Throughout India, Sri Lanka; Tropical Africa, Malay Islands, N. Australia.

Parts Used : Whole plant. Seeds.

Flowering & Fruiting : May – December.

Folk Use : Plant and seeds used as a nutritive tonic.

References : Ambasta (1986), Bhalla & Bhalla (2002), Kirtikar & Basu (1984).

502. *Inula racemosa* Hook. f.

Family: Asteraceae.

*Hindi and Sanskrit Names:***Hindi :** Pahakarmul, Pusharmul; **Sans. :** Pauskaram, Pushkaram.

Distribution : Temperate and alpine W. Himalayas: 1,700-4,650m.

Parts Used : Root.

Flowering & Fruiting : July – August.

Folk Use : Roots used as tonic and stimulant.

Active Constituents : Alantolactone, isoalantolactone, dihydroisoalantolactone (*root*).

References : Kapur & Singh (1996), Kirtikar & Basu (1984), Watt (1972).

503. *Iochroma fuchsioides* (HBK) Meirs, Hooker

Family: Solanaceae.

English Name: **Eng. :** Red Iochroma.

Distribution : Sub-tropical and mild temperate regions of India; Columbia – Argentina, Peru and Ecuador.

Parts Used : Bark. Leaf.

Flowering & Fruiting : Throughout the year.

Folk Use : Tea prepared from bark and leaves hallucinogenic.

Reference : Schultes (1993).

504. *Ipomoea batatas* (L.) Lam.

Syn.: Batatas edulis Chois.; *Convolvulus batatas* L.; *C. edulis* Thunb.; *C. esculentus* Salisb.

Family: Convolvulaceae.

English, Hindi and Sanskrit Names: **Eng. :** Pichhila, Pindula, Sweet Potato; **Hindi :** Mitha Alu, Sakkarkand; **Sans. :** Kandagranthi, Pindaluh, Raktaluh.

Distribution : Throughout India, cultivated.

Parts Used : Root.

Flowering & Fruiting : June – March.

Folk Use : Roots used as a tonic.

Active Constituents : Amylase, protease, invertase, catalase, laccase, arabinase, galactonase, polygalacturonase, peroxidase, monophenolase, catecholase, cytochrome c – oxidase, phosphorylase and phosphatase have been isolated from *root*.

Reference : Prajapati *et al.* (2006).

505. *Ipomoea carnea* ssp. ***fistulosa*** (Mart. *ex* Choisy) Austin

Syn.: Batatas crassicaulis Benth.; *Ipomoea crassicaulis* (Benth.) Robinson; *I. fistulosa* Mart. *ex* Choisy.

Family: Convolvulaceae.

Hindi Name: **Hindi :** Hakim.

Distribution : Native of tropical America, naturalized as a hedge plant throughout India.

Parts Used : Aerial Parts.

Flowering & Fruiting : Flowering and fruiting almost throughout the year.

Folk Use : Extract of aerial plant parts considered CNS depressant.

Active Constituents : Known to contain tricontane, saponin, ipamotocin, n- tritriacontane agroclavine and α – dihydrolysergol (*leaf*).

References : Dhiman (2005), Kaushik & Dhiman (2000).

506. *Ipomoea digitata* auct. non Linn.

Syn.: Batatas paniculata Chois.; *Convolvulus paniculata* L., *Ipomoea gossypifolia* Willd.; *I. mauritiana* Jacq.; *I. paniculata* (L.) R. Br., non Burm.

Family: Convolvulaceae.

English, Hindi and Sanskrit Names: **Eng. :** Giant Potato; **Hindi :** Bidar Kand, Bilaikand, Bhuyikohada; **Sans. :** Bhumikushmanda, Kasiravidari, Vidari.

Distribution : Moist places throughout tropical parts in India.

Parts Used : Tuberous Roots.

Flowering & Fruiting : July – December.

Folk Use : Roots used as a tonic and stimulant.

Active Constituents : β – sitosterol and other sterols isolated (*tuberous root).*

References : Ambasta (1986), Anonymous (1959), Dhiman (2005), Dwivedi (2007), Kirtikar & Basu (1984), Prajapati *et al.* (2006), Sharma (2004), Watt (1972).

507. *Ipomoea pes – caprae* (L.) Sweet

Syn.: Batatas maritima Bojer.; *Convolvulus bilobatus* Roxb.; *C. pes-carpae* L.; *Ipomoea biloba* Forsk.; *I. maritima* R. Br.

Family: Convolvulaceae.

English, Hindi and Sanskrit Names: **Eng.** : Bayhops, Beach Morning Glory, Goat's Foot Creeper, Railroad Vine; **Hindi** : Dopatilata; **Sans.** : Maryadavalli, Sagaramekhala.

Distribution : Throughout India, especially along sandy shores.

Parts Used : Whole Plant. Leaf.

Flowering & Fruiting : October – March.

Folk Use : Plant and leaves used as a tonic.

Active Constituents : These are pentatriacontane, triacontane, a sterol, behenic acid, melissic acid, butyric acid and myristic acid, quercetin–3–glucoside, -3– galactoside and – 3 – (6" – O – acetyl) – β – D – glucopyranoside (*plant*), malic, citric, maleic, tartaric, succinic, fumaric acids, β- damscenoneand – phytol (*leaf*).

References : Ambasta (1986), Anonymous (1959), Nayak *et al.* (2003), Prajapati *et al.* (2006).

508. ***Ipomoea sepiaria*** Roxb.

Syn.: Convolvulus marginatus Lamk.; *C. maximus* Lamk.; *C. sepiarius* Wall.; *Ipomoea heynii* Wall.; *I. maxima* auct. non (L. f.) Sweet.; *I. striata* Roth.

Family: Convolvulaceae.

English, Hindi and Sanskrit Names: **Eng.** : Purple Heart Glory; **Hindi** : Bankalmi; **Sans.** : Laksmana, Manjika, Nagini, Tulini.

Distribution : Hedges near streams and tanks up to 150m throughout India.

Parts Used : Whole Plant.

Flowering & Fruiting : Throughout the year.

Folk Use : Plant used as a tonic.

References : Bhattacharjee (2001), Kirtikar & Basu (1984), Prajapati *et al.* (2006).

509. ***Ipomoea violacea*** L.

Syn.: I. rubrocaerulea Hook.; *I. tricolor* Cav.

Family: Convolvulaceae.

English Names: **Eng.** : Beach Moonflower, Heavenly Blues, Morning glory, Sea Moonflower.

Distribution : Meghalaya, U.P., Kerala, Tamil Nadu; also grown exclusively for showy flowers in Indian gardens.

Parts Used : Seed.

Flowering & Fruiting : Flowering during July – October.

Folk Use : Seeds hallucinogenic.

Active Constituents : Ergoline alkaloids (*seeds*).

Reference : Schultes (1993).

510. *Ixora coccinea* L.

Syn.: I. bandhuca Roxb.; *I. grandifolia* Br.; *I. incarnata* DC.; *I. obovata* Hegne.; *Pavetta bandhuca* Miq.; *P. coccinea* Blume.; *P. incarnata* Blume.

Family: Rubiaceae.

English, Hindi and Sanskrit Names: **Eng. :** Flame of the Woods, Jungle Flame Ixora, Sacred Ixora; **Hindi :** Rajana, Rangan, Rookmini; **Sans. :** Bandhujivaka, Bandhuka, Paranti, Raktata.

Distribution : Throughout the West coast, in forest lands and road cutting; also cultivated as ornamental plant.

Parts Used : Root.

Flowering & Fruiting : Rainy season.

Folk Use : Roots used as sedative.

Active Constituents : Methyl ester of palmitic, stearic, oleic and linoleic acids (*root oil*).

References : Ambasta (1986), Parrotta (2001), Prajapati *et al.* (2006), Sharma (2004).

511. *Ixora parviflora* Vahl.

Syn.: I. arborea Roxb.; *I. decipieus* DC.; *I. pavetta* Andr.

Family: Rubiaceae.

English, Hindi and Sanskrit Names: **Eng. :** Torch Tree, Torch-wood Ixora; **Hindi :** Kotagandhal, Makrichijhar, Nevari; **Sans. :** Iswara, Nevali.

Distribution : Upper Gangetic plains, Bengal, Chittagong, Myanmar, Nicobars, Bombay, in almost all districts Chennai, Sri Lanka.

Parts Used : Bark.

Flowering & Fruiting : March.

Folk Use : Decoction of bark used as a tonic.

Reference : Ranjan (2003).

512. *Jasminum arborescens* Roxb.

Syn.: J. glabellum Wall.; *J. lotifolium* Roxb.; *J. punctatum* Wall.; *J. roxburghianum* Wall.

Family: Oleaceae.

English, Hindi and Sanskrit Names: **Eng. :** Tree Jasmine, Wild Jasmine; **Hindi :** Bela, Chameli, Navamallika, Saptala; **Sans. :** Navamallica, Saptola.

Distribution : Throughout India, in forests up to 1,200m.

Parts Used : Leaf.

Flowering & Fruiting : April – May.

Folk Use : Leaves used as a tonic.

References : Ambasta (1986), Anonymous (1959), Farooq (2005), Kirtikar & Basu (1984), Prajapati *et al.* (2006), Watt. (1972).

513. *Jasminum auriculatum* Vahl.

Family: Oleaceae.

English, Hindi and Sanskrit Names: **Eng. :** Needle Flower Jasmine; **Hindi :** Chameli, Mogra, Juhi, Juyi; **Sans. :** Ganika, Jai, Magadhi, Sucimallika, Yuthika.

Distribution : Throughout South India, on dry slopes of Western Ghats.

Parts Used : Flower.

Flowering & Fruiting : Flowering throughout the year.

Folk Use : Flowers used as cardio tonic.

References : Prajapati *et al.* (2006), Reddi *et al.* (2005).

514. *Jasminum grandiflorum* L.

Syn.: J. aureum Don.; *J. officinale* (L.) forma *grandiflorum* (L.) Kobuski;

Family: Oleaceae.

English, Hindi and Sanskrit Names: **Eng. :** Catalonion Jasmine, Common Jasmine, Spanish Jasmine; **Hindi :** Chameli, Jati, Juhi; **Sans. :** Bahugandha, Chetaki, Janesha, Jati, Malati, Mallika, Manodna, Punyagandha, Sukemari, Surupa, Veshika.

Distribution : Throughout India, in forests up to 2,500m; also extensively cultivated.

Parts Used : Whole Plant.

Flowering & Fruiting : April – September.

Folk Use : Plant used as a tonic.

Active Constituents : Linolyl acetate, benzyl benzoate, geraniol, neral di – linolool, β, γ – hexenol, farnesol, nerolidol, eugenol, p – cresol, cresol, benzaldehyde, jasmine, benzoic acid, methyl anthranilate and indole (*plant*).

References : Prajapati *et al.* (2006), Reddi *et al.* (2005).

515. *Jasminum multiflorum* (Burm. f.) Andr. (Pl. 7C)*

Syn.: J. bracteatum Willd.; *J. congestum* Wall.; *J. hirsutum* Willd.; *J. pubescens* Willd.; *Mogorium pubescens* Lam.; *Nyctanthe multiflora* Burm.f.; *N. pubescenes* Retz.

Family: Oleaceae.

English, Hindi and Sanskrit Names: **Eng. :** Common Jasmine, Downy Jasmine, Musk Jasmine; **Hindi :** Chameli, Kundo,

Kundphul; **Sans.** : Atimukha, Damana, Kundah, Mahamoda, Makaranda, Monohara, Monomoda, Muktapushpa, Tarapushpa, Vanahasa.

Distribution : Forests of Western Ghats and Sub-Himalayan tracts up to 1,500m.

Parts Used : Flower.

Flowering & Fruiting : November – April.

Folk Use : Flowers used as cardio tonic.

References: Parrotta (2001), Prajapati *et al.* (2006), Reddi *et al.* (2005).

516. *Jateorhiza palmata* (Lam.) Miers

Syn.: Cocculus palmatus DC.; *Jateorhiza columba* Miers.

Family: Menispermaceae.

English and Hindi Names: **Eng.** : Calumba; **Hindi** : Kalam-ki-jar.

Distribution : Native to rainforests of east Africa, cultivated in some places, roots imported into India from Africa.

Parts Used : Root.

Flowering & Fruiting : Flowering: spring season & fruiting during January – April.

Folk Use : Roots used as a source of a bitter tonic.

Active Constituents : Palmatine, jatrorrhizine and columbamine, berberine, coptisine, protoberberine have been isolated (*root*).

References : Ambasta (1986), Harborne & Baxter (2001), Prajapati *et al.* (2006).

517. *Jatropha curcas* L.*

Syn.: J. moluccana Herb.

Family: Euphorbiaceae.

English, Hindi and Sanskrit Names: **Eng.** : Physic Nut, Poison Nut, Purging Nut; **Hindi :** Chandarjot, Jungli Errand, Pahari Errand, Ratanjot, Safed Arand; **Sans.** : Akhuparnika, Dravanti, Kananaerand, Nayagrodhi, Paravatay eranda.

Distribution : Plains and in hedges throughout India.

Parts Used : Aerial Parts. Seeds.

Flowering & Fruiting : February – April.

Folk Use : Seeds used as a digestive tonic. Aerial parts used as CNS depressant.

Active Constituents : Oleic, linoleic, myristice, palmitic, stearic, arachidic acids and sitosterol, glucose, arabinose, xylose and rhamnose (*seeds*), vitexin, isovitexin, campesterol, β – sitosterol - 3β, 7β – diol and 1 – triacontanal (*leaf*).

References : Prajapati *et al.* (2006), Singh & Srivastava (2000).

518. *Jatropha gossypifolia* L.

Family: Euphorbiaceae.

English, Hindi and Sanskrit Names: **Eng. :** Bellyache Bush, Tua-tua; **Hindi :** Bherenda, Verenda; **Sans. :** Raktaeramada.

Distribution : Native of Brazil; naturalized in many parts of India.

Parts Used : Whole Plant.

Flowering & Fruiting : February – May.

Folk Use : Plant used as CNS depressant.

Active Constituents : Properties accredited to vitexin, isovitexin and apigenin (*leaf*), jatrophone, 2α – hydroxyjatrophone, 2β - hydroxyjatrophone, and 2β – hydroxyl – 5, 6 – isojatrophone (*root*), β – amyrin, β – sitosterol, threonine, proline, glycine, alanine, valine, isoleucine, leucine, arginine, glutamic acid, xylose, arabinose, rhamnose and myristic, palmitic, stearic, aspartic, oleic and linoleic acids (*seeds*).

References : Reddi *et al.* (2005), Singh & Srivastava (2000).

519. *Juglans cinerea* L.

Family: Juglandaceae.

English Names: **Eng. :** Butternut, White Walnut.

Distribution : Native of North America forests; cultivated for its timber in temperate regions of India.

Parts Used : Inner Bark.

Flowering & Fruiting : April – May.

Folk Use : Inner bark used as a tonic.

Active Constituents : Napthoquinones, a fixed and a volatile oil and tannins (*bark*).

Reference : Prajapati *et al.* (2006).

520. *Juglans regia* L.*

Syn.: J. arguta Wall.; *J. regia* L. var. *kumaonica* DC.

Family: Juglandaceae.

English, Hindi and Sanskrit Names: **Eng. :** Common Walnut, European Walnut, Persian Walnut; **Hindi :** Akhor, Akhrot, Khor; **Sans. :** Akhoda, Akschoda, Aksotah.

Distribution : Temperate Himalayas (1,000-3,350m), wild and cultivated; Khasia hills, cultivated; Ava Hills, Baluchistan, N. Persia, Caucasus and Armenia.

Parts Used : Leaf. Fruit.

Flowering & Fruiting : April – October.

Folk Use : Leaves, oil cakes and fruits given as a general tonic. Also, fruits valued as tonic for improving memory.

Active Constituents : Reported to possess caffeic, cholrogenic, p – coumaric, ferulic, gallic, gentisic, p – hydroxybenzoic, p – hydroxyphenylacetic, protocatechuic, sinapic, salicylic, vannilic and syringic acids, hydrocarbons, esters, aldehydes and fatty acids (*leaf*), primary alcohol, hydrocarbons, esters, aldehydes and fatty acids (*fruit*).

References : Ambasta (1986), Kapur & Singh (1996), Kirtikar & Basu (1984), Pande & Joshi (2001), Prajapati *et al.* (2006), Reddi *et al.* (2005), Viswanathan & Singh (1996), Watt (1972).

521. *Juniperus communis* L.

Syn.: J. argaea Bal. *ex* Parl.; *J. borealis* Salisb.; *J. combrassa* Hort. *ex* Car.

Family: Cupressaceae.

English, Hindi and Sanskrit Names: **Eng. :** Common Juniper; **Hindi :** Aaraar, Abhal, Haubera; **Sans. :** Aparajita, Habusha, Hapush, Kanchhugha, Plihahantri, Vapusha, Visra.

Distribution : Known to occur wild in Western Himalayas from Kumaon westwards between 3,810 – 4,267m.

Parts Used : Fruit.

Flowering & Fruiting : August – September (cone formation); and ripening in second year.

Folk Use : Fruits used as stimulant and tonic.

Active Constituents : Found to contain junionone, Desoxypodophyllotoxin, α – pinene, Myrcene, cineole, p – cymene, α – and γ – cardiene, limonene, β – pinene, y – terpinene, terpinene – 4 – ol, α – thujene, borneol, caryophyllene, epoxydihydrocaryophyllene & β – elemem – 7 – α – ol (*fruit*).

References : Ambasta (1986), Bhattacharjee (2001), Dhiman (2005), Kapur *et al.* (1996), Kirn *et al.* (2000), Kirtikar & Basu (1984), Sharma & Sood (1997), Watt (1972).

522. *Juniperus marcopoda* Boiss.

Syn.: J. excelsa Brand.; *J. gossaainthaneana* Loddig.; *J. polycarpos* K. Koch; *Sabina seravschanica* (Kom.) Nevski.

Family: Cupressaceae.

English and Hindi Names: **Eng. :** Indian Juniper, The Himalayan Pencil Cedar; **Hindi :** Dhup, Dhum Padam.

Distribution : Himalayas westwards up to 4,267m.

Parts Used : Fruit.

Flowering & Fruiting : September – March.

Folk Use : Fruits considered stimulant.

Active Constituent : Hypolactin-7-glucoside (*fruit*).

Reference : Dhiman (2005).

523. *Jurinea dolomiaea* Boiss.

Syn.: Dolomaea marcocephalo DC.; *Jurinea marcocephalo* (Royle) C. B. Clarke.

Family: Asteraceae.

Hindi Name: **Hindi :** Dhoop, Jari-dhoop.

Distribution : Himalayas: Kashmir – Kumaon, (3,650-4,650m).

Parts Used : Root.

Flowering & Fruiting : September – October.

Folk Use : Roots used as stimulant.

Reference : Kaul (1997).

524. *Justicia adhatoda* L. (Pl. 7D)*

Syn.: Adhatoda vasica Nees.; *A. zeylanica* Med.

Family: Acanthaceae.

English, Hindi and Sanskrit Names: **Eng. :** Malabar Nut; **Hindi :** Adusa, Arusha, Banasaa; **Sans.:** Bhisangmata, Raktappittaghni, Vasaka.

Distribution : Throughout the Sub-Himalayan tracts up to 1,219m.

Parts Used : Leaf. Root. Flower.

Flowering & Fruiting : December – April (flowering); February – May (fruiting).

Folk Use : Flowers, leaves and roots used as tonic.

Active Constituents : Characteristic properties due to vascicine, vasinone, vascicone, vaccine, β- sitosterol, tritriacontane, vascicine, vascicinol, 4 new quinazoline alkaloids – vascicoline, adhatodine, vasciolinone, anistone (*leaf*), and vasicinone (*flower*).

References : Parrotta (2001), Rao (1996), Sharma (2003).

525. *Justicia beddomei* (Clarke) Bennet

Syn.: Adhatoda beddomei C. B. Clarke.

Family: Acanthaceae.

English, Hindi and Sanskrit Names: **Eng. :** Malabar Nut; **Hindi :** Adusa, Arusa, Arush, Bansa; **Sans. :** Adaroosha, Basak, Bhishangmata, Shwetavasa, Simhika, Simhasya, Sinhaparni, Vajidanta, Vasa, Vishnu, Vrisha.

Distribution : Southern parts of India.

Parts Used : Whole Plant.

Flowering & Fruiting : November – March.

Folk Use : Plant used as a tonic.

Reference : Prajapati *et al.* (2006).

526. *Kaempferia galanga* L.

Syn.: Alpinia sessilis Kon.

Family: Zingiberaceae.

English, Hindi and Sanskrit Names: **Eng. :** Black Thorn, East India Galingale; **Hindi :** Chandramula, Kacora, Sugandhavacha; **Sans. :** Chandramulika, Sugandhavacha.

Distribution : Hotter parts of India; much cultivated in gardens.

Parts Used : Whole Plant. Rhizome.

Flowering & Fruiting : Flowering during August – September.

Folk Use : Plant used as hallucinogen. Rhizomes used as a stimulant.

Active Constituents : Found to possess p – methoxystrene, p – coumaric acid, ethyl p – methoxy – trans – cinnamate (*rhizome*).

References : Ambasta (1986), Anonymous (1959), Asolkar *et al.* (1992), Behura & Rout (2003), Rahman & Yusuf (1996), Sharma (2003, 2004).

527. *Kalanchoe integra* (Medic.) Kuntze

Syn.: K. acutiflora Kurz.; *K. brasiliensis* Cambess.; *K. nudicaulis* Ham.; *K. spathulata* DC.; *K. varians* Haw.

Family: Crassulaceae.

English and Hindi Names: **Eng. :** Never-die; **Hindi :** Haiza, Rungru, Tatora.

Distribution : Tropical Himalayas : Kashmir – Bhutan up to 2,000m.

Parts Used : Leaf.

Flowering & Fruiting : September – November.

Folk Use : Juice of leaves used as a tonic.

Active Constituents : Glycoside-patuletin-3,7-di-O-rhamnoside (*leaf*).

References : Ambasta (1986), Anonymous (1959), Farooq (2005).

528. *Krameria argentea* Mart.

Family: Krameriaceae.

English Names: **Eng. :** Brazilian Rhatany, Para Rhatany.

Distribution : Found across America; native to the tropical regions.

Parts Used : Root.

Flowering & Fruiting : October – December.

Folk Use : Dried roots used as tonic.

References : Ambasta (1986), Asolkar *et al.* (1992).

529. ***Krameria triandra*** Ruiz & Pav.

Family: Krameriaceae.

English Name: **Eng. :** Rhatany.

Distribution : Europe, Peru and Bolivia on Western slopes of the Andes at 900-3,000m.

Parts Used : Whole Plant.

Flowering & Fruiting : Throughout the year.

Folk Use : Plant used as a tonic.

References : Ambasta (1986), Watt (1972).

530. ***Lactuca indica*** L.

Syn.: Lactuca brevirestris Champ.

Family: Asteraceae.

English Name: **Eng. :** Indian Lettuce.

Distribution : Wild in wasteplaces, Sikkim Himalayas, Assam and Khasia hills, and extending further into Japan, China and Philippines.

Parts Used : Leaf.

Flowering & Fruiting : Throughout the year.

Folk Use : Leaves used as tonic.

Reference : Ambasta (1986).

531. ***Lactuca serriola*** L.

Syn.: Lactuca scariola L.

Family: Asteraceae.

English and Hindi Names: **Eng. :** Garden Lettuce, Prickly Lettuce; **Hindi :** Kahu, Khas, Salad.

Distribution : Wild in Western Himalayas up to 3,000m; sometimes cultivated as a pot herb.

Parts Used : Whole Plant. Leaf. Seeds.

Flowering & Fruiting : April – October.

Folk Use : Seeds used as hypnotic. Leaves, juice of plant and seeds considered as sedative and nerve tonic.

Active Constituents : Linoleic acid, oleic acid and saturated acids (*seeds*).

References : Ambasta (1986), Bhattacharjee (2001), Kirtikar & Basu (1984), Thakur *et al.* (1989).

532. ***Lagenaria siceraria*** (Mol.) Standley

Syn.: Cucurbita siceraria Molina; *Lagenaria leucantha* (Duchesne) Rusby; *L. vulgaris* Ser.

Family: Cucurbitaceae.

English, Hindi and Sanskrit Names: **Eng.** : Bitter Bottle Gourd, Bitter Calbash Gourd, Bottle-gourd, White Pumpkin; **Hindi** : Alkaddu, Golkaddu, Kadvilauki, Lau, Laue, Lauka, Titalauki, Tumri; **Sans.** : Alabu, Katutumbi, Tiktalabu.

Distribution : Indigenous to India; occurs throughout India, wild or cultivated.

Parts Used : Seeds.

Flowering & Fruiting : Flowering and Fruiting throughout the year but mostly during winter.

Folk Use : Seeds used as a brain tonic and nerve stimulant.

Active Constituents : Attributed properties due to rhamnose, fructose, glucose, galactose, sucrose, raffinose, stachyose, avenasterol, codisterol, clerosterol, isofucosterol, stigmasterol, campesterol, sitosterol, spinasterol, 25(27)-dehydroporiferasterol, 25(27)-dehydrofungisterol, 25(27)-dehydrochondrillasterol, 24β-ethyl-25(27)-dehydrolathosterol, 24ζ-methyllathasterol and 22-dihydrospinasterol (*seeds*).

References : Kaushik & Dhiman (2000), Prajapati *et al.* (2006), Trivedi (2003).

533. *Lagerstroemia indica* L.

Syn.: L. elegans Wall.; *Velaga globosa* Gaertn.

Family: Lythraceae.

English and Hindi Names: **Eng.** : China Privet, Common Crape Myrtle, Indian Lilac; **Hindi** : Pharash, Saoni, Telingachina.

Distribution : Cultivated throughout India in gardens for its beautiful flowers.

Parts Used : Bark. Seeds.

Flowering & Fruiting : Flowering and Fruiting during rainy season.

Folk Use : Bark used as stimulant and seeds valued as narcotic.

References : Ambasta (1986), Anonymous (1962), Kaushik & Dhiman (2000), Kirtikar & Basu (1984).

534. *Lagerstroemia speciosa* Pers.

Syn.: L. flosreginae Retz.; *L. reginae* Roxb.

Family: Lythraceae.

English, Hindi and Sanskrit Names: **Eng.** : Pride of India, Queen Crape Martle; **Hindi** : Arjuna, Jarul; **Sans.** : Jarula, Kramuka.

Distribution : Distributed more of less throughout India, especially in Assam, Bengal and Deccan Peninsula.

Parts Used : Root. Seeds.

Flowering & Fruiting : May – September.

Folk Use : Seeds used as narcotic. Roots considered stimulant.

References : Parrotta (2001), Raju (2000), Sharma (2003).

535. *Lamium album* L.

Family: Lamiaceae.

English Names: **Eng. :** Bee Nettle, Blind Nettle, Deaf Nettle, Dumb Nettle, White Deadnettle.

Distribution : Native to Europe and Central and Northern Asia; found in wasteplaces and roadsides in Western Himalayas (Kashmir – Kumaon): 1,500-3,000m.

Parts Used : Flower.

Flowering & Fruiting : Flowering during May – September.

Folk Use : Flowers used as tonic.

Active Constituents : Flowers rich in lamalbide, three sapanosides, four aglycones, caffeic acid, cysteine, asparatic acid, glutamic acid, serine, asparagine, glycine, threonine, alanine, γ-aminobutynic acid, methionine, valine, leucine, proline, ornithine and lysine.

References : Ambasta (1986), Farooq (2005), Prajapati *et al.* (2006).

536. *Lantana camara* auct. (Pl. 7E)*

Syn.: Lantana aculeate L.; *L. camara* L. var. *aculeate* (L.) Moldenke; *L. myxa* L.

Family: Verbenaceae.

English, Hindi and Sanskrit Names: **Eng. :** Lantana Weed, Wild Sage; **Hindi :** Caturang, Desi Lanatan, Ghaneri; **Sans. :** Caturangi, Vanacchedi.

Distribution : Native of tropical America, naturalized in many parts of India as a troublesome prickly weed.

Parts Used : Whole Plant. Leaf.

Flowering & Fruiting : May – November.

Folk Use : Plant and decoction of leaves used as a tonic. Also, tea brewed from leaves should be taken before going to bed for enhancing memory.

Active Constituents : Reported to possess catalase, amylase, invertase, lipase, tannase & glucosidase, caryophyllene, l-α-phellandrene, lancamarone and lantanine (*plant*), arachidic, linoleic, myristic, palmitic acids, 1-triacontanol, 1-8-cineole, citral, eugenol, furfural, geranial, linalool, α-phellandrene, phellandrone, α-amyrin, β-sitosterol and lantadene (*leaf*).

References : Khatoon & Geol (2010), Kirtikar & Basu (1984), Nayak *et al.* (2003), Prajapati *et al.* (2006), Subramaniam (2000).

537. *Lathyrus aphaca* L.

Family: Fabaceae.

English and Hindi Names: **Eng. :** Yellow-flowered Pea, Yellow Vetchling; **Hindi :** Jangli Matar.

Distribution : Throughout N. India; met as a weed ascending up to 2,100m; Bengal, M.P., Nilgiris in the South; W. Asia, N. Africa, Europe.

Parts Used : Seeds.

Flowering & Fruiting : April – May.

Folk Use : Seeds have narcotic effect.

References : Anonymous (1962), Kirtikar & Basu (1984), Manandhar (2003).

538. *Laurus nobilis* L.

Family: Lauraceae.

English Names: **Eng. :** Bay Laurel, Sweet Bay, True Laurel.

Distribution : Native to Mediterranean countries, cultivated in Indian gardens.

Parts Used : Fruit. Leaf.

Flowering & Fruiting : Flowering in spring season.

Folk Use : Leaves and fruits used as stimulant and narcotic.

Active Constituents : Known to possess rutin, costunolide, artemorin, verlotorin, santamarine, reynosin, (+) catechin, (-) epicatechin, (-) epigallocatechin, (+) gallocatechin, procyanidins B2, B4, B5 and B7 (*leaf*); [4R, 7S, 11ζ, 1 (10)E, 5E] 12-acetoxygermacra-1(10), 5-dien-4,11-diol, eremanthin, dehydrocostuslactone, zaluzanin D, 10-hydroxyoctacosanyl tetradecanonate, 1-docosanyl tetradecanonate and 11-hydroxytriacontan- 9-one (*fruit*).

References : Ambasta (1986), Anonymous (1962), Bhattacharjee (2001), Parrotta (2001), Prajapati *et al.* (2006).

539. *Lawsonia inermis* L.

Syn.: L. alba Lam.; *L. pupurea* Lamk.; *L. spinosa* L.

Family: Lythraceae.

English, Hindi and Sanskrit Names: **Eng. :** Cypress Shrub, Egyptian Priven, Henna; **Hindi :** Henna, Mehndi; **Sans. :** Madayantika, Medhini, Nakharanjani, Ranjaka, Sakachara.

Distribution : Dry deciduous forests; widely cultivated as a hedge plant.

Parts Used : Flower. Leaf.

Flowering & Fruiting : April – July.

Folk Use : Leaves used as liver tonic and flowers valued as cardio tonic. Also, flowers considered good as brain tonic (Yunani).

Active Constituents : Lawsone, laxathone II, luteolin, β-sitosterol and its 3-O-glucoside, gallic acid, 2-hydroxyl-1, 4-napthoquinone, 1, 4-dihydroxynapthalene, 1, 4-naothoquinone, fraxetin, scopoletin, esculetin, β-sitosterol, stigmasterol, 1, 2-dihydroxy-4'-glucosyloxynapthalene, apigenin-7-glucoside, apigenin-4'-glucoside, luteolin-7-glucoside and luteolin-3'-glucoside (*leaf*).

References : Parrotta (2001), Prajapati *et al.* (2006).

540. *Leea hirta* Roxb. *ex* Hornem

Syn.: L. aequata L.*; L. kurzii* Cl.

Family: Vitaceae.

Hindi and Sanskrit Names: **Hindi** : Kakajangha; **Sans.** : Kakajangha.

Distribution : Assam, Sikkim Himalaya, Khasia mountains and Andaman Islands.

Parts Used : Root. Stem.

Flowering & Fruiting : June – September.

Folk Use : Roots and stem valued as stimulant.

Reference : Chatterjee & Pakrashi (1997).

541. *Lens culinaris* Medikus

Syn.: Ervum lens L.; *Lens esculenta* Moench.

Family: Fabaceae.

English, Hindi and Sanskrit Names: **Eng.** : Lentil; **Hindi** : Masser, Masur, Masuri; **Sans.** : Gurubija, Masurah, Ragadoli, Sura.

Distribution : Throughout colder parts of India, cultivated.

Parts Used : Seeds.

Folk Use : Seeds used as a tonic.

Active Constituents : Imidazole, p – coumaric and ferulic acid (*seeds*).

References : Kirtikar & Basu (1984), Prajapati *et al.* (2006), Reddi *et al.* (2005).

542. *Leonurus cardiaca* L.

Family: Lamiaceae.

English and Hindi Names: **Eng.** : Common Motherwort; **Hindi** : Guma.

Distribution : Temperate Western Himalayas (Kashmir – Kumaon): 2,000-3,350m; Europe, N. Asia and Asia Minor.

Parts Used : Aerial parts.

Flowering & Fruiting : Flowering during June – August.

Folk Use : Aerial parts considered stimulant and tonic.

Active Constituents : Therapeutic properties due to quercitrin, isoquercitrin, rutin, hyperoside, quercetin-7-O-β-D-glucoside, kaempferol-3-O-β-D-glucoside, apigenin-5, 7- and 4'-O-β-D-glucosides, quinqueloside and leocardin (*aerial parts*).

References : Anonymous (1962), Bhattacharjee (2001).

543. *Leonurus japonicus* Houtt.

Syn.: L. heterophyllus Sweet; *L. sibiricus* L.; *L. tartaricus* Burm.

Family: Lamiaceae.

English and Hindi Names: **Eng. :** Siberian Motherwort; **Hindi :** Guma, Lal Halkucha.

Distribution : Plains of India from Bengal and Sylhet – Coorg, but doubtfully indigenous; also found in tropical Asia, Africa and America.

Parts Used : Whole Plant.

Flowering & Fruiting : June – September.

Folk Use . Whole plant used as a tonic.

Active Constituents : Leourine nitrate, leosibirin, isoleosibrin, leosibiricin and marrubiin (*plant*).

References : Kirtikar & Basu (1984), Sharma (2004).

544. *Lepidium sativum* L.

Family: Brassicaceae.

English, Hindi and Sanskrit Names: **Eng. :** Garden Grass, Water Cress; **Hindi :** Chamasur, Chandrasur, Halim, Hurf; **Sans. :** Asalic, Asalika, Chandrashura, Chandrika, Vaspushpa.

Distribution : Cultivated throughout India.

Parts Used : Leaf. Seeds.

Flowering & Fruiting : March – May.

Folk Use : Leaves used as mild stimulant and seeds as a tonic.

Active Constituents : Characteristic properties due to lepidine, sinapic acid ethyl ester, N, N' – dibenzylthiourea and N, N' – dibenzylurea (*seeds*).

References : Ambasta (1986), Anonymous (1962), Kirtikar & Basu (1984), Parrotta (2001), Prajapati *et al.* (2006), Reddi *et al.* (2005), Sharma (2004), Watt (1972).

545. *Leptadenia reticulata* W. & A.

Syn.: Asclepias tuberosa Roxb.; *Cynanchum asthmaticum* Ham.; *C. reticulatum* Retz.; *Gymnema aurantiaca* Wall.; *Leptadenia brevipes* Wight.; *L. imberbe* Wight.

Family: Asclepiadaceae.

English, Hindi and Sanskrit Names: **Eng. :** Meda; **Hindi :** Dodi, Dori; **Sans. :** Arkapushpi, Jivanti, Shakashreshtha, Svarnajvanti.

Distribution : Distributed in India in Punjab, Konkan, Poona and Gujarat, also in the Western Peninsula.

Parts Used : Whole Plant. Fruit.

Flowering & Fruiting : June – September.

Folk Use : Plant used as a stimulant and a tonic. Fruits prescribed as a tonic for better eye-sight.

Active Constituents : Stigmasterol, tocopherols, α-amyrin, β-sitosterol, diosmetin, luteolin and leptaculatin (*plant*).

References : Anonymous (1962), Bhatt (2002), Bhatt *et al.* (2006), Farooq (2005), Kirtikar & Basu (1984), Prajapati *et al.* (2006), Ranjan (2000).

546. *Leucas aspera* (Willd.) Link.

Syn.: L. dimidiata Benth.; *L. plukenetii* (Roth.) Spreng.; *Pholmis aspera* Willd.; *P. esculenta* Roxb.; *P. obliqua* Ham.; *P. plukenetii* Roth.

Family: Lamiaceae.

English, Hindi and Sanskrit Names: **Eng. :** Thumbe; **Hindi :** Chota-italkusa, Chotalalkura; **Sans. :** Drona Pushpi.

Distribution : Plains of India; Sikkim and Bihar – Punjab and southwards – Cape Comorin; Mauritius, Java, and Philippine Islands.

Parts Used : Whole Plant.

Flowering & Fruiting : June – October.

Folk Use : Plant used as a stimulant and tonic.

Active Constituents : Galactose, oleanolic acid, ursolic acid and β-sitosterol (*plant*).

Reference : Subramaniam (2000).

547. *Leucas cephalotus* Spreng.

Syn.: L. capitata Desf.; *Phlomis cephalotus* Roth.

Family: Lamiaceae.

English, Hindi and Sanskrit Names: **Eng. :** Goma; **Hindi :** Delotona, Dhurpisag, Guma, Motapati; **Sans. :** Chhatraka, Dronapushpi, Kaundinya, Palindi.

Distribution : Himalayas (Kashmir – Sikkim) up to 1,828m, southwards throughout India.

Parts Used : Whole Plant.

Flowering & Fruiting : Flowering: July – October; Fruiting: December – January.

Folk Use : Herb used as a stimulant.

Active Constituents : β-sitosterol and glucose (*plant*).

References : Anonymous (1962), Bhattacharjee (2001), Dhiman (2005), Farooq (2005), Kapur & Singh (1996), Kaushik & Dhiman (2000), Kirtikar & Basu (1984), Kumar & Lal (2003), Parrotta (2001), Sharma (2003), Watt (1972).

548. *Leucas lavandulaefolia* Rees.

Syn.: Leucas linifolia Spreng.; *Phlomia linifolia* Roth.; *P. zeylanica* Roxb.

Family: Lamiaceae.

Hindi and Sanskrit Names: **Hindi :** Guma, Halkusa, Kumbha; **Sans.** : Dronapushpi, Kumbhi, Rudrapushpa.

Distribution : More or less throughout India, Mauritius, and Malaya.

Parts Used : Leaf.

Folk Use : Decoction of leaves used as a sedative.

Reference : Ambasta (1986).

549. *Leucas stelligera* Wall.

Syn.: L. oriostoma var. *longifolia* Hook. f.

Family: Lamiaceae.

English Name: **Eng. :** Starry Leucas.

Distribution : Saurashtra, Deccan, Konkan, Western Ghats, hills of S. Kanara and Mysore up to 900m.

Parts Used : Whole Plant.

Flowering & Fruiting : November – December.

Folk Use : Herb used as stimulant.

References : Anonymous (1962), Kirtikar & Basu (1984), Srivastava *et al.* (2003b).

550. *Ligusticum elatum* C.B. Cl.

Family: Apiaceae.

Distribution : N.W. Himalayas: 2,350-2,700m, Lahul.

Parts Used : Fruit.

Flowering & Fruiting : July – October.

Folk Use : Fruits used as stimulant.

Reference : Kapur (1996a).

551. *Lilium polyphyllum* D. Don

Family: Liliaceae.

English, Hindi and Sanskrit Names: **Eng. :** Ksheerakakoli; **Hindi :** Ksirakakoli; **Sans. :** Ksirakakoli.

Distribution : Western temperate Himalayas, up to 3,600m.

Parts Used : Bulb.

Flowering & Fruiting : May.

Folk Use : Bulbs used as a tonic.

Active Constituents : Linalool and a-terpineol (*bulb*).

Reference : Prajapati *et al.* (2006).

552. *Limnophila gratioloides* R. Br.

Syn.: L. indica (L.) Druce; *L. racemosa* Benth.

Family: Scrophulariaceae.

English, Hindi and Sanskrit Names: **Eng. :** Indian Ambulia; **Hindi :** Kuttra; **Sans. :** Ambuja, Amragandhaka.

Distribution : Throughout India ascending to 1,200m in the Himalayas; Baluchistan, Sri Lanka, Malay Archipelago, China, Australia and tropical Africa.

Parts Used : Whole Plant.

Flowering & Fruiting : March – November.

Folk Use : Plant used as a brain tonic by Paharia and Santhal tribes of Jharkhand.

References : Khatoon & Goel (2010), Kumar & Goel (1998).

553. *Limnophila rugosa* (Roth) Merr.

Syn.: L. roxburghii G. Don.

Family: Scrophulariaceae.

English and Hindi Names: **Eng. :** Limnophila; **Hindi :** Bhringraj.

Distribution : Found in aquatic situations and moist lands almost throughout India ascending up to 1,800m in the Himalayas.

Parts Used : Whole Plant.

Flowering & Fruiting : August – November.

Folk Use : Whole plant used in the preparation of medicated oil which is considered good for brain.

Active Constituents : Ursolic acid, hentriacontanol, 5-hydroxyl-7, 2', 4'- trimethoxyflavone, 3β-hydroxyl-lup-20 (29)-en-27-oic acid, nevadensin, demthoxysudachitin, betulin and betulinic acid (*plant*).

Reference : Khatoon & Goel (2010).

554. *Linum usitatissimum* L.

Syn.: L. trinervium Roth.

Family: Linaceae.

English, Hindi and Sanskrit Names: **Eng. :** Blaebows, Flax Plant, Lin, Lint Bells, Lint Bow, Linseed, Lyne, Vlix; **Hindi :** Alsi, Tisi; **Sans. :** Atasi, Chanaka, Devi, Haimwati, Kshuma, Madagandha, Sunila, Uma.

Distribution : Cultivated throughout India, occasionally found in wild.

Parts Used : Flower. Seeds.

Flowering & Fruiting : November – April.

Folk Use : Flowers used as cardio and brain tonic. Seeds also used as tonic.

Active Constituents : Active properties due to cholesterol, campesterol, stigmasterol, sitosterol, 5-dehydro -avenasterol, cycloartenol, 24-methylenecycloartanol, oleic, linoleic and linolenic acid (*seeds*).

References : Kirtikar & Basu (1984), Parrotta (2001), Prajapati *et al.* (2006), Reddi *et al.* (2005), Watt (1972).

555. *Liparis paradoxa* Rchb. f.

Family: Orchidaceae.

English Name: **Eng. :** Wild Lily Orchid.

Distribution : Distributed in Uttaranchal, Naga and Khasia hills, Sikkim, Chhatisgarh and Karnataka.

Parts Used : Whole Plant.

Flowering & Fruiting : November – December.

Folk Use : Plant used as a general tonic.

References : Bhattacharjee (2001), Sharma (2003).

556. *Liparis rostrata* Rchb. f.

Family: Orchidaceae.

English Name: **Eng. :** Beaked Widelip Orchid.

Distribution : Mussorie and Chakrata: 2,100-2400m; Kashmir, Pakistan.

Parts Used : Bulb.

Flowering & Fruiting : Flowering: July – August.

Folk Use : Bulbs of this plant used as tonic.

Reference : Bhattacharjee (2001).

557. *Liquidambar orientalis* Mill.

Syn.: L. imberbe Aiton.

Family: Hamamelidaceae.

English, Hindi and Sanskrit Names: **Eng. :** Asiatic Storax, Oriental Sweet Gum, Liquid Amber; **Hindi :** Silaras; **Sans. :** Silhaka, Turuskah.

Distribution : Native of Asia Minor; large forests of the plant found in the southwestern parts of the India.

Parts Used : Whole Plant.

Flowering & Fruiting : May (flowering); October – November (fruiting).

Folk Use : Plant used as a stimulant and tonic.

Active Constituents : Known to yield cinnamic acid, cinnnamyl, cinnamate or styracine, phenylpropyl cinnamate and small amounts of ethyl cinnamate, benzyl cinnamate, styrene, vanillin, phenylpropyl, cinnamyl alcohols and cinnamic acid (*plant*).

References : Ambasta (1986), Prajapati *et al.* (2006).

558. *Liriodendron tulipifera* L.

Family: Magnoliaceae.

English Names: **Eng. :** American Tulip Tree, Tulip Poplar, Tulip Tree, Yellow Poplar.

Distribution : Native of eastern North America; planted as an ornamental species in gardens in the hilly regions of North India.

Parts Used : Root Bark.

Flowering & Fruiting : April – June (flowering); September – October (fruiting).

Folk Use : Root bark used as a stimulant.

Active Constituents : γ-liridenolide, epitulipdienolide, α- and β-liridenolides (*root bark*).

References : Ambasta (1986), Bhattacharjee (2001).

559. *Liriope platyphylla* Wang et Tang

Syn.: L. graminifolia (L.) Baker.

Family: Liliaceae.

English Names: **Eng. :** Bigblue Lilyturf, Border Grass, Lilyturf, Monkey Grass.

Distribution : N.W. Himalayas; native of China, Japan and Korea.

Parts Used : Tuber.

Flowering & Fruiting : Flowering during winters and fruiting in autumn.

Folk Use : Tubers used as tonic.

Active Constituents : Accredited properties due to saponins : ophiogonins A, B, C, D etc; Montoterpenoids: borneol derivative (*tuber*).

Reference : Prajapati *et al.* (2006).

560. *Litchi chinensis* (Gaertn.) Sonn.*

Syn.: Nephelium litchi Camb.

Family: Sapindaceae.

English and Hindi Names: **Eng. :** Litchi, Lychee; **Hindi :** Lichi.

Distribution : Native of China, widely cultivated throughout India, chiefly in warmers parts like North Bihar, Muzaffarpur, Saharanpur, Dehradun, Haridwar, Muzaffarnagar, Gorakhpur and West Bengal.

Parts Used : Fruit.

Flowering & Fruiting : Flowering: February – April; Fruiting: May – June.

Folk Use : Fruits used as tonic.

Active Constituents : Fruits rich in cyanidin-3-rutinoside, cyaniding-3-glucoside and malvidin -3-acetylglucoside.

References : Farooq (2005), Kaushik & Dhiman (2000), Kirtikar & Basu (1984), Singh & Srivastava (2000).

561. *Lithospermum officinale* L.

Syn.: Magarospermum officinale Dene.

Family: Boraginaceae.

English Names: **Eng. :** Common Gromwell, European Gromwell, Gromwell.

Distribution : Kashmir and Kumaon: 1,500-2,700m; Nepal, Bhutan, Afghanistan and Russia.

Parts Used : Leaf.

Flowering & Fruiting : June – August.

Folk Use : Infusion of leaves used as sedative.

Active Constituents : Caffeic, chlorogenic and ellagic acids and amino acids (*leaf*).

Reference : Ambasta (1986).

562. *Litsea glutinosa* (Lour.) C. B. Robinson

Syn.: Laurus involucrata Koenig.; *Litsea chinensis* Lamk.; *L. mutiflora* Blume.; *L. sebifera* Pers.; *L. tetranthera* Pers.; *Neolitsea chinensis* (Lam.) Chun.; *Sebifera glutinosa* Lour.; *Tomex apetala* Roxb.; *T. capitata* Herb.; *T. laurifolia* Roxb.; *T. tetraanthera* Willd.

Family: Lauraceae.

English, Hindi and Sanskrit Names: **Eng. :** Common Tallow Laurel, Vasa; **Hindi :** Garbijaur, Menda, Singrauf; **Sans. :** Drava, Jivani, Medasaka, Medovati, Rasa, Shreshtha.

Distribution : Found throughout India ascending up to 1,350m in the Himalayas.

Parts Used : Root.

Flowering & Fruiting : Flowering during summer season and fruiting in rainy season.

Folk Use : Roots considered as tonic.

References : Anonymous (1962), Dhiman (2005), Kirtikar & Basu (1984).

563. *Litsea monopetala* (Roxb.) Pers.

Syn.: Litsea polycantha Juss.; *Tetranthera alnoides* Miq.; *T. fruticosa* Ham.; *T. macrophylla* Wall.; *T. monopetala* Roxb.; *T. pubescens* Willd.

Family: Lauraceae.

English, Hindi and Sanskrit Names: **Eng. :** Common Grey Mango Laurel; **Hindi :** Karkawa, Kerauli, Meda, Patoia, Randkari, Singran; **Sans. :** Gajapippali.

Distribution : Throughout north, east and central India up to 1,200m.

Parts Used : Bark.

Flowering & Fruiting : July – August (flowering); October – November (fruiting).

Folk Use : Bark used as stimulant.

Active Constituents : β-sitosterol and actinodaphnine (*bark*).

References : Dhiman (2005), Kirtikar & Basu (1984), Watt (1972).

564. *Lodoicea maldivica* (Poir.) Pers.

Syn.: L. seychellarum Comm. & Labill.

Family: Arecaceae.

English, Hindi and Sanskrit Names: **Eng. :** Double Coconut Palm, Love Nut, Sea Coconut Palm, Seychelles Nut; **Hindi :** Darya-ka-nariyal, Dariya-l-nariyal; **Sans. :** Abdhinarikelah, Agnijarah, Arkaragah.

Distribution : In Seychelles Islands; cultivated in India.

Parts Used : Kernel.

Flowering & Fruiting : Throughout the year.

Folk Use : Kernels used as a tonic and cardio tonic.

References : Kirtikar & Basu (1984), Prajapati *et al.* (2006), Reddi *et al.* (2005).

565. *Luffa acutangula* (L.) Roxb.

Syn.: Cucumis acutangulus Linn.; *Luffa acutangula* var. *amara* (Roxb.) Cl.; *L. kleinii* W. & A.; *L. umbellate* (Klein *ex* Willd.) Roem.

Family: Cucurbitaceae.

English, Hindi and Sanskrit Names: **Eng. :** Ribbed Gourd, Ridged Gourd; **Hindi :** Jinga, Kali Tori, Tori; **Sans. :** Dhamargavah, Koshataki, Svadukosataki.

Distribution : Cultivated throughout India.

Parts Used : Fruit.

Flowering & Fruiting : August – December.

Folk Use : Fruits used as a tonic.

Active Constituents : Protein, fats, carbohydrates, calcium, phosphorus, iron, carotene (as vitamin A), iodine and phytin, arginine, glycine, threonine, glutamic acid, leucine, serine, alanine, γ-amino butyric acid and pipecolic acid (*fruit*).

References : Kaushik & Dhiman (2000), Prajapati *et al.* (2006), Watt (1972).

566. *Luffa cylindrica* (L.) M. Roem.*

Syn.: Bryonia cheirophylla Wall.; *Luffa aegyptiaca* Mill.; *L. clavata* Roxb.; *L. gosa* Wall.; *L. hedaracea* Wall.; *L. parvala* Wall.; *L. pentandra* Roxb.; *L. petola* Ser.; *L. racemosa* Roxb.; *L. satipatia* Wall.; *Momordica cylindrica* L.; *M. luffa* L.

Family: Cucurbitaceae.

English, Hindi and Sanskrit Names: **Eng. :** Sponge Gourd, Vegetable Sponge, Wash Sponge; **Hindi :** Ghiya Tori, Nenua, Purula; **Sans. :** Dhirghapatolika, Hastiparna, Mahakosataki, Rajakoshatoki, Sapitaka.

Distribution : Wild in wastelands especially along the coastal areas throughout India; also cultivated.

Parts Used : Fruit.

Flowering & Fruiting : August – December.

Folk Use : Fruits used as a tonic.

Active Constituents : Proteins, carbohydrates and ash, Calcium, phosphorus, iron, carotene (as vitamin A), thiamine, riboflavin, niacin, ascorbic acid, choline, phytin, lysine, arginine, aspartic acid, glycine, threonine, glutamic acid, alanine, tryptophan, phenylalanines and leucines have been isolated from *fruits*.

References : Parrotta (2001), Prajapati *et al.* (2006).

567. *Lycopersicon esculentum* Mill. (Pl. 7F)*

Syn.: L. lycopersicon (L.) Britton & Brown; *L. lycopersicum* (L.) Karsten; *Solanum humboldtii* Willd.; *S. lycopersicum* L.; *S. pseudolycopersicum* Jacq.

Family: Solanaceae.

English, Hindi and Sanskrit Names: **Eng.** : Love-apple, Tomato; **Hindi** : Kakjangha, Tamatar, Timoti; **Sans.** : Raktamaci.

Distribution : Throughout India, cultivated.

Parts Used : Fruit.

Flowering & Fruiting : Throughout the year.

Folk Use : Fruits used as a tonic.

Active Constituents : Principle constituents are ascorbic acid, biotin, folic, nicotinic and panthothenic acids, riboflavin, thiamine, vit. K, fructose, glucose, a ketoheptose, raffinose, sucrose, eugenol, β-ionene, essential amino acids (except tryptophan), cystine, asparatic acid and glutamic acids, glycine, histamine, pipecolic acid, serine, tyrosine, ethanolamine, serotonine, tyramine, arachic, acetic, citric, lactic, malic, succinic, caffeic and hydroxyl-cinnamic acids (*fruit*).

References : Kaushik & Dhiman (2000), Prajapati *et al.* (2006).

568. *Lycopodium clavatum* L.

Syn.: L. inflexum Sw.

Family: Lycopodiaceae.

English Names: **Eng.** : Club Moss, Groundpine, Stag's-horn Clubmoss, Wolf's-foot Clubmoss.

Distribution: Throughout regions of the northern hemisphere; common on mountains and in moorland.

Parts Used : Moss.

Flowering & Fruiting : July – September.

Folk Use : Moss used as sedative.

Active Constituents : Clavatol, lycoflexine, lycopodine, clavatine, lycodoline, dihydrolycopodine, clavolonine, N-demethyl-α-obscurine and nicotine (*moss*).

Reference : Prajapati *et al.* (2006).

569. *Madhuca indica* J. F. Gmel.

Syn.: Bassia latifolia Roxb.; *B. longifolia* Koen.; *Madhuca latifolia* Macbr.; *M. longifolia* (Koen.) Macbr.; *M. longifolia* (Koen.) Macbr. var. *latifolia* (Roxb.) Chev.

Family: Sapotaceae.

English, Hindi and Sanskrit Names: **Eng.** : Butter Tree, Illipe, Mahua, Mowra; **Hindi** : Mahua, Mauwa, Mohwa; **Sans.** : Dalaphala, Garudapu, Lodhrapushpa, Madhava, Madhu, Madhuka, Priyala, Vanaprastha.

Distribution : Plains and lower hills of India up to 1,200m; common throughout W. Bengal, Bihar, M.P., Orissa, Punjab and U.P. including plains of Uttaranchal.

Parts Used : Bark. Flower. Fruit.

Flowering & Fruiting : Flowering: April – May; Fruiting: June – July.

Folk Use : Bark, flowers and fruits used as tonic. Also, flowers valued as stimulant.

Active Constituents : These are leucodelphindin (*bark*), vitamin A and C, ethyl cinnamate, sesquiterpene alcohol, α-terpeneol, 3-β-monocaprylic ester of erythrodiol and 3β-capryloxy oleanolic acid and α- and β-amyrin (*fruit*), L-arabinose, D-galactose, glucose, D-glucceronic acid, fructose, maltose, L-rhamnose, D-xylose, biotin, folic acid, inositol, niacin, pantothenic acid, riboflavin, thiamine, amylase, catalase, emulsion, invertase, maltose, oxidase, malic acid and succinic acid (salts), anthocyanin, betaine and protein (*flower*).

References : Ambasta (1986), Bhattacharjee (2001), Chauhan *et al.* (2007), Dhiman (2005), Dwivedi (2007), Kaushik & Dhiman (2000), Kumar & Lal (2003), Parrotta (2001), Prajapati *et al.* (2006).

570. *Maerua arenaria* Hook. f. & Thoms.

Syn.: Maerua oblongifolia (Forsk.) A. Rich.; *M. ovalifolia* Cambess.; *Niebuhria oblongifolia* Royle.

Family: Capparaceae.

English, Hindi and Sanskrit Names: **Eng.** : Necklace Berried Caper; **Hindi** : Hemkanda; **Sans.** : Murahari.

Distribution : Abundant in Western Himalaya and Central India; also found in dry areas in Punjab, Upper Gangetic plains, M.P., Rajasthan, Gujarat, Deccan and Carnatic districts from Godavari southwards.

Parts Used : Root.

Flowering & Fruiting : February – June.

Folk Use : Roots used as stimulant and tonic.

References : Ambasta (1986), Anonymous (1962), Kirtikar & Basu (1984), Thakor (2001).

571. *Magnolia grandiflora* L. (Pl. 8A)

Syn.: M. angustifolia Millais; *M. elliptica* Link.; *M. foetida* (L.) Sarg.; *M. lanceolata* Link.; *M. rotundifolia* Millois.

Family: Magnoliaceae.

English and Hindi Names: **Eng.** : Bull Bay, Southern Magnolia; **Hindi** : Andachampa, Him Champa.

Distribution : Indigenous to southeastern United States (N. Carolina, Florida, Texas); cultivated in Indian gardens up to 2,100m in the Himalayas and Nilgiris.

Parts Used : Bark.

Flowering & Fruiting : May – August.

Folk Use : Bark used as stimulant and tonic.

Active Constituents : Magnoflorine, magnolenioc, syringing and acanthoside B (*bark*).

References : Ambasta (1986), Anonymous (1962), Singh & Srivastava (2000).

572. *Majorana hortensis* Moench.

Syn.: Origanum majorana L.; *O. wallichianum* Benth.

Family: Lamiaceae.

English, Hindi and Sanskrit Names: **Eng.** : Sweet-knotted Marjoram, Sweet Marjoram; **Hindi** : Marua, Murwa; **Sans.** : Ajanmasura-bhipatra, Bahuvirya, Kharapatra, Maru, Phani, Suravha.

Distribution : Commonly grown in Indian gardens.

Parts Used : Whole Plant.

Flowering & Fruiting : August – November.

Folk Use : Infusion of plant used as a stimulant and tonic for liver. Wine containing this plant consumed to sharpen memory.

Active Constituents : Known to yield luteolin-7-β-D-diglucoside, apigenin-7-β-D-glucoside, diosmetin-7-β-D-glucuronide, arbutin, methylarbutin and their aglycones, hydroquinone and hydroquinone mono methyl ether (*plant*).

References : Ambasta (1986), Anonymous (1962), Kaushik & Dhiman (2000), Khatoon & Goel (2010), Kirtikar & Basu (1984), Rajan *et al.* (2003), Watt (1972).

573. *Malaxis acuminata* D. Don

Syn.: Microstylis wallichi Lindl.

Family: Orchidaceae.

English, Hindi and Sanskrit Names: **Eng.** : Jeevak; **Hindi** : Jivak; **Sans.** : Jivakah.

Distribution : Throughout India, on hills up to 3,000m.

Parts Used : Green swollen stem base.

Flowering & Fruiting : June – August.

Folk Use : Stem base used as tonic.

Active Constituents : Isolated components are β-sitosterol, ceryl alcohol, glucose, rhamnose and choline (*pseudo bulbs*).

References : Negi *et al.* (1999), Prajapati *et al.* (2006), Singh (2001).

574. *Malaxis muscifera* (Lindley) Kuntze

Syn.: Microstylis muscifera (Lindley) Ridley.

Family: Orchidaceae.

English, Hindi and Sanskrit Names: **Eng. :** Adder's Mouth Orchid; **Hindi :** Rsabhak; **Sans. :** Rsabhakah.

Distribution : Throughout India, in hills up to 4,000m.

Parts Used : Stem Base. Root

Flowering & Fruiting : July – August.

Folk Use : Roots and stem base used as a tonic.

References : Bhattacharjee (2001), Hemlata *et al.* (2007), Kala (2003), Prajapati *et al.* (2006), Reddi *et al.* (2005), Sharma (2003), Singh (2001).

575. *Mallotus philippinensis* (Lam.) Muell. – Arg. (Pl. 8B)*

Syn.: Croton cascarilloides Rauesch. *ex* Steud; *C. coccineus* Vahl.; *C. distans* Wall.; *C. montanus* Willd.; *C. philippinensis* (Lam.) Muell. - Arg.; *Rattlera affinis* Hassk.; *R. aurantiaca* Hook & Arn.; *R. montana* Wall.; *R. tinctoria* Roxb.

Family: Euphorbiaceae.

English, Hindi and Sanskrit Names: **Eng. :** Kamala Tree, Monkey Face Tree; **Hindi :** Kamila, Rohni, Rula; **Sans. :** Kampillaka, Lohitanga, Nadivasa, Pikaksha, Rechanaka.

Distribution : Widely distributed throughout tropical India, ascending up to 1,500m in the outer Himalayas; Sri Lanka, Myanmar, Malay islands, Australia, China.

Parts Used : Root.

Flowering & Fruiting : April – October.

Folk Use : Root powder used as tonic.

Reference : Subramaniam (2000).

576. *Malva parviflora* L.

Family: Malvaceae.

English and Hindi Names: **Eng. :** Cheese Weed, Small Whorled; **Hindi :** Panirak.

Distribution : Bengal, Kashmir, Punjab, Sind, Baluchistan, Bombay, Deccan, Mysore and Madura. Afghanistan, Persia, Arabia, Mediterranean and Nubia.

Parts Used : Leaf.

Flowering & Fruiting : May – September.

Folk Use : Infusion of leaves taken as a nervine tonic.

References : Ambasta (1986), Anonymous (1962).

577. *Mandragora autumnalis* Spreng.

Syn.: Mandragora officinarum L.; *M. microcarpa* Bertol.

Family: Solanaceae.

English, Hindi and Sanskrit Names: **Eng. :** Mandrake, The Official Mandrake; **Hindi :** Luckmuna, Lufah; **Sans. :** Lufah.

Distribution : Found in Mediterranean region; roots sold in the bazaars of N. India.

Parts Used : Root.

Flowering & Fruiting : Flowering during autumn and fruiting in winters.

Folk Use : Roots used as sedative and hypnotic.

Active Constituents : Hyoscyamine, scopolamine, atropine, cuskhygrine and scopine (*root*).

References : Ambasta (1986), Anonymous (1962).

578. *Mangifera indica* L. (Pl. 8C)

Syn.: M. domestica Gaertn.

Family: Anacardiaceae.

English, Hindi and Sanskrit Names: **Eng. :** Cuckoo's Joy, Cupid's Favourite, Mango Tree, Spring Tree. **Hindi :** Am, Amb; **Sans. :** Adisairrabha, Alipriya, Amra, Kamanga, Koshi, Manmathalaya, Parapushtamahotsava, Priyambu, Rasala, Sripriya.

Distribution : Cultivated throughout tropical regions from Kumaon – Assam, W. Bengal, Bihar, Orissa and southwards to Peninsular India.

Parts Used : Fruit. Rind.

Flowering & Fruiting : March – July.

Folk Use : Fruits regarded as restorative tonic. Rind valued as stimulating tonic in debility of stomach.

References : Anonymous (1962), Bhattacharjee (2001), Kirtikar & Basu (1984), Kumar *et al.* (2003), Parrotta (2001), Sood *et al.* (2009), Subramaniam (2000), Watt (1972).

579. *Manilkara hexandara* (Roxb.) Dubard

Syn.: Mimusops hexandra Roxb.

Family: Sapotaceae.

English, Hindi and Sanskrit Names: **Eng.** : Milk Tree, Obtuse Leaved Minusops; **Hindi :** Khair, Khirni; **Sans. :** Rajadanah.

Distribution : Evergreen forests throughout India; also cultivated.

Parts Used : Bark. Fruit.

Flowering & Fruiting : Flowering: December – January; Fruiting: March – April.

Folk Use : Fruits and bark used as a tonic.

Active Constituents : Known to possess amino acids, ascorbic acid, nicotinic acid, riboflavin and thiamine, carotene and glucose (*fruit*), α-spinasterol, α- and β-amyrin acetates, α- and β-amyrin, cinnamates, taraxerol, tarraxeryl acetate, ursolic acid and two unidentified triterpenoids, β-sitosterol, glucoside, quercitol, quercitin and dihydroquercitin (*bark*).

References : Anonymous (1962), Kaushik & Dhiman (2000), Prajapati *et al.* (2006).

580. *Manilkara kauki* (L.) Dubard

Syn.: Mimusops kauki L.

Family: Sapotaceae.

English, Hindi and Sanskrit Names: **Eng.** : Obtused Leaved Mimusops; **Hindi :** Khirni; **Sans. :** Khirni.

Distribution : Planted in Indian gardens.

Parts Used : Seeds.

Flowering & Fruiting : April – May.

Folk Use : Seeds used as tonic.

References : Ambasta (1986), Anonymous (1962).

581. *Maranta arundinacea* L.

Family: Marantaceae.

English, Hindi and Sanskrit Names: **Eng. :** Arrowroot, West Indian Arrowroot; **Hindi :** Tikhor; **Sans. :** Tavaksiri, Tugaksiri.

Distribution : Cultivated throughout India; native of tropical America.

Parts Used : Rhizome.

Flowering & Fruiting : June – December.

Folk Use : Rhizome used as a tonic.

Active Constituents : Moisture, crude protein, fat, starch, dextrin and sugars, crude fiber and ash (*rhizome*).

Reference : Prajapati *et al.* (2006).

582. *Marrubium vulgare* L.

Family: Lamiaceae.

English and Hindi Names: **Eng. :** East Indian Peppermint, Hoarhound, White Horehound; **Hindi :** Pahari Gandana.

Distribution : Wild in temperate Western Himalayas, chiefly in Kashmir: 1,524-2,438m.

Parts Used : Whole Plant.

Flowering & Fruiting : July – August.

Folk Use : Plant known to possess tonic properties.

Active Constituents : Herb rich in marrubin, marrubenol, flavonoids and alkaloids, betonicin, choline, premarrubiin, vulgarol, apigenin, its 7-glucoside, luteolin, its 7-glucoside, quercetin-3-glucoside and quercetin-3-rhamnoglucoside.

References : Anonymous (1962), Dhiman (2005), Jain & Sharma (2000), Kapur & Singh (1996), Kirtikar & Basu (1984), Prajapati *et al.* (2006), Watt (1972).

583. *Marsilea minuta* L.

Syn.: M. quadrifolia L.

Family: Marsileaceae.

English, Hindi and Sanskrit Names: **Eng. :** Caupatiya; **Hindi :** Chaupatia, Sunsuniya; **Sans. :** Catuspatri, Sunisannak.

Distribution : Grows as a weed in marshy places, edges of canals and ponds throughout India.

Parts Used : Whole Plant. Leaf.

Flowering & Fruiting : Fruiting during November – January.

Folk Use : Plant used as a sedative in insomnia, and also as a tonic after fever. Leaves also valued as sedative.

Active Constituents : Marsilin, 3-hydroxyl triacontan-11-one, hentriacoutan-6-al, methylamine, β-sitosterol, marsileagenin A, flavonal, O-mono and diglycoside, C-glucosyl flavones and C-glucosylxanthones (*plant*).

References : Dhiman (2005), Kaushik & Dhiman (2000), Kumar & Goel (1998), Parrotta (2001), Prajapati *et al.* (2006), Sharma (2004), Singh & Viswanathan (2006).

584. *Matricaria chamomilla* L.

Syn.: M. recutita L.; *M. suaveolens* L.

Family: Asteraceae.

English and Hindi Names: **Eng. :** German Chamomile, Wild Chamomile; **Hindi :** Babuna, Bauna, Sanamoti.

Distribution : Mainly in gardens or as an escape in Punjab, H.P. and Upper Gangetic plains, cultivated in U.P.

Parts Used : Whole Plant. Flower. Root.

Flowering & Fruiting : August – September.

Folk Use : Plant used as a sedative. Extract of flowers soaked in water taken orally and also for washing feet to retain good memory. Flowers and roots valued as stimulant.

Active Constituents : These are patulitrin, luteolin-7-maonoglucoside, quercimaritrin, apigenin-7-glucoside, 2"- and 6"-acetates of apigenin-7-glucoside, apigenin-7-β-D-glucoside and (-) α-bisabolal oxides A and B (*plant*), glucose, fructose, arabinose, xylose, rhamnose, galactouronic acid, herniarin, umbelliferone and epigenin (*flower*).

References : Ambasta (1986), Anonymous (1962), Kaul (1997), Kayal *et al.* (2003), Khatoon & Goel (2010), Sharma (2003), Thakur, puri, Husain (1989), Watt (1972).

585. *Matthiola flavida* Boiss.

Family: Brassicaceae.

Distribution : Pakistan – Kashmir : 2,700-4,500m; Common in Ladakh.

Parts Used : Whole Plant.

Flowerıng & Fruiting : June – July.

Folk Use : Powdered plant used locally as a tonic.

Reference : Singh *et al.* (1996).

586. *Matthiola incana* R. Br.

Family: Brassicaceae.

English Names: **Eng. :** Common Stock, Gilli Flower, Hoary Stock.

Distribution : Indigenous to Mediterranean region and Western Europe; cultivated in gardens of N. India.

Parts Used : Seeds.

Flowering & Fruiting : Flowering: May – July; fruiting in August.

Folk Use : Seeds used as tonic.

Active Constituents : Dactylin characterized as 5, 7-dihydroxy-3'-methoxy-flavone-3, 4'-diglucoside (*seeds*).

References : Ambasta (1986), Anonymous (1962).

587. *Mazus japonicus* (Thunb.) Kuntze

Syn.: Lindernia japonica Thunb.; *Mazus bicolor* Benth.; *M. rugosus* Lour.; *Tittmannia oborata* Bunge.

Family: Scrophulariaceae.

English Name: **Eng.** : Japanese Figwort.

Distribution : Temperate and sub-tropical Himalayas, plains of Northern India; Kashmir – Bhutan, ascending to 2,350m, Khasia hills, 1,350-2,000m, Myanmar, Upper Assam, Son river.

Parts Used : Whole Plant.

Flowering & Fruiting : Flowering during July – October.

Folk Use : Infusion of plant used as tonic.

References : Ambasta (1986), Anonymous (1962).

588. *Meconopsis aculeata* Royle

Syn.: M. nepalensis Honigberger.

Family: Papaveraceae.

English and Hindi Names: **Eng.** : Blue Poppy; **Hindi :** Kanta, Vanita.

Distribution : Kashmir, Garhwal, Kumaon, 3,650-5,000m.

Parts Used : Aerial Parts. Seeds. Fruit. Root.

Flowering & Fruiting : June – October.

Folk Use : Roots, fruits and seeds used as narcotic. Powdered aerial parts (2-3g) given thrice a day as a tonic for general weakness of the body.

References : Ambasta (1986), Jain & Goel (2005), Kirtikar & Basu (1984), Negi *et al.* (1999), Samant *et al.* (2001), Sood *et al.* (2001), Watt (1972).

589. *Medicago sativa* L.

Family: Fabaceae.

English and Hindi Names: **Eng.** : Alfalfa, Lucerne; **Hindi** : Lasunghar, Wilayti-gawuth.

Distribution : Kashmir, Ladakh, Kunawar, 1,700-4,350m; Afghanistan, Orient and all through Europe.

Parts Used : Whole Plant.

Flowering & Fruiting : Flowering during May – July; Fruiting: July – October.

Folk Use : Believed to improve memory.

Active Constituents : 10-methoxymedicarpin, 7-hydroxyl-2', 3', 4'-trimethoxyisoflavan, 7, 5'-dihyroxy - 2', 3', 4'-trimethoxyisoflavan (*plant*).

Reference : Khatoon & Goel (2010).

590. *Melia azedarach* L.

Syn.: M. bukayun Royle; *M. sempervirens* Sw.

Family: Meliaceae.

English, Hindi and Sanskrit Names: **Eng. :** Barbados Liliac, Bead Tree, China Berry, Hoop Tree, Persion Liliac, Pride of India; **Hindi :** Bakain, Bakorja, Drek; **Sans. :** Himadruma, Mahadroksha, Mahanimba, Parvata-nimbu-vrikshaha.

Distribution : Commonly cultivated, wild in the forest of hotter parts of India.

Parts Used : Root. Root Bark. Fruit. Seeds.

Flowering & Fruiting : Flowering during February – March; fruiting in October – December.

Folk Use : Fruits and roots used as tonic. Root bark used as bitter tonic. Seeds valued as brain tonic.

Active Constituents : Active properties due to azaridine, bakayanin, bakalactone, margosine, azadirone, azadiradione, epoxyazadiradione, ohchinol, ohchimin, ohchinolal, ohchinalides A and B, nimbolinin B, 1 – desacetylnimboline B, nimbolidine A and B, triterpene B, meliacins A_1, A_2, B_1 and B_2 , sendanin, sendenal, 1 – cinnamoylmelianolone, meliontriol, meliantrial, melianane, melianol, lupeol, β-sitosterol, catechin, vanillin and cinnamic acid (*fruit*).

References : Ambasta (1986), Farooq (2005), Kirtikar & Basu (1984), Sharma (2004), Siwakoti & Varma (1996).

591. *Melia composite* Willd.

Syn.: M. dubia Hiern; *M. robusta* Roxb.

Family: Meliaceae.

English, Hindi and Sanskrit Names: **Eng. :** Malabar Neem Wood, White Cedar; **Hindi :** Nimbara; **Sans. :** Arangaka.

Distribution : E. Himalayas up to 2,000m, Assam, W. Ghats and below the Ghats to Tinnevelly, N. Circars, Ganjam, Deccan, Sri Lanka, Malay Peninsula, China, Java, Australia.

Parts Used : Fruit.

Flowering & Fruiting : September – December.

Folk Use : Fruits eaten raw as a liver tonic.

Active Constituent : Salannin (*fruit*).

Reference : Chhetri (2005).

592. *Melissa officinalis* L.

Family: Lamiaceae.

English Names: **Eng. :** Balm, Balm Mint, Blue Balm, Dropsy Plant, Garden Balm, Lemon Balm, Sweet Balm.

Distribution : Native to S. Europe, W. Asia and N. Africa; lemon balm now grows throughout the world.

Parts Used : Whole Plant. Aerial Parts. Leaf. Fruit

Flowering & Fruiting : June – September.

Folk Use : Plant considered stimulating. Leaves valued as nervine tonic. Aerial parts used as sedative. Fruits regarded as brain tonic.

Active Constituents : Flavonoids, triterpenes, polyphenols and tannins (*leaf*).

References : Bhattacharjee (2001), Harborne & Baxter (2001), Khatoon & Goel (2010), Prajapati *et al.* (2006), Rajan *et al.* (2003).

593. *Melissa parviflora* Benth.

Family: Lamiaceae.

English and Hindi Names: **Eng. :** Arabian Balm, Gentle Balm; **Hindi :** Bililotan.

Distribution : Temperate Himalayas: Garhwal – Sikkim and Mishmi, Khasia Hills - Java.

Parts Used : Fruit.

Flowering & Fruiting : August.

Folk Use : Fruits considered as brain tonic.

References : Ambasta (1986), Anonymous (1962), Kirtikar & Basu (1984).

594. *Melodinus monogynus* Roxb.

Family: Apocynaceae.

English Name: **Eng. :** Seh Barb.

Distribution : Sikkim Himalayas, Assam, Sylhet, Khasia mountains; Malay Peninsula, Timor.

Parts Used : Leaf. Root. Wood.

Flowering & Fruiting : March – June.

Folk Use : Leaves, wood and roots used as narcotic.

Active Constituents : Digitoxigenin, medigenin, its glucoside - medinin, seven other cardenolines, butenolides – jaintigenin, pisigenin, medigenin, digitoxigenin, harinin, jaintin, gynin and pisidin (*root*).

Reference : Ambasta (1986).

595. *Melothria maderaspatana* (L.) Cogn.

Syn.: Mukia maderaspatana (L.) Roem.

Family: Cucurbitaceae.

Hindi and Sanskrit Names: **Hindi :** Agumaki, Bilari; **Sans. :** Krtarandhrah, Trikosaki.

Distribution : Throughout India up to 1,800m.

Parts Used : Whole Plant.

Flowering & Fruiting : September – October.

Folk Use : Plant used as a tonic.

Reference : Prajapati *et al.* (2006).

596. *Memecylon augustifolium* Wight.

Family: Melastomataceae.

Distribution : Southern India, Western Ghats, Sri Lanka.

Parts Used : Bark.

Folk Use : Bark used as tonic.

References : Ambasta (1986), Anonymous (1962), Kirtikar & Basu (1984).

597. *Menispermum canadense* L.

Family: Menispermaceae.

English Names: **Eng. :** Common Moonseed, Maple Vine, Yellow Parilla.

Distribution : Native to eastern North America, from Southern Canada south to North Florida in thickets, moist woods and the banks of streams.

Parts Used : Root.

Flowering & Fruiting : Flowering: May – July; fruiting in September – November.

Folk Use : Roots used as tonic.

Reference : Bhattacharjee (2001).

598. *Mentha aquatica* L.

Family: Lamiaceae.

English Names: **Eng. :** Marsh Mint, Winter Mint.

Distribution : Occurs in Indian gardens, and as escapes.

Parts Used : Whole Plant.

Flowering & Fruiting : July – August.

Folk Use : Plant used as stimulant.

Active Constituents : Attributed properties due to luteolin, its 7-rutinoside, 7-β-D-glucopyranoside, hesperidin acacetin, its 7-rutinoside, apigenin, its 7-rutinoside, 7-β-D-glucopyranoside, eriodictyol-7-rutinoside and eriodictyol-7-β-D-glucopyranoside (*plant*).

Reference : Ambasta (1986).

599. *Mentha arvensis* L.

Family: Lamiaceae.

English, Hindi and Sanskrit Names: **Eng. :** Corn Mint, Field Mint, Japanese Mint; **Hindi :** Podina, Pudina; **Sans. :** Putiha.

Distribution : Wild in the mountains and cultivated everywhere; W. Himalayas, Kashmir, 1,700-3,350m. Europe, N. & W.Asia–China

Parts Used : Whole Plant. Leaf.

Flowering & Fruiting : August – December.

Folk Use : Plant and oil used as stimulant. Leaves used as sedative, stimulant and cardio tonic.

Active Constituents : Active principles are L-menthol, menthhyl acetate, L-menthone, L-a-pinene, limonene (*plant*).

References : Ambasta (1986), Anonymous (1962), Bhattacharjee (2001), Kirtikar & Basu (1984), Reddi *et al.* (2005), Sharma (2003), Sharma (2004), Siwakoti & Varma (1996), Watt (1972).

600. *Mentha longifolia* (L.) Nathh.*

Syn.: M. longifolia var. *royleana* Rech. f.; *M. royleana* Wall. *ex* Benth.; *M. spicata* L. var. longifolia L.; *M. sylvestris* L.

Family: Lamiaceae.

English, Hindi and Sanskrit Names: **Eng. :** Horsemint; **Hindi :** Jungli Pudina, Podina; **Sans. :** Ajirnahara, Pudina, Rochani, Sugandhipatra, Vantihara.

Distribution : Known to occur in Kashmir, Punjab, Garhwal and Kumaon hills up to 4,000m.

Parts Used : Whole Plant. Flower. Leaf.

Flowering & Fruiting : Flowering during July – October.

Folk Use : Plant, leaves and flowers used as stimulant.

Active Constituents : Diospheral, piperitenone oxide, diosphenolene, piperitone, piperitenone, limonene and cineol (*plant*).

References : Ambasta (1986), Anonymous (1962), Farooq (2005), Kapur & Singh (2000), Kaul (1997), Kaushik & Dhiman (2000), Watt (1972).

601. *Mentha piperita* L. emend. Huds.

Family: Lamiaceae.

English and Hindi Names: **Eng. :** Brandy Mint, Pepper Mint; **Hindi :** Gamathi Pudina, Paparaminta.

Distribution : Cultivated in Kashmir, Dehra Dun, Delhi, Nilgiris and Mysore.

Parts Used : Whole Plant.

Flowering & Fruiting : July – September.

Folk Use : Plant used as a stimulant.

Active Constituents : Menthol, menthone, menthyl acetate, isomenthone, neomenthal, cineole, limonene, pipertione and pulegone (*plant*).

References : Ambasta (1986), Anonymous (1962), Kirtikar & Basu (1984), Rajan *et al.* (2003), Watt (1972).

602. *Mentha pulegium* L.

Family: Lamiaceae.

English Names: **Eng. :** English Pennyroyal, European Pennyroyal, Pudding Grass.

Distribution : Native of Europe and Western Asia; introduced in various parts of India.

Parts Used : Whole Plant.

Flowering & Fruiting : July – September.

Folk Use : Plant stimulant and digestive tonic.

Active Constituents : Pulegone, isopulegone, menthol and other terpenoids (*plant*).

References: Ambasta (1986), Anonymous (1962), Prajapati *et al.* (2006).

603. *Mentha spicata* L.

Syn.: M. spicàta var. *virdis* L.; *M. virdis* L.

Family: Lamiaceae.

English, Hindi and Sanskrit Names: **Eng. :** Garden Mint, Lamb Mint, Spearmint; **Hindi :** Pahari Pudina, Pudina; **Sans. :** Putiha, Rocani.

Distribution : Indigenous to England; naturalized and cultivated throughout India in gardens.

Parts Used : Whole Plant.

Flowering & Fruiting : July – December.

Folk Use : Herb regarded as stimulant.

Active Constituents : Protein, fat, carbohydrates, calcium, phosphorus, iron, carotene, nicotinic acid, riboflavin and thiamine (*leaf*), d-carvons, careen, d-sylvestrene and citranellol (*essential oil*).

References : Ambasta (1986), Anonymous (1962), Bhattacharjee (2001), Kaushik & Dhiman (2000), Sharma (2003, 2004).

604. *Menyanthes trifoliata* L.

Family: Gentianaceae.

English Names: **Eng. :** Buckbean, Goat's Bean, Marsh Trefoil, Water Shamrock, White Fluff.

Distribution : Kashmir and Temperate regions of India; also found in Europe, Caucasus, N. Asia, Amurland, Japan and N. America.

Parts Used : Whole Plant. Leaf. Rhizome.

Flowering & Fruiting : May – November.

Folk Use : Plant, leaves and rhizomes used as tonic and sedative.

Active Constituents : These are gentianine, gentianidine, gentialutine and gentiatibetine (*leaf*), coumarin, foliamenthin, loganin, dihydrofoliamenthin, menthiafoline, scopoletin and sweroside (*plant*), a major saponin – menyanthoside (*rhizome*).

References : Anonymous (1962), Bhattacharjee (2001), Kirtikar & Basu (1984), Watt (1972).

605. *Meriandra bengalensis* Benth.

Syn.: Salvia bengalensis Roxb.; *S. dianthera* Roth.

Family: Lamiaceae.

English and Hindi Names: **Eng. :** Bengal Sage; **Hindi :** Kafurkapat.

Distribution : Native of Abyssinia, cultivated in India.

Parts Used : Leaf.

Folk Use : Infusion of leaves used as tonic.

References : Ambasta (1986), Anonymous (1962), Watt (1972).

606. *Mesua ferrea* L.

Syn.: M. coromandeliana Wight.; *M. nagassarium* (Burm. f.) Kosterm; *M. pedunculata* Wight.; *M. speciosa* Chois.

Family: Clusiaceae.

English, Hindi and Sanskrit Names: **Eng. :** Ceylon Iron-wood, Iron-wood of Assam, Iron-wood Tree, Mesua; **Hindi :** Nagakesar, Naghar; **Sans. :** Champeya, Hema, Kanchana, Nagakesarah, Nagapushpah, Nogaya, Pinjara, Rukma.

Distribution : Evergreen forests throughout India up to 1,500m.

Parts Used : Flower.

Flowering & Fruiting : May – June.

Folk Use : Flowers used as cardio tonic.

References : Kirtikar & Basu (1984), Prajapati *et al.* (2006), Reddi *et al.* (2005).

607. *Michelia champaca* L.

Syn.: Champaca velutina Kuntze; *M. aurantiaca* Wall.; *M. daldsopa* Ham.; *M. rheedii* Wight.; *M. rufinervis* DC.

Family: Magnoliaceae.

English, Hindi and Sanskrit Names: **Eng. :** Champak, Golden Champa, Yellow Champa; **Hindi :** Champa, Champaca; **Sans. :**

Anjana, Champaka, Deepapushpa, Hemapushpa, Kanchana, Peetapushpa, Svarnapushpa.

Distribution : Evergreen forests throughout India, also cultivated.

Parts Used : Bark. Leaf. Flower. Fruit.

Flowering & Fruiting : April – September.

Folk Uses: Known so far as tonic (leaves, flowers, fruits) and stimulant (bark, flowers, fruits).

References: Ambasta (1986), Anonymous (1962), Bhattacharjee (2001), Kirtikar & Basu (1984), Parrotta (2001), Raju (2000), Ranjan (2000, 2003), Sharma (2003), Watt (1972).

608. *Mimosa hamata* Willd.

Syn.: M. armata Rottl.

Family: Mimosaceae.

English, Hindi and Sanskrit Names: **Eng. :** Hooked Mimosa; **Hindi :** Bander-ki-rakhi.

Distribution : Plains of Punjab, Central and South India.

Parts Used : Seeds.

Flowering & Fruiting : July – September (flowering); November – January (fruiting).

Folk Use : Seeds boiled with milk used as tonic.

Reference : Deshmukh *et al.* (2000).

609. *Mimusops elengi* L.

Syn.: Imbricaria perroudii Montrouz.; *Mimusops lucida* Poir.; *M. parvifolia* R. Br.

Family: Sapotaceae.

English, Hindi and Sanskrit Names: **Eng. :** Bullet-wood Tree; **Hindi :** Bakul, Bolsari, Maulser, Maulsiri; **Sans. :** Angaka, Bakulah, Chirapushpa, Dhanvi, Karuka, Madhupanjara, Mukula, Sharadika, Surabhi, Tailanga, Vakula.

Distribution : Evergreen forests throughout South India and Andaman Islands.

Parts Used : Bark. Flower. Fruit. Seeds.

Flowering & Fruiting : Flowering: March – April; Fruiting: April – June.

Folk Use : Bark and fruits used as tonic. Powdered dried flowers used as brain tonic. Flower oil considered stimulant. About 2g powdered seeds and ½ tsp powder given to women after delivery as general tonic.

Active Constituents : Known to yield quercetin, β-sitosterol and its glucoside and D-mannitol (*flower*), β-amyrin, basic and betulinic acids, lupeol acetate, quercitol, α-spinasterol, lupeol, taraxerol and taraxerone (*bark*), quercitol, ursolic acid, glucose, a triterpene alcohol, quercetin, dihydroquercetin and β sitosterol glucoside (*fruit, seeds*).

References : Anonymous (1962), Bhattacharjee (2001), Jain & Singh (1997), Kaushik & Dhiman (2000), Kirtikar & Basu (1984), Parrotta (2001), Prajapati *et al.* (2006), Sharma (2003, 2004), Siwakoti & Varma (1996), Watt (1972).

610. *Momordica balsamina* L.

Family: Cucurbitaceae.

English, Hindi and Sanskrit Names: **Eng. :** Balsam Apple; **Hindi :** Mokha; **Sans. :** Jangli Karela.

Distribution : Sind, Gujarat, Deccan, Punjab plain; also found in Malaya, Australia, W. Asia, Africa; introduced in America.

Parts Used : Whole Plant.

Flowering & Fruiting : August – November.

Folk Use : Plant used as tonic.

Active Constituents : Verbascoside, calceolarioside and rosmarinic acid (*aerial parts*).

References : Ambasta (1986), Anonymous (1962), Jain & Sharma (2000).

611. *Momordica charantia* L.*

Syn.: Cucumis africans Bat.; *Momordica humilis* Wall.; *M. muricata* DC.; *M. senegalensis* Lamk.

Family: Cucurbitaceae.

English, Hindi and Sanskrit Names: **Eng. :** Balsam Apple, Bitter Gourd, Carilla Fruit; **Hindi :** Karela, Kareli, Karola; **Sans. :** Ambuvallika, Karavella, Karvellak, Sushavi, Visakantaki.

Distribution : Common in coastal thickets, along creeks and streams and lowland forest margins; also commonly cultivated.

Parts Used : Fruit.

Flowering & Fruiting : Flowering and fruiting chiefly during April – November.

Folk Use : Fruits used as a tonic.

Active Constituents : Found to possess charantin, β-sitosterol-glucoside, stigmast-5, 25-diene-3β-O-glucoside, a mixture of two acylglycosyl sterols – 3-O-[6'-O-palmitoyl-β-D-glucosyl] stigmast-5, 25 (27)-diene (major) and 3-O-[6'-O-stearyl-β-D-glucosyl] stigmasta 5, 25 (27)-diene, benzyl alcohol, myrtenol, cis-3-hexenol, trans-2-hexenol, 1-penten-3-ol and cis-2-penten-1-ol (*fruit*).

References : Ambasta (1986), Anonymous (1962), Kaushik & Dhiman (2000), Parrotta (2001), Watt (1972).

612. *Monochoria vaginalis* Presl.

Syn.: M. plantaginea (Roxb.) Solms.; *Pontederia vaginalis* Burm. f.

Family: Pontederiaceae.

English and Sanskrit Names: **Eng. :** Heartleaf False Pickerel Weed, Oval-leaved Pondweed, Marshy Betelvine; **Sans. :** Indivarah.

Distribution : Ponds and marshes throughout India up to 1,500m.

Parts Used : Leaf. Root.

Flowering & Fruiting : July.

Folk Use : Roots and leaves used as a tonic.

Reference : Prajapati *et al.* (2006).

613. *Morinda citrifolia* L.

Family: Rubiaceae.

English, Hindi and Sanskrit Names: **Eng. :** Indian Mulberry; **Hindi :** Ach, Ak, Borraal; **Sans. :** Achchhuka, Ashyuka.

Distribution : Indigenous in the Darjeeling Terai and outer hills and on the Andaman and along the Konkan coast; cultivated largely in India and Myanmar; China Formosa.

Parts Used : Leaf.

Flowering & Fruiting : Throughout the year.

Folk Use : Leaves used as a tonic.

References : Anonymous (1962), Kirtikar & Basu (1984), Mukherjee (1986), Watt (1972).

614. *Morinda pubescens* J. E. Smith

Syn.: M. coreia Buch.-Hom.; *M. tinctoria* Roxb.; *M. tomentosa* Roth.

Family: Rubiaceae.

English, Hindi and Sanskrit Names: **Eng. :** Morinda Tree, Togari Wood of Madras; **Hindi :** Acc, Ach, Al; **Sans. :** Achchhuka, Paphanah.

Distribution : Bengal, Central and South India.

Parts Used : Leaf. Root.

Flowering & Fruiting : Flowering during April – July; fruiting in November – April.

Folk Use : Roots and leaves used as a tonic.

Active Constituents : Reported to yield morindone, damnacanthal, nordamnacanthal, β-sitosterol (*root*), and ursolic acid (*leaf*).

Reference : Prajapati *et al.* (2006).

615. *Moringa oleifera* Lam.

Syn.: Guilandina moringa L.; *Hyperanthera decandra* Willd.; *H. moringa* Vahl.; *Moringa polygana* DC.; *M. pterygosperma* Gaertn.; *M. zeylanica* Pers.

Family: Moringaceae.

English, Hindi and Sanskrit Names: **Eng.** : Drumstick Tree, Horse Radish Tree; **Hindi** : Mungna, Sahinjna, Seyla, Shajna; **Sans.** : Akshiba, Chalusha, Jalaproya, Sainjna, Shigru, Shobhanjana, Tilashigru.

Distribution : Indigenous to N.W. India; also cultivated throughout India.

Parts Used : Flower. Root. Fruit.

Flowering & Fruiting : Flowering during February – April; fruiting in May – June.

Folk Use : Flowers and fruits used as tonic and stimulant. Roots valued as stimulant, cardiac and circulatory tonic.

Active Constituents : Known to possess quercetin and kaempferol, lysine, sucrose, glucose (*flower*), and benzyl isothiocyanate (*root*).

References : Ambasta (1986), Anonymous (1962), Banerjee (1996, 2000b), Bhattacharjee (2001), Dhiman (2000), Karnick *et al.* (1981), Kaushik & Dhiman (2000), Kirtikar & Basu (1984), Mughal *et al.* (2000), Parrotta (2001), Sharma (2003), Siwakoti & Varma (1996).

616. *Moschosma polystachyum* Benth.

Syn.: Ocimum polystachyum L.; *Plectranthus parviflorus* Br.; *P. micranthus* Spreng.

Family: Lamiaceae.

English Name: **Eng.** : Musk Basil.

Distribution : Wet places in Bengal; Dacca; Deccan Peninsula, Myanmar and Tenasserim; Sri Lanka, Java, China, Philippine Islands, Tropical Africa and Australia.

Parts Used : Leaf.

Folk Use : Decoction of leaves used as sedative.

References : Ambasta (1986), Anonymous (1962).

617. *Mucuna monosperma* DC.

Syn.: Carpopogon monospermum Roxb.; *Mucuna anguina* Wall.; *M. corymbosa* Grah.; *M. cristata* Ham.

Family: Fabaceae.

English and Sanskrit Names: **Eng.** : Negro Bean; **Sans.** : Dadhipushpi, Khatavangi, Kupa, Vanshya.

Distribution : E. Himalayas, Khasia, Assam, Chittagong, Konkan and Sri Lanka.

Parts Used : Seeds.

Flowering & Fruiting : October – May.

Folk Use : Seeds used as sedative and tonic.

References : Ambasta (1986), Kirtikar & Basu (1984), Raju (2000), Watt (1972).

618. *Mucuna prurita* Hook.

Syn.: Caryopogon pruriens Roxb.; *Dolichos pruriens* L.; *Mucuna pruriens* Baker, non DC.; *M. utilis* Wall.; *Stizolobium pruriens* (L.) Medicus.

Bean;*Family:* Fabaceae.

English, Hindi and Sanskrit Names: **Eng. :** Common Cowitch, Cowhage, Velvet Bean; **Hindi :** Goncha, Kaunch, Kiwach; **Sans. :** Atamagupta, Badari, Chanda, Kapikacchu, Romalu.

Distribution : Occurs wild almost throughout India; also in Andaman and Nicobar Islands.

Parts Used : Leaf. Root. Seeds.

Flowering & Fruiting : Flowering: September – October; fruiting in December – March.

Folk Use : Leaves, roots and seeds used as tonic. Seeds and roots used as nervine tonic. Also, roots valued as stimulant.

Active Constituents : Associated properties due to dihydroxyphenylalanine, lecithin, galic acid, prurieninine and pruienidine; mucunine, mucundine, tryptamine and 5-hydroxy-tryptamine (*seeds*), 6-methoxyharman (*leaf*), resin, tannin, fat and traces of manganese (*root*).

References : Ambasta (1986), Anonymous (1962), Bhattacharjee (2001), Dhiman (2005), Dwivedi (2000), Farooq (2005), Kaushik & Dhiman (2000), Kirtikar & Basu (1984), Kumar & Lal (2003), Lalramnghinglova (2003), Manandhar (2003), Parrotta (2001), Prajapati *et al.* (2006), Ranjan (2000), Reddi *et al.* (2005), Sharma (2003), Singh (2001), Singh *et al.* (2007), Sood *et al.* (2009), Siwakoti & Varma (1996), Watt (1972).

619. *Mundulea sericea* (Willd.) A. Cheval

Syn.: M. suberosa Benth.

Family: Fabaceae.

English Names: **Eng. :** Cork Bush, Rhodesian Silver-leaf, Silver Bush.

Distribution : Dry forests and rocky hills of West and South India, in Konkan, Circars, Deccan and Carnatic to Tinnevelly, up to 1,200m.

Parts Used : Root Bark.

Flowering & Fruiting : October – January (flowering); February – April (fruiting).

Folk Use : Paste of its grounded root bark with stem bark of *Sapindea emarginatus* and *Azadirachta indica* and garlic in equal quantities used as a general tonic.

Active Constituents : Munetone and mundulone (*root bark*).

Reference : Kumar & Pullaiah (1998).

620. *Murraya koenigii* (L.) Spreng.

Syn.: Bergera koenigii L.

Family: Rutaceae.

English, Hindi and Sanskrit Names: **Eng. :** Curry Leaf Tree; **Hindi :** Barsanga, Gandhela, Kathnim, Kurry Patta, Mitha Neem; **Sans. :** Krishna-nemba, Raman, Surabhininiba.

Distribution : Throughout India, ascending to 1,650m.

Parts Used : Whole Plant. Bark. Leaf. Root.

Flowering & Fruiting : Flowering during February – April; fruiting in cold season.

Folk Use : Plant, leaves, roots and bark used as tonic whereas bark, root and leaves also valued as stimulant.

Active Constituents : These are pyranocarbazole, alkaloid – girinimbine and murraganine (*bark*), a carbazole carboxylic acid – mukoeic acid (*stem bark*), mahanimbine, Koenimbine, koenigicine, cyclomahanimbine, mahanimbidine, mahanine, koenine, koeniginen koenigicine koenidine, koenimbine and muconicine (*leaf*), girinimbine, mahanimbine and murrayacine (*root*).

References : Ambasta (1986), Anonymous (1962), Bhattacharjee (2001), Farooq (2005), Kaushik & Dhiman (1984), Kirtikar & Basu (1984), Ranjan (2000), Ranjan *et al.* (2003), Sharma (2003), Siwakoti & Varma (1996), Subramaniam (2000).

621. *Murraya paniculata* (L.) Jack.

Syn.: Chalcas exotica (L.) Millsp.; *C. paniculata* L.; *C. sumatrana* Roem.; *Murraya exotica* L.; *M. sumatrana* Roxb.

Family: Rutaceae.

English, Hindi and Sanskrit Names: **Eng. :** Orange Jasmine, The China Box Tree; **Hindi :** Bibsar, Juti, Kamini, Marchula; **Sans. :** Ekangi, Surabhininiba.

Distribution : Commonly found in the shady ravine slopes up to 1,500m in Uttaranchal Himalayas.

Parts Used : Bark. Leaf. Root.

Flowering & Fruiting : March – September.

Folk Use : Leaves used as stimulant. Leaves, bark and roots also possess tonic properties.

Active Constituents : Known to contain phebalosin, hibiscetin, heptamethyl ether, paniculatin, murragone, phebalosin hydrate, hentriacontane, octacosanol, aurapten, murralongin, murrangatin, meragin hydrate, exozoline, murraxonin, murraxocin, colensanone, coumurrin, murpaniculol, paniculol and paniculin, peroxyauraptenol, cis – dehydroosthol, murraol, murranganon, murrangatin acetate, isomurranganon, senecioate, isomurralonginol acetate, chloticol, isomurralonginol nicotinate, panial, cisosthenon, peroxymurraol and paniculonal (*leaf*), mexolide, 3-(1, 1 - dimethylallyl) xanthyletin, aurantiamide acetate, sitosterol-β-D-galactoside (*bark*), yuehchukene (*root*).

References : Anonymous (1962), Bhattacharjee (2001), Kirtikar & Basu (1984), Parrotta (2001), Ranjan (2003), Sharma (2003).

622. *Musa paradisiaca* L. (Pl. 8D)*

Syn.: Musa sapientum L.

Family: Musaceae.

English, Hindi and Sanskrit Names: **Eng. :** Adam's Fig, Banana, Fig of India, Kadali, Plantain; **Hindi :** Maoz-Kula, Kach-Kula, Kela; **Sans. :** Alapu, Ambusara, Bhanuphala, Bhanupriya, Kadali, Nisara, Rambha, Sukumara, Tatpatri, Varavriksha.

Distribution : Indigenous in Bihar and the E. Himalayas up to 1,350m, Sri Lanka; cultivated throughout India and the tropics.

Parts Used : Root. Stem. Fruit.

Flowering & Fruiting : Flowering and fruiting almost throughout the year.

Folk Use : Fruits, roots and stem used as tonic.

Active Constituents : Protein, glutelin, prolamines and proteoses. Essential amino acids are arginine, histidine, isoleucine, leucine, lysine, methionine, phenylalanine, threonine, tryptophan and valine. Besides, it also contains Ca, Fe, K, Mg, Na, P, S, Mn, Cu, Cl, Si, I, Al, Zn, Co and As and vitamin A, B and C (*fruit*).

References : Kaushik & Dhiman (2000), Prajapati *et al.* (2006), Saikia & Nath (2003b), Sharma (2003), Watt (1972).

623. *Mussaenda frondosa* L.

Family: Rubiaceae.

Hindi and Sanskrit Names: **Hindi :** Bebina, Bedina; **Sans. :** Rajatarih, Sriparnah.

Distribution : Throughout India in Forests, also cultivated.

Parts Used : Whole Plant.

Flowering & Fruiting : July – October.

Folk Use : Plant used as cardio tonic.

Reference : Prajapati *et al.* (2006).

624. *Myrica esculenta* Buch. – Ham. *ex* D. Don

Syn.: M. integrefolia Roxb.; *M. nagi* Hook. f.; *M. rubra* Sieb. & Zucc.; *M. sapida* Wall.

Family: Myricaceae.

English, Hindi and Sanskrit Names: **Eng. :** Box Berry, Box Myrtle; **Hindi :** Kaephal, Kaiphal; **Sans. :** Aranya, Katphala, Laghukashmarya, Nasanu, Rohini.

Distribution : Sub-tropical Himalayas from Ravi eastwards to Assam as well as in Lushar, Khasia, Jaintia and Naga hills between 900-2,100m.

Parts Used : Bark. Fruit.

Flowering & Fruiting : Flowering: October – November; and fruiting later.

Folk Use : Bark considered as tonic. Fruits used as sedative.

Active Constituents : Active principles due to β-sitosterol, taraxerol, myricadiol, myricanone, glucoside, myricanone, myricanol-5-O-β-D-glucopyranosyl (1→3)-β-D-glucopyranoside, myricanol-5-O-α-L-arabinofuranosyl (1→6)-β-D-glucopyranoside (*bark*).

References : Anonymous (1959), Dhiman (2005), Kirtikar & Basu (1984), Sharma (2003), Watt (1972).

625. *Myristica fragrans* Houtt.*

Family: Myristicaceae.

English, Hindi and Sanskrit Names: **Eng. :** Nutmeg Tree; **Hindi :** Jaiphal, Javitri, Jayapatri, Jotri; **Sans. :** Jaiphala, Javitri.

Distribution : Native of Moluccas Islands; grown with success at various places in Kerala, Tamil Nadu and West Bengal.

Parts Used : Seed (nutmeg).

Folk Use : Nutmeg used as a stimulant tonic.

Active Constituents : Activities due to dodeconoic acid, tetradecenoic acid, hexadeconoic acid, octadecenoic acid, octadecadienoic acid, α-terpinene, α- and β-pinene, myrcene, camphene, sabinene, p-menth-1,4-diene, p-menth-1, 8-diene, α-phellandrene, β-phellandrene, 3-carene, thujone, toluene, p-cymene, linalyl acetate, bornyl acetate (*seeds*).

References : Rajan *et al.* (2003), Thakur *et al.* (1989), Watt (1972).

626. *Myroxylon balsamum* Harms.

Syn.: Myroxylon toluiferum H. B. & K.

Family: Fabaceae.

English Names: **Eng. :** Quina-quina, Tolu Balsam Tree.

Distribution : Indigenous to Venezuela, Colombia and Peru; grown in Lal Bagh Botanic Garden (Bangalore) and Kallar (Nilgiris); some areas in S. India and eastern Himalayas between 450-900m considered suitable for its cultivation.

Parts Used : Trunk.

Flowering & Fruiting : July – September (flowering); October – November (fruiting).

Folk Use : Trunk oil called, oil of Tolu Balsam used as a stimulant.

Active Constituents : Known to possess (±) 7-hydroxyl-4'-methoxyisoflavnone (I), (±)-7, 3'-dihydroxy - 4' - methoxyisoflavnone (II) and a benzofuran – 2-(2', 4'-dihydroxyphenyl)-5, 6-dimethoxybenzofuran (III) (*trunk wood*).

References : Ambasta (1986), Anonymous (1962).

627. *Myrtus communis* L.

Family: Myrtaceae.

English and Hindi Names: **Eng. :** Common Myrtle, Myrtle; **Hindi :** Murad, Vilayatimehndi.

Distribution : Indigenous from Mediterranean – N.W. Himalayas; often grown in gardens throughout India.

Parts Used : Leaf. Fresh Shoots. Fruit.

Flowering & Fruiting : Flowering during summers.

Folk Use : Shoots and leaves used as nerve sedative. Fruits used as a tonic to brain and heart (Yunani).

Active Constituents : Reported to possess myricetin, kaempferol, ellagic acid and gallic acid, myricetin-3-galactoside, myricetin-rhamnoside, quercetin-3-galactoside and -3-glucoside, esculin, caffeic acid, 3-rhamnoglucoside and 3, 3'-digalactoside (*leaf*).

References : Kirtikar & Basu (1984), Rajan *et al.* (2000).

628. *Myxopyrum serratulum* A. W. Hill

Syn.: M. smilacifalium C. B. Clarke.

Family: Oleaceae.

Hindi and Sanskrit Names: **Hindi :** Pilacameli; **Sans. :** Caturdharalata, Hemamalati.

Distribution : Evergreen forests throughout Kerala at 600-900m.

Parts Used : Leaf.

Folk Use : Leaves used as a tonic.

Reference : Prajapati *et al.* (2006).

629. *Nardostachys jatamansi* DC.

Syn.: N. grandiflora DC.; *Valeriana jatamansi* auct non. Jones.

Family: Valerianaceae.

English, Hindi and Sanskrit Names: **Eng. :** Indian Nard, Spikenard; **Hindi :** Bal-chir, Bhoot Jata, Jalalasi, Jatamansi; **Sans. :** Amritajata, Bhutajata, Gauri, Hinsra, Jatamansi, Jatila, Keshi, Kiratini, Nadala, Peshi.

Distribution : Distributed in alpine Himalayas: 3,352-4,572m extending eastwards from Kumaon – Sikkim up to 5,181m.

Parts Used : Rhizome. Root.

Flowering & Fruiting : February – May (flowering); May – June (fruiting).

Folk Use : Rhizomes used as a tonic and stimulant. Roots valued as tranquilizer, sedative and nervine tonic.

Active Constituents : Found to contain new terpenic caumarin – jatamansin, oroselol, α-pinene, β-pinene, 3-carene, β-endesmol, elemol a, C30-hydrocarbom, β-sitosterol, angelicin, jatamansinol and a new diethenoid bicyclic ketone – nardostachone (*root*), jatamansone and actinidine (*rhizome*).

References : Ambasta (1986), Anonymous (1962), Arya (2004), Bhattacharjee (2001), Chhetri (2005), Dhiman (2005), Farooq (2005), Gupta (1997), Gautam *et al.* (2007), Kirtikar & Basu (1984), Negi *et al.* (1999), Prakash & Mehrotra (1996), Raju (2000), Rawat *et al.* (2007), Reddi *et al.* (2005), Samant *et al.* (2001), Shah (2006), Sharma (2003, 2004), Sharma & Sood (1997), Thakur *et al.* (1989), Watt (1972).

630. *Nasturtium officinale* R. Br.

Syn.: Cardamine fontana Lam.; *Nasturtium fontanum* Aschers.; *Sisymbrium nasturtium - aquaticum* L.

Family: Brassicaceae.

English Names: **Eng. :** Shamrock, Stertion, Water Cress.

Distribution : Native of Europe, N. Africa and W. Asia; naturalized in many countries including India, and is commonly found in ditches, pools and margins of shallow streams up to 2,100m.

Parts Used : Whole Plant. Leaf.

Flowering & Fruiting : April – September.

Folk Use : Leaves used as stimulant. Plant taken boiled or raw as a tonic.

References : Ambasta (1986), Anonymous (1966), Chhetri (1994), Kirtikar & Basu (1984).

631. *Nelsonia campestris* R. Br.

Syn.: Justicia hirsuta Vahl.; *J. lamifolia* Koen.; *J. tomentosa* Wall.; *Nelsonia canescens* Nees.; *N. lamiifolia* Spreng.; *N. origanoides* Roem. & Sch.; *N. tomentosa* Nees.

Family: Acanthaceae.

English Name: **Eng. :** Reinwardtia.

Distribution : Throughout India, up to 1,350m, Himalayas – Sri Lanka and Malacca, common; not in Western desert of India. Also, found in S. E. Asia, Australia, Africa, and America.

Parts Used : Root.

Flowering & Fruiting : November – May.

Folk Use : About 4cm of root grounded with water taken as a tonic.

Reference : Shah & Singh (1990).

632. *Nelumbo nucifera* Gaertn.

Syn.: Cyamus nelumbo Smith.; *Nelumbo indica* Poir.; *N. nelumbo* Druce.; *N. speciosum* Willd.

Family: Nymphaeaceae.

English, Hindi and Sanskrit Names: **Eng. :** Chinese Water Lily, Egyptian Beans, Indian Lotus, Pythagorean Bean, Sacred Lotus; **Hindi :** Ambuj, Kamal, Kanval; **Sans. :** Ambhoja, Asyapatra, Drishopadma, Indiralaya, Padama, Pushkara, Varisona.

Distribution : Throughout warmer parts of India in ponds; also cultivated for its sweet - scented flowers.

Parts Used : Flower. Fruit.

Flowering & Fruiting : Flowering: summer – autumn; & fruiting in winters.

Folk Use : Fruits and carpel taken as tonic. White flowers valued as good tonic for heart and brain.

Active Constituents : Known to exhibit isoquercetin and glucoluteolin (*petals, stamens*), Kaempferol – 3 – galactorhamnoglucoside (*flower*), quercetin and luteolin (*receptacle of flower*).

References : Chauhan *et al.* (2007), Kaul (1997), Kaushik & Dhiman (2000), Kirtikar & Basu (1984), Maya *et al.* (2003), Sharma (2003, 2004).

633. *Nepeta cataria* L.

Family: Lamiaceae.

English Names: **Eng. :** Catmint, Catnip.

Distribution : Native of Europe; found in W. temperate Himalayas from Dalhousie – Kashmir, up to 1,500m.

Parts Used : Aerial Parts. Leaf. Flower.

Flowering & Fruiting : Flowering during July – September & fruiting : September – October.

Folk Use : Leaves and flowering tops used as tonic. Aerial parts used as sedative.

Active Constituents : These are nepetoglucosyl ester (*leaf*), and an iridoid glucoside – nepetaside (*aerial parts*).

References : Ambasta (1986), Anonymous (1966), Harborne & Baxter (2001), Prajapati *et al.* (2006).

634. ***Nepeta discolor*** Royle ex Benth.

Family: Lamiaceae.

Distribution : Western temperate and Alpine Himalayas; from Kashmir – Garhwal and Western Tibet: 3,350-5,000m, Afghanistan.

Parts Used : Whole Plant.

Flowering & Fruiting : July – September.

Folk Use : Plant used as a tonic.

Reference : Singh *et al.* (1996).

635. ***Nepeta flocosa*** Benth.

Syn.: Marmoritis rotundifolia Benth.; *Nepeta rotundifolia* Benth.

Family: Lamiaceae.

English Name: **Eng. :** Woolly Catmint.

Distribution : Kashmir and Ladakh: 2,500-6,000m; Western Tibet: 2,350-3,650m.

Parts Used : Whole Plant.

Flowering & Fruiting : May – September.

Folk Use : Plant used as tonic.

Reference : Shah (2010).

636. ***Nepeta hindostana*** (Roth.) Haines

Syn.: Glechoma hindostana Roth.; *Nepeta ruderalis* Buch.-Ham. *ex* Benth.

Family: Lamiaceae.

English and Hindi Names: **Eng. :** North Indian Catmint; **Hindi :** Billilotan.

Distribution : Practically throughout India, chiefly in hilly parts of Punjab, Bengal, Bihar, U. P., Rajputana, Central India, Deccan and Konkan.

Parts Used : Whole Plant. Flower. Leaf.

Flowering & Fruiting : Throughout the year.

Folk Use : Plant taken as cardiac tonic. Leaves and flowering tops used as stimulant.

Active Constituents : Principle constituents are nepitrin, stigmasterol, β-sitosterol, dinatin, nepetin, 6-hydroxyluteolin, a new flavone – 7, 4'-O-dimethylscutellarein, nehipetol, nehipediol, α- and β-amyrins, eicosane, heneicosane, hexacosane, hentriacontane, triacontane, dotriacontane (*plant*).

References : Anonymous (1966), Bhattacharjee (2001), Kaushik & Dhiman (2000), Kirtikar & Basu (1984), Sharma (2003), Watt (1972).

637. *Nerium indicum* Mill. (Pl. 8E)

Syn.: Nerium latifolium Mill.; *N. odoratum* Lamk.; *N. odorum* Soland.; *N. oleander* L.

Family: Apocynaceae.

English, Hindi and Sanskrit Names: **Eng. :** Indian Oleander, Sweet-scented Oleander; **Hindi :** Karber, Kuruvira, Lal Kaner; **Sans. :** Asvamaraca, Chandata, Hayamaraca, Karavira, Sataprasa.

Distribution : Western Himalayas (Nepal westwards, ascending to 2,000m); extensively cultivated throughout the greater part of India as well as in China and Japan.

Parts Used : Leaf. Root.

Flowering & Fruiting : April – November.

Folk Use : Leaves used as cardiac tonic. Roots used as tonic and cardiac tonic.

Active Constituents : Accredited properties due to dambonitol, oleandrigenin-β-D-glucopyranosyl-β-D-diginopyranoside, digitoxigenin-α-L-oleandroside, 5α-adynerin, quercitrin, isoquercitrin, kaneric acid, neriucoumaric and isoneriucoumaric acids, oleanderol and betulinic ursolic and oleanolic acids, kanerocin, oleandrolic acid, kanerodione, adynerigenin-β-odorotrioside, 16-dehydroadynerigenin-β-odotrioside, 16-dehydro-adynerigenin-β-gentiobiosyl – β – D - sarmentoside, olendrigenin-β-neritrioside, oleandrigenin-β-D-glucosyl-β-D-sarmentoside, digitoxigenin-β-oleatrioside, 8-β-hydroxydigitoxigenin-β-neritrioside, 8β-hydroxy-16-dehydrodigitoxigenin-β-neritrioside and 8β-hydroxy-16-dehydrodigitoxigenin-β-odorobioside (*leaf*), six cardenolide glycosides – digitoxigenin-β-gentiotriosyl (1→4)-β-D-digitaloside, uzarigenin-β-gentiobiosyl (1→4)-β-D-digitaloside, 5α-oleandrigenin, its β-D-digitaloside, β-D-glucosyl (1→4)-β-D-diginoside and β-D-glucosyl (1→4)-β-D-digitaloside and two pregnanolone glycosides - 5α-pregnanolone-O-β-D-glucosyl [(2→1)-β-D-glucosyl] (6→1) –β-D-glucoside (III) and pregnenolone-β-D-apiosyl (1→6)-β-D-glucoside (IV) (*root*).

References : Anonymous (1966), Parrotta (2001), Verma & Chauhan (2006).

638. *Nervilia aragoana* Gaud.

Syn.: Pogonia carinata Wt.; *P. flabelliformis* Lindl.

Family: Orchidaceae.

Hindi and Sanskrit Names: **Hindi :** Sthalakamal, Sthalapadma; **Sans. :** Padmacarini.

Distribution : Himalayas from Garhwal eastwards : 1,200-1,500m, Pulney hills, Konkan, N. Kanara and Travancore.

Parts Used : Whole Plant.

Flowering & Fruiting : June – July.

Folk Use : Plant used as a tonic.

Active Constituents : Phytol, cycloeucolenol, stigmasterol, linolenic acid, linoleic acid, L-norleucine and a glycine ester have been characterized from the *plant*.

References : Prajapati *et al.* (2006), Reddi *et al.* (2005).

639. *Nicotiana tabacum* L. (Pl. 8F)*

Family: Solanaceae.

English, Hindi and Sanskrit Names: **Eng. :** American Tobacco, Tobacco, Virginian Tobacco; **Hindi :** Bujjerbhang, Tamaku, Tambaku; **Sans. :** Dhumrapatrika, Kalanja, Krimighni, Tamakhuh, Vajrabhringi.

Distribution : Cultivated in hotter parts of India.

Parts Used : Leaf.

Flowering & Fruiting : December – April.

Folk Use : Leaves used as sedative, stimulant and tonic; also chewed by Soligas as narcotics.

Active Constituents : Nicotine, nornicotine, Anabasine, l-nicotine, nicotyrine, nicotimine, l-noricotine, d-nornicotine, piperidine, pyrrolidine, N-methylpyrroline, 2, 3'-dipyridyl, l-anabasine, N-methyl-l-anabasine, l-anatabine, N-methyl-l-anatobine, nicotoine, necatalline, myosmine (*leaf*).

References : Ambasta (1986), Anonymous (1966), Farooq (2005), Kirtikar & Basu (1984), Kshirsagar & Singh (2007), Parrotta (2001), Prajapati *et al.* (2006), Ranjan (2003), Sharma (2004), Watt (1972).

640. *Nigella sativa* L.

Syn.: N. indica Roxb. *ex* Flem.

Family: Ranunculaceae.

English, Hindi and Sanskrit Names: **Eng.** : Black Cumin, Small Fennel; **Hindi** : Kalajira, Kalonji, Mugrela; **Sans.** : Bashpika, Karavi, Krsnajraka, Musani, Upakuncika.

Distribution : Punjab, Bihar; often cultivated and an occasional weed of cultivation.

Parts Used : Seeds.

Flowering & Fruiting : September – January.

Folk Use : Seeds used as stimulant.

Active Constituents : Known to possess nigellimine, cholesterol, campasterol, stigmasterol, β-sitosterol, α-spinasterol, nigellimine N-oxide and nigellicine (*seeds*).

References : Ambasta (1986), Anonymous (1966), Farooq (2005), Kirtikar & Basu (1984), Pal & Paul (2003), Parrotta (2001), Prajapati *et al.* (2006), Ranjan *et al.* (2003).

641. *Nyctanthes arbor - tristis* L.

Syn.: Parilium arbor-tristis Gaertn.; *Scabrita scabra* L.; *S. triflora* L.

Family: Oleaceae.

English, Hindi and Sanskrit Names: **Eng.** : Coral Jasmine, Indian Mourner, Night Jasmine, Sorrowful Tree; **Hindi** : Binari, Ganga Sedi, Har Singar, Karosli, Kuri, Nibari, Parja; **Sans.** : Atyuha, Kharopatraka, Nalakunkuma, Parijatah, Rajanikasa, Sophalika.

Distribution : Wild in outer Himalayan range from Kashmir – Sikkim, Assam, Bengal, C. India and southwards to Godavari; also cultivated throughout India.

Parts Used : Leaf.

Flowering & Fruiting : Flowering: August – November & fruiting in cold season.

Folk Use : Juice of leaves used as a tonic, cardiac tonic and mild bitter tonic.

Active Constituents : Isolated constituents are arborsides A, B and C, arbortristoside A, ascorbic acid, carotene, hentriacontane, 6β-hydroxyloganin, fructose, glucose, mannitol, benoic acid, methylsalicylate, tannic acid, β-amyrin, friedelin, lupeol, oleanolic acid and β-sitosterol glucoside (*leaf*).

References : Dhiman (2005), Kaushik & Dhiman (2000), Sharma (2003), Srivastava *et al.* (2003a).

642. *Nymphaea alba* L.

Syn.: N. cachemeriana Cambess.

Family: Nymphaeaceae.

English and Hindi Names: **Eng.** : European White Water-lily, Lotus Lily, Water Rose; **Hindi** : Kumuda.

Distribution : Native to Europe; found in ponds and still water in lakes, rivers and canals (lakes of Kashmir, below 1,800m).

Parts Used : Rhizome.

Flowering & Fruiting : May – August.

Folk Use : Rhizomes used as a mild sedative and narcotic.

Active Constituents : Nymphaeine, resin, glycosides and tannins (*rhizome*).

References : Ambasta (1986), Kirtikar & Basu (1984), Prajapati *et al.* (2006), Ranjan (2003), Watt (1972).

643. *Nymphaea nouchali* Burm. f.

Syn.: N. lotus Hooks.; *N. pubescens* Willd.

Family: Nymphaeaceae.

English, Hindi and Sanskrit Names: **Eng. :** Indian Water-lily; **Hindi :** Chhota-sundi, Kanval, Kokka, Rakta Chandana; **Sans. :** Alagandha, Alipriya, Alohita, Alpapatra, Bhadra, Halika, Indivara, Kamuda, Kumudan, Ravipriya, Rochana.

Distribution : Throughout the warmer parts of India, in tanks, ponds and ditches.

Parts Used : Flower. Rhizome.

Flowering & Fruiting : Throughout the year.

Folk Use : Rhizome and flowers used as a tonic and cardio tonic.

References : Anonymous (1966), Gogoi & Das (2003), Kirtikar & Basu (1984), Prajapati *et al.* (2006), Reddi *et al.* (2005), Sharma (2004), Srivastava *et al.* (2003a), Watt (1972).

644. *Nymphaea stellata* Willd.

Syn.: Castalia stellata (Willd.) Blume.

Family: Nymphaeaceae.

English, Hindi and Sanskrit Names: **Eng. :** Blue Water-lily, Indian Blue Water-lily; **Hindi :** Lilophal, Nilkamal, Nilpadma; **Sans. :** Asistotpala, Indiwar, Kandotra, Mridutpala, Nilopala, Utpala.

Distribution : Tropical Asia and Africa; warmer parts of India.

Parts Used : Flower.

Flowering & Fruiting : August – November.

Folk Use : Decoction of flowers considered narcotic.

Active Constituents : A hot water – soluble polysaccharide consisted of xylose (64.5%) and arabinose (35.5%) (*flower*).

Reference : Ambasta (1986).

645. *Ochna jabotapita* L.

Syn.: O. lucida Lamk.; *O. squarrosa* L.

Family: Ochnaceae.

English and Sanskrit Names: **Eng.** : Golden Champak; **Sans.** : Kanakchampa.

Distribution : Assam, Bihar, Burma, Orissa and the Deccan Peninsula, W. Peninsula, Sri Lanka.

Parts Used : Bark.

Flowering & Fruiting : March – May.

Folk Use : Bark used as tonic and digestive tonic.

References : Ambasta (1986), Anonymous (1966), Kirtikar & Basu (1984).

646. *Ocimum americanum* L.

Syn.: O. africanum Lour.; *O. canum* Sims.

Family: Lamiaceae.

English, Hindi and Sanskrit Names: **Eng.** : Common Basil, Hoary Basil; **Hindi** : Babutulsi, Bahari, Mamri, Kala Tulsi, Vantulsi; **Sans.** : Ajaka, Aranyatulasi, Gambhira, Kuthera, Vanabarbarika.

Distribution : In fields and wastelands throughout India.

Parts Used : Leaf. Flower. Seeds.

Flowering & Fruiting : April – September.

Folk Use : Seeds used as tonic. Leaves and flowers considered useful as cardio tonic.

Active Constituents : Nevadensin, Salvigenin, uraolic acid and oleanolic acid (*leaf*).

References : Ambasta (1986), Anonymous (1966), Bhattacharjee (2001), Parrotta (2001), Prajapati *et al.* (2006), Sharma (2003), Singh *et al.* (2003), Srivastava *et al.* (2003a).

647. *Ocimum basilicum* L.*

Syn.: O. album L.; *O. caryophyllatum* Roxb.; *O. ciliatum* Hornem.; *O. hispidum* Lamk.; *O. methaefolium* Benth.; *O. minimum* Burm.; *O. pilosum* Willd.; *Plectranthus barrelieri* Spreng.

Family: Lamiaceae.

English, Hindi and Sanskrit Names: **Eng.** : Common Basil, Sweet Basil; **Hindi** : Babui Tulsi, Ban Tulsi, Gulal Tulsi, Kali Tulsi, Marua; **Sans.** : Arjakan, Asurasa, Karahi, Munjariki, Surasa, Tungi, Varvara.

Distribution : Cultivated throughout the greater parts of India, Sri Lanka and Myanmar.

Parts Used : Whole Plant. Flower.

Flowering & Fruiting : June – September.

Folk Use : Plant and flowers used as a stimulant.

Active Constituents : Methyl cinnamate, linalool, methyl chaviocol, ocimene, linalyl acetate, eugenol and trans-anethole (*plant*).

References : Ambasta (1986), Anonymous (1966), Jain & Sharma (2000), Kirtikar & Basu (1984), Ranjan (2003), Singh *et al.* (2007).

648. *Ocimum gratissimum* L.*

Syn.: O. citronatum Ham.; *O. robustum* Heyne.; *O. virids* Willd.

Family: Lamiaceae.

English, Hindi and Sanskrit Names: **Eng. :** Large Basil, Lemon Basil, Shrubby Basil, Tea Bush; **Hindi :** Ban Tulsi, Malatulsi, Ram Tulsi; **Sans. :** Ajaka, Nidralu, Shophahari, Vriddhutulsi.

Distribution : Almost throughout India; often cultivated in gardens.

Parts Used : Whole Plant.

Flowering & Fruiting : September – October.

Folk Use : Plant used as tonic and stimulant.

Active Constituents : Citral, geraniol, citronelol, geranyl acetate and gratissimol (*plant*).

References : Ambasta (1986), Anonymous (1966), Kirtikar & Basu (1984), Parrotta (2001), Thakur *et al.* (1989).

649. *Ocimum sanctum* L.*

Syn.: O. inodorum Burm; *O. monochorum* L.; *O. tenuiflorum* L.

Family: Lamiaceae.

English, Hindi and Sanskrit Names: **Eng. :** Holy Basil, Sacred Basil; **Hindi :** Baranda, Kala-Tulsi, Tulsi; **Sans. :** Ajaka, Brinda, Manjari, Sursa, Tulsi.

Distribution : Cultivated throughout India and often runs wild.

Parts Used : Leaf.

Flowering & Fruiting : Flowering and fruiting almost throughout the year, chiefly in winter.

Folk Use : Juice and infusion of leaves considered stimulant. Juice of leaves mixed with honey and a few drops of juice extracted from rhizome of *Zingiber officinale* Rosc. taken as tonic.

Active Constituents : Known to possess cardiene, β-carotene, ursolic acid, eugenol and limoneme, stenocerol, hexaatinocentine, eugenol, orientin, vicenin, bornyl acetate, cadinene, camphene, camphor, carvacrol, β-caryophyllene, 1, 8-cineole,deacylaldehyde, humelene, limonene, methyl charical, nerol, α- and β-pinene, apigenin, molludustin and ursolic acid (*leaf*).

References : Anonymous (1966), Farooq (2005), Kaushik & Dhiman (2000), Parrotta (2001), Raju (2000), Saikia & Nath (2003b), Sarma *et al.* (2002), Sharma (2003), Siwakoti & Varma (1996).

650. *Oenothera biennis* L.

Family: Onagraceae.

English Name: **Eng. :** Evening Primrose.

Distribution : Native to N. America; introduced in Indian gardens.

Parts Used : Flower. Leaf. Stem Bark.

Flowering & Fruiting : Flowering: July – October & fruiting in September – November.

Folk Use : Flowers, leaves and stem bark regarded as sedative.

Reference : Prajapati *et al.* (2006).

651. *Onosma bracteatum* Wall.

Syn.: O. macrocephala Don

Family: Boraginaceae.

English, Hindi and Sanskrit Names: **Eng. :** Sedge; **Hindi :** Gaozaban, Shankhahuli; **Sans. :** Gaozaban, Gojihva.

Distribution : Occurs wild in Kashmir and Kumaon upto 3,352m.

Parts Used : Whole Plant. Flower. Leaf.

Flowering & Fruiting : Flowering: July – August and Fruiting at a later stage.

Folk Use : Dried leaves and whole plant used as tonic. Flowers used as stimulant and cardiac tonic.

References : Ambasta (1986), Farooq (2005), Kirtikar & Basu (1984), Sharma (2003), Watt (1972).

652. *Operculina turpethum* (L.) Silva Manso

Syn.: Ipomoca turpethum R. Br.; *Merremia turpethum* (L.) G.L. Shah & R.G. Bhat.

Family: Convolvulaceae.

English, Hindi and Sanskrit Names: **Eng. :** Indian Jalap, Turpeth; **Hindi :** Nisotar, Nisoth, Pitohri; **Sans. :** Aruna, Kalameshi, Kalaparni, Kalavika, Kali, Nandi, Rechani, Shayama, Suvaha, Trivrt.

Distribution : Throughout India upto 900m; occasionally cultivated.

Parts Used : Root.

Flowering & Fruiting : August.

Folk Use : Roots useful as stimulant.

Active Constituents : Glycosidic resin, turpethin, α- turpethein and β-turpethein (*root*).

Reference : Prajapati *et al.* (2006).

653. *Ophioglossum reticulatum* L.

Family: Ophioglossaceae.

English Name: **Eng. :** Adder's Tongue.

Distribution : Mussoorie, Bihar, Bengal, Assam and S. India upto 1,800m.

Parts Used : Frond.

Folk Use : Fronds used as tonic.

References : Saini (2008), Sharma (2003).

654. *Ophioglossum vulgatum* L.

Family: Ophioglossaceae.

English Names: **Eng. :** Adder's Tongue, Christ's Spear.

Distribution : Sikkim, 1,350m; below Darjeeling 700m, Europe, Africa, America, Japan, Australia and New Zealand.

Parts Used : Frond.

Folk Use : Fronds used as a tonic.

References : Kirtikar & Basu (1984), Singh (2000b).

655. *Ophiorrhiza mungos* L.

Family: Rubiaceae.

English, Hindi and Sanskrit Names: **Eng. :** Indian Snake Root, Mongoose Plant; **Hindi :** Sarahati; **Sans. :** Bhiyangakshi, Chhatrika, Gandali, Nadikalapa, Nakuli, Rasna, Sarpaksi, Sarpari, Sugan.

Distribution : Shady places on plains and hills of low elevations throughout India.

Parts Used : Root.

Flowering & Fruiting : April – August.

Folk Use : Roots used as sedative and tonic.

Active Constituents : Known to contain starch, β-sitosterol, 5α-ergost-7-en-3β-ol and 5α-ergost-8-en-3β-ol (*root*).

References : Anonymous (1966), Prajapati *et al.* (2006), Watt (1972).

656. *Origanum vulgare* L.

Syn.: O. laxiflora Royle; *O. normale* Don.

Family: Lamiaceae.

English and Hindi Names: **Eng. :** Argans, Wild Marjoram, Orgament, Pot Marjoram; **Hindi :** Mirzanjosh, Sathra.

Distribution : Native to Europe; temperate Himalayas: (Kashmir – Sikkim): 2,350-4,000m.

Parts Used : Whole Plant. Seeds.

Flowering & Fruiting : July – September.

Folk Use : Plant and oil from seeds used as stimulant and tonic in diarrhoea.

Active Constituents : Accredited properties due to α-pinene, p-cymene, γ-terpinene, thymol, carvacrol, β-caryophyllene, β-bisabolene, terpinen-4-ol and germacrene D (*oil*), cinnamic, p-hydroxybenzoic, vanillic, syringic, protocatechuic, caffeic and chlorogenic acids, leptosidin, naringin, 5-hydroxyl-3, 3', 4', 7-tetramethoxyflavone, luteolin, apigenin, peonidin and catechol (*plant*).

References : Ambasta (1986), Anonymous (1966), Bhattacharjee (2001), Farooq (2005), Gupta (1997), Kapur & Singh (1996), Kaul (1997), Kirtikar & Basu (1984), Rajan *et al.* (2003), Rana *et al.* (1996), Sharma (2003), Watt (1972).

657. *Ormocarpum cochinchinensis* (Lour.) Merrill

Syn.: O. glabrum Teijsm. & Binn. *O. sennoides* DC.

Family: Fabaceae.

Sanskrit Names: **Sans. :** Kananashekhara, Kananashigru.

Distribution : W. India, Sri Lanka, Siam, Philippines, Polynesia, tropical Africa.

Parts Used : Root.

Flowering & Fruiting : June – September (flowering); September – October (fruiting).

Folk Use : Roots used as tonic and stimulant.

References : Ambasta (1986), Anonymous (1966), Kirtikar & Basu (1984).

658. *Oroxylum indicum* Vent.

Syn.: Bignonia indica L.; *B. petandra* Lour.; *Calosanthes indica* Blume.; *Spathodea indica* Pers.

Family: Bignoniceae.

English, Hindi and Sanskrit Names: **Eng. :** Indian Trumpet Tree; **Hindi :** Arlu, Pharkhat, Sonapatha, Syona, Ullu; **Sans. :** Adhvantashatrava, Aratu, Bhalluka, Kandarpa, Nata, Shyonaka, Tuntukah, Vatu.

Distribution : Found throughout India except in the Western drier area ascending up to 1,219m in the Himalayas.

Parts Used : Bark. Root Bark.

Flowering & Fruiting : Flowering: June – July; Fruiting: cold season.

Folk Use : Root bark used as a tonic. Infusion of bark useful as a bitter tonic.

Active Constituents : Recorded earlier to possess oroxylin-A, baicalein and chrysin (*root bark, bark*); alkaloid, tannic acid, sitosterol and galactose (*bark*).

References : Anonymous (1966), Baruah & Sarma (1987), Dhiman (2005), Farooq (2005), Kaushik & Dhiman (2000), Kirtikar & Basu (1984), Parrotta (2001), Raju (2000); Reddi *et al.* (2005); Sharma (2003), Watt (1972).

659. *Oryza sativa* L.*

Syn.: Oryza plena (Prain.) Chowdhuryi.

Family: Poaceae.

English, Hindi and Sanskrit Names: **Eng. :** Paddy, Rice; **Hindi :** Caval, Dhan; **Sans. :** Dhanyah, Sali, Tandula, Vrihi.

Distribution : Cultivated throughout India.

Parts Used : Grains.

Flowering & Fruiting : July – September.

Folk Use : Grains used as tonic. Rice beer taken as a cooling light tranquilizer.

References : Kirtikar & Basu (1984), Kumar & Rao (2001), Prajapati *et al.* (2006).

660. *Osmunda regalis* L.

Family: Osmundaceae.

English Names: **Eng. :** Buckhorn, Fern Brake, Hartshorn Bush, Herb Christopher, King's Fern, Royal Fern, Water Fern.

Distribution : N. India, Kumaon, moist parts of the Himalayas, Khasia hills and the Western Ghats at altitudes of 1,500-3,000m.

Parts Used : Whole Plant. Root.

Flowering & Fruiting : Flowering occurs in April.

Folk Use : Roots used as tonic and stimulant. Plant also used as a tonic.

Active Constituents : Caffeic, p-hydroxybenzoic acid and p-coumaric acid (*root*).

References : Ambasta (1986), Anonymous (1966), Jha *et al.* (2003), Kirtikar & Basu (1984), Singh & Viswanathan (1996).

661. *Ostrya virginiana* (Mill.) K. Koch.

Family: Betulaceae.

English Names: **Eng. :** American Horphorn Bean, Iron Wood, Lever Wood.

Distribution : Sikkim and N. W. Himalayas; Native to eastern North America, southeast to northern Florida and southwest to eastern Texas and northeastern Mexico.

Parts Used : Bark.

Flowering & Fruiting : Flowering: March – April and fruiting during summers.

Folk Use : Bark of this plant used as a tonic.

Reference : Bhattacharjee (2001).

662. *Ougeinia oojeinensis* (Roxb.) Hochr.

Syn.: Dalbergia ougeinensis Roxb.; *Ougeinia dalbergioides* Benth.

Family: Fabaceae.

English, Hindi and Sanskrit Names: **Eng. :** Chariot Tree, Sandan; **Hindi :** Panan, Panjan, Sandan, Tinnas; **Sans. :** Akshaka, Chakri, Meshri, Nemi, Rathika, Ratna, Shakata, Syandanah, Tinisah, Vanjula.

Distribution : Outer Himalayas and Sub-Himalayan tract upto 1,500m.

Parts Used : Bark.

Flowering & Fruiting : March – May.

Folk Use : Bark used as a stimulant.

Active Constituents : Lupeol, hydroxylupeol and betulin (*bark*).

Reference : Prajapati *et al.* (2006).

663. *Ouratea serrata* (Gaertn.) Robson

Syn.: Gomphia angustifolia Vahl.; *Ouratea angustifolia* (Vahl.) Baill; *Ouratea zeylanica* (Lam.) Alston.

Family: Ochnaceae.

Distribution : Bombay Konkan, W. Coast and lower slopes of W. Ghats from S. Kanara to Tinnevelly up to 1,000m in evergreen forests, Sri Lanka.

Parts Used : Leaf. Root.

Flowering & Fruiting : January - June

Folk Use : Decoction of roots and leaves used as tonic.

References : Ambasta (1986), Kirtikar & Basu (1984).

664. *Oxalis corniculata* L. (Pl. 9A)

Syn.: O. foliosa Blatt.; *O. pusilla* Salisb.; *O. repens* Thunb.; *O. villosa* Bieb.

Family: Oxalidaceae.

English, Hindi and Sanskrit Names: **Eng. :** Indian Sorrel, Yellow Oxalis; **Hindi :** Amrul, Anboti, Chuka tripati, Seh; **Sans. :** Amlika, Amlotaja, Kangeri, Shuklika.

Distribution : Common throughout India in the warmer regions.

Parts Used : Whole Plant. Leaf.

Flowering & Fruiting : April – November.

Folk Use : Hot water extract of leaves with salt and pepper used as a tonic for nursing mothers. Whole plant juice valued as liver tonic.

Active Constituents : Known to contain tartaric acid and citric acid (*leaf*).

References : Abraham (1997), Chhetri (2005), Reddi *et al.* (2005).

665. *Oxydendrum arboratum* (L.) DC.

Syn.: Andromeda arborea L.

Family: Ericaceae.

English Names: **Eng. :** Sorrel Tree, Sourwood.

Distribution : India, Pakistan, Sri Lanka and Myanmar.

Parts Used : Leaf.

Folk Use : Leaves considered as tonic.

Reference : Bhattacharjee (2001).

666. *Paederia foetida* auct.

Syn.: P. ovata Miq.; *P. scandens* (Lour.) Merr.

Family: Rubiaceae.

English, Hindi and Sanskrit Names: **Eng. :** Chinese Moon-creeper, King's Tonic, Strinking Opalberry; **Hindi :** Gandhali, Gundali, Somaraji; **Sans. :** Balya, Chandraparni, Gandhali, Prabala, Prasarani, Sharani.

Distribution : Wild in some areas of Himalayas, Bihar, West Bengal, Orissa and Assam, up to 1,800m.

Parts Used : Leaf.

Flowering & Fruiting : July – August.

Folk Use : Leaves used as tonic.

Active Constituents : Isolated constituents are hentriacontane, hentriacontanol, methyl mercaptan, ceryl alcohol, palmitic acid, sitosterol, stigmasterol, campesterol, ursolic acid and iridoid glycosides – asperuloside, paederoside and scandoside (*leaf*).

References : Ambasta (1986), Anonymous (1966), Kirtikar & Basu (1984), Raju (2000), Thakur *et al.* (1989).

667. *Paeonia emodi* Wall. ex Royle.

Family: Paeoniaceae.

English and Hindi Names: **Eng. :** Himalayan Paeony, Paeony Rose; **Hindi :** Udsalap.

Distribution : N.W. Himalayas from Kashmir – Uttaranchal: 1,600-3,000m.

Parts Used : Whole Plant.

Flowering & Fruiting : May.

Folk Use : Plant used as CNS depressant.

Reference : Sharma (2003).

668. *Panax bipinnatifidus* Seem.

Family: Araliaceae.

English Name: **Eng. :** Wild False-ginseng.

Distribution : Sikkim Himalayas: 2,800-3,850m.

Parts Used : Root.

Folk Use : Root possesses tonic properties.

Reference : Prajapati *et al.* (2006).

669. *Panax pseudo - ginseng* Wall.

Syn.: Aralia pseudo - ginseng Benth.

Family: Araliaceae.

English Name: **Eng. :** False Ginseng.

Distribution : Endemic in Sikkim Himalayas: 1,800-3,600m.

Parts Used : Root.

Folk Use : Roots used as tonic and stimulant.

Active Constituents : Found to contain triterpenoid saponins, oleanolic acid, amino acids – leucine, valine, proline, oxyproline, histidine, lysine and cysteine (*root*).

References : Chhetri (2005), Jamir (1997), Prajapati *et al.* (2006), Sharma (2003).

670. *Panax quinquefolium* L.

Family: Araliaceae.

English Name: **Eng. :** American Ginseng.

Distribution : Imported into India.

Parts Used : Root.

Folk Use : Roots used as sedative having effect on cerebrum and mildly stimulant action on vital centres.

References : Ambasta (1986), Harborne & Baxter (2001).

671. *Pandanus odoratissimus* L. f.

Syn.: P. amaryllifolius Roxb.; *P. fascicularis* Lam.; *P. laevis* Kunth.; *P. tectorius* auct. non Soland. ex Parkinson.

Family: Pandanaceae.

English, Hindi and Sanskrit Names: **Eng. :** Screw Pine, Umbrella Tree; **Hindi :** Gagandhul, Kedgi, Kevda; **Sans. :** Chamarapushpa, Halina, Jambuka, Ketaki.

Distribution : Throughout India, along the sea coast and water courses.

Parts Used : Bracts. Leaf. Root.

Flowering & Fruiting : February – August.

Folk Use : Roots and leaves used as a tonic. Oil of Otto obtained from bracts considered as stimulant.

Active Constituents : Accredited properties due to pandamarine, pandamarilactone 1, pandamarilactone 31 and pandamarilactone 32 (*leaf*).

References : Kirtikar & Basu (1984), Prajapati *et al.* (2006), Watt (1972).

672. *Panicum miliaceum* L.

Syn.: Milium esculentum Moench; *M. panicum* Mill.; *Panicum asperrimum* Fisch.; *P. callosum* Hochst. ex A.; *P. densepilosum* Steud.; *P. striatum* Spreng.

Family: Poaceae.

English and Hindi Names: **Eng. :** Common Millet, Hog Millet, Prosa Millet; **Hindi :** Chena, Chin.

Distribution : India, Africa and other hot countries; in India, chief producing regions are A.P., Maharashtra, Mysore, Chennai, Bihar, M.P., U.P. and Punjab.

Parts Used : Seeds.

Flowering & Fruiting : July – October.

Folk Use : Seeds used as narcotic in Arunanchal Pradesh.

Reference : Anonymous (1986).

673. *Panicum sumatrense* Roth. ex Roem. & Schult.

Syn.: P. attenuatum Willd.; *P. menieri* Koen. ex Nees; *P. miliare* Lamk.; *P. psilopodium* Wight; *P. simplex* Rattl. ex Trin.

Family: Poaceae.

English and Hindi Names: **Eng. :** Little Miliare; **Hindi :** Kodra, Kungu, Kutki, Shavan.

Distribution : Throughout India up to 2,100m in the Nilgiris.

Parts Used : Whole Plant.

Flowering & Fruiting : June – October.

Folk Use : Grass considered as nerve stimulant and also it exhibits little narcotic property.

Reference : Srivastava & Singh (2003).

674. *Papaver rhoeas* L.

Family: Papaveraceae.

English, Hindi and Sanskrit Names: **Eng. :** Blind Eyes, Canker Rose, Collin Hood, Corn Poppy, Red Poppy, Thunder Bolts, Yedwork; **Hindi :** Lalpost, Post, Postekebija; **Sans. :** Raktaposta, Raktapostavrikshaha.

Distribution : Native to Europe, N. Africa and temperate regions of Asia; thrives on cultivated land and on road verges.

Parts Used : Flower.

Folk Use : Petals used as sedative.

Active Constituent : Rhoeadine (*flower*).

References : Ambasta (1986), Anonymous (1966), Kirtikar & Basu (1984), Prajapati *et al.* (2006).

675. *Papaver somniferum* L. (Pl. 9B)*

Family: Papaveraceae.

English, Hindi and Sanskrit Names: **Eng. :** Joan Silverpin, Opium Poppy, White Poppy; **Hindi :** Afim, Afyun, Kashkash, Post; **Sans. :** Ahifen, Chosa, Khasa.

Distribution : Cultivated in the warmer parts of the country.

Parts Used : Whole Plant. Capsule (latex). Seeds.

Flowering & Fruiting : February – April.

Folk Use : Plant used as stimulant and tonic. Opium obtained from capsules used as sedative, narcotic and hypnotic. Seeds also used as tonic.

Active Constituents : Reported to possess morphine, codeine, thebaine, noscapine, papaverine and narceine, minor alkaloids lioke aporeine, codamine, cryptopine, gnoscopine, hydrotarnine, lanthopine, laudanidine, laudanine, laudanosine, meconidine, narcotoline, neopine, papaveramine, prophyroxine, protopine, reticuline, rhoeadine and xanthaline (*latex*).

References : Ambasta (1986), Anonymous (1966), Bhattacharjee (2001), Harborne & Baxter (2001), Kayal *et al.* (2003), Kirtikar & Basu (1984), Prajapati *et al.* (2006), Prakash (2001), Rajan *et al.* (2000), Ranjan (2000), Rao (1996), Sharma (2003), Thakur *et al.* (1989), Watt (1972).

676. *Parmelia perlata* (Huds.) Ach.

Family: Permeliaceae.

English Name: **Eng. :** Stone Flower.

Distribution : Commonly found growing on rocks and on the tracks and branches of oak tree between 1,600 – 3,200m.

Parts Used : Whole Plant.

Folk Use : Herb used as a tonic.

Active Constituents : Known to yield lecanoric acid, atronorin and a yellow crystalline matter called chrysophanic acid (*herb*).

Reference : Sharma (2003).

677. *Parnassia palustris* L.

Family: Saxifragaceae.

English Names: **Eng.** : Bog-star, Marsh Grass-of-Parnassus, Northern Grass-of-Parnassus.

Distribution : Kashmir (Baltistan) at 2,250m.

Parts Used : Whole Plant.

Flowering & Fruiting : Flowering: July – September; fruiting during October.

Folk Use : Decoction of plant used as a sedative.

References : Ambasta (1986), Anonymous (1966).

678. *Parthenium hysterophorus* L. (Pl. 9C)

Family: Asteraceae.

English and Hindi Names: **Eng.** : Congress Grass, Santa Maria Feverfew, Whitetop Weed; **Hindi** : Chatak Chandni, Gajar Ghas.

Distribution : Native of America, naturalized throughout India.

Parts Used : Whole Plant. Root.

Flowering & Fruiting : Flowering and fruiting during greater part of the year.

Folk Use : Decoction of roots and plant used as tonic.

Active Constituents : Parthenin, hysterin, ambrosin, 6-hydroxykaempferol-3, 7-dimethyl ether, kaempferol, quercetagetin-3,7-dimethyl ether, a quercetin-3-O-glycoside, coronopilin and dihydroisoparthenin (*plant*).

References : Anonymous (1966), Farooq (2005), Kaushik & Dhiman (2000), Sood & Thakur (2004).

679. *Paspalum scrobiculatum* L.

Syn.: P. commersonii Lam.; *P. scrobiculatum* Lam.

Family: Poaceae.

English, Hindi and Sanskrit Names: **Eng.** : Kodo Millet; **Hindi** : Kodo, Kodra; **Sans.** : Kodravah, Madanagraka, Uddala.

Distribution : Wild as well as cultivated throughout India.

Parts Used : Whole Plant. Grains.

Flowering & Fruiting : May – November.

Folk Use : Plant used as tonic. Grains used as sedative, tonic and narcotic.

Active Constituents : Hentriacontanol, hentriacontanone, sitosterol and campesterol (*seeds*).

References : Anonymous (1966), Kirtikar & Basu (1984), Prajapati *et al.* (2006), Srivastava & Singh (2003).

680. *Passiflora foetida* L.

Family: Passifloraceae.

English and Sanskrit Names: **Eng. :** Hispid Granadilla, Stinking Passion Flower, Tagua Passion Flower; **Sans. :** Mukkopara.

Distribution : Widely cultivated in India.

Parts Used : Whole Plant. Leaf.

Flowering & Fruiting : July – January.

Folk Use : Whole plant considered as sedative. Leaf powder mixed with milk taken to increase memory power.

Active Constituent : Harman (*plant*).

References : Prajapati *et al.* (2006), Thomas & Britto (2000).

681. *Passiflora incarnata* L.

Family: Passifloraceae.

English Names: **Eng. :** Maypop, Purple Passionflower, True Passionflower, Wild Apricot, Wild Passion Vine.

Distribution : Native of S.E. United States of America; grown frequently in gardens in India.

Parts Used : Flower. Fruit.

Flowering & Fruiting : July – October.

Folk Use : Flowers and fruits used as sedative and narcotic.

Active Constituent : Harman (*fruit*).

References : Ambasta (1986), Anonymous (1966), Harborne & Baxter (2001).

682. *Pavetta indica* L. ssp. ***tomentosa*** (Roxb. ex Smith) Bennet

Syn.: Ixora pavetta Roxb.

Family: Rubiaceae.

English, Hindi and Sanskrit Names: **Eng. :** Indian Pellet Shrub, White Pavetta; **Hindi :** Karnikara, Kankra, Kathachampa, Papari; **Sans. :** Kakachedi, Papata, Tiriakphala.

Distribution : Throughout India, up to 1,500m, and Andamans.

Parts Used : Flower. Root.

Flowering & Fruiting : July – October.

Folk Use : Roots used as tonic. 8 – 10 flowers pounded with milk recommended as an aphrodisiac tonic.

References : Anonymous (1966), Siwakoti & Varma (1996), Sood *et al.* (2009), Subramaniam (2000).

683. *Pedalium murex* L.

Syn.: P. microcarpum Decne.; *P. muricatum* Salisb.; *Rogeria microcarpa* Klotzsch.

Family: Pedaliaceae.

Hindi and Sanskrit Names: **Hindi :** Bara Gokhru; **Sans. :** Gokshura, Sthulagoksurah, Tiktagoksurah.

Distribution : Throughout India, especially in the coastal areas of South India.

Parts Used : Whole Plant. Leaf. Stem. Fruit.

Flowering & Fruiting : March – December.

Folk Use : 2 tsp of powdered dried fruits, plant and infusion of stem and leaves soaked in water used as a tonic.

Active Constituents : Heptatriacontan-4-one, hexatriacontanoic acid, ursolic acid, pentatriacontane, sitosterol and vanillin (*plant*).

References : Anonymous (1966), Prajapati *et al.* (2006), Ram *et al.* (2002), Singh & Pandey (1998).

684. *Pedicularis pectinata* Wall. ex Benth.

Family: Scrophulariaceae.

English Name: **Eng. :** Pectinate-leaved Lousewort.

Distribution : Western temperate Himalayas (Kashmir – Kumaon): 2,350-3,650m; Afghanistan.

Parts Used : Whole Plant. Inflorescence. Stem.

Flowering & Fruiting : July – August.

Folk Use : Plant, inflorescence and stem used as sedative.

References : Aman & Dar (2003), Jain & Goel (2005).

685. *Peganum harmala* L.

Family: Rutaceae.

English, Hindi and Sanskrit Names: **Eng. :** Foreign Henna, Harmal, Syrian Rue, Wild Rue; **Hindi :** Harmal, Lahouri-hurmul, Isband-lahouri; **Sans. :** Harmal, Hurmal.

Distribution : Native to Middle East, N. Africa and S. Europe; found almost throughout N.W. India and drier regions of Deccan.

Parts Used : Seeds.

Flowering & Fruiting : Flowering: June – July and fruiting in September – October.

Folk Use : Seeds used as narcotic and hypnotic.

Active Constituents : Harmine, harmaline, harmalol, vasicine, oxodeoxypeganine, oxopeganine, lanosterol, kryptogenin and β-sitosterol (*seeds*).

References : Ambasta (1986), Dhiman (2005), Ballabh & Chaurasia (2006), Bhattacharjee (2001), Kaul (1997), Sharma (2003), Singh *et al.* (1996), Watt (1972).

686. *Pennisetum americanum* (L.) Leeke

Syn.: Panicum americanum L.; *Pennisetum alopecroides* Spreng.; *P. cereale* Trin.; *P. cylindrium* Sw.; *P. giganteum* Ten.; *P. typhoides* (Burm. f.) Stapf. *et* Hubb.; *P. typhoideum* Rich.; *P. violaceum* Rich.

Family: Poaceae.

English, Hindi and Sanskrit Names: **Eng. :** Pearl Millet, The Bulrush, The Cumboo, The Spiked Millet; **Hindi :** Bajra, Lahra, Kara-jonar; **Sans. :** Agradhyanya, Nali, Sajaka, Vajarika.

Distribution : Grown as a crop plant throughout drier parts of India.

Parts Used : Grains.

Flowering & Fruiting : June – July.

Folk Use : Grains considered as tonic.

Active Constituents : Found to yield proteins, fats, carbohydrates, calcium, phosphorus, iron and vitamins, carotene, thiamine, nicotinic acid, riboflavin and chlorine (*grains*).

References : Kaushik & Dhiman (2000), Kirtikar & Basu (1984), Parrotta (2001).

687. *Pennisetum flaccidum* Griseb.

Syn.: Gymnothrix flaccida Munro ex Aitchis.; *Pennisetum mongolicum* Franch.

Family: Poaceae.

English and Hindi Names: **Eng. :** Flaccid Grass, Himalayan Fountain Grass; **Hindi :** Phark, Phirki.

Distribution : Temperate and alpine Himalayas: 1,700-3,650m, (Kashmir – W. Nepal and Western Tibet), Afghanistan, Central Asia.

Parts Used : Whole Plant.

Flowering & Fruiting : July – October.

Folk Use : Plant extract used as tonic.

Reference : Kapur (1996b).

688. *Peperomia tetraphylla* (Forst. f.) Hk. & Arn.

Syn.: P. reflexa (L. f.) A. Dietr.; *Piper tetraphyllum* G. Forst.

Family: Piperaceae.

English, Hindi and Sanskrit Names: **Eng. :** Four-leaved Peperomia.

Distribution : Punjab – Assam, ascending to 2,100m in the Himalayas and in the hills of A.P., Orissa, Chennai and Kerala, usually above 1,200m.

Parts Used : Whole Plant.

Flowering & Fruiting : May – November.

Folk Use : Whole plant used as a tonic in Darjeeling.

References : Anonymous (1966), Farooq (2005), Mukherjee & Roy (1987).

689. *Perilla frutescens* (L.) Britton

Syn.: Melissa maxima Arduin; *Mentha perilloides* Willd.; *Ocimum frutescens* L.; *P. ocimoides* L.

Family: Lamiaceae.

English and Hindi Names: **Eng. :** Perilla; **Hindi :** Bhanjira.

Distribution : Almost throughout the Himalayas up to 3,500m, and in the hills of Assam and Bihar.

Parts Used : Whole Plant.

Flowering & Fruiting : August – December.

Folk Use : Herb used as a sedative.

Active Constituents : Therapeutic properties due to perilla – aldehyde (4-isoprophenyl-1-cyclohexen-7-al), limonene, a-pinene, dihydrocumine, oleic acid, linoleic acid, linolenic acid and amino acids; arginine, histidine, leucine, lysine and valine (*plant*).

References : Ambasta (1986), Anonymous (1966), Kaushik & Dhiman (2000).

690. *Petroselinum crispum* (Mill.) Nym. & A. W. Hill

Syn.: Apium crispum Mill.; *A. petroselinum* L.; *Carum petroselinum* (L.) Benth.; *P. hortense* Hoffm.; *P. sativum* Hoffm.

Family: Apiaceae.

English and Hindi Names: **Eng. :** Parsley; **Hindi :** Achu.

Distribution : Native of the Mediterranean region; occasionally grown in kitchen gardens in India.

Parts Used : Root.

Flowering & Fruiting : June – July.

Folk Use : Roots of the plant considered stimulant. Also, dried roots soaked in good wine believed to improve memory

Active Constituents : Imperatorin, isopimpinellin, psoralen, 5-methoxypsaralen, 8-methoxypsoralen, oxypeucedanin, apiole, bergapten, camphene, myristicin, α-pinene, psoralen and xanthotoxin (*root*).

References : Bhattacharjee (2001), Khatoon & Goel (2010).

691. *Peucedanum grande* C. B. Clarke

Syn.: Pastinaca grandis Dals. & Gibs.

Family: Apiaceae.

English, Hindi and Sanskrit Names: **Eng. :** Hill Carrot, Wild Carrot; **Hindi :** Daku, Duku; **Sans. :** Baspika, Ela.

Distribution : Occurs gregariously on the Western Ghats and the hills of Deccan Plateau.

Parts Used : Herb. Fruit.

Flowering & Fruiting : July – November.

Folk Use : Herb and fruits used as tonic and stimulant.

References : Ambasta (1986), Anonymous (1966), Kirtikar & Basu (1984).

692. *Peucedanum nagpurense* (Clarke) Prain

Family: Apiaceae.

Hindi Names: **Hindi :** Bhojraj, Tejraj.

Distribution : Bihar and Orissa.

Parts Used : Root.

Folk Use : Roots used as general tonic.

Reference : Jain *et al.* (2003).

693. *Phaseolus radiatus* L.

Syn.: P. aurens Roxb.; *Vigna radiata* (L.) Wilczek var. *radiata*

Family: Fabaceae.

English and Sanskrit Names: **Eng. :** Dord, Thikiri, Urid; **Sans. :** Baladhya, Dhanyavira, Masha, Pitrya.

Distribution : Extensively cultivated all over India.

Parts Used : Whole Plant. Root. Seeds.

Flowering & Fruiting : Throughout the year.

Folk Use : Plant and seeds used as tonic. Roots considered narcotic.

Active Constituents : γ-glutamyl-γ-glutamylmethionine, 3-(2 - furayl) alanine, L-pipecolic acid (*seeds*), Stigmasterol, campesterol, β-sitosterol (*underground parts*), scopoletin, isoscopoletin, umbelliferone, rutin, quercetin, kaempferol, chlorogenic acid and neochlorogenic acid, doidzin, formononetin, ononin, quercetin-3-O-sophoroside and 3-O-β-D-glucuronide, isoquercitrin and kaempferol-3-O-β-D-glucuronide (*aerial parts*), three γ-glutamyl peptides-γ-glutamyl-S-methylcysteinyl-β-alanine, γ-glutamyl-Nδ-acetylornithine and γ-glutamyl-γ-glutamyl-S-methylocysteine, homoglutathione and γ-glutamyl derivatives of glutamic acid, aspartic acid, phenylalanine, leucine and isoleucine (*seeds*).

References : Kirtikar & Basu (1984), Saikia & Nath (2003b), Watt (1972).

694. *Phoenix pusilla* Gaertn.

Syn.: P. farinifera Roxb.

Family: Arecaceae.

English, Hindi and Sanskrit Names: **Eng. :** Small Wild Date-palm; **Hindi :** Palavat; **Sans. :** Parusakah.

Distribution : Dry forests of Kerala, Karnataka and Tamil Nadu at low elevations.

Parts Used : Fruit.

Folk Use : Fruits used as cardio tonic.

Active Constituents : Galactose and galacturonic acid (*fruit*).

Reference : Prajapati *et al.* (2006).

695. *Phoenix sylvestris* Roxb.

Syn.: Elate sylvestris L.; *E. versicolor* Salisb.; *Katou indel* Ham.; *Phoenix humilis* Royle.

Family: Arecaceae.

English, Hindi and Sanskrit Names: **Eng. :** Date Sugar Palm, Indian Wine Palm, Wild Date Palm; **Hindi :** Khajur, Salma, Thakil; **Sans. :** Bhumikhajurika, Kakakaekati, Kharjur, Skandhappula.

Distribution : Wild or more often cultivated up to 1,500m and throughout the drier parts of India.

Parts Used : Bud. Fruit.

Flowering & Fruiting : April – May (flowering); September – October (fruiting).

Folk Use : Buds used as cooling tonic. Fruits used as tonic and cardio tonic. Fruit mixed with almonds, quince seeds, pistachio nuts, spices and sugar forms a tonic.

References : Dhiman (2005), Farooq (2005), Hebbar *et al.* (2003), Kaushik & Dhiman (2000), Kirtikar & Basu (1984), Sood *et al.* (2009).

696. *Pholidota articulata* Lindl.

Family: Orchidaceae.

English Name: **Eng. :** Rattle Snake Orchid.

Distribution : Uttaranchal, Sikkim, Manipur and Khasia hills between 700-1,300m.

Parts Used : Whole Plant.

Flowering & Fruiting : Flowering: June – August; Fruiting: October – December.

Folk Use : Whole plant used as a tonic.

Active Constituents : Two 9, 10-dihydrophenanthrenes – isoflavidin and iso-oxoflavidin, flavidin (*plant*).

References : Bhattacharjee (2001), Sharma (2003).

697. *Phyllostachys bambusoides* Sieb. & Zucc.

Syn.: Bambusa bifolia Sieb. mss. ex Munro.; *B. reticulata* Rwpr.; *Phyllostachys megastachya* Steud.

Family: Poaceae.

English Name: **Eng. :** Giant Timber Bamboo.

Distribution : Upper Assam; Mishmi Hills (without flowers). China, Japan.

Parts Used : Root.

Flowering & Fruiting : Flowering during summers.

Folk Use : Roots used as tonic.

References : Ambasta (1986), Anonymous (1969).

698. *Physalis minima* L.

Syn.: P. divaricata Don.; *P. hermanii* Dunal.; *P. parviflora* Br.;. *P. pubescens* Wight.; *P. rothiana* Roem. & Sch.; *P. villosa* Roth.

Family: Solanaceae.

English, Hindi and Sanskrit Names: **Eng. :** Country Gooseberry; **Hindi :** Bandhapariya, Cirpoti, Tulati Pati; **Sans. :** Chirapotha, Tankaree.

Distribution : As a weed on cultivated land throughout India, up to 2,300m.

Parts Used : Whole Plant. Fruit.

Flowering & Fruiting : August – January.

Folk Use : Powdered plant recommended as a tonic (½ tsp twice daily for 12-15 days). Fruits also used as a tonic.

Active Constituents : Reported to yield – 5β, 6β-epoxyphysalin B, withaphysalis A, B and C, physalin B, new dihydroxyphysalin B, quercewtin – 3 – O – galactoside (*leaf*), withaminimin, 5-methoxy-6, 7, methylenedioxyflavone, 5, 6, 7-trimethoxyflavone, withaphysalin D, withaphysalin A (*plant*).

References : Hebbar *et al.* (2003), Kaushik & Dhiman (2000), Kirtikar & Basu (1984), Maya *et al.* (2003), Nayak *et al.* (2003), Prajapati *et al.* (2006), Sahu (2004), Srivastava *et al.* (2003a), Watt (1972).

699. *Physochlaina praealta* (Decne.) Miers.

Syn.: Hyoscyamus praealtus Walp.; *Physochlaina grandiflora* Hook. f.; *Scopolia praealta* Dunal.

Family: Solanaceae.

English Name: **Eng. :** Tall Physochlaina.

Distribution : Roadside weed often forming patches running and covered large area in Kashmir, H.P. and Western Tibet at altitudes of 3,300-4,600m.

Parts Used : Leaf. Seeds.

Flowering & Fruiting : June – August.

Folk Use : Leaves and seeds used as a narcotic.

Active Constituents : Tropane, hyoscyomine, atropine, hyoscine and sucrose (*leaf*).

References : Ballah & Chaurasia (2006), Kaul (1997), Sharma (2003), Singh *et al.* (1996).

700. *Physostigma venenosum* Balf.

Family: Fabaceae.

English Name: **Eng. :** Ordeal Bean of Calabar.

Distribution : Coastal areas of Southern India and Nilgiris hills.

Parts Used : Seeds.

Folk Use : Seeds used as sedative.

Active Constituents : Starch, albuminoids and several alkaloids like physostigmine, eserdine, calabarine and physovenine isolated from *seeds*.

References : Bhattacharjee (2001), Sharma (2003).

701. *Phytolacca acinosa* Roxb.

Syn.: P. latbenia (Moq.) Walter; *Pircunia latbenia* Moq.

Family: Phytolaccaceae.

English and Hindi Names: **Eng. :** Indian Poke, Sweet Belladona; **Hindi :** Matazor.

Distribution : Forest shades of temperate Himalayas: Kashmir – Bhutan and in Assam (2,300-2,800m).

Parts Used : Whole Plant.

Flowering & Fruiting : June – September.

Folk Use : Plant used as narcotic.

Active Constituents : Known to yield supergulagenic acid, jailigonic acid, acinosolic acid, β-sitosterol, myristic acid, n-pentacosane, lignoceryl palmitate, 16-hentriacontanol, ursolic acid and its galactoside (*fruit*), phytolaccanol, epiacetylaleuritolic acid, sitosterol, spergulagenic acid A, isophytolaccinic acid A, isophytolaccagenin A and 3-O-methyl spergulagenate (*plant*), phytolaccagenin and demethylphytolaccagenin (*root*).

References : Anonymous (1969), Kapur & Nanda (1996), Sharma (2003).

702. *Picrorhiza kurrooa* Benth.

Syn.: P. scrophulariflora Penell.

Family: Scrophulariaceae.

English, Hindi and Sanskrit Names: **Eng.** : Picrorhiza; **Hindi** : Katki, Karu, Kutki; **Sans.** : Anjani, Chakranga, Katuka, Katurohini, Katvi, Shakuladani, Vantida.

Distribution : Known to occur wild in alpine Himalayas (Kashmir – Sikkim): 2,743-4,572m.

Parts Used : Rhizome. Root.

Flowering & Fruiting : Flowering during May – June and fruiting later.

Folk Use : Roots of the plant used as a cardio tonic and brain tonic. Rhizomes used as bitter tonic.

Active Constituents : Attributed properties due to picrorhizin, picrorhizetin, kutkin, apocynin, α-sitosterol (*rhizome*), caccurbitacin glycoside, 6-feruloyl catalpol, veronicoside, minecoside, picein, androsin, 2β-(glucosyloxy)-16, 20, 22-trihydroxy-9-methyl-19-norlanosta-5, 24-dien-3, 11-dione (VIII), 25-deacetyl derivative of (VII) and arvenin III, 25-(acetoxy)-2β-(β-D-glucopyranosyloxy)-3, 16, 20-trihydroxy-9-methyl-19-norlanosta-5, 23 (Z)-dien-22-one (X), 25-(acetoxy)-2β-(β-D-glucopyranosyloxy)-3, 16-dihydroxy-9-methyl-19-norlanosta-3,23-dien-22-one (IX), 2-(β-D-glucopyranosyloxy)-3, 16-dihydroxy-4, 4, 9, 14-tetramethyl -19-norpregn-5-en-20-one (XII), 2, 3, 16, 20, 25-pentahydroxy-9-methyl-19-norlanost-5-ene-22-one (XIII) and 2β-(6-O-cinnamoyl-β-D-glucopyranosyloxy)-3, 16, 20, 25-tetrahydroxy-9-methyl-19-norlanost-5-en-22-one (XIV), bitter principle named kutkin, non bitter product named kurrin and later identified as D-mannitol, vanillic acid, alcohol kutikial and sterol kutkisterol (*root*).

References : Aman & Dar (2003), Bhattacharjee (2001), Dhiman (2005), Dwivedi (2007), Farooq (2005), Haridasan *et al.* (2001), Kapur *et al.* (1996), Kirtikar & Basu (1984), Prajapati *et al.* (2006), Uniyal & Malhotra (1989).

703. *Pimenta dioica* (L.) Merrill

Syn.: Eugenia pimenta DC.; *Pimenta officinalis* Lindl.

Family: Myrtaceae.

English Names: **Eng. :** Allspice Tree, Jamaica Pepper Tree, Pimenta Tree.

Distribution : Grown in gardens throughout India, also found in Bangalore, Bihar, Orissa, Kerala and West Bengal.

Parts Used : Fruit.

Folk Use : Fruits used as stimulant and tonic.

Active Constituents : Eugenol, limonene, b-caryophyllene, 1-8-cineole, a-phelandrene, terpinolene, b-selinene, a-humulene and p-cymene (*fruit*).

References : Ambasta (1986), Anonymous (1969), Bhattacharjee (2001), Rajan *et al.* (2003), Sharma (2003).

704. *Pimpinella anisum* L.

Family: Apiaceae.

English, Hindi and Sanskrit Names: **Eng. :** Anise, Aniseed; **Hindi :** Badian, Saunf, Sawonf; **Sans. :** Shetapushapa.

Distribution : Native to eastern Mediterranean, Western Asia and North Africa and to a less extent in Mexico and South America; grown to a small extent as a culinary herb in India.

Parts Used : Fruit.

Flowering & Fruiting : August – December.

Folk Use : Dried fruits used as stimulant.

Active Constituents : Recorded to possess luteolin, luteolin-7-O-glucoside and luteolin-7-O-xyloside (*fruit*).

References : Ambasta (1986), Anonymous (1969), Bhattacharjee (2001), Sharma (2003).

705. *Pinellia ternate* (Thunb.) Breitenb.

Family: Araceae.

English Name: **Eng. :** Three-leaf Pinellia.

Distribution : Cultivated fields and roadsides throughout India; also found in Korea, China and Japan.

Parts Used : Corm.

Folk Use : Corms used as sedative.

Active Constituents : Corm possesses amino acids, polysaccharides, homogentisic acid, ephedrine, 3,4-dihydroxybenzaldehyde-diglucoside, pinellin and phytosterols.

Reference : Prajapati *et al.* (2006).

706. *Pinus gerardiana* Wall. ex D. Don

Syn.: Pinus gerardii Forbes.

Family: Pinaceae.

English, Hindi and Sanskrit Names: **Eng. :** Chilgoza Pine, Edible Pine; **Hindi :** Chilgoza, Gunober, Neoza; **Sans. :** Bhutamari, Chidha, Darugandha, Grahajita, Mangalya, Tara, Taruni.

Distribution : N.W. Himalayas: 1,800-3,000m.

Parts Used : Seeds.

Flowering & Fruiting : Flowering: May – June; cones ripen during September – October of the following year.

Folk Use : Seeds used as stimulant.

References : Ambasta (1986), Anonymous (1969), Kirtikar & Basu (1984), Watt (1972).

707. *Pinus roxburghii* Sarg.

Syn.: Pinus longifolia Roxb. *ex* Lamb.

Family: Pinaceae.

English, Hindi and Sanskrit Names: **Eng. :** Chir Pine, Himalayan Long-leaved Pine, Three-leaved Pine; **Hindi :** Chir, Cid, Cil, Sarala; **Sans. :** Bhadradaru, Dhupavrikshaka, Manojna, Putikashtha, Saralan, Surabhidaruka.

Distribution : Himalayas (Kashmir – Sikkim) and in the Shiwalik hills: 450-2,400m.

Parts Used : Root. Wood.

Flowering & Fruiting : March – May.

Folk Use : Roots used as a stimulant. Wood also used as a stimulant and liver tonic.

Active Constituents : b-pinene and car-3-ene (*wood oil*).

References : Dhiman (2005), Kirtikar & Basu (1984), Prajapati *et al.* (2006), Singh (2001), Watt (1972).

708. *Piper betle* L.*

Syn.: Chavica betle Miq.; *C. siriboa* L.; *Piper chavya* Ham.; *P. peepuloides* Wall.; *P. siriboa* L.

Family: Piperaceae.

English, Hindi and Sanskrit Names: **Eng. :** Betel Leaf Pepper, Betel Leaf Vine, Betel Pepper; **Hindi :** Pan, Tambuli; **Sans. :** Bhakshyapatra, Divabhishta, Nagavalli, Saptashira, Tambula, Vitika.

Distribution : Cultivated in many areas, especially in South India.

Parts Used : Leaf.

Flowering & Fruiting : Throughout the year.

Folk Use : Leaves used as a tonic to brain; also leaves chewed with areca nuts to produce a stimulating effect on central nervous system.

Active Constituents : Chavibetal (53.1), chavibetal acetate (15.5), caryophyllens (3.79), allylpyrocatachol diacetate (0.71), camphene (0.48), eugenol (0.32), α-pinene (0.71), β-pinene (0.21), α-limonene (0.14), safrole (0.11), 1, 8-cineole (0.04), p-cymene (0.11), and allylpyrocatechol monoacetate (*leaf*).

References : Ambasta (1986), Anonymous (1969), Bhattacharjee (2001), Kirtikar & Basu (1984), Saikia & Nath (2003b), Watt (1972).

709. *Piper brachystachyum* Wall.

Syn.: Chavica sphaerostachya Miq.; *P. mullesua* D. Don.; *P. vasculosum* Wall.

Family: Piperaceae.

English, Hindi and Sanskrit Names: **Eng. :** Wild Pepper; **Hindi :** Cab, Pahari Pan, Pahari Pipar; **Sans. :** Cavyah.

Distribution : Sub-tropical Himalayas: 700-1,700m, Khasia Hills: 1,000-1,700m, Nilgiri hills.

Parts Used : Fruit.

Flowering & Fruiting : May – July.

Folk Use : Dried fruits used as tonic.

Reference : Bhattacharjee (2001).

710. *Piper cubeba* L. f.

Syn.: Cubeba officinalis L.

Family: Piperaceae.

English, Hindi and Sanskrit Names: **Eng. :** Cubebs, Java Pepper, Tailed Pepper; **Hindi :** Kabab Chini; **Sans. :** Kankolam.

Distribution : Native of Indonesia; cultivated in Karnataka.

Parts Used : Fruit.

Flowering & Fruiting : July – September.

Folk Use : Fruits used as stimulant, sedative and cardio tonic.

Active Constituents : Cubebin, cubebal and cubebic acid (*fruit*).

References : Ambasta (1986), Anonymous (1969), Prajapati *et al.* (2006).

711. *Piper longum* L.

Syn.: Chavica roxburghii Miq.; *C. sarmentosa* Miq.; *Piper latifolium* Hunter.; *P. sarmentosum* Wall.

Family: Piperaceae.

English, Hindi and Sanskrit Names: **Eng.** : Indian Long Pepper; **Hindi** : Gazpipal, Pipal, Pipli, Piplamul; **Sans.** : Kana, Pippali, Shaundi, Upakalya.

Distribution : Hotter parts of India from Central Himalayas – Assam, Khasi, Mikir hills, lower hills of W. Bengal and evergreen forests of W. Ghats from Konkan – Kerala, and Car Nicobar Islands.

Parts Used : Root. Fruit.

Flowering & Fruiting : Flowering during rainy season and fruiting in autumn.

Folk Use : Roots used as sedative. Unripe fruits used as a tonic and ripened ones as tonic to liver.

Active Constituents : These are 1-undecyl-3, 4-dimethylene-dioxybenzene, sesamin, piperine, piplartine and caryophyllene (*fruit*), alkaloids piperine, piplartine, piperlongumins, piperloguminine, triacontane, dihydrostigmasterol, a sterol, reducing sugars and glycosides (*root*).

References : Ambasta (1986), Anonymous (1969), Dhiman (2005), Farooq (2005), Kaushik & Dhiman (2000), Kirtikar & Basu (1984), Sen & Batra (1997), Sharma (2003), Siwakoti & Varma (1996), Thakur *et al.* (1989), Watt (1972).

712. *Piper nigrum* L.

Syn.: Muldera wightiana Miq.; *Piper trioicum* Roxb.

Family: Piperaceae.

English, Hindi and Sanskrit Names: **Eng.** : Black Pepper, Common Pepper, Pepper; **Hindi** : Chocamirch, Golki, Golmirch, Habush, Kali Mirch; **Sans.** : Dharmavarttana, Kolaka, Maricha, Pavita, Ruksha, Vara, Yavanapriya.

Distribution : Hot and moist parts of Southern India.

Parts Used : Fruit. Leaf.

Flowering & Fruiting : July – September.

Folk Use : Fruits used as stimulant. Leaves valued as narcotic.

Active Constituents : Known to possess a new amide-pipericide, characterized as isobutylamide of 11-(3, 4-methylenedioxyphenyl)-2E, 4E, 10E-undecatrienoic acid, guineensine, N-isobutyl-2E, 4E, 8Z-eicosatrienamide, N-isobutyl-2E, 4E-octadecadienamide, pellitorine, new phenolic amides – N-trans-feruloyl-piperdine, feruperine and dihydroferuperine (*fruit*).

References : Anonymous (1986), Bhattacharjee (2001), Mukherjee & Roy (1987), Sharma (2003, 2004), Watt (1972).

713. *Piper puberulum* Benth. ex Seemann

Family: Piperaceae.

English Name: **Eng. :** Downy Pepper.

Distribution : Common in disturbed forests, along streams, beach thickets, forest margins, plantations and wasteareas from sea – level to over 1,000m.

Parts Used : Leaf.

Folk Use : Decoction of leaves taken as a tonic after childbirth.

Reference : Prajapati *et al.* (2006).

714. *Piper retrofractum* Vahl.

Syn.: P. chaba Hunter, non Blume; *P. officinarum* DC.

Family: Piperaceae.

English, Hindi and Sanskrit Names: **Eng. :** Bengal Pepper, Java Long Pepper; **Hindi :** Chab, Chavi; **Sans. :** Chawika, Gandhanakuli, Kola, Nakuli, Tejovati, Ushana, Vashira.

Distribution : Cultivated in various parts of India and Malay Islands.

Parts Used : Fruit.

Flowering & Fruiting : Throughout the year.

Folk Use : Fruits used as stimulant.

Active Constituents : Found to contain a new amide – N-isobuty-lridica-13 (3,4-methylenedioxyphenyl) 2, 4, 12-trienaamide (I), filfiline *(fruit).*

References : Ambasta (1986), Anonymous (1969), Kirtikar & Basu (1984).

715. *Piscidia piscipula* (L.) Sarg.

Syn.: P. erythrina L.

Family: Fabaceae.

English Name: **Eng. :** Jamaica Logwood.

Distribution : Native of tropical America, introduced into Indian gardens.

Parts Used : Bark. Root Bark.

Folk Use : Bark and root bark used as sedative.

Active Constituents : Bark rich in glucosides, an amorphous alkaloid, piscidic acid, rotenone, ichthynone and 5 new aromatic substa-nces viz jamaicin, lesetin and traces of substances designated D, E and G.

References : Ambasta (1986), Anonymous (1969), Harbone & Baxter (2001).

716. *Pistacia chinensis* Bunge ssp. ***integerrima*** (Stewart) Rech. f.

Syn.: P. integerrima Stewart; *P. khinjuk* Stocks.; *Rhus integerrima* Wall.; *R. kakrasingee* Royle.

Family: Anacardiaceae.

English, Hindi and Sanskrit Names: **Eng.** : Crab's Claw; **Hindi** : Kakra, Kakrashingi; **Sans.** : Ghosha, Karkata, Karkata sringi, Shiknari, Vakra.

Distribution : Wild in N.W. Himalayas (Indus – Kumaon): 350-2,500m.

Parts Used : Gall.

Flowering & Fruiting : March – May.

Folk Use : Galls of the plant used as a tonic.

Active Constituents : Reported to possess tetracyclic triterpenes – pistacigerrimones D, E and F, three tetracyclic triterpenoids – pistacigerrimones A, B and C – characterized as 20 (R), 24 (R) 3-oxo-9β-lanosta-1(2), 7, 24-trien-26-oic acid, 20 (R), 24 (R) 3-oxo-9β-lanosta-1(2), 8, 24-trien-26-oic acid and 20 (R), 24 (R) 3-oxo-tirucalla-1 (2) and 24-dien-26-oic acid respectively (*gall*).

References : Dhiman (2005), Kirtikar & Basu (1984), Prajapati *et al.* (2006), Watt (1972).

717. *Pistacia lentiscus* L.

Family: Anacardiaceae.

English Name: **Eng.** : Mastic Tree.

Distribution : Native to Mediterranean region; grows wild in scrub and on wastegrounds.

Parts Used : Whole Plant.

Folk Use : Plant used as stimulant.

Active Constituents : These are masticadienonic, dihydromasticadienonic, oleanonic, masticadienolic, dihydromasticadienolic, oleanolic, 3-epimasticadienolic, dihydro-3-epimasticadienolic and epioleanolic acids isolated as methyl esters, β-amyrin, β amyrone, oleanolic aldehyde, 28-hydroxyl-β-amyrone, dammarenediol, erythrodiol and masticadiendiol (*gall*), α-and β-pinenes, myrcene, limonene, thymol and carvacol (*fruit*).

References : Ambasta (1986), Anonymous (1969), Watt (1972).

718. *Pittosporum floribundum* Wight. & Arn.

Syn.: Calastrus verticileata Roxb.; *Senacia nepalensis* DC.

Family: Pittosporaceae.

English Name: **Eng.** : Himalayan Pittosporum.

Distribution : Sub-tropical Himalayas from Punjab – Sikkim up to 1,700m, Ganjam, Konkan, W. Ghats, Nilgiris and southwards, hills of S. Arcot and Salem.

Parts Used : Bark.

Flowering & Fruiting : May – October.

Folk Use : Bark used as narcotic.

References : Ambasta (1986), Anonymous (1969), Kirtikar & Basu (1984).

719. *Plantago major* L.

Syn.: P. asiatica L.; *P. erosa* L.; *P. longiscopa* Jacq.; *P. maxima* Ruching.; *P. officinarum* Crantz.;

Family: Plantaginaceae.

English, Hindi and Sanskrit Names: **Eng. :** Englishman's Foot, Great Plantain, Ribgrass, Waybroad; **Hindi :** Isbgol, Luhuriya; **Sans. :** Ashvagola.

Distribution : Temperate and Alpine Himalayas, Assam, Myanmar, Baluchistan, Konkan, W. Ghats, Nilgiris, Pulneys, Sri Lanka, Malaya, Afghanistan and westwards to the Atlantic; wild or introduced in these and many other countries.

Parts Used : Whole Plant. Seeds.

Flowering & Fruiting : April– October.

Folk Use : Seeds used as stimulant and also as a tonic in diarrhoea and dysentery whereas whole plant used as tonic and stimulant.

Active Constituents : Baicalein, baicalin, scutellarein, chlorogenic and neochlorogenic acids isolated; apigenin, luteolin-7-O-β-D-glucoside and luteolin-7-O-β-D-glucuronide (*leaf*), aucubosides (*leaf, seeds*), plantamajoside, oleanolic acid, acetoside, luteolin, its 7-O-glucoside, 6-hydroxyluteolin and its 7-O-glucoside, β-sitosterol, campesterol, stigmasterol, four new iridoid glycosides – plantarenaloside, ixoroside, asperuloside and melampyroside (*plant*).

References : Ambasta (1986), Anonymous (1969), Jain & Sharma (2000), Kirtikar & Basu (1984), Sood & Thakur (2004), Watt (1972).

720. *Plantago ovata* Forsk.*

Syn.: P. decumbens Forsk.; *P. ispaghul* Roxb.; *P. lanata* Wall.

Family: Plantaginaceae.

English, Hindi and Sanskrit Names: **Eng. :** Blond Psyllium, Ispaghula, Spogel Seeds; **Hindi :** Isabgal, Isabgul, Issufgul; **Sans. :** Ishadgola, Snigdhabijah, Snigdhajirakah.

Distribution : Punjab plains and low hills from the Sutlej westwards, Sind, Baluchistan westwards to Spain and the Canary islands.

Parts Used : Seeds.

Folk Use : Seeds used as tonic.

Active Constituents : Known to contain linolenic acid, oleic acid, palmitic, stearic, lignoceric acids, aucuboside and catalpal (*seeds*).

References : Kirtikar & Basu (1984), Prajapati *et al.* (2006).

721. *Platycodon grandiflorum* A. DC.

Syn.: Campanula grandiflora Jacq.

Family: Campanulaceae.

English Names: **Eng. :** Ballon Flower, Chinese Bell Flower, Japanese Bell Flower.

Distribution : Native of China and Japan; edges of ponds and streams or in swamps in India, China, Japan, Korea and Soviet Union.

Parts Used : Root.

Folk Use : Roots used as a tonic.

Active Constituents : Saponin and platycodin (*root*).

References : Ambasta (1986), Anonymous (1969), Prajapati *et al.* (2006).

722. *Plectranthus mollis* (Ait.) Spr.

Syn.: Ocimum mayporense Roth.; *O. molle* Ait.; *P. divaricatus* Weinm.; *P. incanus* Link.; *P. secundus* Roxb.

Family: Lamiaceae.

English Names: **Eng. :** Soft-stem Mintleaf.

Distribution : Hills of continental India, from the Himalayas (Shimla – Sikkim), ascending to 1,350m; Khasia Mts. and Central India, to the Anamallays (absent in Sri Lanka and in Malay Peninsula).

Parts Used : Seeds.

Flowering & Fruiting : August – October.

Folk Use : A teacup of seed extract taken as tonic in early morning on empty stomach by Jenukunibas.

Reference : Kshirsagar & Singh (2000).

723. *Plumbago indica* L.

Syn.: P. rosea L.; *Thela coccinea* Lour.

Family: Plumbaginaceae.

English, Hindi and Sanskrit Names: **Eng. :** Fire Plant, Rosy-flowered Leadwort; **Hindi :** Chitra, Lal-chita, Rakta-chitra; **Sans. :** Chitraka, Dahanah.

Distribution : Throughout India, in moist situations, as well as cultivated; native of Sikkim and Khasia hills.

Parts Used : Root.

Flowering & Fruiting : February – May.

Folk Use : Roots used for stimulating action on Central Nervous System and narcotic.

Active Constituents : Plumbagin, stigmasterol acetate, friedelinol, lupanone and sitosterone (*root*).

References : Ambasta (1986), Anonymous (1969), Prajapati *et al.* (2006).

724. *Plumbago zeylanica* L.

Syn.: P. auriculata Blume; *Thela alba* Lour.

Family: Plumbaginaceae.

English, Hindi and Sanskrit Names: **Eng. :** Ceylon Leadwort, White Flowered Leadwort; **Hindi :** Chitarak, Chiti, Chitta; **Sans. :** Agni, Anala, Analanamia, Bhali, Chitraca, Dahaka, Jataveda, Pavaka, Vaishavan-ara.

Distribution : Throughout India, ascending up to 1,650m.

Parts Used : Root.

Flowering & Fruiting : April – September.

Folk Use : Roots said to be stimulant and tonic.

Active Constituents : Accredited properties due to plumbagin, 3-chlorplumbagin and a new substance – 3, 3'-biplumbagin, a new binaphthaquinone – chitranone, zeylinone, isozeylinone, elleze-glanone and plumbagic acid (*root*).

References : Kirtikar & Basu (1984), Swami & Gupta (1996), Watt (1972).

725. *Plumeria acuminata* Ait.

Syn.: P. acutifolia Poir.; *P. rubra* L. var. *acutifolia* Bailey.

Family: Apocynaceae.

English, Hindi and Sanskrit Names: **Eng. :** Frangipani Plant, Pagoda Tree, Temple Tree; **Hindi :** Chameli, Goburchamp, Golainchi; **Sans. :** Devagangagalu, Gosampige, Kshira Champaka.

Distribution : Almost all over the India in the vicinity of religions shrines and burial grounds.

Parts Used : Bark.

Flowering & Fruiting : June – September.

Folk Use : Bark used as stimulant.

Active Constituents : Pluierides and fulvoplumierin and α-amyrin acetate (*bark*).

References : Ambasta (1986), Anonymous (1969), Parrotta (2001), Raju (2000), Sharma (2004).

726. *Podophyllum hexandrum* Royle

Syn.: P. emodi Wall. *ex* Hk. f.

Family: Berberidaceae.

English, Hindi and Sanskrit Names: **Eng. :** Indian Podophyllum, May Apple; **Hindi :** Nibishi, Papra, Pilijati; **Sans. :** Giriparpata.

Distribution : Open slopes, forest clearings, moist humid places from Kashmir – Uttaranchal in temperate Himalayas between 2,600-4,500m.

Parts Used : Fruit. Rhizome. Root.

Flowering & Fruiting : May – September.

Folk Use : Rhizomes and roots considered stimulant. Fruits used as a tonic whereas roots valued used as bitter tonic.

Active Constituents : Reported to contain podophyllotoxin (*rhizome*), podophyllotoxin, deoxypodophyllotoxin, dehydropodophyllotoxin and picropodophyllotoxin, four cytptoxic lognans – (1R, 2R, 3R)-deoxypodophyllotoxin, 4'-demethylpodophyllotoxon (*root*).

References : Anonymous (1969), Chauhan (2001), Kapur & Nanda (1996), Kapur & Srivastava (1996), Kaul (1997), Rawat *et al.* (2007), Sharma (2003).

727. *Pogostemon benghalense* (Burm. f.) Kuntze

Syn.: Mentha fruticulosa Roxb.; *M. secunda* Roxb.; *Origanum benghalense* Burm.; *O. indicum* Roth.; *Pogostemon plectranthoides* Desf.

Family: Lamiaceae.

English and Hindi Names: **Eng. :** Bengal Pogostemone; **Hindi :** Phangla.

Distribution : Western Himalayas, lower Bengal, Bihar, Gujarat, Karnataka and Deccan.

Parts Used : Root.

Flowering & Fruiting : November – February.

Folk Use : Roots used as stimulant.

References : Kaushik & Dhiman (2000), Parrotta (2001).

728. *Pogostemon parviflorus* Benth.

Syn.: P. frutescens Grah.; *P. intermedius* Wall.; *P. pubescens* Benth.; *P. purpuricaulis* Dalz.

Family: Lamiaceae.

English Name: **Eng. :** Pogostemon.

Distribution : Himalayas from Kumkaon – Bhutan and in the hills of Assam and W. Ghats, ascending to 1,200m.

Parts Used : Leaf. Root.

Folk Use : Leaves and roots used as stimulant.

References : Ambasta (1986), Anonymous (1969).

729. *Polemonium caeruleum* L.

Family: Polemoniaceae.

English and Hindi Names: **Eng. :** Jacob's Ladder; **Hindi :** Jhal.

Distribution : Alpine Western Himalayas: 3,000-4,000m; from Kashmir – Kumaon, frequently in Europe, N. and Central Asia, N. America.

Parts Used : Root.

Flowering & Fruiting : July – August.

Folk Use : Roots used as sedative.

Active Constituents : Resin (1.3%), organic acid and essential oil, a saponin complex constituted of acetic, angelic, tiglic, α-methylbutyric, propionic, isobutyric acids and polemoniumgenin A, a diester of camelliagenin E with acetic, angelic or tigelic acids have been isolated from *roots*.

References : Ambasta (1986), Anonymous (1969), Kapur (1996b).

730. *Polygonatum cirrhifolium* (Wall.) Royle

Family: Asparagaceae.

English, Hindi and Sanskrit Names: **Eng. :** Meda; **Hindi :** Meda; **Sans. :** Meda.

Distribution : Himalayas: 1,200-4,200m.

Parts Used : Rootstock.

Flowering & Fruiting : May.

Folk Use : Rootstock used as a tonic.

Active Constituents : Steroidal saponins – sibiricoside A (PS - 1) and sebiriciside B (PS - IV) (*rootstock*).

References : Anonymous (1969), Prajapati *et al.* (2006).

731. *Polygonatum kingianum* Coll. et Hemsl.

Family: Asparagaceae.

English Name: **Eng. :** Solomons Seal.

Distribution : Wild in the highlands above 1,200m.

Parts Used : Rhizome.

Flowering & Fruiting : April – September.

Folk Use : Rhizomes considered as tonic.

Active Constituents : Mucilage, starch and sugar (*rhizome*).

Reference : Prajapati *et al.* (2006).

732. *Polygonatum verticillatum* (L.) All.

Syn.: Campydorum verticillatum Salisb.; *Convallaria verticillata* L.; *Evallaria verticillata* Neck.; *Polygonatum jacquemontianum* Kunth.; *P. leptophyllum* Royle.

Family: Asparagaceae.

English, Hindi and Sanskrit Names: **Eng. :** Mahameda; **Hindi :** Mahameda, Mitha Dudia; **Sans. :** Mahameda.

Distribution : Himalayas from Kashmir – Sikkim: 2,000-4,200m.

Parts Used : Rhizomes. Root.

Flowering & Fruiting : April.

Folk Use : Roots and rhizomes used as a tonic and nervine tonic.

Active Constituents : Diosgenin, four saponosides A, B, C and D (*rhizome*).

References : Gupta (1997), Prajapati *et al.* (2006), Rawat *et al.* (2007), Sharma (2003), Singh (2001).

733. *Polygonum amplexicaule* D. Don

Syn.: P. ambiguum Meissn.; *P. oxyphyllum* Wall.; *P. petiolatum* Don.

Family: Polygonaceae.

Hindi Names: **Hindi :** Durpa Tandar, Kukar Makri.

Distribution : Distributed in temperate Himalayas (Kashmir – Sikkim): 1,800-4,800m.

Parts Used : Root.

Folk Use : Roots used as stimulant. Tea prepared from roots valued as stimulant.

Active Constituents : β-amyrin, chrysophanol, emodin, physcion and stigmasterol (*root*).

References : Kaul (1997), Kumar & Rao (2001), Swami & Gupta (1996).

734. *Polygonum aviculare* L.

Family: Polygonaceae.

English, Hindi and Sanskrit Names: **Eng. :** Allseed, Bian Xu, Blood Wort, Doorweed, Hogweed, Iron-grass, Knotgrass, Knot Wort, Mantie, Pig-weed, Red Legs, Sparrow Tongue, Way Grass, Wireweed; **Hindi :** Bannalia, Ban-natia, Hunraj, Machoti; **Sans. :** Nisomali.

Distribution : Wastegrounds and along shorelines in temperate regions throughout the India.

Parts Used : Whole Plant.

Flowering & Fruiting : June – October.

Folk Use : Herb used as a tonic.

Active Constituents : Quercetin, avicularin, quercitrin, umbelliferone, scopoletin, kaempferol-3-arabinoside, luteolin, quercetin-3-galactoside, rhamnetin-3-galactoside, vitexin, isovitexin, aviculin, juglanin, astragalin and betmidin (*herb*), pyrogallic tannins, formic, acetic, pyruvic, valeric, glucuronic and galacturonic acids (*aerial parts*).

References : Ambasta (1986), Anonymous (1969), Kirtikar & Basu (1984), Sharma (2003).

735. *Polygonum barbatum* L.

Syn.: Persicaria barbatum (L.) Hara.; *Polygonum fluviatile* Herb.; *P. horemanni* Meisn.; *P. marmoramae* Herb.; *P. rivulare* Koen.; *P. stagninum* Buch.-Ham. *ex* Meissn.

Family: Polygonaceae.

English and Hindi Names: **Eng. :** Jointweed, Knotgrass, Smart-weed; **Hindi :** Jal Bahar.

Distribution : Known to occur throughout hotter parts of India.

Parts Used : Seeds.

Flowering & Fruiting : August – December.

Folk Use : Seeds used as tonic.

References : Ambasta (1986), Anonymous (1969), Dhiman (2005), Kaushik & Dhiman (2000), Kumar & Lal (2003), Srivastava *et al.* (2003a).

736. *Polygonum chinensis* L.

Syn.: Coccoloba crispata Ham.; *Polygonum auriculatum* Meissn.; *P. brachiatum* Poir.; *P. corymbosum* Willd.; *P. cymosum* Roxb.; *P. ovatum* Heyne.; *P. patens* Don.; *P. polycephalum* Wall.

Family: Polygonaceae.

English Name: **Eng. :** Chinese Knotweed.

Distribution : Sub-tropical and temperate Himalayas (Shimla – Bhutan): 1,350-2700m; Assam, the Khasia Mts., Silhet, Cachar, Chittagong, Rangoon and Myanmar – Malacca, Bihar, top of Parusnath, Deccan Peninsula; on the Western Ghats from the Concan southwards. Sri Lanka, Sumatra, Java, Philippine Islands, China and Japan.

Parts Used : Whole Plant.

Flowering & Fruiting : Almost throughout the year.

Folk Use : Juice of plant used as tonic.

Active Constituents : Kaempferol, quercetin, kaempferol-7-O-glucoside, kaempferol-7-O-glucuronide, β-sitosterol and ellagic, gallic and 3-O-methyl ellagic acids (*leaf*).

References : Anonymous (1969), Lalramnghinglova (1996).

737. *Polygonum hydropiper* L.

Syn.: P. flaccidum Meissn.; *P. mite* Wall.

Family: Polygonaceae.

English and Hindi Names: **Eng. :** Biting Pepper, Pepper Wort, Smart Weed, Water Pepper; **Hindi :** Machoti.

Distribution : Damp places, ditches, wetlands, shallow water ponds from N.W. India – East.

Parts Used : Herb. Root.

Flowering & Fruiting : June – October.

Folk Use : Used as stimulant, tonic (herb, root) and bitter tonic (root).

Active Constituents : Active properties due to a new isocoumarin – polygonolide, elucidated as 3, 4-dimethyl-6-methoxy-8-hydroxyisocoumarin (*root*), pyrogallic tannins (5.5 %) and formic, acetic, pyruvic, valeric, glucuronic and galacturonic acids (*aerial parts*), quercetin, isoquercetin, 7, 4'-dimethyl-quercetin, 3'-methylquercetin, quercetin-3-sulphate, isorhamnetin-3, 7-disulphate and tamarixetin-3-glucosido-7-sulphate (*leaf*).

References : Ambasta (1986), Anonymous (1969), Kaul (1997), Kirtikar & Basu (1984), Sharma (2003, 2004), Srivastava *et al.* (2003a).

738. *Polygonum multiflorium* Thunb.

Family: Polygonaceae.

English Name: **Eng. :** Many-flowered Knotweed.

Distribution : Native of China and Japan; cultivated in gardens in India.

Parts Used : Root.

Folk Use : Roots used as tonic and nervine tonic.

Active Constituents : Rhponticin (rhapontin, ponticin), chrysophanic acid, emodin, physcion, rhein and polygonic acid (*root*).

References : Anonymous (1969), Gupta (1997), Prajapati *et al.* (2006).

739. *Polygonum viviparum* L.

Syn.: P. affine Wall.; *P. angustifolium* Don.; *P. bistorta* Gaercke.; *P. bracteatum* Spreng.; *P. bulbiferum* Royle.

Family: Polygonaceae.

English Names: **Eng. :** Alpine Knotweed, Viviparous Bistort.

Distribution : Kashmir – Sikkim, up to 5,000m, Alpine N. and arctic Europe, Asia and America.

Parts Used : Root.

Flowering & Fruiting : June – October.

Folk Use : Roots used as tonic.

References : Ambasta (1986), Anonymous (1969), Kirtikar & Basu (1984).

740. *Pongamia pinnata* (L.) Pierre

Syn.: Cystisus pinnatus L.; *Derris indica* (Lam.) Bennett; *Pongamia glabra* Vent.; *P. mitis* (L.) Kurz.; *P. xerocarpa* Hassk.; *Pterocarpus flavus* Lour.; *Robinia mitis* L.

Family: Fabaceae.

English, Hindi and Sanskrit Names: **Eng. :** Indian Beech, Karanj, Pongam Oil Tree; **Hindi :** Kanji, Karanji, Papar; **Sans. :** Angaravalli, Badhaphala, Chirabilva, Gaura, Karanjah, Vishari, Vrittapornia.

Distribution : Native of Western Ghats, chiefly found along the banks of streams and rivers near sea coast almost throughout India, up to 1,200m.

Parts Used : Bark. Rind (pod), Seeds.

Flowering & Fruiting : March – June.

Folk Use : Powdered seed and rind of the pod considered as tonic. Bark valued as stimulant.

Active Constituents : Known to yield karanjin, pongapin, pongaglabrone, kanjone, isolonchocarpin, karanjachromene and isopongachromene, pongamol, ovaletenone, glabrin, desmethoxykanugin, glabrachalcone [4], glabrachromene [5], isopongaflavone [6], pongol [7] and 2'-methoxyfurano [2", 3" :7, 8] flavones [8] (*seeds*).

References : Anonymous (1969), Sharma (2003), Thakur *et al.* (1989).

741. *Populus alba* L.

Family: Salicaceae.

English Names: **Eng. :** Able, White Poplar.

Distribution : Native of Central Europe – Central Asia; found in N.W. Himalayas: 1,200-3,000m, Kashmir and Pakistan.

Parts Used : Bark.

Flowering & Fruiting : May - September

Folk Use : Bark used as tonic.

Active Constituents : Salicin and populin (*bark*).

References : Ambasta (1986), Anonymous (1969), Bhattacharjee (2001), Watt (1972).

742. *Populus ciliata* Wall. ex Royle

Syn.: P. pyriformis Royle; *P. rotundifolia* Griff.

Family: Salicaceae.

English and Hindi Names: **Eng.** : Himalayan Poplar; **Hindi** : Paharipipal, Poplar, Sharphara.

Distribution : Throughout Himalayas up to 3,500m.

Parts Used : Bark.

Flowering & Fruiting : March – April (flowering); May – June (fruiting).

Folk Use : Bark used as tonic and stimulant.

References : Ambasta (1986), Anonymous (1969), Bhattacharjee (2001), Farooq (2005), Kapur & Nanda (1996), Kapur & Singh (1996), Kaushik & Dhiman (2000), Kirtikar & Basu (1984), Singh *et al.* (1996), Watt (1972).

743. *Potentilla anserina* L.

Syn.: Argentina anserina Rydberg.; *P. pseudoanserina* Best.

Family: Rosaceae.

English Names: **Eng.** : Cinguefoil, Silverweed.

Distribution : Western Himalayas: 2,100-4,800m; N. Asia, Orient, Europe, N. America, Australia.

Parts Used : Whole Plant.

Flowering & Fruiting : June – October.

Folk Use : Plant used as tonic.

Active Constituents : Tannins (2-10%); quercitin, quercitrin, tormentol, an essential oil and vitamin E, scopoletin and umbelliferone (*plant*).

References : Anonymous (1969), Kirtikar & Basu (1984), Vishwanathan (2000).

744. *Potentilla argyrophylla* Wall.

Syn.: P. argyrophylla var. *atrosanguinea* Hook. f.; *P. atrosanguinea* var. *argyrophylla* (Hook. f.) Grierson & Long; *P. insignis* Royle ex Lindl.

Family: Rosaceae.

English Names: **Eng.** : Himalayan Cinquefoil, Silver-leaved Cinquefoil.

Distribution : Western and Central Himalayas, from Kashmir (2,400-4,800m) – Nepal.

Parts Used : Root.

Folk Use : Roots used as tonic.

Reference : Kapur & Srivastava (1996).

745. *Potentilla fruticosa* L.

Syn.: P. arbuscula Don.; *P. rigida* Wall.

Family: Rosaceae.

English Name: **Eng. :** Shrubby Cinguefoil.

Distribution : Kashmir (2,750-4,000m) – Sikkim (4,000-5,350m); N. Asia, Europe.

Parts Used : Leaf.

Folk Use : Leaf used as a stimulant by tribals in Arunachal Pradesh.

Reference : Anonymous (1976a).

746. *Potentilla sibbaldi* Hook. f.

Family: Rosaceae.

English Names: **Eng. :** Creeping-glow-wort, Creeping Sibbaldi, Prostrate Sibbaldia.

Distribution : Himalayas (Kashmir – Sikkim): 4,200-4,800m.

Parts Used : Root.

Folk Use : Roots used as a tonic.

Reference : Kapur & Srivastava (1996).

747. *Potentilla supina* L.

Syn.: Comarum flavum Roxb.; *Potentilla cana* Wall.; *P. denticulata* Ser.; *P. heynii* Roth.; *P. ruthenia* Ham.

Family: Rosaceae.

English Name: **Eng. :** Balkan Cinquefoil.

Distribution : Throughout the warmer parts of India, ascending to 2,600m.

Parts Used : Root.

Flowering & Fruiting : December – March.

Folk Use : Roots used as tonic.

References : Anonymous (1969), Kapur *et al.* (1996), Kapur & Singh (1996), Srivastava *et al.* (2003a), Watt (1972).

748. *Prangos pabularia* Lindl.

Family: Apiaceae.

English and Hindi Names: **Eng. :** Prangos, Yugan; **Hindi :** Komal.

Distribution : Kashmir and Himachal Pradesh: 1,800-4,000m.

Parts Used : Herb. Fruit.

Flowering & Fruiting : June (flowering); August – September (fruiting).

Folk Use : Herb used as stimulant. Fruits valued as a stimulant and liver tonic.

Active Constituents : Osthol, prangolarin, imperatorin, prangogenidin, bergapten, oxypeucodanin, an essential oil, a fatty oil, resin, quercetrin and alkaloids including prangosine (*fruit*).

References : Ambasta (1986), Anonymous (1969), Kaul (1997), Kirtikar & Basu (1984), Sharma (2003), Uniyal & Malhotra (1989), Watt (1972).

749. *Premna obtusifolia* R. Br.

Syn.: P. corymbosa Rottl.; *P. integrifolia* L.; *P. serratifolia* L.

Family: Verbenaceae.

English, Hindi and Sanskrit Names: **Eng. :** Headache Tree; **Hindi :** Agetha, Arani, Ustabunda; **Sans. :** Agnimanthah, Ganakasika, Jayanti, Nadegi, Pavaka, Vanhimula.

Distribution : Throughout Indian plains.

Parts Used : Root.

Folk Use : Roots and its decoction used as tonic and cardio tonic.

Active Constituents : Attributed principles are premnine, ganiarine, ganikarine, 6-C-β-D-glucopyranosyl-8-C-β-D-xylopyranosyl apigenin, two new diterpenoids – 11, 12, 16-trihydroxy-abieta-5, 8, 11, 13-tetraen-7-one and 11, 14-dihydroxy-12, 16-epoxy-abieta-5, 8, 11, 13-tetraen-7-one (*root*).

References : Anonymous (1969), Kirtikar & Basu (1984), Nayak *et al.* (2003), Prajapati *et al.* (2006).

750. *Primula rosea* Royle

Syn.: P. elegans Duby.; *P. rosea* var. *elegans* (Duby) Hook. f.

Family: Primulaceae.

English Name: **Eng. :** Himalayan Meadow Primerose.

Distribution : Western Himalayas (Kullu – Kashmir): 3,350-4,000m; Afghanistan.

Parts Used : Whole Plant.

Flowering & Fruiting : May – August.

Folk Use : Plant used as stimulant.

Active Constituents : Glycosides of peonidin and a new anthocyanidin named rosinidin, 14β, 26-epoxy-C-homo-pentacyclic triterpenes (I, II and III), 14β, 26-epoxyserratane-3β-21α-diol (I) (*plant*).

Reference : Kapur (1996b).

751. *Primula veris* L.

Syn.: P. officinalis Jacq.

Family: Primulaceae.

English Name: **Eng.** : Cowslip.

Distribution : Grown in Indian gardens.

Parts Used : Flower. Leaf.

Flowering & Fruiting : Flowering during April – May.

Folk Use : Infusion of leaves and flowers said to be sedative.

Active Constituents : A flavonal glycoside named primulaflavonaloside have been isolated (*flower*).

Reference : Bhattacharjee (2001).

752. *Prunella vulgaris* L.

Syn.: Brunella vulgaris L.

Family: Lamiaceae.

English Names: **Eng.** : Blue Curls, Carpenter's Herb, Hercules Wound Wort, Hook Heal, Self Heal, Sicklewort.

Distribution : N.W. Himalayas (Kashmir – Kumaon), also cultivated in gardens; Europe, Asia and North America.

Parts Used : Whole Plant.

Flowering & Fruiting : June – October.

Folk Use : Plant used as a tonic and cerebral tonic.

Active Constituents : These are flavonoids, coumarins, tannins, esculetin, scopoletin, umbelliferone, cinaroside, luteoline and homooreientin (*aerial parts*), isoquercetin, hyperin, rutin, β-amyrin, daucosterol, oleanolic acid, ursolic acid, and 1α, 2α-dihydroursolic acid (*plant*).

References : Kaul (1997), Rana *et al.* (1996).

753. *Prunus amygdalus* Batsch.

Syn.: Amygdalus communis L.; *Prunus communis* Fritsch.; *Prunus dulcis* (Miller) D. A. Webb.

Family: Rosaceae.

English, Hindi and Sanskrit Names: **Eng.** : Almond; **Hindi** : Badam; **Sans.** : Badamitte, Netroamaphala, Suphala, Vatadah, Vatamah.

Distribution : Cultivated in Kashmir and Punjab at an altitude of 760-2,400m.

Parts Used : Seeds.

Flowering & Fruiting : May – September.

Folk Use : 5-10 seeds (soaked in water overnight) prescribed with 'misri' and butter in the morning on empty stomach as brain tonic. Also, its oil used as sedative.

Active Constituents : Known to possess myristic, palmitic, stearic, oleic and linoleic acid (*seed oil*), A new biflavone - 4'''-methoxy-2'', 3''-dihydroflavonyl-4', 4''-oxymethyl-5, 7-dimethoxy-flavone (I), coumaric acid, 2-methylnonacosan-3-one, n-octacosanol, n-triacontane, β-sitosterol and procyanidin dimer (II) (*seed coat*).

References : Ambasta (1984), Anonymous (1969), Kirtikar & Basu (1986), Prajapati *et al.* (2006), Sood *et al.* (2009).

754. *Prunus armeniaca* L.

Syn.: Armeniaca vulgaris Lam.; *Prunus americana* Marsh.

Family: Rosaceae.

English, Hindi and Sanskrit Names: **Eng. :** Apricot, Common Apricot, Mishmus, Moon of The Faithful; **Hindi :** Chilu, Khubani, Zardalu; **Sans. :** Aluka.

Distribution : Cultivated in N.W. India.

Parts Used : Seeds.

Flowering & Fruiting : May – August.

Folk Use : Seeds used as stimulant and nervine tonic. Seed oil also used as tonic.

Active Constituents : Reported to contain 24-dehydrox-cholesterol, cholesterol, estrone, a-estradiol (estradiol-b-17-ol), oleic acid and linoleic acid (*seeds*).

References : Kumar & Singh (1996), Sood *et al.* (2009).

755. *Prunus cerasoides* D. Don

Syn.: Cerasus phoshia Ham.; *C. puddum* Wall.; *Prunus puddum* (Wall.) Roxb. ex Brandis.; *P. sylvatica* Roxb.

Family: Rosaceae.

English, Hindi and Sanskrit Names: **Eng. :** Bird Cherry, Himalayan Wild Cherry; **Hindi :** Paddam, Patmakath, Patmakh; **Sans. :** Charu, Hima, Padmagandhi, Padmakah.

Distribution : Himalayas: 900-2,300m.

Parts Used : Heartwood. Stem Bark.

Flowering & Fruiting : November – May.

Folk Use : Heartwood and stem bark possess tonic properties.

Active Constituents : Flavonone sakuranetin, flavones genkwanin, isoflavone prunetin, isoflavonone and padmakastin (*stem bark*).

References : Prajapati *et al.* (2006), Sharma (2003).

756. *Prunus cerasus* L.

Syn.: Cerasus caproniana DC.; *C. vulgaris* Miller.

Family: Rosaceae.

English and Hindi Names: **Eng. :** Dwarf Cherry, Red Cherry, Sour Cherry, Wilol Cherry; **Hindi :** Alubalu.

Distribution : Indigenous to W. Asia; cultivated in N.W. Himalayas.

Parts Used : Kernel. Fruit.

Flowering & Fruiting : April – June.

Folk Use : Fruits used as tonic to the brain. Kernels used as nervine tonic.

Active Constituents : Quercetin-3-O-rutinosyl-7-3'-O-bigucoside, kaempferol-3-O-rutinosyl-4'-di-O-glucoside and two isomers of quercetin-3-O-qutinosyl-4'-di-O-glucoside (*fruit*).

References : Ambasta (1986), Anonymous (1969), Kirtikar & Basu (1984), Watt (1972).

757. *Prunus domestica* L.

Syn.: Druparia prunus Clairv.; *P. communis* Huds.; *P. insititia* L.

Family: Rosaceae.

English, Hindi and Sanskrit Names: **Eng. :** Common Plum, Prune Tree; **Hindi :** Alubukhara. Alucha; **Sans. :** Arukam.

Distribution : Cultivated at high altitudes in W. Himalayas, Nilgiris and Himachal Pradesh.

Parts Used : Fruit.

Flowering & Fruiting : January – March (flowering) and April – March (fruiting).

Folk Use : Fruits used as tonic.

Active Constituents : Found to possess D-galactose, D-mannose, L-arabinose, D-xylose, L-rhamnose, glucusronic acid and traces of 4-O-methyl-glucuronic acid, Ca, k, Cu, vitamin B complex, vitamin A and C and mallic and citric acids (*fruit*).

References : Kaushik & Dhiman (2000), Prajapati *et al.* (2006).

758. *Prunus virginiana* L.

Family: Rosaceae.

English Names: **Eng. :** Bitter Berry, Common Chokecherry, Virginia Bird Cherry.

Distribution : Native to North America.

Parts Used : Stem Bark.

Flowering & Fruiting : Flowering during late spring.

Folk Use : Tea prepared from stem bark used as sedative.

Reference : Bhattacharjee (2001).

759. *Pseudarthria viscida* (L.) Wight. & Arn.

Family: Fabaceae.

Hindi and Sanskrit Names: **Hindi :** Chapakno; **Sans. :** Sanaparni.

Distribution : Throughout South India, up to 900m in the hills, also in Gujarat.

Parts Used : Root.

Flowering & Fruiting : September – February.

Folk Use : Roots used as tonic, nervine tonic and cardio tonic.

Reference : Prajapati *et al.* (2006).

760. *Psidium guajava* L. (Pl. 9D)*

Syn.: P. pyriferum L.

Family: Myrtaceae.

English, Hindi and Sanskrit Names: **Eng. :** Common Guava, Guava Tree; **Hindi :** Amrud, Safed Safari; **Sans. :** Aprithaktvacha, Mausala, Parevata.

Distribution : Widely cultivated throughout India.

Parts Used : Bark. Fruit.

Flowering & Fruiting : Flowering: April – June and fruiting in July – October.

Folk Use : Bark extract and fruits used as tonic.

Active Constituents : These are acetaldehyde ethyl cis-hexenylacetal, 1, 1-diethoxy-methane, 1, 1-diethoxyethane, 1, 1-diethoxyhexene, β-bisabolene, cadinene, β-copaene, curcumine, farnesene, α- and β- humulene, α- and β-selinene, citric (E)-cinnamic, (Z)-3-hexenoic, 1-malic, oxalic, quinic, succinic and tartaric acids, folic and nicotinic acids, riboflavin, thiamine, vitamin A, α-, β-, γ- and ζ-carotenes, cis-γ-carotene, 5, 6, 5', 6'-diepoxy-β-carotene, 5,8-epoxy-3,3-4-trihydroxy-β-carotene, β-cryptoxanthin, lycopene, xanthophylls, 5, 8-epoxy-zienoxanthin zeinoxanthin, cyaniding diglucoside, glucose, fructose, rhamnose, sucrose, starch, xylose, arabin, composed of d-galacturonic acid, d-galactose, l-arabinose ester of ellagic and hexahydroxydiphenic acids, gallic acid, polyphenol oxidase, protein, cholesterol, saponin containing oleanolic acid, zeatin glucoside and zeatin riboside (*fruit*).

References : Ambasta (1986), Anonymous (1969), Farooq (2005), Kaushik & Dhiman (2000), Kirtikar & Basu (1984), Parrotta (2001), Subramaniam (2000).

761. *Pteris vittata* L.

Syn.: P. longifolia L.

Family: Pteridaceae.

English and Hindi Names: **Eng. :** Brake Fern, Chinese Brake, Chinese Ladder Brake, Ladder Brake; **Hindi :** Lingru.

Distribution : Tropical and sub-tropical regions of India.

Parts Used : Fronds.

Flowering & Fruiting : Throughout the year.

Folk Use : Fronds used as a tonic.

Active Constituents : Known to possess palmitic acid, β-sitosteryl palmitate, β-sitosteryl acetate and n-triacontanyl hexacosanoate (*plant*).

References : Kapur (1996b), Singh (2000b), Singh & Viswanathan (1996).

762. *Pterocarpus marsupium* Roxb.

Syn.: P. bilobus Roxb.

Family: Fabaceae.

English, Hindi and Sanskrit Names: **Eng. :** Indian Kino Tree, Malabar Kino Tree; **Hindi :** Bija, Bijasar, Vijayasar; **Sans. :** Asanah, Bijakah, Pitasara.

Distribution : Hilly regions throughout Southern India, extending to Gujarat, M.P., U.P., Bihar and Orissa.

Parts Used : Bark.

Flowering & Fruiting : Flowering during rainy season and fruiting in summer.

Folk Use : Fomentation of stem bark taken as tonic to overcome mother's general weakness and fatigue by Bhils in Rajasthan.

Active Constituent : l-epicatechin (*bark*).

References : Dhiman (2005), Kothari & Londhe (2000), Prajapati *et al.* (2006).

763. *Pterocarpus santalinus* L. f.

Family: Fabaceae.

English, Hindi and Sanskrit Names: **Eng. :** Red Sandal Wood, Red Sanders, Ruby Wood; **Hindi :** Lalchandan, Raktha Chandan, Undum; **Sans. :** Agarugandha, Harichandana, Lohita, Raktasara, Tamrabhra.

Distribution : Restricted to southern most districts of A.P., Karnataka, northern Tamil Nadu between 150 – 900m.

Parts Used : Wood.

Flowering & Fruiting : May – June.

Folk Use : Wood used as tonic.

Active Constituents : Reported to yield scentalin, desoxysantalin, santal, pterostilbene, pterocarpin and homopterocarpin, pterocarptriol, isopterocarpolone, pterocarpodiolone, β-endesmol, pterocarpol and cryptomeridiol (*heartwood*), acetyloleanolic acid, acetyloleanolic aldehyde and erythrodiol (*sapwood*).

References : Anonymous (1969), Gopal & Sinha (2000), Kirtikar & Basu (1984), Parrotta (2001), Watt (1972).

764. *Pterospermum acerifolium* Willd.

Syn.: P. aceroides Wall.

Family: Malvaceae.

English, Hindi and Sanskrit Names: **Eng. :** Hathipaila; **Hindi :** Kanakchampa, Kaniar, Muchkund; **Sans. :** Karnikara, Padapotpala, Privyadha.

Distribution : Sub-Himalayan tracts up to 1,200m, extending southwards to Western Ghats. Also, cultivated along roadsides.

Parts Used : Bark. Leaf. Flower.

Flowering & Fruiting : Flowering during March – June and fruiting in cold season.

Folk Use : Bark, leaves and flowers considered as tonic.

Active Constituents : These are n-triacontanol, 3, 7-diethyl-7-methyl-1, 5-pentacosanolide, η hexacosan-1, 26-diol and its dilignocerate, 24 β-ethyl-chalest-5-en-3β-O-α-cellobioside, β-sitosterol, β-amyrin, friedelan-3α- & 3β- ols, kaempferol, kaempferide-7-O-β-D-glucopyranoside, fatty acids, arachidic, arachidonic, behenic, linoleic, linolenic, lignoceric, myristic, palmitic and stearic acids (*flower*), bauerenal, betulin, friedelin, kaempferol-3β-D-galactoside, β-sitosterol, luteolin, its 7β-D-glucoside and 7β-D-glucuronide (*leaf*).

References : Anonymous (1969), Banerjee (1996, 2000b), Dhiman (2005), Kaushik & Dhiman (2000), Kirtikar & Basu (1984), Tiwari & Tiwari (1996).

765. *Pterygota alata* (Roxb.) R. Br.

Syn.: Sterculia alata Roxb.

Family: Sterculiaceae.

English and Hindi Names: **Eng. :** Narikel; **Hindi :** Paglaper.

Distribution : Occurs wild in Western Peninsula, Chittagong, Sylhet and in the Andamans.

Parts Used : Seeds.

Flowering & Fruiting : Flowering: February – March; Fruiting: November – January.

Folk Use : Seeds used as a substitute for opium as narcotics.

Active Constituents : Hydrolysis of fatty acids esters yield sterculic and malvalic acids, 4.0 and 5.8% respectively (*seeds*).

References : Anonymous (1969), Dhiman (2005).

766. *Pueraria tuberosa* (Roxb. ex Willd.) DC.

Syn.: Hedysarum tuberosum Roxb.

Family: Fabaceae.

English, Hindi and Sanskrit Names: **Eng. :** Indian Kudzu; **Hindi :** Patalkand, Sural, Tirra, Vidari Kaud; **Sans. :** Bhumi Kusmanda, Gajeshta, Triparna, Vidari.

Distribution : W. Himalayas – Sikkim, up to 1,350m, Kumaon, lower hills of the Punjab, Mt. Abu, hilly tracts of Bengal and S. India.

Parts Used : Root. Tuber.

Flowering & Fruiting : Flowering: February – April; Fruiting: June – August.

Folk Use : Roots and tubers used as a tonic.

Active Constituents : Major constituents are puerarostan, pterocarpan-tuberosin, hydroxytuberosone, anhydrotuberosin ,3-O-methylhydrotuberosin and puerarone (*tuber*), β-sitosterol, stigmasterol, daidzein, puerarin, a new isoflavone - C-glycoside-4', 6''-diacetylpuerarin and tuberosin (*root*).

References : Farooq (2005), Kirtikar & Basu (1984), Parrotta (2001), Sharma (2004), Watt (1972).

767. *Pulsatilla koreana* Nakai

Family: Ranunculaceae.

English Names: **Eng. :** Korean Pasque Flower, Grand Mother Flower.

Distribution : Native of Korea; open wastelands and grassy places.

Parts Used : Root.

Folk Use : Roots used as cardio tonic.

Active Constituents : Protoanemonin, anemonin, saponins, hederagenin monodesmoside, hederagenin bisdesmoside (*root*).

Reference : Prajapati *et al.* (2006).

768. *Punica granatum* L.*

Syn.: P. nana L.

Family: Punicaceae.

English, Hindi and Sanskrit Names: **Eng. :** Pomegranate Tree; **Hindi :** Anar, Dhalim, Dharimb; **Sans. :** Bijapura, Dadima, Granat, Pindapushpa, Vrittaphala.

Distribution : Cultivated throughout India.

Parts Used : Aerial Parts. Fruit. Seeds.

Flowering & Fruiting : April – July.

Folk Use : Aerial parts CNS depressant. Fruits and seed juice considered useful as a cardiac tonic.

Active Constituents : Reported to possess glucose, fructose, maltose, starch, carotene, thiamine, riboflavin, nicotinic acid, vitamin, pectin and amino acids (*seeds*), punicalin, punicalagin, punicafolin, granatin A, granatin B, conlagin, strictinin, 1, 2, 4, 6-tetra-O-galloyl, β-D-glucose (*leaf*), cyaniding-3-glucoside, delphinidin-3-glucoside and 3, 5-diglucoside (*seed coat*), 28 % gallotannic acid together with a yellow colouring matter (*rind*), Glucose, fructose, maltose, starch, carotene, thiamine, riboflavin, nicotinic acid, vitamin, pectin and amino acids (*fruit*).

References : Kaul (1997), Kirtikar & Basu (1984), Reddi *et al.* (2005), Singh & Srivastava (2000), Sood & Thakur (2004).

769. *Pygmaeopremna herbacea* (Roxb.) Moldenke

Syn.: Premna herbacea Roxb.

Family: Verbenaceae.

Hindi and Sanskrit Names: **Hindi :** Bharangi; **Sans. :** Bhumijambu.

Distribution : Forests of Kumaon and Terai of subtropical Himalayas (200-1,000m), also in Deccan Peninsula.

Parts Used : Root.

Flowering & Fruiting : March – May.

Folk Use : Roots considered as stimulant and tonic.

Active Constituents : Pygmacoherin, diterpenoid, bharangin, pygmaeocine B, C and E, sirutekkone and 7-isopropyl-3,3-dimethyl-2,3-dihydro-4, 9-dihydroxyphenalen-1-one (*root*).

Reference : Kaushik & Dhiman (2000).

770. *Pyrus communis* L.

Syn.: P. pyrifolia Nakai var. *culta* Nakai; *P. sinensis* Hort.

Family: Rosaceae.

English, Hindi and Sanskrit Names: **Eng. :** Pear Tree, Wild Pear; **Hindi :** Nashpati; **Sans. :** Amritaphala.

Distribution : Distributed in the temperate regions of India.

Parts Used : Fruit.

Flowering & Fruiting : March – April (flowering); July – September (fruiting).

Folk Use : Fruits used as sedative.

References : Farooq (2005), Kaushik & Dhiman (2000).

771. *Quassia indica* Noateboom

Syn.: Samadera lucida Wall.; *S. indica* Gaertn.

Family: Simaroubaceae.

English Name: **Eng. :** Surinam Quassia.

Distribution : Evergreen forests in West coast and along back waters and sandy places in Kerala, India, Madagascar, Sri Lanka, Myanmar, Indo – China, Thailand, Peninsular Malaysia – New Guinea.

Parts Used : Wood.

Folk Use : Infusion of wood used as bitter tonic.

References : Ambasta (1986), Anonymous (1969).

772. *Quercus velutina* Lamb.

Family: Fagaceae.

English Name: **Eng. :** Black Oak.

Distribution : Native to eastern N. America from southern Ontaria south to northern Florida.

Parts Used : Bark.

Flowering & Fruiting : April – May (flowering); August – October (fruiting).

Folk Use : Bark used as a tonic.

Active Constituent : Quercitron (*bark*).

Reference : Bhattacharjee (2001).

773. *Randia spinosa* Poir.

Syn.: R. brandisii Gamble; *R. dumetorum* Poir.; *R. longispina* Wight & Arn.; *R. tomentosa* Wight & Arn.; *Xeromphis spinosa* Keay.

Family: Rubiaceae.

English, Hindi and Sanskrit Names: **Eng. :** Common Emetic Nut; **Hindi :** Arar, Mainphal; **Sans. :** Bastishodhana, Madana, Ramachhardanaka.

Distribution : Throughout India up to 1,350m in the hills.

Parts Used : Bark.

Flowering & Fruiting : April – September.

Folk Use : Bark used as sedative.

Active Constituents : Randialic acid-A and randialic acid-B, mannitol, saponins, a new coumarin glycoside (I) and a new aliphatic diol-randiol (tetratriaconta-12, 21-diol) (*bark*).

References : Ambasta (1986), Anonymous (1969), Mukherjee & Namhata (1990), Prakash & Mehrotra (1991).

774. *Ranunculus laetus* Wall. ex D. Don

Syn.: R. brevirastris Edgew.; *R. distans* Royle.

Family: Ranunculaceae.

English Name: **Eng. :** Cheerful Buttercup.

Distribution : Temperate Himalayas, common in the inner ranges of Sikkim.

Parts Used : Leaf.

Flowering & Fruiting : March – April.

Folk Use : Leaves used as tonic.

Reference : Kala (2003).

775. *Ranunculus sceleratus* L.

Syn.: Ranunculus carnosus Wall.; *Ranunculus digitatus* Gilib.; *Ranunculus umbellatus* Roxb. ex Willd.

Family: Ranunculaceae.

English, Hindi and Sanskrit Names: **Eng. :** Blister Buttercup, Celery-leaved Crowfoot, Marsh Crowfoot, Water Celery; **Hindi :** Jaldhania; **Sans. :** Kandir.

Distribution : Warm valleys of the Himalayas from Assam – Kashmir up to 1,500m and in the plains of India.

Parts Used : Whole Plant, Seeds.

Flowering & Fruiting : Flowering and fruiting during winter season.

Folk Use : Plant used as stimulant. Seeds used as a tonic.

Active Constituents : Protoanemonin (93.0%) and anemonin (7.0%) (*oil*).

References : Ambasta (1986), Anonymous (1969), Kaushik & Dhiman (2000).

776. *Rauvolfia serpentina* Benth. ex Kurz.

Syn.: Ophioxylon album Gaertn.; *O. serpentium* L.; *O. trifoliatum* Gaertn.; *Rauvolfia trifoliata* (Gartn.) Boill.

Family: Apocynaceae.

English, Hindi and Sanskrit Names: **Eng. :** Rauvolfia Root, Serpentine, Serpentine Root; **Hindi :** Chandrabhaga, Chota-chand; **Sans. :** Chundrika, Sarpagandha.

Distribution : Wasteplaces and in shady forest in different parts of India from Punjab eastwards to Sikkim and Assam, also in some parts of Central India and Western Ghats.

Parts Used : Root.

Flowering & Fruiting : Chiefly in April – May.

Folk Use : Roots used as sedative, tranquilizer and hypnotic.

Active Constituents : A number of indoles and dihydrole indole alkaloids as indoles: (ajamalicidine, ajmalacine, rauimbine, deserpidine, indobine, indobinine, rescinnamidine, rescinnamine, sarpagine, serpentine, serpentinine and yohimbine) and indulines (ajmalimine, ajmaline, ajmaliminine, ajmalinine, isoajmaline, rauwolfine, sandioicolidine, sandiwicaline and serpentine) have been characterized from *roots*.

References : Ambasta (1986), Anonymous (1969), Dhiman (2005), Farooq (2005), Haridaran *et al.* (2001), Kaushik & Dhiman (2000), Nayak *et al.* (2003), Parrotta (2001), Prajapati *et al.* (2006), Prakash (2001), Raju (2000), Ranjan (2000), Ranjan (2003), Rao (1996), Reddi *et al.* (2005), Shah (1995), Sharma (2003, 2004), Siwakoti & Varma (1996).

777. *Rauvolfia vomitora* Afz.

Syn.: Hylacium owariense Afzel.; *Rauvolfia senegambiae* A. DC.

Family: Apocynaceae.

English Name: **Eng. :** Swizzle Stick.

Distribution : Native of Cameroon, Democratic Republic of Congo, Ghana, Liberia, Nigeria, Senegal, Sudan and Uganda.

Parts Used : Leaf. Root.

Folk Use : Decoction of leaves and roots or mixture of their powder mixed with honey useful as tranquillizer and tonic.

References : Bhattacharjee (2001), Khatoon & Goel (2010).

778. *Rehmania glutinosa* Gaert.

Family: Scrophulariaceae.

English Names: **Eng. :** Chinese Foxglove, Glutinous Rehmania.

Distribution : Introduced species in many parts of the country.

Parts Used : Root.

Folk Use : Roots used as cardio tonic.

Active Constituents : Glucose, mannite, glucoside, rehmanin and carotene (*root*).

Reference : Prajapati *et al.* (2006).

779. *Rhamnus purpurea* Edgew.

Family: Rhamnaceae.

English Name: **Eng. :** Purple Buchthorn.

Distribution : Pakistan, Kashmir, N. Punjab, Kumaon and Nepal.

Parts Used : Bark.

Folk Use : Bark used as a tonic.

Reference : Raju (2000).

780. *Rhamnus triquetra* Brandis

Family: Rhamnaceae.

Distribution : Pakistan, Western Himalayas from Jhelum to Kumaon, Kashmir.

Parts Used : Bark.

Flowering & Fruiting : July – September.

Folk Use : Bark used as a tonic.

Reference : Ambasta (1986).

781. *Rheum emodi* Wall. ex Meissn.

Syn.: Rheum australe D. Don

Family: Polygonaceae.

English, Hindi and Sanskrit Names: **Eng. :** Himalayan Rhubarb, Indian Rhubarb; **Hindi :** Dolu, Hindirevandchini, Revatchini; **Sans. :** Gandhini, Pitamulika, Revatchini.

Distribution : Wild in the alpine Himalayas at an altitude of about 4,000m.

Parts Used : Rhizome. Root.

Flowering & Fruiting : July – August.

Folk Use : Dried rhizomes and roots constitute the Himalayan rhubarb which is used as a tonic and stimulant.

Active Constituents : Emodin, emodinmanomethyl ether, chrysophanol, aloe-emodin and rhein, gallic acid, glucogallin, epicatechin, gallate and rutin (*root*).

References : Ambasta (1986), Dhiman (2005), Dwivedi (2007), Kirtikar & Basu (1984), Negi *et al.* (1999), Prajapati *et al.* (2006), Rawat *et al.* (2007), Sharma (2003), Sharma & Sood (1997), Singh (2001), Thakur *et al.* (1989), Uniyal & Malhotra (1989).

782. *Rheum webbianum* Royle

Family: Polygonaceae.

English, Hindi and Sanskrit Names: **Eng. :** Gilgiti Rhubarb, Indian Rhubarb, Small Himalayan Rhubarb; **Hindi :** Archa, Atis; **Sans. :** Amlaparni, Revatchini.

Distribution : Central and W. alpine Himalayas: 3,350-4,650m.

Parts Used : Rhizome. Root.

Flowering & Fruiting : Flowering during June – November.

Folk Use : Roots and rhizomes used as tonic.

Active Constituents : Reported to contain 1,8-dihydroxy-3-methyl, 1,8-dihydroxy-3-hydroxymethyl and 1, 6, 8-trihydroxy-3-methyl-anthaquinones (*root*).

Reference : Kapur *et al.* (1996).

783. ***Rhodiola imbricata*** Edgew.

Syn.: Sedum imbricatum Hk. f. & Thoms.; *S. rhodiola* auct.

Family: Crassulaceae.

Distribution : Kashmir, Kumaon, Nepal, Pakistan: 3,650-4,000m.

Parts Used : Root.

Flowering & Fruiting : Flowering during June – August.

Folk Use : Roots used for preparing a tonic to restore memory by Boto tribes in Ladakh.

References : Ballabh & Chaurasia (2006), Khatoon & Goel (2010).

784. ***Rhododendron anthopogon*** Don

Syn.: Azalea lapponica Pall.; *Osmothamnus fragrans* DC.; *Rhododendron aromaticum* Wall.

Family: Ericaceae.

English and Hindi Names: **Eng. :** Dwarf Rhododendron; **Hindi :** Talis, Talisri.

Distribution : Himalayas (Kashmir – Bhutan): 3,000-5,200m.

Parts Used : Leaf.

Flowering & Fruiting : June – July.

Folk Use : Leaves or its decoction used as tonic, sedative, tranqui-lizer and stimulant.

Active Constituents : Active properties due to β-sitosterol, friedelin, ursolic acid, quercetin, quercetin-3-O-α-L-ether and 3-O-glucoside (*leaf*).

References : Chhetri (2005), Kirtikar & Basu (1984), Sarin (1990), Sharma (2003).

785. ***Rhododendron arboreum*** Sm. (Pl. 9E)

Syn.: R. album Sweet; *R. campaelliae* Hook. f.; *R. cinnamomeum* Wall.; *R. nilagiricum* Zenk.; *R. puniceum* Roxb.; *R. windsorii* Nutt.

Family: Ericaceae.

English and Hindi Names: **Eng. :** Rose Tree, Tree Rhododendron; **Hindi :** Burans.

Distribution : Temperate Himalayas: 1,500-3,600m.

Parts Used : Flower.

Flowering & Fruiting : March – May.

Folk Use : Flowers used as cardio tonic.

Active Constituents : Known to contain ursolic acid, quercetin, resin and rhamnoside (*flower*).

Reference : Sood *et al.* (2009).

786. *Rhododendron campanulatum* D. Don

Syn.: R. aeruginosum Hook. f.; *R. edgarii* Gamble; *R. nobile* Wall.

Family: Ericaceae.

English and Hindi Names: **Eng. :** Rhododendron; **Hindi :** Cherailu.

Distribution : Alpine Himalayas (Kashmir – Bhutan): 3,000-4,650m.

Parts Used : Leaf. Flower.

Flowering & Fruiting : May – September.

Folk Use : Leaves and flowers used as tonic.

Active Constituents : Leaves rich in campanulin, epifriedelinol, ursolic acid, quercetin, friedelin and α-amyrin.

Reference : Kaul (1997).

787. *Rhus succedanea* L.

Syn.: R. acuminata DC.

Family: Anacardiaceae.

English, Hindi and Sanskrit Names: **Eng. :** Wild Varnish Tree, Wax Tree; **Hindi :** Kakra-singi; **Sans. :** Karkata Sringee.

Distribution : Mixed forests of Himalayas up to 2,500m, Punjab – Assam and in the Khasia hills.

Parts Used : Galls. Fruit.

Folk Use : Galls and fruits used as tonic and stimulant.

Active Constituents : Succedaenins A and B (*fruit*).

References : Ambasta (1986), Farooq (2005).

788. *Rorippa dufia* Hara.

Syn.: Nasturtium heterophyllum Blume.; *Nasturtium indicum* DC.; *Rorippa indica* Hiern var. *apetala* Hochr.

Family: Brassicaceae.

Distribution : Damp places throughout India, ascending up to 2,100m in the Himalayas.

Parts Used : Whole Plant.

Folk Use : Herb used as stimulant.

Reference : Ambasta (1986).

789. *Rosa centifolia* L.

Family: Rosaceae.

English, Hindi and Sanskrit Names: **Eng. :** Cabbage Rose, Hundred-leaved Rose, Pale Rose; **Hindi :** Gulab, Gulabzurdi; **Sans. :** Atikesara, Kharva, Satapatri, Saumyagandha.

Distribution : Cultivated throughout India.

Parts Used : Flower.

Folk Use : Flowers used as cardio tonic and tonic.

References : Kirtikar & Basu (1984), Prajapati *et al.* (2006).

790. *Rosa damascena* Mill.

Family: Rosaceae.

English, Hindi and Sanskrit Names: **Eng. :** Damask Rose, Otto of Roses; **Hindi :** Fasli Gulab, Gulab, Sudharg; **Sans. :** Atimanjula, Lakshapushpa, Shatapatri.

Distribution : Widely grown in gardens throughout India.

Parts Used : Flower.

Flowering & Fruiting : Flowering and fruiting almost throughout the greater parts of year.

Folk Use : Flowers used as tonic. 'Gulkand', a preserve made from equal parts of sugars and its petals, valued as tonic.

Active Constituents : Essential oils, citronellel, nerol, geraniol, β-phenylethanol and its glucoside, eugenol and methyl eugenol; organic acids, cholorgenic acid, tannin, cyanin, cyanidin and its 3,5-diglucoside, quercetrin, carotene and sugars (*flower*).

References : Kaushik & Dhiman (2000), Kirtikar & Basu (1984), Sharma (2003), Watt (1972).

791. *Rosa laevigata* Michaux.

Syn.: R. sinica Lindl.

Family: Rosaceae.

English Name: **Eng. :** Cherokee Rose.

Distribution : Indigenous to Cao Bang and Lang Son provinces; introduced and naturalized in India.

Parts Used : Fruit.

Flowering & Fruiting : Flowering during May – July.

Folk Use : Fruits regarded as a general tonic.

Active Constituents : 2α, 3β, 19α, 23-tetrahydroxyurs-12-en-28-oic acid, 2α, 3α, 19α, 23-tetrahydroxyurs-12-en-28-oic acid, euscaphic acid, diacetyl-2α,3β-dihydroxy-lupan-28-oic acid, daucosterol and β-sitosterol (*fruit*).

Reference : Prajapati *et al.* (2006).

792. *Roscoea alpina* Royle

Family: Zingiberaceae.

English Name: **Eng. :** Mountain Roscoe Lily.

Distribution : Uttaranchal – South-East Tibet between 1,500-3,000m; found in India, Shimla – Fagu and from Kashmir – Bhutan.

Parts Used : Tuberous Root.

Flowering & Fruiting : Flowering during Spring.

Folk Use : Tuberous roots useful as tonic in general debility.

Reference : Sharma (2003).

793. *Rosmarinus officinalis* L.

Family: Lamiaceae.

English and Hindi Names: **Eng. :** Rosemary; **Hindi :** Rusmari.

Distribution : Native to the Mediterranean; cultivated throughout the world and India.

Parts Used : Whole Plant. Leaf. Flower.

Folk Use : Plant used as nervine tonic. Leaves useful as tonic and stimulant. Flowers also considered stimulant.

Active Constituents : Known to yield volatile oil containing borneal, camphene, camphor, cineole, flavonoids, tannins, rosmarininc acid, diterpenes and rosemaricine (*plant*).

References : Bhattacharjee (2001), Khatoon & Geol (2010), Prajapati *et al.* (2006), Rajan *et al.* (2003), Sharma (2003), Watt (1972).

794. *Rourea minor* (Gaertn.) Alston

Syn.: R. acuminata Hook. f.; *R. caudata* Planch.; *R. commutota* Planch.; *R. pulchella* Planch.; *R. santaloides* Wight. & Arn.; *Santaloides erectum* Schellenb.; *S. floridum* Kuntze; *S. minus* Schellenb.

Family: Cannaraceae.

Hindi and Sanskrit Names: **Hindi :** Kalavidhara, Vidhara; **Sans. :** Antri, Junga, Kotarapushpi, Rukshagandha, Vridha.

Distribution : Konkan, S.M. country, N. Kanara – Travancore.

Parts Used : Root.

Folk Use : Roots used as a tonic.

Active Constituents : β-sitosterol-β-D-glucoside, hentriacontane, meso-inositol, rapanone, leucopelargonidin and β-sitosterol (*root*).

References : Ambasta (1986), Kirtikar & Basu (1984).

795. *Roylea cinerea* (D. Don) Bail.

Syn.: Ballota cinerea D. Don; *Phlomis calycina* Roxb.; *Roylea calycina* (Roxb.) Briq.; *R. elegans* Wall.

Family: Lamiaceae.

Hindi Name: **Hindi :** Patkarru.

Distribution : Subtropical W. Himalayas from Kashmir – Kumaon: 700-1,700m.

Parts Used : Leaf.

Flowering & Fruiting : March – November.

Folk Use : Leaves used as bitter tonic.

Active Constituents : Isolated components are betulin, β-amyrin, β-sitosterol, stigmasterol, cetyl alcohol, glucose, fructose, arabinose and palmitic, stearic, oleic, oxalic and tartaric acids (*leaf*).

References : Farooq (2005), Raju (2000), Watt (1972).

796. *Rubia cordifolia* L. sensu Hook. f.

Syn.: R. alata Wall.; *R. chinensis* Regel & Maack; *R. javana* DC.; *R. mitis* Miq.; *R. mungisth* Desv.; *R. munjista* Roxb.; *R. purpurea* Decne.; *R. scandens* Zoll. & Morr.; *R. secunda* Moon.

Family: Rubiaceae.

English, Hindi and Sanskrit Names: **Eng. :** Indian Madder; **Hindi :** Manjit, Manjith; **Sans. :** Aruna, Bhandi, Chhatra, Gandiri, Harini, Jingi, Kala-meshika, Manjistha, Rohini, Samanga, Tamrika, Vapra, Vikasa, Yojanavallika.

Distribution : Throughout the hilly regions in India.

Parts Used : Leaf. Root.

Flowering & Fruiting : July – September (flowering); November – December (fruiting).

Folk Use : Infusion of leaves and roots recommended as a tonic.

Active Constituents : Reported to possess alizarin, physcion, nor-damnacanthal, lucidine, primevoside, ruberythric acid, munjistin, xnthopurpurin, pseudopurpurin, 1,4-dihydroxy-6-methylanthraquinone, 1-hydroxyl-2-methylanthraquinone, nordamnacanthol, physcion, 1-hydroxyl-2-methoxyanthraquinone, 1,4-dihydroxy-2-methyl-5-methoxyanthraquinone, rubiadin, 1,3-dimethoxy-2-carboxyanthraquinone, 1,4dihydroxy-2-methylanthraquinone, alizarin, mollugin, 1,5-dihydroxy-2-methylanthraquinone, 3-prenyl-5-methoxy-1,4-napthoquinone, 1-hydroxyl-2-methyl -9,10-anthraquinone, 1, 3-dihydroxy-2-ethoxymethyl-9,10-anthraquninoe, lucidin primeveroside, ruberythric acid, 1,3-dihydroxy-2-methoxymethylanthraquninoe, 1-methoxy-2-methoxymethyl -3-hydroxyanthraquinone, 4-hydroxyl-2-carboxyanthraquinone, 1,4-dihydroxy-2-hydroxymethylanthraquinone and 1-hydroxyl-2-hydroxymethylanthraquinone (*root*).

References : Ambasta (1986), Dhiman (2005), Farooq (2005), Haridaran *et al.* (2001), Prajapati *et al.* (2006), Sharma & Sood

(1997), Sood & Thakur (2004), Sood *et al.* (2009), Thakur *et al.* (1989), Tiwari & Tiwari (1996).

797. *Rubia tinctorum* L.

Family: Rubiaceae.

English Names: **Eng. :** Alizari, European Madder.

Distribution : Native to S. Europe, W. Asia and N. Africa; flourishes on wastegrounds in hedgerows and amid rubbles in Kashmir.

Parts Used : Root.

Folk Use : Roots used as a tonic.

Active Constituents : Reported to possess alizarin, purpurin, purpurin carboxylate, quinizarine, christofin, nordamnalanthoic acid methyl ester, nordamnalanthol, tectoquinone and 7-hydroxytectoquinone (*root*).

Reference : Ambasta (1986).

798. *Rubia wallichiana* Decne.

Syn.: R. asperrima Decne.

Family: Rubiaceae.

English Name: **Eng. :** Madder.

Distribution : Himalayas to S. China.

Parts Used : Whole Plant.

Folk Use : Plant used as tonic.

Reference : Joshi (2009).

799. *Rubus coreanus* Miq.

Family: Rosaceae.

English Names: **Eng. :** Korean Black Raspberry, Korean Bramble.

Distribution : Thickets on mountain slopes of Korea, Japan and China.

Parts Used : Fruit.

Folk Use : Fruits used as a tonic.

Active Constituents : Polysaccharides, Triterpenoids : nagaichigoside F1, F2, Suavissimoside R1, Coreanoside F1, (+)-catechin, (-) - epicatechin, polycyanidin B-4, sanguiin H-4 (*fruits*).

Reference : Prajapati *et al.* (2006).

800. *Rubus niveus* Thunb. (Pl. 9F)

Syn.: R. albescens Roxb.; *R. bijugus* Focke; *R. distans* Don; *R. furfuracens* Wall.; *R. horsefieldii* Miq.; *R. indicus* Heyne; *R. lasiocarpus* Sm.; *R. mysorensis* Heyne; *R. rosaeflorus* Roxb.; *R. rotundifolius* Royle.

Family: Rosaceae.

English and Hindi Names: **Eng. :** Mahabaleshwar Raspberry, Mysore Raspberry, Raspberry; **Hindi :** Kala Hinsalu.

Distribution : Throughout Himalayas: 1,350-3,500m.

Parts Used : Root.

Flowering & Fruiting : March – May.

Folk Use : Nepali and Gorkhali (migrating tribes collecting medicinal plants) make their hallucinogenic drinks from the root extract of this plant.

Reference : Negi *et al.* (1999).

801. *Ruellia patula* Jacq.

Syn.: Dipteracanthus patulus (Jacq.) Nees.

Family: Acanthaceae.

English Name: **Eng. :** Wild Petunia.

Distribution : S.W. India; Pakistan, Sri Lanka, Tropical Africa and Arabia.

Parts Used : Root.

Flowering & Fruiting : January – April.

Folk Use : 3g per dose of powdered roots given for 8 – 10days with cow's milk as a tonic for children in the age group 5 – 10 years.

Reference : Sood & Thakur (2004).

802. *Rumex hymenosephalus* Torr.

Syn.: R. arizonicus Britton.; *R. hymenosephalus* var. *salinus* (A. Nelson) Rech.; *R. salinus* A. Nelson.; *R. saxei* Kellogg.

Family: Polygonaceae.

English Names: **Eng. :** Arizona Dock, Canaigre, Ganagra, Red Dock, Tanner's Dock, Wild Rhubarb, Wild Pie Plant.

Distribution : Native to U.S. and Mexico and southwestern California and Mojave desert.

Parts Used : Root.

Flowering & Fruiting : January – June.

Folk Use : Extract of roots used as tonic.

Reference : Bhattacharjee (2001).

803. *Rumex nepalensis* Spreng. (Pl. 10A)

Syn.: R. eregrinus Boiss.; *R. hamatus* Trevir.; *R. ramulosus* Meisn.; *R. roxburghianus* Schult. f.; *R. tuberosus* Roxb.; *R. uncinatus* Hort.

Family: Polygonaceae.

English Name: **Eng. :** Sorrel.

Distribution : Temperate Himalayas: 1,350-4,000m, W. Ghats, Nilgiris and Pulney Hills.

Parts Used : Root.

Flowering & Fruiting : May – October.

Folk Use : Infusion of root used as a liver tonic.

Active Constituents : Chief ingredients are chrysophanol, emodin, physcion, lupeol, sitosterol, its glucoside, musizin-1-O-β-D-glucopyranoside, orientalone, two new compounds – 3-methoxy-5,6-methylenedioxybenzaldehyde (I) and chrysophanol-8-O-β – D-galactopyranoside (*root*).

Reference : Chhetri (2005).

804. *Ruta chalepensis* Wall.

Syn.: R. angustifolia Pers.

Family: Rutaceae.

English, Hindi and Sanskrit Names: **Eng. :** Common Rue, Fringed Rue, Garden Rue; **Hindi :** Pismaram, Sadab, Satari; **Sans. :** Gucchapatra.

Distribution : Cultivated in Indian gardens; native to Eurasia and North Africa.

Parts Used : Whole Plant.

Folk Use : Plant used as a tonic.

Active Constituents : These are rutamarin, kokusaginine, xanthotoxin, chalepensin, skimmianine, isoimperatorin, isotaifine, 8-methoxytaifine, γ-fagarine, rutamarin, bergapten, xanthotoxin, isopimpinellin, graveoline, chalepin (*plant*), moskachans A, B, C, D, chalepensin, chalepin, coumarin, bergapten, isopimpinellin, xanthotoxin, chalepensin, heliettin, rutamarin, kokusaginine, skimmianine and arborinine (*aerial parts*).

References : Kayal *et al.* (2003), Prajapati *et al.* (2006), Watt (1972).

805. *Ruta graveolens* L.

Family: Rutaceae.

English, Hindi and Sanskrit Names: **Eng. :** Ave Grace, Herb of Grace, Rue Plant; **Hindi :** Pismarum, Sadab, Satari; **Sans. :** Guchhapatra, Somalata; Vishopaha.

Distribution : Cultivated throughout India westwards to the Canaries.

Parts Used : Whole Plant.

Folk Use : Plant used as narcotic, stimulant and tonic.

Active Constituents : Reported to possess kokusaginine, skimmianine, graveoline, graveolinins, rutin, isoimperatorin, psoralen, ribolinium, rutalinium and N-methyl platydesmin (*plant*), suberenone, gravacridon-chorine, gravacridonalchlorine, gravacridonediol and its monoethyl ether, rutaculotin, rutamarin alcohol, daphnorin, scopoletin, gravacridonetriol, furacridone, 1-hydroxyl-3-methoxy-N-methylacridone (IV) and isogravacridone-chlorine (*root*), 2-nonanone, 2-nonyl acetate and 2-undecyl acetate (*leaf oil*), gossypetin-7-methylether-3-rutinoside (*flower*), 1-methyl-2n-nonyl-4-quinollone, lictamnine, γ-fagarine, kokusaginine, skimmianine and graveoline (*leaf, shoots, flower*).

References : Bhattacharjee (2001), Kirtikar & Basu (1984), Parrotta (2001).

806. *Saccharum spontaneum* L.

Syn.: S. aegyptiacum Willd.; *S. balense* Boga; *S. biforum* Forsk.; *S. canaliculatum* Roxb.; *S. chinense* Nees.; *S. glaza* Reinw; *S. insulare* Borngn.; *S. propinguum* Steud.; *S. punctatum* Schumach; *S. semidecumbens* Roxb.

Family: Poaceae.

English, Hindi and Sanskrit Names: **Eng. :** Thatch Grass, Wild Sugarcane; **Hindi :** Kas, Kus, Kagara, Kansi, Kar, Kosa; **Sans. :** Kasah, Kasalu, Khaggara.

Distribution : Warmer parts of India, Sri Lanka and ascending to 2,000m in the Himalayas.

Parts Used : Root.

Flowering & Fruiting : June – October.

Folk Use : Roots used as a tonic.

References : Prajapati *et al.* (2006), Watt (1972).

807. *Salacia reticulata* Wight.

Family: Celastraceae.

Sanskrit Names: **Sans. :** Pitika,Vairi.

Distribution : Found throughout India in forests.

Parts Used : Root.

Folk Use : Roots used as a tonic.

Reference : Prajapati *et al.* (2006).

808. *Salix babylonica* L.

Family: Salicaceae.

English and Hindi Names: **Eng. :** Weeping Willow; **Hindi :** Majnun.

Distribution : Cultivated in Indian plains and Himalayas up to 2,473m, and elsewhere in gardens.

Parts Used : Bark. Leaf.

Flowering & Fruiting : Flowering during February - March and fruiting later in summer.

Folk Use : Bark and leaves of the plant considered as tonic.

Active Constituents : Known to possess a new benzyl ester of gentisic-2'-O-acetyl-β-D-glucoside, apigenin-7-O-galactoside, kaempferol-7-O-glucoside, luteolin-4'-O-glucoside, trichocarpin, salicin acid and an ester of terephthalic acid (*leaf*), sucrose, glucose and salicin (*bark*).

References : Kaushik & Dhiman (2000), Kirtikar & Basu (1984), Watt (1972).

809. *Salix caprea* L.

Family: Salicaceae.

English and Hindi Names: **Eng. :** Sallow, Goat Willow; **Hindi :** Bedmushk.

Distribution : Cultivated in Rohil Khand and N.W. India, W. Asia and Europe.

Parts Used : Leaf.

Folk Use : Oil obtained from leaves or its decoction used as tonic.

Active Constituents : A new flavones bioside – salicarpene, luteolin-7-β-D-glucopyranoside, (+) epigallocatechin and (+) gallocatechin dimmers, I and II (*leaf*).

Reference : Ambasta (1986).

810. *Salix clegans* Wall.

Syn.: S. denticulata Anderss; *S. kumaoneusis* Lindl.

Family: Salicaceae.

English Name: **Eng. :** Elegant Willow.

Distribution : Western Himalayas (Nepal – Marui): 2,000-3,650m, and Gilgit.

Parts Used : Leaf.

Flowering & Fruiting : April – July.

Folk Use : Leaves used as a tonic.

Reference : Kapur (1996b).

811. *Salvadora persica* L.

Syn.: Cissus arborea Forsk.; *Salvadora indica* Wight.; *S. wightiana* Planck.

Family: Salvadoraceae.

English, Hindi and Sanskrit Names: **Eng. :** Mustard Tree, Salt Bush, Tooth-brush Tree; **Hindi :** Jhak, Kharjal, Pilu; **Sans. :** Brihatpilu, Maha vriksha, Pilu, Piluh.

Distribution : Saline lands and black cotton soil of Peninsular India and Sri Lanka.

Parts Used : Leaf. Root Bark. Seeds.

Flowering & Fruiting : Throughout the year.

Folk Use : Leaves, seeds and decoction of root bark used as tonic and stimulant.

Active Constituents : Known to possess β-sitosterol, elemental γ-monoclinic sulphur (S8) and glucopaeolin (*root*); salvadoricine (*leaf*).

References : Ambasta (1986), Kirtikar & Basu (1984), Prajapati *et al.* (2006), Singh *et al.* (2007).

812. *Salvia officinalis* L.*

Family: Lamiaceae.

English and Hindi Names: **Eng. :** Garden Sage, Red Sage, Sage; **Hindi :** Salbiasetakuss.

Distribution : Native to Mediterranean; cultivated in India in gardens.

Parts Used : Whole Plant. Leaf. Flower.

Folk Use : Plant and leaves used as tonic. Flowers mixed with sugar used for strengthening weak memories.

Active Constituents : Picrosalvin, genkwanin, 6-methoxygenkwanin, luteolin, 6-methoxyluteolin and its 7-methyl ether and hispidulin, viridifloral, 7 – ethylrosmanol, cirsimarilin, luteolin-7-O-glucoside, apigenin-7-O-glucoside, ursolic acid, oleanolic acid, β-sitosterol and saccharose (*leaf*), salvin and its monomethyl ether (*flower*).

References : Bhattacharjee (2001), Khatoon & Goel (2010), Kirtikar & Basu (1984), Prajapati *et al.* (2006), Rajan *et al.* (2003), Watt (1972).

813. *Samadera indica* Gaertn.

Syn.: Niota penta - petala Poir.

Family: Simaroubaceae.

English, Hindi and Sanskrit Names: **Eng. :** Niepa Bark Tree; **Hindi :** Lokhanti; **Sans. :** Gucchakaranjah.

Distribution : Evergreen forests and along backwaters of South India.

Parts Used : Bark. Wood.

Folk Use : Bark and infusion of wood used as a tonic.

Active Constituents : Taraxerone, stigmostanone, stigmasterol, 2, 6-dimethoxybenzoquninone, samaderine A, samaderine B, samaderine C, indaquassins A, B, C, D, E and F, dihydrosamaderine B, brucein D, soulameolide, cedronin, cedronin and canthin-3, 6-dione (*bark*).

References : Kirtikar & Basu (1984), Prajapati *et al.* (2006), Watt (1972).

814. *Sambucus canadensis* L.

Family: Caprifoliaceae.

English Name: **Eng. :** American Elder.

Distribution : Native of central and eastern Canada and U. S. A., introduced into gardens of India and Nepal.

Parts Used : Whole Plant.

Folk Use : Plant used as stimulant.

Active Constituents : Cyanidin-3-O-(6"-O-E-p-carmaroyl) glucosyl (2→1)-β-D-xyloside-5-O-β-D-glucoside (I), cyanidin-3-sambubioside-5-glucoside, cyanidin-3-sambubioside and cyaniding-3-glucoside (*fruit*).

Reference : Ambasta (1986).

815. *Sambucus nigra* L.

Family: Caprifoliaceae.

English Names: **Eng. :** Common Elder, European Elder.

Distribution : Native of Europe and Asia Minor; cultivated in India to a small extent in botanical gardens, and occur wild in Kangra District.

Parts Used : Bark. Flower.

Flowering & Fruiting : May - September.

Folk Use : Bark of the plant possesses narcotic properties. Infusion of dried flowers used as a sedative.

Active Constituents : Found to possess α-amyrenone, α-amyrin, betulin, oleanolic acid, β-sitosterol, ursolic acid, ceryl alcohol, n-heptacosane (*bark*), α- and β- amyrins, lupeol, cycloartenol, 24-methylenecycloartanol, cycloeuclenol, cholesterol, campesterol and stigmasterol in free as well as esterified forms (*flower*).

Reference : Bhattacherjee (2001).

816. *Sandorium kortjape* (Burm.f.) Merrill

Syn.: S. indicum Cav.; *S. nervosum* Blume; *S. ternatum* Blanco; *Trichilia nevosa* Vahl.; *T. venosa* Spreng.

Family: Meliaceae.

English Names: **Eng. :** Malay Sandal, Red Sental, Yellow Sental.

Distribution : Tropical forests of Pegu and Tenasserim; largely cultivated in Myanmar, the Malay Peninsula, W. Peninsula and Sri Lanka.

Parts Used : Root.

Folk Use : Roots used as tonic.

Reference : Ambasta (1986).

817. *Sansevieria trifasciata* Prain.

Syn.: S. jacquinii N. E. Br.; *S. laurentii* De Wild.; *S. trifasciata* var. *laurentii* (De Wild.) N. E. Br.; *S. zeylanica* var. *laurentii* (De Wild.) L. H. Bailey.

Family: Liliaceae.

English Names: **Eng. :** African Bowstring Hemp, Mother-in-law's Tongue.

Distribution : A common ornamental Indian garden plant.

Parts Used : Root.

Flowering & Fruiting : June – July.

Folk Use : Roots used as tonic and stimulant.

Reference : Ambasta (1986).

818. *Sansevieria zeylanica* Willd.

Family: Liliaceae.

English, Hindi and Sanskrit Names: **Eng. :** Bowstring Hemp; **Hindi :** Morwa; **Sans. :** Mumva.

Distribution : Eastern coast of India from Bengal – Chennai.

Parts Used : Root.

Folk Use : Decoction of roots used as tonic.

Reference : Sharma (2004).

819. *Santalum album* L.

Syn.: Sandalum album Rumph.; *S. myrtifolium* Roxb.; *S. verum* L.; *Sirium myrtifolium* Roxb.

Family: Santalaceae.

English, Hindi and Sanskrit Names: **Eng.** : Sandal Tree, White Sandal Tree; **Hindi :** Chandal, Safed Chandan, Santal; **Sans. :** Bhadrasri, Candanah, Malayaja, Pitachandana, Srikhandam.

Distribution : Southern regions of India, especially in Karnataka, Tamil Nadu and Kerala, ascending up to1,200m.

Parts Used : Heartwood.

Flowering & Fruiting : February - May.

Folk Use : Heartwood used as a tonic and cardiotonic.

Active Constituents : α-santalenes, β-santalenes, santenol and teresantanol (*heartwood*).

References : Kirtikar & Basu (1984), Parrotta (2001), Prajapati *et al.* (2006), Watt (1972).

820. *Sapindus mukorossi* Gaertn.

Syn.: S. abruptus Lour.; *S. detergens* Roxb.; *S. laurifolia* Vahl.

Family: Sapindaceae.

English, Hindi and Sanskrit Names: **Eng. :** Soapnut Tree of North India; **Hindi :** Dodan, Kanwar, Ritha; **Sans. :** Arishtak, Phenilah, Urista.

Distribution : Dehradun and in the valleys of the Himalayas up to 1,219m, also cultivated throughout N.W. India, Bengal and Assam.

Parts Used : Fruit.

Flowering & Fruiting : Flowering during May – June; and fruiting in cold season.

Folk Use : Fruits used as tonic.

Active Constituents : Saponin emarginatoside, saponin, mucorosin, glucose and pectin (*fruit shell*), and saponin have been isolated (*pericarp*).

References : Dhiman (2005), Kaushik & Dhiman (2000), Prajapati *et al.* (2006), Sharma (2004).

821. *Sapindus trifoliatus* L.

Syn.: S. abstergens Roxb.; *S. accetus* Roxb.; *S. emarginatus* Vahl.; *S. laurifolius* Vahl. var. *emarginatus* (Vahl.) Cooke.

Family: Sapindaceae.

English, Hindi and Sanskrit Names: **Eng. :** Soapnut Tree; **Hindi :** Reetha; **Sans. :** Arishta, Phenila, RithaKaranja,

Distribution : Aravalli hills and in the Western Peninsula; also cultivated in Bengal and North India.

Parts Used : Fruit.

Flowering & Fruiting : Flowering during October – December and fruiting in March – April.

Folk Use : Fruits and its pulp used as tonic.

Active Constituents : Reported to yield two new hederagenin glycosides – emargintosides B and C, A new saponin – emarginatoside, whose acid hydrolysis yielded hederagenin, rhamnose, xylose and glucose (*pericarp*), saponin A, sapindoside B, saponin C and sapindoside A (*rind*).

References : Ambasta (1986), Dhiman (2005), Kirtikar & Basu (1984), Watt (1972).

822. *Sapium insigne* (Royle) Benth. ex Hook.

Syn.: Carumbium insigne Kurz.; *Falconeria insignis* Royle; *F. wallichiana* Royle.

Family: Euphorbiaceae.

English and Hindi Names: **Eng. :** Chinese Tallow, Tiger's Milk Spruce; **Hindi :** Khinna, Khirum, Lendwa, Lienda.

Distribution : Sub-tropical Himalayas: 500-1,800m.

Parts Used : Leaf.

Flowering & Fruiting : January – March.

Folk Use : ½ - 1tsp of its powdered leaves given twice daily for 10-15 days as a tonic.

Reference : Sood & Thakur (2004).

823. *Sapium sebiferum* Roxb.

Syn.: Carumbium sebiferum Kurz; *Croton sebiferus* L.; *Excoecaria sebifera* Muell.; *Stillingfleetia sebifera* Bojer; *Stillingia sebifera* Michaux; *S. sinesis* Baill.

Family: Euphorbiaceae.

English, Hindi and Sanskrit Names: **Eng. :** Chinese Tallow-tree; **Hindi :** Pahari-shisham, Pippalyang, Tarcharbi, Vilyati Shisham; **Sans. :** Toyapippali.

Distribution : Native of China; cultivated in various parts of India and elsewhere in warm countries.

Parts Used : Root Bark.

Flowering & Fruiting : Flowering during rainy season and fruiting in winters.

Folk Use : Decoction of root bark used as tonic.

Active Constituents : 3-epimoretenol, moretenol, moretinone, phloracetophenone -2, 4-dimethylether and xanthoxylin (*root bark*).

References : Ambasta (1986), Farooq (2005).

824. *Saraca asoca* (Roxb.) De Wilde

Syn.: Jonesia asoca Roxb.; *J. pinnata* Willd.; *S. arborescens* Burm.; *S. indica* auct. non L.; *S. minor* Miq.; *S. zollingeriana* Miq.

Family: Fabaceae.

English, Hindi and Sanskrit Names: **Eng. :** Ashoka Tree; **Hindi :** Ashok, Asok; **Sans. :** Ashoka, Gatasokah, Hempuspha.

Distribution : Along streams or in shady evergreen forest up to 750m in Central & Eastern Himalayas as well as in the Khasia, Garo and Lushai hills.

Parts Used : Bark. Flower.

Flowering & Fruiting : Flowering during March – April and fruiting in August – September.

Folk Use : Flower buds (5-7 in number) eaten raw by females as uterine tonic to prevent miscarriage. Decoction of bark also used as uterine tonic and tonic.

Active Constituents : Quercetin, kaempferol-3-O-β-D-glucoside, quercetin-3-O-β-D-glucoside, apigenin-7-O-β-D-glucoide, pelargonidin-3, 3-diglucoside and cyanidin -3,3-diglucoside (*flower*); 24-

methylcholesterol-5-en-3β-ol, (22E)-24-ethylcholesta-5, 22-dien-3β-ol and 24-ethylcholest-5-en-3β-ol (*bark*).

References : Ambasta (1986), Banerjee (1996), Dhiman (2005), Hemlata *et al.* (2007), Parrotta (2001), Prajapati *et al.* (2006), Sarma *et al.* (2002), Sood (2009), Sharma (2004), Thakur *et al.* (1989).

825. *Sarcostemma acidum* (Roxb.) Voigt

Syn.: Asclepias aphylla Roxb.; *Sarcotemma brevistigma* Wight. & Arn.

Family: Asclepiadaceae.

English, Hindi and Sanskrit Names: **Eng. :** Moon Creeper, Moon Plant, Sour Creeper; **Hindi :** Somlata; **Sans. :** Chandravalli, Dhanurlata, Indulekha, Mahagulma, Padmakastha, Soma, Somavalli, Yajnashreshtha, Yajnavalli.

Distribution : Bihar, Bengal and South India in dry rocky places up to 1,350m.

Parts Used : Whole Plant. Stem.

Folk Use : Plant used as a narcotic. Stem extract valued as tonic.

Active Constituents : Malic acid, succinic acid, reducing sugar, sucrose, traces of tannin, an alkaloid, a phytosterol, α- and β-sitosterol (*plant*).

References : Prajapati *et al.* (2006), Upadhyay (2003).

826. *Sarcostemma brunonianum* Wt. & Arn.

Family: Asclepiadaceae.

Distribution : South India, Sri Lanka.

Parts Used : Fruit.

Flowering & Fruiting : Flowering during September – April.

Folk Use : Fermented fruits used as narcotic.

Reference : Subramaniam (2000).

827. *Sassafras albidum* (Nutt.) Nees

Syn.: Sassafras officinalis Nees & Eberm.; *Sassafras variifolium* Salisb.

Family: Lauraceae.

English Names: **Eng. :** Ague Tree, Sassafras.

Distribution : Native of the forests of North America, from Canada – Florida; grown in Indian gardens.

Parts Used : Root.

Folk Use : Roots used as stimulant and tonic. Oil extracted from root also considered stimulant.

Active Constituents : 80% safrole and little quantity of eugenol, pinene, phellendrene, sesquiterpene and d-camphor (*root oil*).

References : Ambasta (1986), Bhattacharjee (2001), Watt (1972).

828. *Satyrium nepalense* D.Don

Family: Orchidaceae.

English Name: **Eng. :** Nepal Satyrium.

Distribution : Kashmir, Khasia hills, Darjeeling, Manipur and hilly regions of Southern India between 2,700-3,000m.

Parts Used : Root.

Flowering & Fruiting : July – September.

Folk Use : Roots considered as a tonic.

Reference : Sharma (2003).

829. *Saussurea lappa* C.B.Clarke

Syn.: Aplataxis lappa Decne.; *Sassurea costus* (Falc.) Lipsch.

Family: Asteraceae.

English, Hindi and Sanskrit Names: **Eng. :** Costus, Kuth; **Hindi :** Kot, Kur, Kut, Pachak; **Sans.** : Agada, Bhasura, Kushtha, Pavana.

Distribution : Cultivated in Lahul and Spiti district and along Indo – Tibetan border between: 2,500-3,500m.

Parts Used : Root.

Folk Use : Roots used as tonic, stimulant and narcotic.

Active Constituents : Costunolide, dehydrocostuolactone, α-cyclostunolide and costic, palmitic and linoleic acids, besides β-sitosterol, costunolide, dehydro-costuslactone, β-cyclocostunolide and isolantolactone (*root*).

References : Ambasta(1986), Chauhan (2001), Kirtikar & Basu (1984), Prajapati *et al.* (2006), Sharma (2003), Thakur *et al.* (1989), Watt (1972).

830. *Saussurea sacra* Edgew.

Syn.: S. simpsoniana Lipsch.

Family: Asteraceae.

Distribution : Alpine Himalayas: Garhwal – Sikkim (4,650-6,000m).

Parts Used : Whole Plant.

Flowering & Fruiting : July – September.

Folk Use : Plant used as tonic and stimulant.

Active Constituents : Saussurol, colchicine, limonene, 1-cadinol and β-amyrenol (*plant*).

Reference : Kapur *et al.* (1996).

831. ***Saxifraga oppositifolia*** L.

Syn.: S. asiatica Hayek.

Family: Saxifragaceae.

English Name: **Eng. :** Common Mountain Saxifrage.

Distribution : Western Tibet (5,000-5,650m); Sassar; Alps of Europe and Asia, Arctic Regions.

Parts Used : Whole Plant.

Folk Use : Plant extract considered as tonic.

Reference : Chaurasia *et al.* (2000).

832. ***Schefflera octophylla*** (Lour.) Harms

Family: Araliaceae.

English Name: **Eng. :** Octophyllous Aralia.

Distribution : Grows wild in secondary forest in mountainous regions.

Parts Used : Root Bark. Stem Bark.

Folk Use : Root bark and stem bark possess tonic properties.

Reference : Prajapati *et al.* (2006).

833. ***Schinus terebinthifolius*** Raddi

Family: Anacardiaceae.

English Name: **Eng. :** Brazilian Pepper Tree.

Distribution : Native of temperate and tropical regions of South America, introduced into Indian gardens as ornamental. Also, found in Argentina, Paraguay, Brazil and adjacent territory.

Parts Used : Bark.

Folk Use : Bark used as tonic and stimulant.

Active Constituents : Two new triterpene ketones – bauerenone and terebenthifolic acid, α-amyrin and α-amyrenone have been isolated (*bark*).

Reference : Ambasta (1986).

834. ***Schleichera oleosa*** (Lour.) Oken

Syn.: S. trijuga Willd.

Family: Sapindaceae.

English, Hindi and Sanskrit Names: **Eng. :** Ceyolon Oak, Lac Tree; **Hindi :** Kausum, Kusum; **Sans.** : Mukulakah, Laksavrksah, Raktamrah.

Distribution : Semi-evergreen and deciduous forests throughout India.

Parts Used : Fruit. Seeds.

Folk Use : Fruits and seeds used as a tonic.

Active Constituents : Phenols and polyene pigments (*seeds*).

References : Prajapati *et al.* (2006), Upadhye & Kumbhojkar (1998).

835. *Schweinfurthia sphaerocarpa* A.Br.

Syn.: Antirrhinum glaucum Stocks; *A. papilionaceum* Burm; *Linaria sphaerocarpa* Benth.; *S. papilionacea* Boiss.

Family: Scrophulariaceae.

Hindi and Sanskrit Names: **Hindi :** Nepal Nimb, Sanipat; **Sans. :** Nepala-nimba, Sanipat.

Distribution : Rocky places in Sind and distributed to Baluchistan and Afghanistan.

Parts Used : Fruit. Leaf. Stem.

Flowering & Fruiting : October – December.

Folk Use : Broken pieces of fruits together with powdered leaves and portion of dried stems used as tonic.

Reference : Ambasta (1986).

836. *Scindapsus officinalis* (Roxb.) Schott.

Syn.: Pothos officinalis Roxb.

Family: Araceae.

Hindi and Sanskrit Names: **Hindi :** Baripippli, Gajapipal, Pipalijhamgha; **Sans. :** Chavyaphala, Gajapippali, Ibhoshana.

Distribution : Tropical Himalayas from Sikkim eastwards, Bengal, Chittagong, Myanmar, Andaman Islands.

Parts Used : Fruit. Inflorescence.

Folk Use : Dried inflorescence used as stimulant. Fruits used as stimulant, tonic and cardio tonic.

Active Constituents : Scindapsin A, scindapsin B, a sterol, a mixture of sugars (*fruit*).

References : Farooq (2005), Kirtikar & Basu (1984), Prajapati *et al.* (2006), Siwakoti & Varma (1996), Watt (1972).

837. *Scutellaria lateriflora* L.

Family: Lamiaceae.

English Names: **Eng. :** Mad Dog, Skullcap, Virginian Skullcap.

Distribution : Native of N. America. It still grows wild in much of the US and Canada.

Parts Used : Seeds.

Flowering & Fruiting : Flowering during summers.

Folk Use : Seeds used as sedative and nervine tonic.

Active Constituents : Flavonoids, bitter iridoids, volatile oil tannins (*seeds*).

References : Bhattacharjee (2001), Prajapati *et al.* (2006).

838. *Scutellaria prostrata* Jacq. ex Benth.

Family: Lamiaceae.

English Names: **Eng. :** Madweed, Skullcap.

Distribution : Western Himalayas (Kashmir – Kumaon): 2,700-4,650m.

Parts Used : Shoot.

Flowering & Fruiting : August – September.

Folk Use : Shoots used as nervine tonic.

Reference : Kala (2003).

839. *Securinega leucopyrus* (Willd.) Muell – Arg.

Syn.: Fluggea leucpyrus Willd.

Family: Euphorbiaceae.

English, Hindi and Sanskrit Names: **Eng. :** Spinous Fluggea; **Hindi :** Ainta, Hartto; **Sans. :** Bhuriphali, Panduraphali.

Distribution : Dry localities throughout India.

Parts Used : Whole Plant.

Folk Use : Plant used as a tonic.

Reference : Prajapati *et al.* (2006).

840. *Selaginella bryopteris* (Sw.) Spring.

Family: Selaginellaceae.

English and Hindi Names: **Eng. :** Selaginella; **Hindi :** Sanjivini.

Distribution : Throughout India.

Parts Used : Whole Plant.

Folk Use : Plant used as a tonic.

Reference : Jha *et al.* (2003).

841. *Selinum vaginatum* C.B. Clarke

Family: Apiaceae.

Distribution : W. Himalayas: 1,800-3,800m; Temperate and alpine moist pastures in Chota – Bara Bhangal of Kangra, Hali range of Chamba, Parvati valley and Rohtang area in Kullu, Rampur and Rohru division of Shimla district.

Parts Used : Root.

Folk Use : Roots used as a nervine sedative.

Active Constituents : Known to possess selinidin, vaginidin, angelicin, oroselol and lomatin (*root*).

References : Ambasta (1986), Prajapati *et al.* (2006), Sharma (2003).

842. *Selinum wallichianum* (DC.) Raizada & Saxena *ex* Raizada

Syn.: Selinum tenuifolium Wall.

Family: Apiaceae.

Hindi Names: **Hindi :** Bhootheshi, Kher.

Distribution : Temperate Himalayas: 2,000-4,000m, usually found in temperate pasture lands, forests shades and alpine pasture in Kangra, Kullu, Chamba, Kinnaur, Sirmour and Shimla districts between 2,400-4,200m.

Parts Used : Root.

Folk Use : Roots used as nervine sedative and tranquillizer.

Active Constituents : Bergapten, heraclenin, heraclenol, angelicin and xanthotoxol (*root*).

References : Kapur (1996b), Prajapati *et al.* (2006), Sarin (1990), Sharma (2003).

843. *Semecarpus anacardium* L. f.

Syn.: Anacardium latifolium Lamk.; *A. longifolius* Lam.; *A. officinarum* Gaertn.; *A. orientale* auct. *ex* Steud.; *Cassuvium anacardium* Kuntze; *C. longifolium* Kuntze; *Semecarpus latifolius* Pers.

Family: Anacardiaceae.

English, Hindi and Sanskrit Names: **Eng. :** Marking Nut Tree, Oriental Cashew, Varnish Tree; **Hindi :** Bhela, Bhilara, Belatak; **Sans. :** Arusharah, Bhallatakah, Shailabija, Tapona, Vatori.

Distribution : Throughout India, in semievergreen and moist deciduous forests.

Parts Used : Flower. Fruit. Kernel.

Flowering & Fruiting : Flowering during May – June and fruiting in December – March.

Folk Use : Fruits used as liver tonic and cardiotonic. Kernels valued as nervine tonic and eaten to improve memory. Also, extract of flowers taken 2-3 times for a week for good memory.

Active Constituents : Anacardic acid, cardol, catechol, anacardol and fixed oil and semicarpol bhilawanol (*fruit*).

References : Khatoon & Goel (2010), Kirtikar & Basu (1984), Prajapati *et al.* (2006), Reddi *et al.* (2005), Reddy (1996).

844. *Senecio jacquemontianus* (DC.) Benth.

Syn.: Senecillis jacquemontiana Decne.

Family: Asteraceae.

English Name: **Eng. :** Jacquemont's Senecio.

Distribution : Western Himalayas: 3,350-4,650m.

Parts Used : Root.

Folk Use : Fresh root extract used as nervine tonic.

References : Gupta (1997), Kirtikar & Basu (1984), Singh (1996).

845. *Serenoa serrulata* (Michx.) Hook. f.

Syn.: S. repens (Bartram) Small.

Family: Arecaceae.

English and Hindi Names: **Eng.** : Dwarf Palm, Saw-palmetto; **Hindi** : Sabal.

Distribution : South California to Florida and Louisania, South eastern N. America; planted in Indian gardens.

Parts Used : Fruit.

Folk Use : Dried fruits said to be sedative.

Reference : Bhattacharjee (2001).

846. *Sesamum indicum* L.*

Syn.: S. luteum Retz.; *S. occidentale* Heer & Regel.; *S. orientale* L.

Family: Pedaliaceae.

English, Hindi and Sanskrit Names: **Eng.** : Gingelly, Sesame; **Hindi:** Bariktel, Til, Tilkatel; **Sans.** : Homadhanya, Pavitra, Snehaphala, Tila, Tilah, Vanodbhava.

Distribution : Cultivated throughout Indian plains and up to 1,200m.

Parts Used : Whole Plant. Seeds.

Flowering & Fruiting : Flowering during September – October and fruiting in November – December.

Folk Use : Plant and seeds used as a tonic.

Active Constituents : These are sesamolinol, sesaminol, pinoresinol, pinoresinol-4'-O-β-D-glucosyl (1→6)-β-D-glucoside (KP1), pinoresinol-4'-O-β-D-glucosyl (1→2)-β-D-glucoside (KP2) (*seeds*).

References : Farooq (2005), Jain & Sharma (2000), Kirtikar & Basu (1984), Kumar *et al.* (2003), Parrotta (2001), Prajapati *et al.* (2006), Saikia & Nath (2003b), Sharma (2003), Singh *et al.* (2007), Siwakoti & Varma (1996), Watt (1972).

847. *Sesbania grandiflora* Pers.

Syn.: Agati grandiflora Desv.; *Coronilla grandiflora* Willd.

Family: Fabaceae.

English, Hindi and Sanskrit Names: **Eng.** : Agathi, Sesban, Swamp Pea; **Hindi** : Agasti, Bak, Basna, Hatiya; **Sans.** : Agasti, Agastyah, Agati, Anari.

Distribution : Cultivated in Southern and Eastern India, the Ganges Doab and Burma.

Parts Used : Bark. Leaf.

Flowering & Fruiting : Flowering throughout the year and fruiting in March – May.

Folk Use : Bark and leaves used as a tonic.

Active Constituents : Oleanolic acid, galactose, rhamnose and glucuronic acid (*leaf*).

References : Ambasta (1986), Kirtikar & Basu (1984), Parrotta (2001), Prajapati *et al.* (2006), Raju (2000).

848. *Sesbania roxburghii* Merrill

Syn.: S. aculeata Pers. var. *paludosa* Baker.

Family: Fabaceae.

English and Hindi Names: **Eng. :** Sesban; **Hindi :** Kathsola.

Distribution : Indigenous to south – eastern Asia and Malaysia; grown as garden fence in West Bengal and in Marshes on the W. Coast.

Parts Used : Seeds.

Flowering & Fruiting : September – November.

Folk Use : Seeds used as stimulant.

Active Constituents : Methyl linoleate, methyl linolenate, methyl oleate and methyl stearate (*seeds*).

Reference : Ambasta (1986).

849. *Sesbania sesban* (L.) Merrill

Syn.: Aeschynomene sesban L.; *Caronilla sesban* Willd.; *Sesbania aegyptiaca* Pers.

Family: Fabaceae.

English, Hindi and Sanskrit Names: **Eng. :** Common Sesban, Egyptian Rattle Pod; **Hindi :** Jainti, Jait, Rawasan; **Sans. :** Jayanti, Jayantika, Nadayi, Raja.

Distribution : Throughout India up to 1,200m.

Parts Used : Seeds.

Flowering & Fruiting : Flowering during August – September and fruiting in December – January.

Folk Use : Seeds used as stimulant.

Active Constituents : Stigmasta-5, 24 (28) diene-3β-O-β-D-galactopyranoside, galactomannan (*seeds*).

References : Ambasta (1986), Kaushik & Dhiman (2000), Kirtikar & Basu (1984), Parrotta (2001), Raju (2000), Watt (1972).

850. *Seseli diffusum* (Roxb. *ex* Sm.) Santapau & Wagh

Syn.: Ligusticum diffusum Roxb.; *Seseli indicum* Wight. & Arn.

Family: Apiaceae.

Sanskrit Name: **Sans. :** Vanayamani.

Distribution : Sub-Himalayan tracts from Dehra Dun – Gorakhpur, Bundel Khand, Assam, Central Bengal and Coromandel Coast.

Parts Used : Whole Plant. Fruit.

Flowering & Fruiting : August – December.

Folk Use : Whole Plant and fruits used as stimulant.

References : Ambasta (1986), Siwakoti & Varma (1996).

851. *Shorea robusta* Gaertn. f.

Syn.: Vatica robusta Steud.

Family: Dipterocarpaceae.

English, Hindi and Sanskrit Names: **Eng. :** Sal Tree, Indian Dammer; **Hindi :** Damar, Dhuna, Ral, Sal; **Sans. :** Asvakarna, Guggilam, Koushi-kaha, Salah.

Distribution : N.E. and Central India up to 1,700m.

Parts Used : Shoot. Bark. Flower. Fruit. Seeds.

Flowering & Fruiting : March – June.

Folk Use : Juice of young shoots or bark and fruits used as a tonic. Flowers and seeds pasted with rice and grass (*Cynodon dactylon* (L.) Pers.) eaten as tonic.

Active Constituents : These are shoreaphenol (*bark*), corilogin, ellagic, chebulinic and gallic acids (*seeds*).

References : Banerjee (1996), Banerjee (2000b), Kirtikar & Basu (1984), Prajapati *et al.* (2006).

852. *Sida acuta* Burm. f.

Syn.: Sida carpinifolia Mast.; *S. lanceolata* Roxb.; *S. scoparia* Lour.; *S. stauntoniana* DC.; *S. stipulata* Cav.

Family: Malvaceae.

English, Hindi and Sanskrit Names: **Eng. :** Hornbean Leaved Sida; **Hindi :** Bariara, Kareta, Kharenta; **Sans. :** Bala, Pitberela, Rajbala.

Distribution : Wasteplaces throughout the hotter parts of India.

Parts Used : Root.

Flowering & Fruiting : September – March.

Folk Use : Roots used as tonic and bitter tonic.

Active Constituents : α-amyrin, ecdysterone, ephedrine, cryptolepine and oxalic acids (*root*).

References : Ambasta (1986), Dhiman (2005), Gogoi & Das (2003), Kirtikar & Basu (1984), Parrotta (2001), Ranjan (2000), Shukla & Verma (1996), Srivastava *et al.* (2003b), Subramaniam (2000), Watt (1972).

853. *Sida cordata* (Burm. f.) Barssum

Syn.: Melochia cordata Burm.f.; *Sida humilis* Cav.; *S. humilis* var. *veronicaefolia* (Lamk.) Masters; *S. multicaulis* Cav.; *S. veronicaefolia Lamk.*

Family: Malvaceae.

Hindi and Sanskrit Names: **Hindi :** Banaiyar, Bhyunli, Farid buti; **Sans. :** Bhumibala, Nagabala, Rajbala.

Distribution : Hotter parts of India up to 1,500m.

Parts Used : Whole Plant.

Flowering & Fruiting : Flowering and fruiting almost throughout the year.

Folk Use : Plant decoction used as a tonic in heart diseases.

Active Constituents : N-methyl ψ-ephedrine, ephedrine, N-methylephedrine, β-phenylthylamine, ψ-ephedrine, vasicinol, vasicinone, vasicine, choline and betaine (*plant*).

References : Dhiman (2005), Farooq (2005), Kaushik & Dhiman (2000), Parrotta (2001), Prajapati *et al.* (2006), Singh & Kumar (2000a).

854. *Sida cordifolia* L.

Syn.: S. herbacea Cav.; *S. rotundifolia* Cav.

Family: Malvaceae.

English, Hindi and Sanskrit Names: **Eng. :** Country-mallow; **Hindi :** Bariyar, Kharenti, Kungyi; **Sans. :** Baladhya, Kanaka, Motapati, Samanga, Variga.

Distribution : Wild throughout the tropical and sub-tropical India, ascending upto 1,050m.

Parts Used : Whole Plant. Root. Seeds.

Flowering & Fruiting : August – December.

Folk Use : Plant and roots used as tonic. Seeds valued as tonic for improving sexual strength.

Active Constituents : Found to yield ephedrine (*root, stem*), palmitic, stearic, hexacosanoic acids and β-sitosterol (*aerial parts*).

References : Ambasta (1986), Bhattacharjee (2001), Dhiman (2005), Kaushik & Dhiman (2000), Kirtikar & Basu (1984), Meena *et al.* (2003), Prajapati *et al.* (2006), Sharma (2003), Watt (1972).

855. *Sida ovata* Forssk.

Family: Malvaceae.

English and Hindi Names: **Eng. :** Mapungubwe Sida; **Hindi :** Bal, Dabi.

Distribution : Gujarat; occasionally in forest edges.

Parts Used : Root.

Flowering & Fruiting : October – January.

Folk Use : Roots boiled in water taken orally as a tonic.

Reference : Singh & Pandey (1998).

856. *Sida rhombifolia* L. (Pl. 10B)

Syn.: S. canariensis Willd.; *S. compressa* Wall.

Family: Malvaceae.

Hindi and Sanskrit Names: **Hindi :** Bhiunli, Sahadabi, Swetbarela; **Sans. :** Ahikhanda, Brihadagala, Lalbarila, Sahadeva, Vatya.

Distribution : Widely distributed throughout the tropical regions of India.

Parts Used : Root.

Flowering & Fruiting : August – October.

Folk Use : Root used as a tonic.

References : Kirtikar & Basu (1984), Tiwari & Tiwari (1996).

857. *Sida spinosa* L.

Syn.: S. alba L.; *S. alnifolia* L.; *S. boriara* Wall.; *S. glandulosa* Roxb.; *S. retusa* Wight.

Family: Malvaceae.

English, Hindi and Sanskrit Names: **Eng. :** Prickly Sida; **Hindi :** Bariyara, Gulsakari, Jangli Methi; **Sans. :** Avishta, Devadanda, Mahabala.

Distribution : Hotter parts of India.

Parts Used : Root. Root Bark

Flowering & Fruiting : September – January.

Folk Use : Roots and root bark used as a tonic.

References : Dwivedi (2007), Kirtikar & Basu (1984), Prajapati *et al.* (2006), Watt (1972).

858. *Sissymbriun iria* L.

Syn.: S. irioides Boiss.

Family: Brassicaeceae.

English and Hindi Names: **Eng. :** Hedge Mustard, London Rocket; **Hindi :** Khubkalan.

Distribution : Found distributed in Northern India, Rajasthan and Punjab.

Parts Used : Seeds.

Flowering & Fruiting : November – February.

Folk Use : Seeds used as stimulant and tonic.

Active Constituent : Isorhamnetin (*seeds*).

References : Dhiman (2005), Kirtikar & Basu (1984), Watt (1972).

859. *Sium latijugum* C.B. Clarke

Family: Apiaceae.

English Name: **Eng. :** Large Leaflet Water Parsnip.

Distribution : Kashmir and Baltisthan, in ditches, 1,700-3,000m; very common.

Parts Used : Root.

Folk Use : Alcoholic extract of the roots when administered intraperitoneally produce depression of central nervous system.

Reference : Ambasta (1986).

860. *Smilax china* L.

Family: Liliaceae.

English, Hindi and Sanskrit Names: **Eng. :** China Root; **Hindi :** Cobcini; **Sans. :** Dvipantaravaca, Madhusnuhi.

Distribution : China and Japan.

Parts Used : Rhizome.

Folk Use : Rhizomes used as tonic.

Active Constituents : Tannin, resin, cinchonin, smilacin, saponin, flavonoid glycoside, diosgenin, rutin and isoseryl S-methylcysteamine sulphoxide (*rhizome*).

Reference : Prajapati *et al.* (2006).

861. *Smithia conferta* Sm.

Syn.: Smithia germiniflora Roth.

Family: Fabaceae.

English and Hindi Names: **Eng. :** Paired Flower Smithia. **Sans. :** Lakshmana.

Distribution : Throughout India, Sri Lanka, Java, N. Australia.

Parts Used : Whole Plant.

Flowering & Fruiting : August – December.

Folk Use : Herb used as a tonic.

References : Ambasta (1986), Parrotta (2001).

862. *Solanum dulcamara* L.

Syn.: S. laxum Royle; *S. lyratum* Thunb.; *S. persicum* Willd.

Family: Solanaceae.

English, Hindi and Sanskrit Names: **Eng. :** Bitter Sweet, Dulcamara, Woody Nightshade; **Hindi :** Bhalumas; **Sans. :** Kakmachi.

Distribution : Wild in Kashmir, H.P. and U.P.: 1,219-2,438m.

Parts Used : Branches.

Flowering & Fruiting : July – September.

Folk Use : Infusion of dried branches used as sedative, stimulant and narcotic.

Active Constituents : Soladulcidine, tigogenin, new steroidal glucuronides – (25R and 25S) spirost-5-ene-3β-O-α-L-rhamnopyranosyl (1→2)-β-D-glucuronopyranoside(II and III), (22ξ, 25R and 25S) furost-5-en-22-methoxy-3β-O-α-L-rhamnopyranosyl (1→2) –β-D-glucuronopyranosyl-26-O-β-D-glucopyranoside (IV and V), and tigogenin-3-O-β-D-glucopyranoside (*branches*).

References : Ambasta (1986), Dhiman (2005), Kirtikar & Basu (1984), Prajapati *et al.* (2006).

863. *Solanum indicum* auct. non L.

Syn.: S. anguivi Lam.; *S. cuneatum* Moench.; *S. cunescens* Blume; *S. himalense* Dunal.;*S. heynii* Roem. & Sch.; *S. junghuhnii* Miq.; *S. pinnatifidum* Roth.; *S. violaceum* Jacq.

Family: Solanaceae.

English, Hindi and Sanskrit Names: **Eng.** : Poison Berry; **Hindi :** Barhauta, Birhatta, Kattarha; **Sans.** : Akranta, Brhati, Hinguli, Raktapaki, Simhi, Torani, Vyaghri.

Distribution : Throughout Indian plains and at low elevations.

Parts Used : Herb. Root.

Flowering & Fruiting : June – August.

Folk Use : Herb possesses tonic properties. Decoction of roots used as cadio tonic.

Active Constituents : Herb rich in solanine, solanidine, solasonine, solasodine, diosgenin, β-sitosterol, lanosterol and solamargine.

References : Parrotta (2001), Prajapati *et al.* (2006), Sharma (2003).

864. *Solanum melongena* L. (Pl. 10C)

Family: Solanaceae.

English, Hindi and Sanskrit Names: **Eng.** : Brinjal, Egg Plant; **Hindi** : Badanjan, Baingan, Bhanta; **Sans.** : Bhantaki, Hingoli, Jukutam, Vartaku, Vatigama, Vatigana.

Distribution : Cultivated throughout India.

Parts Used : Fruit. Leaf. Root. Seeds.

Flowering & Fruiting : Flowering and fruiting almost throughout the year.

Folk Use : Roots and seeds used as a stimulant. Leaves used as narcotic. Unripe fruits used as cardio tonic and ripened ones as hypnotic.

Active Constituents : Found to possess arginine, aspartic acid, γ-aminobutyric acid, histidine, isoleucine, γ-hydroxyglutamic acid, leucine, lysine, pipecolic acid, phenylalanine (*fruit*), lanost-8-en-3β-ol, lanosterol (tr), 24-methylene lanost-8-en-3β-ol (tr), cycloortanol, cycloarterol, 24-methylenecycloartanol, lupeol, β-amyrin, daturaolone and daturadiol (*seeds*), 4-ethylcatechol, trans-caffeic, hydrocaffeic, protocatechuic and chlorogenic acids (*leaf*).

References : Ambasta (1986), Farooq (2005), Kaushik & Dhiman (2000), Kirtikar & Basu (1984), Prajapati *et al.* (2006), Watt (1972).

865. *Solanum nigrum* L. (Pl. 10D)

Syn.: S. pupureilineatum Sabins & Bhatt; *S. rubrum* Mill.; *S. villosum* Mill. ssp. *villosum* Edmonds.

Family: Solanaceae.

English, Hindi and Sanskrit Names: **Eng.** : Black Nightshade, Hound's Berry, Morelle, Petty Morel; **Hindi** : Gurkkamai, Kabaiya, Makoi; **Sans.** : Dhvankshamachi, Ghanaghana, Kakamachi, Rasayani, Vayasi.

Distribution : Wasteplaces, roadsides and railway tracks throughout India up to 2,743m in the Western Himalayas.

Parts Used : Whole Plant. Berry.

Flowering & Fruiting : Flowering and fruiting chiefly in February – June.

Folk Use : Berries used as tonic. Plant used as sedative, narcotic, cardio tonic and tonic.

Active Constituents : Reported to possess riboflavin, nicotinic acid, vit.-C, β-carotene and citric acid (*leaf*), glucose, fructose, vit.-C and β-carotene (*fruit*), solamargine and α- and β-solanigrine (*immature green fruit*), diosgenin (*leaf, stem, fruit*).

References : Ambasta (1986), Dhiman (2005), Dhasmana & Rathore (2007), Farooq (2005), Gajurel *et al.* (2006), Hebber *et al.* (2003), Kaushik & Dhiman (2000), Kirtikar & Basu (1984), Parrotta (2001), Prajapati *et al.* (2006), Sharma (2003), Subramaniam (2000), Watt (1972).

866. *Solanum surattense* Burm. f. (Pl. 10E)

Syn.: S. armatum Br.; *S. virginianum* L.; *S. xanthocarpum* Schrad. & Wendl.

Family: Solanaceae.

English, Hindi and Sanskrit Names: **Eng.** : Dwarf Lid Brinjal, Yellow-berried Nightshade; **Hindi** : Kateli, Remgani, Satyanasi; **Sans.** : Dhavanika, Kautakari, Nidigdhika, Putrapada, Sinhi.

Distribution : As a weed on roadsides and wastelands in dry situation throughout India.

Parts Used : Whole Plant. Flower. Fruit.

Flowering & Fruiting : April – August.

Folk Use : Plant, flowers and fruits used as stimulant.

Active Constituents : A new sterol – carpesterol characterized as (22R) 22-hydroxy-6-oxo-4α-methyl-5α-stigmast-7-en-3β-yl benzoate, scopoletin, esculin and esculetin have been isolated from the *plant*.

References : Prajapati *et al.* (2006), Sharma (2004).

867. *Solanum torvum* Swartz. (Pl. 10F)

Syn.: S. ferrugineum Jacq.; *S. pseudosaponaceum* Blume; *S. stramonifolium* Roxb.; *S. wightii* Miq.

Family: Solanaceae.

Hindi and Sanskrit Names: **Hindi** : Bari, Kateli; **Sans.** : Gosthabartaku, Shwet Brihati.

Distribution : Wild throughout the tropical regions of India.

Parts Used : Whole Plant. Fruit.

Flowering & Fruiting : December – June.

Folk Use : Entire plant used as sedative. Fruits valued as tonic.

Active Constituents : Attributed properties due to neochlrogenone, chlorogenin, neochlorogenin, chlorogenone and its 25S-epimer, solasonine, solasodine, sisalagenone, torvogenin, sitosterol-D-glucoside and vit.-A (*fruit*), neochlorogenone, solasodine, paniculogenin, neosolaspigenin, solaspigenin, toronin A and B, 5-hexatriacontanone, octasanyl-tricontanoate, 2, 3, 4-trimethyltriacontane, tetratriacontanoic acid, triacontanol, 3- tritriacontanone, campesterol, sitosterol and sigmasterol (*leaf*), solasidiene (*stem*), 25 S-3β-amino-5α-furostane-22 α, 26-diol-O (26)-β-D glucopyranoside (*root*), α-, β-, γ-carotenes, lutein, cryptoxanthin and lycopene (*plant*).

References : Dhiman (2005), Farooq (2005), Gogoi & Das (2003), Kaushik & Dhiman (2000), Parrotta (2001), Sharma (2009), Subramaniam (2000).

868. *Solanum tuberosum* L.

Family: Solanaceae.

English, Hindi and Sanskrit Names: **Eng.** : Potato; **Hindi** : Alu; **Sans.** : Alukam, Golakaudah.

Distribution : Cultivated in higher elevations in India.

Parts Used : Tuber.

Flowering & Fruiting : Flowering and fruiting during winters.

Folk Use : Tubers used as nervine sedative.

References : Kaushik & Dhiman (2000), Prajapat *et al.* (2006).

869. *Solena amplexicaulis* (Lam.) Ghandhi

Syn.: Melothria heterophylla (Lour.) Cogn.; *Solena heterophylla* Lour.

Family: Cucurbitaceae.

English, Hindi and Sanskrit Names: **Eng. :** Creeping Cucumber, Diversely-leaved Melothria, Clasping-stemmed Solena; **Hindi :** Amantamul, Tarali; **Sans. :** Amlavetasah, Sukakamdah.

Distribution : Throughout India upto 2,100m in the hills.

Parts Used : Tuberous Roots.

Flowering & Fruiting : March – October.

Folk Use : Tuberous roots used as stimulant and cardio tonic.

References : Ambasta (1986), Anonymous (1966), Prajapati *et al.* (2006), Sharma (2004).

870. *Sonchus arvensis* L.

Syn.: S. longifolius Wall.; *S. orixensis* Roxb.; *S. wightianus* DC.

Family: Asteraceae.

English and Hindi Names: **Eng. :** Corn Sow-thistle, Gutweed, Hogweed, Rosemary; **Hindi :** Sadhi, Sahadevibari.

Distribution : More common on the hills; all warm countries, wild or introduced.

Parts Used : Whole Plant.

Flowering & Fruiting : May – July.

Folk Use : Plant used as sedative and hypnotic.

Reference : Bhalla *et al.* (1996).

871. *Sonchus oleraceus* L.

Syn.: S. ciliates Lamk.; *S. royleanus* Wall.

Family: Asteraceae.

English and Hindi Names: **Eng. :** Hare's Lettuce, Hare's Palace, Milk Thistle, Sow Thistle; **Hindi :** Dodak, Dudhi.

Distribution : Sparingly throughout India; Sri Lanka, all temperate and many tropical countries, wild or introduced.

Parts Used : Leaf. Root. Stem.

Flowering & Fruiting : December – March.

Folk Use : Stem of plant used both as tonic and sedative. Roots and leaves exhibit tonic properties.

Active Constituents : Scopoletin, esculetin, ceryl alcohol, α- amyrin and β-sitosterol (*leaf*).

References : Ambasta (1986), Chaudhary & Neogi (2000), Farooq (2005), Kapur & Srivastava (1996), Kirtikar & Basu (1984), Sharma (2009), Singh & Srivastava (2000), Sood *et al.* (2001), Srivastava *et al.* (2003a), Watt (1972).

872. *Soymida febrifuga* A. Juss.

Family: Meliaceae.

English, Hindi and Sanskrit Names: **Eng.** : Indian Red Wood; **Hindi** : Rohan; **Sans.** : Mansarohinee.

Distribution : Forests of Central India, Bihar and hilly areas of India.

Parts Used : Bark.

Flowering & Fruiting : April – July.

Folk Use : Bark used as tonic.

Active Constituents : Active properties due to lupeol and two tetranortriterpenoids having substituted-γ-methoxy-butenolide substituent (*bark*).

Reference : Chatterjee & Pakrashi (1997).

873. *Sphaeranthus indicus* auct. non L.

Syn.: S. hirtus Willd.; *S. senegalensis* DC.

Family: Asteraceae.

English, Hindi and Sanskrit Names: **Eng.** : East Indian Globe-thistle; **Hindi** : Gorakh Mundi, Mundi; **Sans.** : Hapusa, Mahamundi, Mundi, Mundirika.

Distribution : Known to be distributed throughout India ascending up to 1,524m; commonly in dry rice fields.

Parts Used : Herb. Leaf. Flower.

Flowering & Fruiting : December – March.

Folk Use : Herb, flowers and powdered leaves used as a tonic.

Active Constituents : Known to possess 7α – hydroxyeudesm – 4 – en – 6, 12 – olide (I) (7 - hyroxyfrullanolide), 7β – hydroxyendesm – 4 – en – 6, 12 – olide (II) and a new dihydrolactone (III) (*flower*).

References : Ambasta (1986), Dhiman (2005), Bhalla *et al.* (1996), Dwivedi (2007), Kirtikar & Basu (1984), Kumar & Lal (2003), Parrotta (2001), Prajapati *et al.* (2006), Reddi *et al.* (2005), Sahu (2004), Watt (1972).

874. *Spilanthes acmella* Murr.

Syn.: S. mauritiana DC.; *Verbesina acmella* L.

Family: Asteraceae.

English and Sanskrit Names: **Eng.** : Paracress, Toothache Plant; **Sans.** : Akalkada, Akarkara, Pokarmul.

Distribution : Occurring as a weed in rice fields throughout India, ascending to 1,700m.

Parts Used : Capitula.

Flowering & Fruiting : September – December.

Folk Use : Tincture of capitula used as stimulant.

Active Constituents : A eudesmanolide (I), limonene, β-caryophyllene, β-ocimene, germacrene D and myrcene (*capitula*).

References : Ambasta (1986), Srivastava *et al.* (2003a).

875. *Spilanthes paniculata* DC.

Family: Asteraceae.

English Name: **Eng. :** Daisy Cress.

Distribution : Meghalaya, Sikkim and W. Ghats up to 1,500m.

Parts Used : Flower Head.

Folk Use : Flower head used as a tonic.

Reference : Gogoi & Das (2003).

876. *Spondias pinnata* (L. f.) Kurz.

Syn.: Evia amara Comn.; *Mangifera pinnata* Koen.; *Spondias mangifera* Willd.

Family: Anacardiaceae.

English, Hindi and Sanskrit Names: **Eng. :** Bile Tree, Hogplum, Wild Mango; **Hindi :** Ambodha, Amra, Jungli Aam; **Sans. :** Amratakan, Bhringiphalla, Kapitana, Rasadhya, Tanukshiri.

Distribution : Wild as well as cultivated throughout India.

Parts Used : Stem Bark. Fruit.

Folk Use : Mixture of powdered stem bark with a cup of country liquor taken for 3 – 5 days after delivery as a tonic. Ripe fruits also valued as tonic.

Active Constituents : β-amyrin, oeanolic acid, glycine, cystine, serine, alanine and leucine (*fruit*).

References : Jain (1996), Kirtikar & Basu (1984), Prajapati *et al.* (2006).

877. *Stephania glabra* Miers.

Syn.: Cissampelos glabra Roxb.; *Stephania rotunda* Hook. f. & Th.

Family: Menispermaceae.

Distribution : Himalayas from Shimla – Sikkim, Khasia Hills, Assam, Tenasserim.

Parts Used : Tuber.

Folk Use : Tubers used as sedative.

Active Constituents : Stepharine, stepharrotine, (-) tetrahydropalmatine, dehydrocorydalmine chloride, protoberberine alkaloid – stepharanine, palmatrubine, stepharanine, corydalmine and stepholidine (*tuber*).

Reference : Ambasta (1986).

878. *Sterculia urens* Roxb.

Syn.: Cavallium urens Schott. & Endl.

Family: Sterculiaceae.

English, Hindi and Sanskrit Names: **Eng. :** Sterculia; **Hindi :** Gulu, Kabru, Kullu, Tanuku; **Sans. :** Arjuna.

Distribution : Gujarat, Konkan, Deccan, N. Kanara, S.M. Country, dry forests of Chennai, N. and Central India, Chotta Nagpur, Myanmar and Sri Lanka.

Parts Used : Gum.

Flowering & Fruiting : October – June.

Folk Use : Mixture of gum and sugar in water used as a general tonic.

References : Kothari & Rao (2000), Upadhyay & Chauhan (2003).

879. *Stereospermum colais* (Buch.-Ham. *ex* Dillw.) Mabberley

Syn.: S. personatum (Hassk.) Chatterjee

Family: Bignoniaceae.

English, Hindi and Sanskrit Names: **Eng. :** Trumpet Flower Tree, Yellow Snake Tree; **Hindi :** Padri, Paral, Patar; **Sans. :** Patala.

Distribution : Deciduous forests upto 1,200m throughout India.

Parts Used : Flower. Root.

Folk Use : Roots used as cardio tonic whereas flowers useful as tonic.

Active Constituents : Lapachol, dehydro-α-lapachone and dehydrotectol (*root*).

Reference : Prajapati *et al.* (2006).

880. *Stereospermum suaveolens* DC.

Syn.: Bignonia chelonoides L. f.; *B. suaveolens* Roxb.; *Stereospermum chelonoides* (L. f.) DC.

Family: Bignoniaceae.

English, Hindi and Sanskrit Names: **Eng. :** Trumpet Flower; **Hindi :** Padaria, Padiala, Paral; **Sans. :** Abhipriya, Kalavrinta, Madhuduti, Patala, Tamrapushpi, Vasantaduti.

Distribution : Distributed in drier parts throughout India.

Parts Used : Root Bark. Stem Bark.

Flowering & Fruiting : Flowering: May – June and fruiting during cold season.

Folk Use : Root bark and stem bark used as a tonic.

Active Constituents : β-sitosterol and n-triacontanol (*root bark*).

References : Ambasta (1986), Dhiman (2005), Kaushik & Dhiman (2000), Kirtikar & Basu (1984), Wall (1972).

881. *Streblus asper* Lour.

Syn.: Epicarpurus asper Steud.; *E. orientalis* Bl.; *Trophis aspera* Retz.

Family: Moraceae.

English, Hindi and Sanskrit Names: **Eng. :** Siamese Rough-bush; **Hindi :** Cherootpathi, Daheya, Karchanua, Rusa, Siora; **Sans. :** Akshadhara, Gavakshi, Patrollekhataruh, Shakhotaka.

Distribution : Drier parts of India from Rohilkhand, eastwards and southwards to Travancore and in Andaman islands.

Parts Used : Latex.

Flowering & Fruiting : Flowering during January – February and fruiting in May – July.

Folk Use : Latex applied to the temples as a sedative in neuralgia.

References : Ambasta (1986), Dhiman (2005).

882. *Strobilanthes ciliates* Nees

Syn.: Nilgirianthus ciliates (Nees) Bremek; *Ruellia ciliata* Wall.

Family: Acanthaceae.

Hindi and Sanskrit Names: **Hindi :** Kara, Karvi; **Sans. :** Sahacorah, Sairyakah.

Distribution : Evergreen forests of Western Ghats up to 1,200m.

Parts Used : Root.

Folk Use : Roots used as tonic.

Reference : Prajapati *et al.* (2006).

883. *Strychnos ignatii* Bergius

Syn.: Ignatia amara L.

Family: Loganiaceae.

English, Hindi and Sanskrit Names: **Eng. :** St. Ignatius Beans; **Hindi :** Pipita; **Sans. :** Papitaa Vilaayati.

Distribution : Growing in Bohol, Samar and Cebu, islands of the Bisaya group of Phillipines, China where it has been introduced; grown in Indian gardens in south.

Parts Used : Seeds.

Folk Use : Seeds valued as a nervine tonic.

Reference : Ambasta (1986).

884. *Strychnos nux – vomica* L.

Syn.: S. colubrine Wight.; *S. lucida* Wall.

Family: Loganiaceae.

English, Hindi and Sanskrit Names: **Eng. :** Nux-vomica, Snake-wood, Strychnine Tree; **Hindi :** Bailewa, Chibbige, Kajra, Kuchla; **Sans. :** Chipita, Dirghapatra, Gerdruma, Marakatincles, Kakasphurja, Karaskara, Vishamushti.

Distribution : Throughout tropical parts of India, including Bihar, Orissa, West Bengal and U.P. ascending upto 1,350m.

Parts Used : Root. Seeds. Wood.

Flowering & Fruiting : Flowering during March – April and fruiting in winters.

Folk Use : Ripe seeds used as a tonic, stimulant and sedative. Also, root tincture and wood used as tonic.

Active Constituents : Therapeutic properties due to C-mavaccerine (*root*), isobrucine, isobrucine N-oxide, isostrychnine N-oxide, 2-hydroxyl-3-methoxystrychnine, β-colubrine, pseudostrychnine, bruceine, its N-oxide, novacine, icajine and vomicine (*seeds*), a bis-iridoid glucoside – ligustrinoside, loganin, loganetin and loganic acid (*wood*).

References : Ambasta (1986), Dhiman (2005), Dwivedi (2003), Farooq (2005), Harborne & Baxter (2001), Kirtikar & Basu (1984), Parrotta (2001), Prajapati *et al.* (2006), Reddi *et al.* (2005), Sharma (2003), Singh *et al.* (2007), Subramaniam(2000), Watt (1972).

885. *Strychnos potatorum* Willd.

Syn.: S. tettankotta Retz.

Family: Loganiaceae.

English, Hindi and Sanskrit Names: **Eng. :** Clearing-nut Tree, Clearing Nut; **Hindi :** Neimal, Nelmal, Nirmali; **Sans. :** Ambuprasada, Chakshushya, Guchnaphala, Katakatphalam, Lekhanatmaka, Ruchishya, Shadanatmaka, Tikaphala.

Distribution : Known to occur wild in Central India, Konkan, Deccan Peninsula and from Karnataka to south Travancore.

Parts Used : Seeds.

Flowering & Fruiting : Flowering: April – May and fruiting during cold season.

Folk Use : Seeds used as tonic.

Active Constituents : Reported to possess mannogalactan, diaboline, β-sitosterol, stigmasterol, oleanolic acid and its 3β-acetate and a saponin containing oleanolic acid, galactose and mannose (*seeds*).

References : Ambasta (1986), Dhiman (2005), Kirtikar & Basu (1984), Prajapati *et al.* (2006).

886. *Swertia angustifolia* L.

Syn.: Ophelia angustifolia D. Don

Family: Gentianaceae.

English and Hindi Names: **Eng. :** Felworts; **Hindi :** Chiraita.

Distribution : Sub-tropical Himalayas (300-2,000m) from the Chenab – Bhutan.

Parts Used : Whole Plant.

Flowering & Fruiting : August – October.

Folk Use : Entire plant used as a tonic.

Active Constituents : Bellidifolin, β-sitosterol, ursolic acid, fourteen tetra – and five penta – oxygenated xanthones, including seven 1-O-glucosides, isolated and characterized-1, 5, 8-trihydroxy-3-methoxy-, 1, 7, 8-trihydroxy-3-methoxy-, 1, 3, 5, 8-tetrahy-droxy-, 1, 3, 7, 8-tetrahydroxy-1 hydroxy-2, 3, 4, 7-tetramet- hoxy-, 1, 8-dihydroxy-3, 5-dimethoxy-, 1, 8-dihydroxy-3, 7-dimethoxy-1-hydroxyl-3, 7, 8-trimethoxy – (decussatin), 1, 3, 8-trihydroxy-7-methoxy-, 3-methoxy-1-O-β-D-glucopyranosyl-5, 8-dihydroxy-, 3-hydroxy-1-O-β-D-glucopyranosyl-4, 5-dimethoxy-, 8-hydroxyl-1-O-β-D-glucopyranosyl-3, 5-dimethoxy, 1-O-β-D-glucopyranosyl-3, 5, 8-trihydroxy-, 1-O-β-D-glucopyranosyl-3, 4, 5, 8-tetrahydroxy-, 1-O-β-D-glucopyranosyl-3,4,7,8-tetrahydroxy-, 3-methyoxy -1-O-β-D-glucopyranosyl-7, 8-dihydroxy-, 1, 3, 4, 5, 8-pentamethoxy– and 1, 3, 4, 7, 8-pentamethoxy – xanthones (*plant*).

Reference : Islam & Jha (2003).

887. *Swertia chirayita* (Roxb. ex Flem.) Karst.

Syn.: Agathotes chirata D. Don.; *Gentiana chirayita* Roxb.; *Ophelia chirata* Griseb.; *Swertia chirata* Buch.-Ham. *ex* C. B. Clarke.

Family: Gentianaceae.

English, Hindi and Sanskrit Names: **Eng. :** Chiretta; **Hindi :** Chirayita; **Sans. :** Bhunimba, Haima, Jvarantaka, Kirata-tikta, Naipala, Ramasenaka, Viktaka.

Distribution : Temperate Himalayas ascending up to 3,350m from Kashmir – Sikkim and in Khasia hills between 1,200-1,500m.

Parts Used : Whole Plant.

Flowering & Fruiting : Flowering during July – August and fruiting in September – October.

Folk Use : A drug, 'Chirata' obtained from dried plant used as a bitter tonic.

Active Constituents : 1, 5, 8-trihydroxy-3-methoxyxanthone, 1-hyroxy-3, 5, 8-trimethoxyxanthone, deccussatin, mangiferin, 1,8-dihydroxy-3, 7-dimethoxyxanthone, 1, 8-dihyroxy 3, 7-

dimethoxyxanthone, 1, 3, 8-trihyroxy-5-methoxyxanthone, 1, 3, 5, 8-tetrahydroxyxanthone and 1, 3, 7, 8-tetrahydroxyxanthone (*plant*).

References : Ambasta (1986), Dhiman (2005), Farooq (2005), Kirtikar & Basu (1984), Kumar & Lal (2003), Prajapati *et al.* (2006), Rawat *et al.* (2007), Sharma (2003, 2004), Sharma *et al.* (2003), Thakur *et al.* (1989), Watt (1972).

888. *Swertia ciliata* (D. Don ex G. Don) B. L. Burt.

Syn.: Ophelia ciliata G. Don.; *O. dalhousiana* Griseb.; *Swertia purpurascens* Wall.

Family: Gentianaceae.

English Name: **Eng. :** Felworts.

Distribution : Temperate N.W. Himalayas from Kashmir – Kumaon: 1,700-4,000m.

Parts Used : Whole Plant.

Flowering & Fruiting : August – November.

Folk Use : Plant extract used as a tonic.

Active Constituents : Oleanolic acid, sitosterol, swertisin, swertiomarin, gentianine, genticrucine and enicoflavine (*plant*).

Reference : Jain & Singh (2005).

889. *Swertia cordata* Wall. ex Clarke

Family: Gentianaceae.

English Name: **Eng. :** Heart-leaf Swertia.

Distribution : Temperate Himalayas and Western Tibet: 1,350-4,000m (Kashmir and Karakorum – Bhutan), abundant westwards, Khasia Hills: 1,000-1,700m.

Parts Used : Whole Plant.

Flowering & Fruiting : July – October.

Folk Use : Whole plant used as a bitter tonic.

Active Constituents : Two new xanthones isolated and characterized as 1-hydroxyl-3, 5, 7, 8-tetramethoxy-xanthone and 1, 7-dihydroxy-3, 5, 8-trimethoxyxanthone (*plant*).

Reference : Bhattacharjee (2001).

890. *Swertia petiolata* Royle ex D. Don

Syn.: S. petiolata var. *lahulensis* (Kerner) Cl.

Family: Gentianaceae.

Hindi Name: **Hindi :** Mamiran.

Distribution : Western Himalayas (Kashmir: 3,000-4,000m), frequent; Gulmurg, Tragbol and Kabul.

Parts Used : Whole Plant.

Flowering & Fruiting : July – September.

Folk Use : Decoction of whole plant given as health tonic.

Active Constituents : Two new isomeric 2,3-secohopene lactones – swertialactone C and swertialactone D, a new triterpene characterized as 3β-hydroxylup-13 (18)-ene, 3β-hydroxylup-12-en-28-oic acid, ursolic acid, 1, 3-dihydroxy-5, 8-dimethoxyxanthone, 2-hydroxydimethyl terephthalate (*plant*).

Reference : Ballabh & Chaurasia (2006).

891. *Swietenia mahagoni* Jacq.

Family: Meliaceae.

English Names: **Eng. :** Cuban Mahagony Tree, Jamaica Mahagony Tree, Spanish Mahagony Tree.

Distribution : Indigenous to the West Indies and Central Africa; cultivated in Kolkata, N.W. regions, Bombay and Chennai.

Parts Used : Bark.

Folk Use : Bark used as a tonic.

Active Constituents : Furanolactones-Me angolensate and Me6-hydroxyl-angolensate, cycloartenol, cycloswietanol, acacic acid lactone diacetate and cycloswieterol-3-O-β-D-glucopyranoside (*bark*).

Reference : Ambasta (1986).

892. *Symplocos paniculata* (Thunb.) Miq.

Syn.: Lhodra crataegoides Dene.; *S. crataegoides* Hamilton.

Family: Symplocaceae.

English Name: **Eng. :** Sapphire-berry.

Distribution : Himalayas (Indus – Assam): 1,000-2,700m, also found in Khasia mountains and hills of Martaban.

Parts Used : Bark.

Flowering & Fruiting : Flowering during spring season and fruiting in summers.

Folk Use : Bark used as a tonic.

References : Singh (2001), Watt (1972).

893. *Symplocos racemosa* Roxb.

Syn.: S. beddomei C.B. Clarke; *S. candolleana* Braud.; *S. hamiltoniana* Wall.; *S. nervosa* A. DC.

Family: Symplocaceae.

English, Hindi and Sanskrit Names: **Eng. :** China Nora, Lodh Tree; **Hindi :** Lodh; **Sans. :** Bhilli, Galava, Lodhra, Rodhra.

Distribution : Very common throughout Chotta Nagpur, North – east India including Assam & terai of Kumaon upto 762m.

Parts Used : Bark.

Flowering & Fruiting : Flowering during October – January and fruiting in December – May.

Folk Use : Bark used as a tonic.

Active Constituents : Symposide, α- amyrin, β-sitosterol, 2δ-hydroxyl-20α-urs-12, 18 (19)-dien-28 oic acid, 24-hydroxyolean-12-en-3-one, betulin and oleaniolic acid (*bark*).

Reference : Dhiman (2005).

894. *Synadenium grantii* Hook. f.

Family: Euphorbiaceae.

English Name: **Eng. :** African Milkbush.

Distribution : Native of Africa, grown in Indian gardens.

Parts Used : Whole Plant.

Folk Use : Plant used as a stimulant of central nervous system.

Active Constituents : 4-deoxyphorbol-12-tigolate-13-phenylacetate, 12-tiglate-13-isobutyrates, 12-tiglate-13-phenylacetates of 4, 20-dideoxy- 5α-hydroxyphorbol and 20-deoxy-5α-hydroxyphorbol (*plant*).

Reference : Ambasta (1986).

895. *Syzygium aromaticum* (L.) Merrill & Perry

Syn.: Caryophyllus aromaticus L.; *Eugenia aromatica* Kuntze; *E. caryophyllata* Thunb.

Family: Myrtaceae.

English, Hindi and Sanskrit Names: **Eng. :** Clove Tree; **Hindi :** Laung; **Sans. :** Devakusumam, Lavangam.

Distribution : Cultivated in South India.

Parts Used : Floral Buds. Fruit.

Folk Use : Cloves used as stimulant and tonic. Fruits used as a remedy against weakness of the brain.

Active Constituents : Protein, Carbohydrates, tannins, oleanolic acid, eugenol acetate, caryophyllene and eugenol (*fruit*).

References : Ambasta (1986), Dwivedi (2007), Khatoon & Geol (2010), Pal & Paul (2003), Prajapati *et al.* (2006), Rajan *et al.* (2003).

896. *Syzygium cumini* (L.) Skeel*

Syn.: Calyptranthes capitellata Ham; *C. jambolana* Willd.; *Eugenia caryophyllifolia* Lam.; *E. frondosa* Wall.; *E. jambolana* Lam.; *Eugenia moorei* F. Muell. Lam.; *Myrtus cumini* L.; *Syzygium jambolanum* DC.

Family: Myrtaceae.

English, Hindi and Sanskrit Names: **Eng. :** Black Plum, Jaman, Jambolan; **Hindi :** Jaman, Jamun, Jambhal, Paiman, Phalinda; **Sans. :** Jambavam, Jambuh, Rajphala.

Distribution : In forests usually along river banks up to 1,800m throughout India; also cultivated.

Parts Used : Fruit.

Flowering & Fruiting : April – August.

Folk Use : Fruits used as a tonic and stimulant.

Active Constituents : Anthocyanins, delphinidin-3-gentiobioside, malvidine-3-laminaribioside and petunidin-3-gentiobioside (*fruit*).

References : Kumar & Lal (2003), Prajapati *et al.* (2006), Sharma & Singh (2001).

897. *Tabebuia pentaphylla* (L.) Hemsl.

Syn.: Tecoma pentaphylla Juss.

Family: Bignoniaceae.

English Names: **Eng. :** Pink Tecoma, Salvador Pinktrumpet.

Distribution : Native of tropical America; grown in Indian gardens.

Parts Used : Fruit.

Folk Use : Peel of fruit used as hypnotic.

Reference : Ambasta (1986).

898. *Tabernaemontana divaricata* L.

Syn.: Ervatamia coronaria (Jacq.) Stapf.; *E. divaricata* (Linn.) Burkill.; *Nerium divaricatum* L.; *Tabernaemontana coronaria* (Jacq.) Willd.

Family: Apocynaceae.

English, Hindi and Sanskrit Names: **Eng. :** East Indian Rosebay, Wax Flower; **Hindi :** Candni, Sugandabala, Taggar; **Sans. :** Gajapadapa, Kshayataru, Nandivrkshah, Vishnupriya.

Distribution : Throughout India; also cultivated as an ornamental plant.

Parts Used : Root.

Flowering & Fruiting : May – December.

Folk Use : Roots used as a tonic; also used as brain tonic (Yunani).

Active Constituents: Found to possess tabernaemontanine, dregamine, vobasine, voacamine and vocangine (*root*).

References : Kirtikar & Basu (1984), Prajapati *et al.* (2006).

899. *Tagetes minuta* L.

Syn.: Tagetes glanduliflora Schrank.

Family: Astereceae.

English and Hindi Names: **Eng. :** Stinking-roger; **Hindi :** Ganindri, Jangli-gainda.

Distribution : Native to S. America; naturalized near wasteplaces and on dry embankments in the N.W. Himalayas: 1,250-2,500m; found almost as weed all over the state in H.P.

Parts Used : Capitula.

Flowering & Fruiting : September – November.

Folk Use : Capitula used as tranquillizer due to presence of a volatile oil in it.

Active Constituent : Volatile oil (*capitula*).

References : Ambasta (1986), Prajapati *et al.* (2006).

900. *Tamarindus indica* L.

Syn.: T. lumbrosa Salisb; *T. occidentalis* Gaertn.; *T. officinalis* Hook.

Family: Caesalpiniaceae.

English, Hindi and Sanskrit Names: **Eng. :** Tamarind Tree; **Hindi :** Amli, Ampli, Imli, Nuli; **Sans. :** Amlika, Chukra, Cinca, Sukta, Tintrini.

Distribution : Throughout India, particularly in the South India; often cultivated.

Parts Used : Root Bark. Fruit. Seeds.

Flowering & Fruiting : Flowering during summer and fruiting in next summer.

Folk Use : Root bark and seeds used as tonic whereas fruits used as tonic to heart.

Active Constituents: Active principles are tamarindienal characterized as 5-hydroxyl-2-oxo-hexa-3, 5-dienol (*fruit pulp*), (-) epicatechin, 2-hydroxyl-3', 4'-dihydroxyacetophenone (TAO), methyl 3, 4-dihydrobenzoate (TA 1) and 3, 4-dihydroxyphenyl acetate (TA 2) (*seeds*), and hordenine (*bark*).

References : Bhattacharjee (2001), Kaushik & Dhiman (2000), Kirtikar & Basu (1984), Parrotta (2001), Prajapati *et al.* (2006), Rajan *et al.* (2003), Sharma (2004), Singh & Srivastava (2000), Subramaniam (2000), Watt (1972).

901. *Tanacetum vulgare* L.

Syn.: Chrysanthemum vulgare (L.) Bernh.

Family: Asteraceae.

English Name: **Eng. :** Tansy.

Distribution : Native of Europe; found as a weed in Jammu, Srinagar and adjoining localities in the state.

Parts Used : Flower. Leaf.

Folk Use : Decoction of leaves and flowering tops used as a tonic.

References : Ambasta (1986), Kapur (1996 a).

902. *Taraxacum officinale* Weber *ex* wiggers (Pl. 11A)

Syn.: Leontodon taraxaccum L.

Family: Asteraceae.

English, Hindi and Sanskrit Names: **Eng. :** Bitter Wort, Canker, Common Dandelion, Fortune-teller, Irish Daisy, Priest's Crown; **Hindi :** Barau, Dudhli, Dulal, Kanphul; **Sans. :** Dugdhfeni.

Distribution : Throughout the Himalayas: 3,040-5,486m and in Mishmi hills.

Parts Used : Rhizome. Root.

Flowering & Fruiting : March – November.

Folk Use : Rhizomes and root extract taken as a liver tonic and tonic.

Active Constituents : Roots rich in umbelliferone, esculetin and p-coumaric, caffeic, ferulic, p-hydroxybenzoic, protocatechuic, vanillic, β-resorcyclic, syringic and p-hydroxyphenylacetic acids.

References : Ambasta (1986), Chhetri (2005), Dhiman (2005), Bhattacharjee (2001), Chaurasia *et al.* (2000, 2003), Jain & Singh (2005), Kaul (1997), Kirtikar & Basu (1984), Prajapati *et al.* (2006), Sharma (2003), Thakur *et al.* (1989).

903. *Taxus baccata* L.

Syn.: T. contorta Griff.; *T. erecta* Hort. ex Carr.; *T. orientalis* Bertholini.; *T. nepalensis* Jacq.; *T. nucifera* Wall.; *T. stricta* Hort. ex Gord.

Family: Taxaceae.

English, Hindi and Sanskrit Names: **Eng. :** Common Yew; **Hindi :** Birmi, Kash, Talispatta, Thuner, Thuno; **Sans. :** Dhatripatra, Manduparni, Talispatra.

Distribution : Temperate Himalayas: 1,828-3,352m and in Khasia hills.

Parts Used : Leaf. Fruit.

Flowering & Fruiting : Cone formation starts in March – May and matures during September – November.

Folk Use : Fruits and leaves used as sedative.

Active Constituents : These are taxine, two new biflavones – isomers of sciadopitysin (I) and sotetsuflavone (II), betuloside, ginkgetin, sequoiaflavone, taxagifin, amentoflavone and mono- and di-Omethylamento flavones (*leaf*), rhodoxanthin and escholtzxanthone (*fruit*).

References : Ambasta (1986), Dhiman (2005), Kirn *et al.* (2000), Sharma *et al.* (2003), Watt (1972).

904. *Tecoma stans* (L.) H. B. & K. (Pl. 11B)

Syn.: Bignonia stans L.; *Stenolobium stans* Seem.

Family: Bignoniaceae.

English and Hindi Names: **Eng. :** Yellow Trumpet Bush; **Hindi :** Sonapatti.

Distribution : Planted in Indian gardens as ornamental plant; also run wild in some parts.

Parts Used : Root.

Flowering & Fruiting : Flowering and fruiting during summer.

Folk Use : Roots used as a tonic.

References : Ambasta (1986), Kaushik & Dhiman (2000), Parrotta (2001), Raju (2000).

905. *Tectaria cicutaria* (L.) Copel.

Family: Tectariaceae.

English and Hindi Names: **Eng. :** Button Fern; **Hindi :** Harishakarr Pattor.

Distribution : Throughout India.

Parts Used : Rhizome.

Folk Use : Paste of 10g rhizome and 7 black peppers given once daily with cow's milk as tonic.

Reference : Brahman & Saxena (1990).

906. *Tectona grandis* L. f.

Syn.: Jatus grandis (L. f.) Kuntze; *Tectona theka* Lour.; *Theka grandis* (L. f.) Lam.

Family: Verbenaceae.

English, Hindi and Sanskrit Names: **Eng. :** Indian Oak, Ship Tree, Teak; **Hindi :** Sagon, Sagwan, Sakhu; **Sans. :** Anila, Bhumishah, Holimaka, Nau, Sakkah, Shak.

Distribution : Occurs both cultivated and wild almost throughout India with greater concentrations in Central and Peninsular India, Assam and Bengal.

Parts Used : Wood.

Flowering & Fruiting : Flowering during August – September and fruiting in November – December.

Folk Use : Wood of the plant used as a stimulant and sedative.

Active Constituents : Betulinic acid, fatty acids, 2-10 dobenzoic acid, carbohydrates, O-tolymethylether, 1,4-polyisoprene (as and trans), squalene, lapachonone, deoxylapachaol, anthraquinone-2-carboxylic acid, anthraquinone-2-carboxyaldehyde, 2-methylanthraquinone, β – lapachone, 3-methyl-2-hydroxymethyl-3 hydroxy-2-methyl-1-hydroxyl-2-methyl-1, 4-dihydroxy-2-methyl-4, 5-dihydroxymethyl anthroquinones, 9,10-dimethoxy-2-methylanthra-1,4-quinine, tecomaquinone 1,2-(3,3-dimthylalyl)-1,4-napthoquinone and dehydro lapachal, dehydro-α-lapachone, tectal, dehydrotectal, β-lapachone, obtusifolin and tectograndiol (*wood*).

References : Dhiman (2005), Farooq (2005), Kaushik & Dhiman (2000), Parrotta (2001).

907. *Tephrosia purpurea* (L.) Pers.

Syn.: Cracca purpurea L.; *Tephrosia hamiltonii* Drumn. ex Gamble; *T. stricta* Grah.

Family: Fabaceae.

English, Hindi and Sanskrit Names: **Eng. :** Purple Tephrosia, Wild Indigo; **Hindi :** Sarphonka; **Sans. :** Banah, Banapunkhah, Kriti, Nilavriksha, Sharpunkha.

Distribution : All over India, ascending up to 1,850m in the Himalayas.

Parts Used : Whole Plant. Root. Seed.

Flowering & Fruiting : Flowering during rainy season and fruiting in winter.

Folk Use : Whole plant used as a tonic and CNS active. Seeds and roots possess only tonic properties.

Active Constituents : β-sitosterol, lupeal, retin, delphinidin chloride, cyaniding chloride, isolonchocarpin, lanceolatin A and B, pongamal, Karangin, Kangone, 5,7-dimethoxy-8-flavanone, 2-methoxy-3,9-dihydroxycouestone, flevichapparins B and C, methyl karanjic acid and pupurin (*plant*).

References : Dhiman (2005), Kaushik & Dhiman (2000), Kirtikar & Basu (1984), Maya *et al.* (2003), Mishra (2008), Parrotta (2001), Prajapati *et al.* (2006), Raju (2000), Reddi *et al.* (2005), Singh & Srivastava (2000), Singh *et al.* (2007), Subramaniam (2000), Watt (1972).

908. *Tephrosia virginia* (L.) Pers.

Family: Fabaceae.

English Names: **Eng. :** Devil's Shoestings, Goat's Rue, Virginia Tephrosia.

Distribution : Eastern N. America – Florida and Louisiana.

Parts Used : Whole Plant. Root.

Folk Use : Plant as well as its roots considered as tonic and stimulant.

Reference : Bhattacharjee (2001).

909. *Teramnus labialis* Spreng.

Syn.: Glycine labialis L. f.

Family: Fabaceae.

Hindi and Sanskrit Names: **Hindi** : Mashaparni; **Sans.** : Ashvapuchhi, Hansamasha, Mashparni, Pandura, Sulabha.

Distribution : Punjab eastwards to West Bengal; grows wild in southern parts of India.

Parts Used : Fruit.

Folk Use : Fruits used as tonic and nervine tonic.

References : Farooq (2005), Kirtikar & Basu (1984), Reddi *et al.* (2005).

910. *Terminalia arjuna* (Roxb.) Wight. & Arn.

Syn.: Pentaptera angustifolia Roxb.; *P. glabra* Roxb.; *T. berryi* W.&A. ; *T. glabra* W. & A.; *T. ovalifolia* Rottl.

Family: Combretaceae.

English, Hindi and Sanskrit Names: **Eng.** : Arjun Tree, White Murdah; **Hindi** : Arhun, Kahu, Kahua, Kangach; **Sans.** : Arjunah, Dhananjaya, Kakubhah, Karaviraka, Sachi, Savya.

Distribution : Banks of rivers, streams and dry water courses in Subhimalayan tract, Central and South India and West Bengal.

Parts Used : Bark. Fruit.

Flowering & Fruiting : Flowering during April – May and fruiting in September – November.

Folk Use : Bark and fruits used as tonic whereas bark also valued for its heart tonic properties.

Active Constituents : Isolated principles are arjunic acid (2α, 3β, 19α-trihydroxyalean 12-en-28-oic acid) & arjunetin, (2α, 3β, 19α-trihydroxyolean-12-en-28-carboxy-1-β-D-(+)-glucopyranoside) (*bark*), β-sitosterol, ellagic acid, arjunic acid & methyl ester of gallic acid (*fruit*).

References : Ambasta (1986), Chauhan *et al.* (2007), Dhiman (2005, 2007), Dixit & Kumar (2003), Dwivedi (2007), Farooq (2005), Kaushik & Dhiman (2000), Khanna (2002), Kirtikar & Basu (1984), Kumar & Lal (2003), Kumar *et al.* (2003), Parrotta (2001), Prajapati *et al.* (2006), Reddi *et al.* (2005), Sharma (2003, 2004), Sharma & Sood (1997), Sood *et al.* (2009), Subramaniam (2000), Thakur *et al.* (1989), Watt (1972).

911. *Terminalia bellirica* (Gaertn.) Roxb. (Pl. 11C)*

Syn.: Myrobalanus bellirica Gaertn.; *Terminalia bellirica* var. *laurinoides* Cl.

Family: Combretaceae.

English, Hindi and Sanskrit Names: **Eng. :** Beddanuts, Belleric Myrobalan; **Hindi :** Bahera, Bhaira, Sagona; **Sans. :** Aksah, Anilaghnaka, Bahira, Behera, Bibhitaka, Kaliyugalaya, Karshaphala.

Distribution : Grows wild up to 1,000m all over India excepting dry and marshy areas.

Parts Used : Whole Plant. Fruit. Kernel.

Flowering & Fruiting : Flowering during April – June and fruiting in December – January.

Folk Use : Plant used as a tonic. Kernel of the fruit said to possess narcotic effect whereas fruits possess both tonic and brain tonic properties.

Active Constituents : β-sitosterol, gallic acid, ellagic acid, ethyl gallate, galloyl glucose, chebulagic acid, mannitol, glucose, galactose, fructose and rhamnose (*fruit*).

References : Bhattacharjee (2001), Dhiman (2005), Dixit & Kumar (2003), Farooq (2005), Kaushik & Dhiman (2000), Kirtikar & Basu (1984), Kumar & Lal (2003), Parrotta (2001), Prajapati *et al.* (2006), Reddi *et al.* (2005), Samant *et al.* (2001), Sharma & Sood (1997), Singh (2001), Siwakoti & Varma (1996), Sood *et al.* (2009), Subramaniam (2000), Thakur *et al.* (1989), Watt (1972).

912. *Terminalia catappa* L.

Syn.: Badamia comonersoni Gaertn.; *Juglans catappa* Lour.; *Terminalia intermedia* Spreng.; *T. moluccana* Lamk.; *T. myrobalana* Roth.; *T. procera* Roxb.; *T. subcordata* Willd.

Family: Combretaceae.

English, Hindi and Sanskrit Names: **Eng. :** Indian Almond, Malabar Almond, Malay Almond; **Hindi :** Badami, Deshi-badam, Jangli-badam; **Sans. :** Desabadama, Grahdruma, Ingudi.

Distribution : Common along beaches, rocky coats, upper margins of mangrove swamps, low land clearings and secondary forests.

Parts Used : Bark.

Folk Use : Fluid from bark used as a tonic.

Active Constituents : Gallic acid, ellagic acid, corilagin and unidentified flavonoids (*bark*).

Reference : Prajapati *et al.* (2006).

913. *Terminalia chebula* Retz.*

Syn.: *Embryogonia arborea* Teys. & Binn.; *Myrobalanus chebula* Gaertn.; *T. aruta* Ham.; *T. reticulata* Roth.

Family: Combretaceae.

English, Hindi and Sanskrit Names: **Eng.** : Chebulic Myrobalan; **Hindi** : Harad, Harara, Harra; **Sans.** : Abhaya, Haritaki, Harra, Pathya.

Distribution : Abundantly found in N. India from Kangra – Bengal and southwards to the Deccan at 304-914m and up to 1,828m in Travancore; higher forests of Bombay Ghats, Satpuras, Bdgaum and Kanara.

Parts Used : Bark. Fruit.

Flowering & Fruiting : Flowering during April – June and fruiting in January – March.

Folk Use : Known earlier as cardio tonic (bark, fruits) and also valued as tonic and brain tonic (fruits).

Active Constituents : Attributed properties due to palmitic, stearic, oleic, linoleic, arachidic and behenic acid (*fruit kernel*), anthraquinone glycoside, chebulinic acid, tannic acid, terchebin and vitamin C, shikimic acid, gallic, triacontanoic and palmitic acids, β-sitosterol, daucosterol, triethyl ester of chubulic acid and ethyl ester of gallic acid. (*fruit*).

References : Ambasta (1986), Dhiman (2005), Dwivedi (2007), Kaushik & Dhiman (2000), Kayal *et al.* (2003), Khatoon & Geol (2010), Kirtikar & Basu (1984), Parrotta (2001), Prajapati *et al.* (2006), Reddi *et al.* (2005), Sharma (2003), Sharma & Sood (1997), Thakur *et al.* (1989), Verma *et al.* (1995), Watt (1972).

914. *Terminalia crenulata* Roth.

Syn.: *T. tomentosa* Wt. & Arn.

Family: Combretaceae.

English, Hindi and Sanskrit Names: **Eng.** : Black Murdah; **Hindi** : Asaina, Kauha, Sadri; **Sans.** : Chagakarnah, Krishnatvaka, Sajada.

Distribution : Semi-evergreen and mixed deciduous forests throughout India.

Parts Used : Bark.

Flowering & Fruiting : Flowering during June – July and fruiting in February – March.

Folk Use : Bark used as cardio tonic and tonic.

Active Constituents : Arjunic, arjunolic, betulinic, ellagic acids and β-sitosterol (*bark*).

References : Kirtikar & Basu (1984), Parrotta (2001), Prajapati *et al.* (2006), Subramaniam (2000).

915. *Terminalia myriocarpa* Heurck & Muell. – Arg.

Syn.: Pentaptera saja Wall.

Family: Combertaceae.

English Name: **Eng. :** East Indian Almond.

Distribution : E. Himalayas from Nepal eastwards up to 1,700m, Bhutan, Abor country, Assam and Upper Myanmar.

Parts Used : Bark.

Folk Use : Infusion of bark taken as a heart tonic.

Active Constituents : Ellagic, gallic, chebulinic, chebulagic acids, β-sitosterol, fructose and 4, 4', 5, 5', 6, 6' – hexahydroxydiphenic acid, dilactone (*bark*).

Reference : Chhetri (2005).

916. *Terminalia paniculata* Roth.

Syn.: Pentaptera paniculata Roxb.; *Terminalia monoptera* Roth.; *T. trioptera* Heyne.

Family: Combreteceae.

English and Sanskrit Names: **Eng. :** Flowering Murdah; **Sans. :** Asvakarnah.

Distribution : Semi-evergreen and moist deciduous forests of Western and Eastern Ghats up to1,200m.

Parts Used : Bark.

Folk Use : Bark used as cardio tonic.

Active Constituents : β-sitosterol, triterpene, carboxylic acid, a new glucoside and dimethyl ellagic acid (*bark*).

References : Ambasta (1986), Prajapati *et al.* (2006).

917. *Thalictrum foliolosum* DC.

Family: Ranunculaceae.

English, Hindi and Sanskrit Names: **Eng. :** Meadow Rue; **Hindi :** Mamira, Peeli Jadi, Pinjari; **Sans. :** Peetranga.

Distribution : Known to occur throughout Himalayas: 1,540-2,438m and in Khasia hills: 1,219-1,828m.

Parts Used : Rhizome. Root.

Flowering & Fruiting : Flowering during October – November and fruiting in November – December.

Folk Use : Rhizome and roots considered as tonic and bitter tonic.

Active Constituents : Properties due to thalrugosidine, thalrugo-saminine, thalisopine, thalirugidine, oxyberberine (berlamine) and noroxyhdrastinine, thalifendine, palmitine, berberine, columbamine, jatraorrhizine, thalidastine, dehydrodiscretamine, rugosinone, tembetarine, xanthoplanine and magnoflorine (*root*), N, O, O – trimethylsparsiflorine, thalicarpine, thalidasine, thalrugosidine, reticuline, magnoflorine, berberine and palmatine (*rhizome*).

References : Bhattacharjee (2001), Dhiman (2005), Kirtikar & Basu (1984), Prajapati *et al.* (2006), Samant *et al.* (2001), Sharma (2003), Watt (1972).

918. *Thalictrum pedunculatum* Edgew.

Family: Ranunculaceae.

Distribution : Western Temperate Himalayas (Shimla – Kafiristan): 2,000-2,700m.

Parts Used : Root.

Folk Use : Root extract used as a general tonic.

Active Constituent : Berberine (*root*).

Reference : Kapur (1996b).

919. *Theobroma cacao* L.

Family: Sterculiaceae.

English Names: **Eng. :** Cacao, Cocoa, Chocolate Tree.

Distribution : Native to Mexico and Central America; introduced into India and now cultivated in the S. India and Sri Lanka.

Parts Used : Seeds.

Folk Use : Seeds used as nervous system stimulant, heart and kidney tonic.

Active Constituents : Seeds rich in epicatechin (2β→5, 4β→6)-epicatechin, 3-O-β-D-galactosylentepicatechin (2α→7, 4α→8)-epicatechin and 3-O-L-arabinosylentepicatechin (2α→7, 4α→8)-epicatechin.

Reference : Prajapati *et al.* (2006).

920. *Thevetia peruviana* (Pers.) K. Schum.

Syn.: Cascabela thevetia (L.) Lipp.; *Cerbera peruviana* Pers.; *Thevetia neriifolia* Juss.

Family: Apocynaceae.

English, Hindi and Sanskrit Names: **Eng. :** Bastard Oleander, Lucky Nut-tree, Exile Tree, Exile Oil Plant, Yellow Olender; **Hindi :** Kulkephul, Pile Kaner, Zard Kunel; **Sans. :** Ashvaghna, Ashvamaraka, Ashvantaka, Divyapushpa, Haripriya, Shatakunda.

Distribution : Cultivated for its elegant foliage and flowers throughout India.

Parts Used : Bark. Leaf. Fruit.

Flowering & Fruiting : July – September.

Folk Use : Bark used as bitter tonic. Fruits useful as cardio tonic. Leaves considered narcotic.

Active Constituents : These are glucosides thevetin, thevetoxin, a phytosterolin (*bark*), lupeol, α- and β-amyrins, taraxasterol, ψ-taraxasterol (*root*), Glutamic acid, aspartic acid, leucine, glycine, isoleucine, arginine, valine, alanine, proline, phenylalanine, cystine, lysine, serine, tyrosine, histidine, threonine, methionine and tryptophan, 3β-hydroxyl-11α, 12α-epoxy-urs-13β, 28-olide, 3β-O-α-L-thevetose-3β, 14β-dihydroxy-14 (13→12)-abeo- 5β, 12β, 14β-carda-13 (18), 20 (22)-dienolide (*leaf*), Arabinose, galactose, glucose, galacturonic acid and xylose, epiperuviol acetate, hesperitin-7-glucoside, α- and β-amyrin, kaempferol and quercetin (*fruit*).

References : Parrotta (2001), Prajapati *et al.* (2006), Sharma (2003, 2004), Sood *et al.* (2009).

921. *Thlaspi arvense* L.

Family: Brassicaceae.

English Names: **Eng. :** Common Pennycress, Fanweed.

Distribution : A weed throughout the temperate and subalpine Himalayas, ascending to 4,650m. Europe, Asia, always in cultivated places.

Parts Used : Seeds.

Folk Use : Seeds used as a stimulant.

Reference : Ambasta (1986).

922. *Thuja orientalis* L. (Pl. 11D)

Syn.: Biota orientalis Endl.; *Platycladus orientalis* L.; *P. stricta* Spach.*; Thuja acuta* Moench; *T. chengii* Borderes & Gaussen; *T. decora* Salisb.

Family: Cupressaceae.

English and Hindi Names: **Eng. :** Chinese Thuja, Oriental Arborvitae; **Hindi :** Mayurpankh, Morepankhi.

Distribution : Mountain thickets in India, China, Japan and Korea.

Parts Used : Leaf. Seeds.

Flowering & Fruiting : July – October.

Folk Use : Leaves used as a tonic. Seeds valued as liver tonic.

Active Constituents : Known to contain nonacosanol, quercitrin, condensed tannins, apigenin, cupressuflavone, quercitrin, kaempferol-7-O-glucoside, myricetin-3-O-α-L-rhamnoside, quercetin-7-O-rhamnoside, pimaric acid, isopimaric acid, deoxypodophyllotoxin and ethyl p-methoxy-trans-cinnamate (*leaf*), 15, 16-bisnor-13-oxo-8 (17), 11 E-labdadien-19-oic acid (II), 14, 15, 16-trinor-8 (17)-labdene-13, 19-dioic acid (III) and 12 (R), 13-dihydroxycommunic acid (*root*).

References : Ambasta (1986), Bhattacharjee (2001), Farooq (2005), Prajapati *et al.* (2006).

923. *Thymus serpyllum* L.

Syn.: T. linearis Benth.

Family: Lamiaceae.

English and Hindi Names: **Eng. :** Bank Thyme, Brotherwort, Creeping Thyme, Hill-wort, Penny Mountain, Serpell, Wild Thyme; **Hindi :** Banajwain.

Distribution : Himalayas (Kashmir – Nepal): 1,500-4,500m and reported to be grown in the gardens of Western India.

Parts Used : Whole Plant. Leaf.

Flowering & Fruiting : March – October.

Folk Use : Plant possesses stimulating properties. Decoction of the plant and leaves used as tonic.

References : Bhattacharjee (2001), Prajapati *et al.* (2006).

924. *Thymus vulgaris* L.

Family: Lamiaceae.

English and Hindi Names: **Eng. :** Common Thyme; **Hindi :** Banajwain.

Distribution : Native to Southren Europe; grown in Northern India up to 900m.

Parts Used : Whole Plant.

Folk Use : Decoction of plant used as a tonic.

Active Constituents : Luteolin-7-β-D-glucoside, luteolin-7-β-D-diglucoside and apigenin-7-β-D-glucoside (*plant*).

Reference : Sharma (2003).

925. *Tilia europaea* L.

Syn.: T. officinarum Crantz.; *T. intermedia* DC.; *T. vulgaris* Hayne.

Family: Tiliaceae.

English Names: **Eng. :** Common Lime, European Linden.

Distribution : Found in N. America; introduced into India in Himachal Pradesh.

Parts Used : Flower.

Folk Use : Flowers used as sedative.

Reference : Ambasta (1986).

926. *Tinospora cordifolia* (Willd.) Miers ex Hook. f. & Thoms.

Syn.: Cocculus convolvulacrus DC.; *C. cordifolius* DC.; *C. verrucosus* Wall.; *Menispermum cordifolium* Willd.

Family: Menispermaceae.

English, Hindi and Sanskrit Names: **Eng.** : Gulancha Tinospora, Tinospora; **Hindi** : Amrita, Giloy, Gulanca; **Sans.** : Chakranai, Gurunchi, Jivanthika, Madhuparni, Nirjara, Tantri.

Distribution : Found throughout the tropical India ascending up to 300m.

Parts Used : Fruit. Seeds. Stem.

Flowering & Fruiting : Flowering during summers and fruiting in winters.

Folk Use : Pulverized fruits and aqueous extract of dry stems considered as tonic and cardio tonic. Seeds also valued for its tonic and brain tonic properties.

Active Constituents : Reported to contain tinosporin and a furanoid diterpene, giloin, giloinin, berberine and waxy substrate, furanolactone (I), C-6, C-12 epimer of 6-hydroxyarcangelicin, magnoflorine (*stem*).

References : Bedi (1978), Dhiman (2005), Dwivedi (2007), Kaushik & Dhiman (2000), Kirtikar & Basu (1984), Malhotra & Murthy (1973), Prajapati *et al.* (2006), Reddi *et al.* (2005), Saini (1996), Shah *et al.* (1983), Shah & Joshi (1971), Sharma (2003, 2004), Singh (2001), Siwakoti & Varma (1996), Srivastava *et al* (2003a), Watt (1972).

927. *Toddalia asiatica* (L.) Lam.

Syn.: Paulinia asiatica L.; *Toddalia aculeata* Pers.; *T. angustifolia* Miq.; *T. floribunda* Wall.; *T. rubricaulis* Willd.

Family: Rutaceae.

English, Hindi and Sanskrit Names: **Eng.** : Wild Orange-tree, Lopez Tree, Forest-pepper; **Hindi** : Dehan, Jangli Kalimirch, Kanj; **Sans.** : Dahana, Kanchano, Tiksnaksah.

Distribution : Throughout India, up to 2,500m; also in dry places in the plains.

Parts Used : Root Bark.

Folk Use : Infusion of fresh root bark used as stimulant and tonic.

Active Constituents : Toddalinine, pimpinellin, isopimpinellin, skimmianine, chelerythrine and dihydrochelerythrine (*root bark*).

References : Ambasta (1986), Kirtikar & Basu (1984), Parrotta (2001), Prajapati *et al.* (2006).

928. *Toona ciliata* Roem. (Pl. 11E)

Syn.: Cedrela australis Muell.; *C. febrifuga* Blume; *C. hexandra* Wall.; *C. serrata* Royle; *C. toona* Roxb. ex Rottl.

Family: Meliaceae.

English, Hindi and Sanskrit Names: **Eng. :** Happy Tree, Indian Mahogany, Red Cedar, Sandal Neem, Toon; **Hindi :** Lim, Mahanim, Toon, Tuni; **Sans. :** Apina, Kanta, Nandiviriksha, Tunah.

Distribution : Sub-Himalayan tracts from the Indus eastwards, Chittagong, Assam, Chotta Nagpur, Western Ghats of Mumbai – Nilgiris and Annamalais and other hills of Western Peninsula.

Parts Used : Bark.

Flowering & Fruiting : Flowering during March – April and fruiting in June – July.

Folk Use : Bark used as a tonic and cardio tonic.

Active Constituents : Toonacilin, β-secotetranortriterpenoid, its 6-acetoxy derivative, 23-(R, S) - and 21-(R, S)-hydroxyl toonacilids (*bark*).

References : Dhiman (2005), Farooq (2005), Kapur & Singh (1996), Kaushik & Dhiman (2000), Kirtikar & Basu (1984), Parrotta (2001), Prajapati *et al.* (2006), Ranjan (2000), Siwakoti & Varma (1996).

929. *Trachyspermum ammi* (L.) Sprague

Syn.: Carum copticum Hiern.; *Trachyspermum copticum* Link.

Family: Apiaceae.

English, Hindi and Sanskrit Names: **Eng. :** Ajowan, Carum; **Hindi :** Ajowan, Ajwain; **Sans. :** Ajmoda.

Distribution : Cultivated throughout India.

Parts Used : Fruit.

Folk Use : Fruits used as stimulant and tonic.

Active Constituents : Thymol, carvacrol, palmitic acid, petroselenic acid, oleic and linoleic acid (*fruit*).

References : Ambasta (1986), Thakur *et al.* (1989).

930. *Trachyspermum roxburghianum* (DC.) Craib.

Syn.: Carum roxburghianum Benth. *ex* Kurz; *Trachyspermum involucratum* Wolff.

Family: Apiaceae.

English, Hindi and Sanskrit Names: **Eng.** : Ajowan; **Hindi** : Ajmod, Ajmot, Ajmud, Randhuni; **Sans.** : Ajamoda, Ajaoda.

Distribution : Cultivated throughout India.

Parts Used : Fruit.

Folk Use : Fruits used as a cardio tonic and stimulant.

Active Constituents : d-limonene, α-terpinene, d-linalool, dl-terpineol, thymohydroquinone, thymol, dl-piperitons, cuminic acid, cumminaldehyde, unidentified esters d-limonene, dipentene mixture and bergapten (*fruit*).

References : Ambasta (1986), Prajapati *et al.* (2006), Sharma (2004).

931. *Trapa natans* L. var. ***bispinosa*** (Roxb.) Makino.

Syn.: T. bispinosa Roxb.

Family: Trapaceae.

English, Hindi and Sanskrit Names: **Eng.** : Water Chestnut, Caltrops, Singhara Nut; **Hindi** : Simghada, Simghara; **Sans.** : Jalaphalam, Smgtakah, Trikota.

Distribution : Throughout India, in ponds, lakes and tanks; also cultivated.

Parts Used : Fruit.

Folk Use : Fruits used as a tonic.

Active Constituents : Thiamine, riboflavin, nicotinic acid, vitamin C, vitamin A, oxalates, β-amylase and gallic acid (*fruit*).

References : Prajapati *et al.* (2006), Sharma (2004).

932. *Tribulus terrestris* L.

Syn.: T. lanuginosus L.

Family: Zygophyllaceae.

English, Hindi and Sanskrit Names: **Eng.** : Land-caltrops, Puncture-vine; **Hindi** : Chhota Gokhru, Gokhru, Gokshri, Hussuk; **Sans.** : Bahukantaka, Chanadruma, Gokshura, Ikshugandha.

Distribution : Wasteplaces and dry habitats throughout warmer regions of India, ascending up to 3,000m including West Rajasthan and Gujarat.

Parts Used : Herb. Fruit. Leaf. Root.

Flowering & Fruiting : August – December.

Folk Use : Herb, fruits, leaves and roots used as a tonic.

Active Constituents : Known to possess kaempferol, kaempferol-3-glucoside, kaempferol-3-rutinoside and tribuloside (*fruit, leaf*); campesterol, β-sitosterol, stigmasterol, diosgenin and neotigogenin (*root*).

References : Ambasta (1986), Bhattacharjee (2001), Dhiman (2005), Kaul (1997), Kaushik & Dhiman (2000), Kayal *et al.* (2003), Kirtikar & Basu (1984), Parrotta (2001), Prajapati *et al.* (2006), Punjani (2002), Reddi *et al.* (2005), Sharma (2003), Srivastava *et al.* (2003a), Thakur *et al.* (1989), Watt (1972).

933. *Tricholepis glaberrima* DC.

Family: Asteraceae.

Hindi and Sanskrit Names: **Hindi :** Brahmadandi; **Sans. :** Ajadandi, Brahmadandi.

Distribution : W. Rajputana, Mt. Abu, Central India, Konkan, Deccan, W. Ghats, S. Kanara, Coorg and the hills of Mysore.

Parts Used : Whole Plant.

Folk Use : Plant used as a nervine tonic.

Active Constituents : Betulin, spinasterol, stigmasterol, stigma-7-enol, cycloart-23-ene-3β, 25-diol, 11, 13-dihydrodesacylcynaropicrin and cynaropicrin (*plant*).

References : Ambasta (1986), Kirtikar & Basu (1984), Watt (1972).

934. *Trichosanthes cucumerina* L.

Syn.: T. brevibracteata Kundu; *T. laciniosa* Klein.; *T. pachyrrhachis* Kundu.

Family: Cucurbitaceae.

Hindi and Sanskrit Names: **Hindi :** Jangli Chichinda; **Sans. :** Amritaphala, Potola.

Distribution : Distributed throughout India.

Parts Used : Whole Plant. Root.

Flowering & Fruiting : July – December.

Folk Use : Whole plant used as general and cardiac tonic. Roots used as tonic.

Active Constituents : β-sitosterol, its glucoside and luteolin-7-glucoside (*leaf*).

References : Dhiman (2005), Kaushik & Dhiman (2000), Kirtikar & Basu (1984), Srivastava *et al.* (2003a), Thakur *et al.* (1989), Watt (1972).

935. *Trichosanthes lobata* Roxb.

Syn.: T. perrottetiana DC.; *T. villosula* DC.

Family: Cucurbitaceae.

English, Hindi and Sanskrit Names: **Eng. :** Wild Snake Gourd, Bitter Snake Gourd; **Hindi :** Jamgali Cicinda, Tikta Cicinda; **Sans. :** Patola, Tiktapatola.

Distribution : Throughout Indian plains and in lower hills; also cultivated.

Parts Used : Whole Plant.

Folk Use : Plant used as a tonic.

Reference : Prajapati *et al.* (2006).

936. ***Tricodesma indicum*** (L.) Br.

Syn.: Borago indica L.; *B. spinosa* Roxb.; *T. hirsutum* Edgew.; *T. perfoliatum* Wall.

Family: Boraginaceae.

Hindi Name: **Hindi :** Chhotokalpa.

Distribution : Common throughout India excepting Bengal plains. Kabul, Baloochistan, Persia, Mauritius.

Parts Used : Whole Plant.

Folk Use : Whole plant used as a tonic for brain.

Reference : Sahu (2004).

937. ***Trifolium pratense*** L.

Family: Fabaceae.

English Names: **Eng. :** Purple Clover, Red Clover, Brood-leaved Clover.

Distribution : Native to Europe and Asia; naturalized in N. America and Australia; also in India in Kashmir – Garhwal: 1,350-2,700m.

Parts Used : Flower.

Folk Use : Flowers used as sedative.

Active Constituents : Acetophenone, methyl cinnamate and 1-phenylethanol (*flower*).

Reference : Ambasta (1986).

938. ***Trigonella foenum - graecum*** L.*

Family: Fabaceae.

English, Hindi and Sanskrit Names: **Eng. :** Fenugreek, Greek Hayes; **Hindi :** Methi, Muthi; **Sans. :** Asumodhagam, Bahuparni, Chandrika, Dipani, Methika.

Distribution : Wild in Kashmir, Punjab, and upper Gangetic plains; also cultivated in many other parts of India.

Parts Used : Seed.

Flowering & Fruiting : Flowering during January – February and fruiting in March – April.

Folk Use : Seeds used as tonic.

Active Constituents : Known to yield galactomannan, diosgenin, gitogenin, neogitogenin, homoorientin, saponaretin, neogigogenin, and tiogogenin, vicenins 1 and 2, vixtexin, homoorientin, saponoretin, trigonellosidec, 3, 26-biglycoside, trigonelline, & choline besides, 7-acetoxy-4-methyl coumarin, furastanol glycosides and kaempferol (*seeds*).

References : Ambasta (1986), Kaushik & Dhiman (2000), Pal & Paul (2003), Parrotta (2001), Prajapati *et al.* (2006), Rajan *et al.* (2003), Singh *et al.* (2007), Siwakoti & Varma (1996), Watt (1972).

939. *Triticum aestivum* L. (Pl. 11F)*

Syn.: T. amyleum Seringe.; *T. compactum* Heer.; *T. compositum* L.; *T. durum* Desf.; *T. sativum* Lam.; *T. spelta* L.; *T. turgidum* L.; *T. vulgare* Villars.

Family: Poaceae.

English, Hindi and Sanskrit Names: **Eng. :** Common Wheat, Bread Wheat; **Hindi :** Gehun, Kunak; **Sans. :** Arupa, Godhumah, Kshiri, Rasala, Yavana.

Distribution : Widely cultivated in India.

Parts Used : Grains.

Flowering & Fruiting : Plant grown as a crop in October – November.

Folk Use : Grains used as a tonic.

Active Constituents : Found to contain p-aminobenzoic acid, biotin, carotene, choline, folic acid, inositol, nicotinic acid, pantothenic acid, riboflavin, thiamine, α-tocopherol, β-tocopherol, α-tocotrienols, β-tocotrienols (*grains*).

References : Kaushik & Dhiman (2000), Kirtikar & Basu (1984), Prajapati *et al.* (2006).

940. *Triumfetta rhomboidea* Jacq.

Syn.: T. bartramia L.; *T. trilocularis* Roxb.

Family: Tiliaceae.

English, Hindi and Sanskrit Names: **Eng. :** Bur-weed; **Hindi :** Chikti, Chiriyari; **Sans. :** Jhinijharita, Kantaphali.

Distribution : Tropical and subtropical India, Sri Lanka, Malay, Peninsula. Malay Archipalago, China, Africa and America.

Parts Used : Root.

Flowering & Fruiting : Flowering during August – December and fruiting in September – January.

Folk Use : Root used as a tonic.

References : Kirtikar & Basu (1984), Parrotta (2001), Singh & Srivastava (2000).

941. *Tropaeolum majus* L.

Family: Tropaeolaceae.

English Names: **Eng. :** Garden Nastrutium, Indian Cress, Nasturtium.

Distribution : Commonly cultivated in Indian gardens.

Parts Used : Whole Plant.

Flowering & Fruiting : May – November.

Folk Use : Plant used as tonic and stimulant.

Reference : Bhattacharjee (2001).

942. *Tulipa stellata* Hk. f.

Family: Liliaceae.

English Name: **Eng. :** Tulip.

Distribution : Common in the Western Punjab, the Salt Range, the Siwaliks and the outer Himalayas – Kumaon.

Parts Used : Bulb.

Flowering & Fruiting : May – August.

Folk Use : Bulbs used as tonic.

References : Kapur (1996b), Kapur *et al.* (1996).

943. *Turbina corymbosa* (L.) Rafin.

Syn.: Rivea corymbosa Hallier. f.

Family: Convolvulaceae.

English Names: **Eng. :** Ololiuqui, Snake Plant.

Distribution : Native of tropical America, introduced as an ornamental in gardens in Bombay, Pune and Belgaum.

Parts Used : Seeds.

Folk Use : Decoction of seeds used as a narcotic, hallucinogen and stimulant.

Active Constituents : Turbicoryn, a new diterpene diglucoside (I), galactomannan composed of galactose and mannose in ratio of 1:7.2 (*seeds*).

Reference : Ambasta (1986).

944. *Tylophora indica* (Burm. f.) Merrill

Syn.: Asclepias asthmatica Willd.; *A. tunicata* Hort. Calc.; *Cynanchum flavum* Thunb.; *C. indicum* Burm. f.; *C. vomitorium* Lam.; *Tylophora asthmatica* (L.f.) Wight. & Arn.

Family: Asclepiadaceae.

English, Hindi and Sanskrit Names: **Eng. :** Emetic Swallow-wort, Indian Ipecacuanha, Country Ipecacuanha; **Hindi :** Antamul, Jangli-pikvam; **Sans. :** Lataksiri.

Distribution : North and East Bengal, Assam, Orissa, Konkan, Deccan and in Chennai and upto 914m in the hilly parts.

Parts Used : Root.

Flowering & Fruiting : July – October.

Folk Use : Roots used as stimulant and tonic.

Active Constituents : γ-fagarine, skimmianine and tylopharinicine characterised as 14-hydroxytylophorine (*root*).

References : Ambasta (1986), Dhiman (2005), Kaushik & Dhiman (2000), Subramaniam (2000).

945. *Typha angustata* Bory & Chaub.

Family: Typhaceae.

English Name: **Eng. :** Cattail.

Distribution : More or less throughout India; N. Asia, N. Africa.

Parts Used : Pollen grains.

Flowering & Fruiting : July – October.

Folk Use : Pollen grains mixed with jaggery eaten as a tonic.

Active Constituents : Known to possess isorhamnetin, quercetin, isorhamnetin-3-O-rutinoside and an unidentified isorhamnetin glycoside (*infloresence*).

Reference : Kalpesh *et al.* (2004).

946. *Typhonium trilobatum* Schott.

Syn.: Arum orixense Roxb.; *A. trilobatum* L.; *Typhonium roxburghii* Sannders.

Family: Araceae.

English Names: **Eng. :** Bengal Arum, Lobed Leaf Typhonium.

Distribution : Grows wild in wet soils in Indian Peninsula, Sri Lanka, China.

Parts Used : Tuber.

Folk Use : Tubers used as stimulant.

Active Constituents : Thiamine, niacin, carotene, folic acid, sterols, β-sitosterol and inorganic substances Ca, P, I, F, Fe, Na, K (*tuber*).

Reference : Ambasta (1986).

947. *Uraria lagopodioides* (L.) Desv.

Syn.: U. alopecuroides Sw.; *U. repanda* Wall. ex Benth.

Family: Fabaceae.

English Name: **Eng. :** Tick Trefoil.

Distribution : Indian continent; S.E. China, Taiwan, S.E. Asia, Malaysia – Polgnesia, northern Australia.

Parts Used : Whole Plant.

Flowering & Fruiting : June – December.

Folk Use : Plant used as tonic.

Active Constituents : 3-hydroxyl-7,4'-dimethyoxyflavone and five other flavanoids (*plant*).

Reference : Siwakoti & Varma (1996).

948. ***Urginea indica*** (Roxb.) Kunth.

Syn.: Drimia indica (Roxb.) Jessop*; Scilla indica* Roxb.*; U. coromandeliana* (Roxb.) Hook. f.*; U. gobindappal* Boraiah & Fatima*; U. wightiana* Hook.f.

Family: Liliaceae.

English, Hindi and Sanskrit Names: **Eng. :** Indian Squill, Sea-onion, True-squill; **Hindi :** Ban Piyazi, Jangli Piyaz, Kanda; **Sans. :** Kalkanda, Patalu, Vanapandan.

Distribution : Wild in N.W. Himalayas up to 1,500m, Bihar, Konkan and Coramandel coast.

Parts Used : Bulb.

Flowering & Fruiting : Flowering during March – May and fruiting in May – July.

Folk Use : A drug called Indian squill, obtained from bulbs used as a cardio tonic and stimulant.

Active Constituents : These have been hentriacontanal, octacosanic acid and sitosterol, mannose, glucose and xylose (*bulb*), two glycosides scillaren A and B (*fresh squill*).

References : Ambasta (1986), Kaushik & Dhiman (2000), Parrotta (2001), Prajapati *et al.* (2006), Sharma (2003), Thakur *et al.* (1989).

949. ***Urginea maritima*** (L.) Baker

Syn.: Drimia maritima (L.) Stearn; *Scilla maritima* L.; *U. pancration* Steinh.

Family: Liliaceae.

English Names: **Eng. :** Ada Sogani, Ein Sit, Red Squill, Sea Onion, Squill.

Distribution : Native to Southern Spain, the Canary Islands and South Africa; cultivated in Mediterranean region.

Parts Used : Bulb.

Folk Use : Bulbs used as cardio tonic.

References : Bhattacharjee (2001), Prajapati *et al.* (2006).

950. *Urtica dioica* L. (Pl. 12A)

Family: Urticaeae.

English, Hindi and Sanskrit Names: **Eng.** : Nettle, The Stinging Nettle, Nacighty, Man's Plaything; **Hindi :** Bichhu Booti; **Sans. :** Vrscikali.

Distribution : Temperate regions throughout the northern hemisphere, Southern Africa, the Andes and Australia; found in the N.W. Himalayas from Kashmir and the salt range – Shimla and W. Tibet: 2,700-4,000m.

Parts Used : Root.

Flowering & Fruiting : May – October.

Folk Use : Root used as a tonic and stimulant.

Active Constituents : Homovanillyl alcohol, its 4'-monoglucoside, neoolivil-4-O-β-D-glucoside, 9, 9'-bisacetyl-neoolivil and its 4-O-β-D-glucoside, sitosterol, 7α- and 7β-hydroxysitosterols, sitosterol-3-O-β-D-glucoside & 24 (R) ethyl-5α-cholestane-3β, 6α diol (*root*).

References : Prajapati *et al.* (2006), Sharma (2004).

951. *Urtica hyperborea* Jacq.

Family: Urticaceae.

English Name: **Eng. :** Himalayan Nettle.

Distribution : Western Tibet: 4,000-5,850m; Eastern Tibet, north of Sikkim: 5,350-5,850m.

Parts Used : Whole Plant.

Flowering & Fruiting : June – September.

Folk Use : Plant used as tonic.

Reference : Kaul (1997).

952. *Urtica parviflora* Roxb.

Syn.: U. ardens Link.; *U. himalayensis* Wall.; *U. virulenta* Wall.

Family: Urticaceae.

English and Hindi Names: **Eng. :** Nettle; **Hindi :** Bichhu.

Distribution : Temperate Himalayas up to 5,940m.

Parts Used : Leaf. Inflorescence.

Folk Use : Leaves and inflorescence used as a tonic.

Active Constituents : Histamine, 5-hydroxytryptamine, acetylcholine (*leaf*).

References : Ambasta (1986), Kaushik & Dhiman (2000).

953. *Vaccinium myrtillus* L.

Family: Vacciniaceae.

English Names: **Eng. :** Bilberry, Blue Berry, European Blue Berry, Wortle Berry.

Distribution : South India; Europe and N. America.

Parts Used : Fruit.

Folk Use : Fruits used as sedative.

Reference : Bhattacharjee (2001).

954. *Valeriana fouriei* Briq.

Family: Valerianaceae.

English Name: **Eng. :** Fourie's Valerian.

Distribution : Wet, shaded places in woods and thickets in Korea, Japan and China.

Parts Used : Root.

Folk Use : Roots used as a sedative.

Active Constituents : Essential oils, sesquiterpenes, iridoid glycosides (*root*).

Reference : Prajapati *et al.* (2006).

955. *Valeriana hardwicikii* Wall.

Syn.: V. acuminata Royle; *V. elata* Don; *V. javanica* Blume; *V. tenera* Wall.

Family: Valerianaceae.

English and Hindi Names: **Eng. :** Indian Valerian; **Hindi :** Muskbala, Nihani,Taggar.

Distribution : Temperate Himalayas (Kashmir – Sikkim): 1,219-3,657m and in Khasia hills: 1,219-1,828m.

Parts Used : Root.

Flowering & Fruiting : Flowering during July – August and fruiting in September – October.

Folk Use : Roots used as stimulant.

Reference : Dhiman (2005).

956. *Valeriana himalayana* Grubov.

Syn.: V. dioica auct.

Family: Valerianaceae.

English Name: **Eng. :** Himalayan Valerian.

Distribution : Alpine and subalpine Himalayas, from Kumaon – J.K. and Tibet.

Parts Used : Root.

Flowering & Fruiting : June – August.

Folk Use : Roots used for enhancing memory.

Reference : Khatoon & Goel (2010).

957. ***Valeriana jatamansi*** Jones

Syn.: V. spica Vahl.; *V. villosa* Wall.; *V. wallichii* DC.

Family: Valerianaceae.

English, Hindi and Sanskrit Names: **Eng. :** Indian Valerian; **Hindi :** Mushkbala; **Sans. :** Barhana, Hatah, Parthiva, Tagarah.

Distribution : Temperate Himalayas (Kashmir – Sikkim) up to 3,048m and in Khasia hills between 1,219-1,828m.

Parts Used : Rhizome. Root.

Flowering & Fruiting : Flowering during June – July and fruiting in September – October.

Folk Use : Rhizomes and roots used as sedative, stimulant, tonic and hypnotic. Dried root extract used as nervine tonic.

Active Constituents : Isovalexic acid, acetic acid, isocaproic and β-acetoxyvaleric acid, actinidine, carotene, calarene, calarenol, elemol, jatamols A and B, jatamansic acid, jatamansone, nordal, nardostachonal, narseychelanone, seychellane, seychellene, spirojatomol, valeranol, valeranone, virolin, angelicin, jatamansin, jatamansinol and oroselol (*rhizome, root*).

References : Ambasta (1986), Dhiman (2005), Chauhan (2001), Dwivedi (2007), Kaul (1997), Kirtikar & Basu (1984), Prajapati *et al.* (2006), Prakash & Mehrotra (1996), Reddi *et al.* (2005), Sharma (2003), Sharma & Rana (2000), Singh (1996), Thakur *et al.* (1989).

958. ***Valeriana officinalis*** auct. non L.

Syn.: V. jaeschkei Cl. var. *kaschmiriensis* (Gurbov) Nasir; *V. kaschmiriensis* Keryer ex Grubov.

Family: Valerianaceae.

English and Hindi Names: **Eng. :** Capon's Tail, Common Valerian, Valerian; **Hindi :** Kakabala.

Distribution : Confined only to Kashmir: 2,438-2,743m.

Parts Used : Rhizome. Root.

Flowering & Fruiting : Flowering during June – July and fruiting in August – September.

Folk Use : Valerian drug obtained from rhizomes and roots used as CNS depressant, stimulant and sedative.

References : Ambasta (1986), Dhiman (2005), Bhattacharjee (2001), Kirtikar & Basu (1984), Prakash (2001), Sharma (2003).

959. ***Vateria indica*** L.

Syn.: V. malabarica Blume.

Family: Dipterocarpaceae.

English, Hindi and Sanskrit Names: **Eng.** : Indian Copal Tree, Piney Varnich-tree, White Dammar; **Hindi** : Kahruba, Saphed Dammar; **Sans.** : Ajakarnah, Dhupa, Sarjah.

Distribution : Along the streams in Western Ghats, up to 1,200m.

Parts Used : Bark. Resin.

Flowering & Fruiting : March – April.

Folk Use : Bark and resin used as a tonic.

Active Constituents : dl – Epicatechin, fistinidol & afzelechin (*bark*).

References : Kirtikar & Basu (1984), Prajapati *et al.* (2006).

960. *Ventilago madraspatana* Gaertn.

Syn.: V. bracteata Wall.

Family: Rhamnaceae.

English, Hindi and Sanskrit Names: **Eng.** : Red Creeper; **Hindi** : Pitti; **Sans.** : Dinesavalli, Raktavalli.

Distribution : In forests throughout India.

Parts Used : Root Bark.

Folk Use : Root bark used as tonic and stimulant.

Active Constituents : Xanthorin-5-methyl ether, chrysophanol, physcion, isolandicin, emodin, xanthorin, new isofuranonaphthoquinones – ventilones A, B, C, D and E characterized as 4,9-dihydro-5,8-dihydroxy-6,7-methylenedioxy-1-methylnaphtho [2, 3-C] furan-4, 9-dione, its 5-methyl ether, 1, 3, 4, 9-tetrahydro-8-hydroxyl-5-methoxy-6, 7-methylene-dioxy-1-methylnaphtho [2,3-C] furan-4, 9-dione, its 8-methyl and 5, 8-dimethyl ethers respectively, eight new benzisochromanquinones – ventiloquinones A, B, C, D, E, F, G and H, two new naphthalenes – ventilaginone and ventilagol and three naphthoquinones – maderone, cordeauxione and isocordeauxione (*root bark*).

References : Ambasta (1986), Kirtikar & Basu (1984), Kothari & Rao (2000), Parrotta (2001), Prajapati *et al.* (2006), Watt (1972).

961. *Verbascum coromandelianum* (Vohl.) Kuntze

Syn.: Celsia coramandeliana Vohl.; *Verbascum celsioides* Benth.; *V. chinensis* (L.) Sant.

Family: Scrophulariaceae.

Hindi and Sanskrit Names: **Hindi** : Gadartambaku; **Sans.** : Bhutakeshi.

Distribution : Throughout India; from the Punjab – Sri Lanka and Pogu, ascending to 1,700m. Afganistan, Ava, China.

Parts Used : Leaf.

Flowering & Fruiting : April – August.

Folk Use : Juice of leaves used as a sedative.

References : Ambasta (1986), Kothari *et al.* (2003).

962. *Verbascum thapsus* L.

Syn.: V. indicum Wall.

Family: Scrophulariaceae.

English and Hindi Names: **Eng. :** Aaron's Rod, Bullock's Lungwort, Common Mullein, Cow's Lungwort, Golden Grain, High Taper, Old Man's Flannel, Velvet Dock, Virgin Mary's Candle; **Hindi :** Ban Tamaku, Cidar Tamaku, Phulla.

Distribution : Native to Central and Southern Europe and Western Asia; naturalized in temperate Himalayas, from Kashmir – Bhutan (2,000-3,650m); and of W. Tibet; distributed westwards to Britain.

Parts Used : Seeds.

Flowering & Fruiting : April – September.

Folk Use : Seeds used as narcotic.

Active Constituents : Seeds rich in palmitic, stearic, oleic, linoleic, linolenic, arachidic and behenic acids along with β-sitosterol and ergosta-7-en-3β-ol.

References . Aman & Dar (2003), Ambasta (1986), Kaul (1997), Kirtikar & Basu (1984), Sharma (2003).

963. *Verbena officinalis* L.

Syn.: V. domingensis Urb.; *V. macrostachya* F. Muell.

Family: Verbenaceae.

English Names: **Eng. :** Vervian, Ma Bian Cao, Common Vervain, Demetria, Holy Herb, Simpler's Joy, Tears of Juno.

Distribution : Wild throughout much of Europe and North Africa as well as in China and Japan; common in the Himalayas from Kashmir – Bhutan: 300-2,000m, and in the Bengal plains to the Sunderbans.

Parts Used : Aerial Parts. Leaf.

Flowering & Fruiting : April – October.

Folk Use : Aerial parts used as nervine tonic and mild sedative. Leaves used as tonic.

Active Constituents : Arotemetin, aucubin, lupeol, β-sitosterol and ursolic acid (*aerial parts*).

References : Kaul (1997), Kirtikar & Basu (1984), Prajapati *et al.* (2006), Sharma (2003), Watt (1972).

964. *Vernonia cinerea* (L.) Juss.

Syn.: Cacalia rotundifolia Willd.; *Conyza cinerea* L.; *C. incana* Wall.; *C. linifolia* Wall.; *C. mollis* Willd.; *C. ovata* Wall.; *Serratula cinerea* Roxb.; *Vernonia albicans* DC.; *V. conyzoides* DC.; *V. laxiflora* Less.

Family: Asteraceae.

English, Hindi and Sanskrit Names: **Eng. :** Ash Coloured Fleabane; **Hindi :** Dandotpala, Sahadevi, Sodori; **Sans. :** Devasasha, Sahadevi, Vishvadeva.

Distribution : Throughout India. Tropical Africa, Asia and Australia.

Parts Used : Whole Plant.

Flowering & Fruiting : July – September.

Folk Use : Plant used as a tonic.

Active Constituents : β-amyrin acetate, β-amyrin benzoate, lupeol and its acetate, β-sitosterol, stigmasterol and α-spinasterol (*plant*).

References : Gogoi & Das (2003), Kirtikar & Basu (1984), Maya *et al.* (2003).

965. *Veronicastrum sibiricum* (L.) Pennell

Family: Scrophulariaceae.

English Names: **Eng. :** Blackroot, Culver's Root.

Distribution : Grassy areas in mountains and lowlands in Korea and Japan; cultivated in Indian gardens.

Parts Used : Root.

Folk Use : Roots used as a liver tonic.

Active Constituents : Active ingredients are mannitol, luteolin-7-b-neohesperidoside, luteolin-7-b-glucoside, iridoids, minecoside, 6-O-veratryl catapol ester, catalpol, aucubin, 6-deoxy-8-isoferuloyl herpogide (*root*).

Reference : Prajapati *et al.* (2006).

966. *Vetiveria zizanioides* (L.) Nash

Syn.: Andropogon muricatus Retz.; *A. squarrosus* Hook.f. non L. f.; *Anatherum zizanioides* (L.) Hitchcock & Chase.

Family: Poaceae.

English, Hindi and Sanskrit Names: **Eng. :** Khas-khas, Khus-khus, Vetiver; **Hindi :** Bena, Khas-khas; **Sans. :** Sevyah, Sugandhi-mula, Reshira, Usirah.

Distribution : Wild throughout the plains of India and at lower hills, up to 1,200m.

Parts Used : Root.

Flowering & Fruiting : August – January.

Folk Use : Vetiver oil extracted from roots used as a stimulant. Roots also used as tonic.

Active Constituents : Roots rich in khusenol, new tricyclic sesquiterpenoids – zizanol and epizizanol together with α- and β-vetivones, khusimone and bicyclodecenone.

References : Ambasta (1986), Kirtikar & Basu (1984), Prajapati *et al.* (2006), Sharma (2003), Srivastava & Singh (2003), Srivastava *et al.* (2003a), Thakur *et al.* (1989).

967. *Viburnum opulus* L.

Family: Caprifoliaceae.

English Names: **Eng. :** Guelder-rose, Cramp Bark, King's Crown, Red Elder, Snowball Tree.

Distribution : Native to Europe; cultivated in gardens of Kashmir and other cooler parts of India.

Parts Used : Bark. Berry.

Flowering & Fruiting : Flowering during early summer and fruiting in August.

Folk Use : Bark and ripe berries used as tonic.

Reference : Bhattacharjee (2001).

968. *Viburnum stellulatum* Wall.

Syn.: V. mullaha Buch.-Ham. ex D. Don.

Family: Caprifoliaceae.

Distribution : Temperate Himalayas: 2,000-3,650m; from Kashmir – Sikkim.

Parts Used : Fruit.

Flowering & Fruiting : May – July.

Folk Use : Fruits consumed as tonic.

Reference : Sood *et al.* (2009).

969. *Vigna mungo* (L.) Hepper*

Syn.: Phaseolus mungo L.; *P. radiatus* Roxb.

Family: Fabaceae.

English, Hindi and Sanskrit Names: **Eng. :** Black Gram; **Hindi :** Urd; **Sans. :** Masha.

Distribution : Cultivated all over India.

Parts Used : Root. Seeds.

Flowering & Fruiting : September – October.

Folk Use : Roots considered narcotic. Seeds used as tonic and nervine tonic.

Active Constituents : Therapeutic properties due to genistein, Kievitone, Dalbergiodin, cyclokievitone, isoferririn, eurenol, glycinol, demethylvesititol, kievitone hydrate, cyclokievitone hydrate, hemicelluloses A, hemicelluloses pectin, starch and hexosons (*root*), 24-epiclerosterol, 24α- and 24β-methyl-5α-cholest -8 (14)-en-3β-ols, 24α-ethyl-5α-cholest-8 (14)-en-3β-ols, 14α, 24α- and 14α- , 24β-dimethyl-5α-cholest-9 (11)-en-3β-ol, 14α-methyl-24α-ethyl-5α-cholest-9 (11)-en-3β-ol, 14α-methyl- 24β-ethyl-5α-cholesta-9 (11), 25-dien-3β-ol, 24α-ethyl-5α-cholest-9 (11)-en-3β-ol, 24-methylenepoliinasteanol and β-sitosterol (*seeds*).

References : Ambasta (1986), Kaushik & Dhiman (2000), Kirtikar & Basu (1984), Prajapati *et al.* (2006).

970. *Vigna radiata* (L.) Wilczek

Syn.: Phaseolus aurens Roxb.; *P. mungo* var. *radiatus* (L.) Baker; *P. radiates* L.

Family: Fabaceae.

English, Hindi and Sanskrit Names: **Eng. :** Wild Black Gram; **Hindi :** Banurd; **Sans. :** Masaparni.

Distribution : Throughout Indian plains and on hills upto 300m.

Parts Used : Root.

Flowering & Fruiting : September – October.

Folk Use : Roots used as a tonic.

Active Constituents : Stigmasterol, campersterol and β-sitosterol (*root*).

Reference : Prajapati *et al.* (2006).

971. *Vigna trilobata* (L.) Verdcourt

Syn.: Dolichos trilobatus L.; *D. trilobus* L.; *Phaseolus trilobatus* (L.) Schreb.; *P. trilobus* Sensu. Ait.

Family: Fabaceae.

English, Hindi and Sanskrit Names: **Eng. :** African Gram, Jungle Mat Bean; **Hindi :** Mugam, Rakhalkalai, Trianguli; **Sans. :** Hrasva, Koshila, Mudgaparni.

Distribution : Known to occur throughout India, ascending up to 2,100m in the N.E. region.

Parts Used : Leaf.

Flowering & Fruiting : September – November.

Folk Use : Leaves used as sedative and tonic.

References : Ambasta (1986), Dhiman (2005), Kirtikar & Basu (1984).

972. *Vigna unguiculata* (L.) Walp.

Syn.: Dalichos catiang L.; *D. sinensis* L.; *Vigna sinensis* (L.) Hassk.; *V. catjang* (Burm.f.) Walp.

Family: Fabaceae.

English, Hindi and Sanskrit Names: **Eng. :** Cow Pea, Blackeye Pea, Jerusalem Pea; **Hindi :** Chowli, Lobiya, Rausa, Santa; **Sans. :** Chavala, Mahamasah, Rajamasah.

Distribution : Cultivated throughout India.

Parts Used : Seeds.

Folk Use : Seed used as a liver tonic and tonic.

Active Constituents : Phytohemagglutinin, five pur trypsin inhibitors – A2-a, A2-b, A2-C, A2-d and A2-e (*seeds*).

References : Kirtikar & Basu (1984), Prajapati *et al.* (2006).

973. *Vinca minor* L.

Family: Apocynaceae.

English Names: **Eng. :** Lesser Periwinkle, Running Myrtle.

Distribution : Native to Europe, grows in hedgerows and along woodland borders; also cultivated as a garden plant.

Parts Used : Whole Plant.

Folk Use : Extract of plant used as sedative.

Active Constituent : Vicamine (*plant*).

Reference : Ambasta (1986).

974. *Viola odorata* L. (Pl. 12B)*

Family: Violaceae.

English, Hindi and Sanskrit Names: **Eng. :** Sweet Violet, Appel-leaf, Bairwort, March Violet; **Hindi :** Banaphasha; **Sans. :** Jvarapaha, Nilapushpa, Sukshmapatra.

Distribution : Kashmir and temperate western Himalayas above 1,800m; distributed to Northern and West Asia, North Africa and Europe.

Parts Used : Root.

Folk Use : Roots used as a tonic.

Active Constituent : Odaratine (*root*).

References : Kirtikar & Basu (1984), Prajapati *et al.* (2006).

975. *Viscum album* L.

Syn.: V. costatum Gamble; *V. stellatum* Don.

Family: Viscaceae.

English and Hindi Names: **Eng. :** Devil's Fuge, European Mistletoe, Spectre's Wand; **Hindi :** Ban, Banda.

Distribution : Jaunsar and Tehri – Garhwal up to 2,000m; parasitizing chiefly on rosaceous trees, shrubs, elm, walnut and willows in H.P. between 1,200-2,600m.

Parts Used : Whole Plant. Berry.

Flowering & Fruiting : May – October.

Folk Use : Plant used as cardio tonic, nervine sedative and narcotic. Berries used as tonic.

Active Constituents : Known to possess fatty acids, nicotine alkaloids, amines, flavanoids, terpenoids-beta-amyrin, betalinic acid, lupeol, oleanoilic acid resin acid, ursolic acid, beta sitosterol, sterol A, phytosterol, glucoside, viscotoxin, sugars, starch, syringing, lutein, α- and β-carotene and vitamin-C (*plant*), arginine, asparagines, proline and other free amino acids (*fruit*).

References : Ambasta (1986), Farooq (2005), Kaul (1997), Kirtikar & Basu (1984), Prajapati *et al.* (2006), Raju (2000); Sharma (2003).

976. *Vitex negundo* L.

Syn.: V. arborea Desf.; *V. bicolor* Willd.; *V. paniculata* Lamk.

Family: Verbenaceae.

English, Hindi and Sanskrit Names: **Eng. :** Five-leaved Chaste Tree, Indian Privet; **Hindi :** Mewri, Nisinda, Panikisambhala, Sambhalu, Shambalu; **Sans. :** Indrani, Nilapushpa, Nirgundi, Suvaha.

Distribution : Warmer parts throughout India in scrub-jungles and roadsides.

Parts Used : Leaf. Flower. Root. Fruit.

Flowering & Fruiting : Flowering during June – August and fruiting in cold season.

Folk Use : Leaves and roots valued for its tonic and tranqullizing properties. Fruits used as nervine tonic. Flowers useful as cardio tonic. Leaves act as CNS depressant.

Active Constituents : Hentriacontane, β-sitosterol, β-sitosterol acetate and stigmasterol (*root*), 2'-p-hydroxybenzoylmussaenosidic acid, nishindaside and negundoside (*leaf*).

References : Ambasta (1986), Dhiman (2005), Farooq (2005), Kaushik & Dhiman (2000), Kirtikar & Basu (1984), Kumar & Lal (2003), Parrotta (2001), Prajapati *et al.* (2006), Reddi *et al.* (2005), Sanjay *et al.* (2007), Sharma (2003, 2004), Sharma & Sood (1997), Singh & Srivastava (2000), Siwakoti & Varma (1996), Watt (1972).

977. *Vitex trifolia* L.

Syn.: Cayratia carnosa (Wall.) Gangep.; *Vitis carnosa* Wall.

Family: Verbenaceae.

English, Hindi and Sanskrit Names: **Eng.** : Fox Grape, Indian Wild Pepper; **Hindi** : Amalbel, Pani ki Sambhalu, Ramchana; **Sans.** : Aranyavasini, Atyamlaporni, Neelanirgundi.

Distribution : Scattered throughout the tropical and subtropical parts in the plains and lower hills in India.

Parts Used : Leaf. Root. Fruit.

Flowering & Fruiting : March – June.

Folk Use : Fruits and leaves used as a nervine tonic. Roots used as tonic. Leaves good for improving memory.

Active Constituents : Friedelin, β-sitosterol, its glucoside, a long chain hydrocarbon, isoorientin, 7-O-β-D-glucuronide and 3'-O-β-D-glucuronide of luteolin (*leaf*), dulcitol and vannilic acid (*fruit*).

References : Dhiman (2005), Kirtikar & Basu (1984).

978. *Vitis vinifera* L.

Family: Vitaceae.

English, Hindi and Sanskrit Names: **Eng.** : Common Grape Vine, European Grape, Wine-grape; **Hindi** : Angur, Dakh; **Sans.** : Amrutaphala, Draksha, Mrdvika, Yakshmaghni.

Distribution : Cultivated throughout India.

Parts Used : Flower. Fruit.

Folk Use : Fruits used as nervine and cardio tonic. Flowers valued as tonic to liver.

Active Constituents : Cholestrol, β-sitosterol, ergosterol and jasmonic acid (*fruit*).

References : Ambasta (1986), Kirtikar & Basu (1984), Prajapati *et al.* (2006).

979. *Wagatea spicata* Dalz.

Syn.: Caesalpinia mimosoides Heyne.

Family: Fabaceae.

English Name: **Eng.** : Candy Corn Plant.

Distribution : Hills of Western Peninsula.

Parts Used : Tuberous Root.

Folk Use : Tuberous roots consumed by tribals as a source of tonic.

Reference : Tosh (1996).

980. *Walsura trifoliata* Harms.

Syn.: Heynea trifoliata A. Juss.; *Trichilia coriacea* Rottl.; *T. trifoliata* Wall.; *Walsura piscidia* Roxb.

Family: Meliaceae.

Distribution : W. Ghats from N.Kanara – Anamalais, Pulneys and Travancore, Sri Lanka, N. Circars, Carnatic, Deccan.

Parts Used : Bark.

Flowering & Fruiting : March – August.

Folk Use : Bark used as a stimulant.

Reference : Ambasta (1986).

981. *Wedelia chinensis* Merrill

Syn.: Joegeria calendulacea Spreng.; *Verbesina calendulacea* L.; *Wedelia calendulacea* Less. non Rich.

Family: Asteraceae.

English, Hindi and Sanskrit Names: **Eng. :** Bhangra, Bhanra; **Hindi :** Bhangra, Bhanra, Pila-bhangra; **Sans. :** Bhringaraja, Kesaraja, Pitabhringi.

Distribution : Wild in wet places; also cultivated.

Parts Used : Leaf.

Flowering & Fruiting : March – April.

Folk Use : Leaves used as tonic.

Active Constituents : Norwedelic acid, norwedelolactose and β-D-glucopyranosyl -3-O-β-D-xylopyanosyl (1-2)-β-D-glucuronopyranosyl-olecin-12-en-28-oate (*leaf*).

References : Ambasta (1986), Kirtikar & Basu (1984), Watt (1972).

982. *Withania coagulans* Dunal

Syn.: Puneeria coagulans Stocks.

Family: Solanaceae.

English and Hindi Names: **Eng. :** Indian Cheese-maker, Indian Rennet, Vegetable Rennet; **Hindi :** Akri, Binputakah, Punir.

Distribution : Punjab, Sutlej valley, Sind and Baluchistan; Afganistan.

Parts Used : Fruit. Leaf.

Flowering & Fruiting : July – October.

Folk Use : Fruits considered sedative. Infusion of leaves prescribed either alone or conjoined with twigs of *Rhazya stricta* as an excellent bitter tonic.

Active Constituents : 3β-hydroxyl-2, 3-dihydrowithanolide F (I), ergosta-5, 25-dien-3β, 24-ξ-diol (II), sitosterol β-D-glucoside (*fruit*).

References : Ambasta (1986), Kirtikar & Basu (1984), Watt (1972).

983. *Withania somnifera* (L.) Dunal (Pl. 12C)*

Syn.: Physalis arborescens Thumb.; *P. flexuosa* L.; *P. somnifera* Link.; *P. tomentosa* Thumb.

Family: Solanaceae.

English, Hindi and Sanskrit Names: **Eng.** : Winter Cherry; **Hindi** : Asgandh, Punir; **Sans.** : Ashwagandha, Gandhapatri, Hoya, Priyakari, Turangi-gandha, Varahakareni.

Distribution : Throughout the drier and subtropical parts of India.

Parts Used : Root. Fruit. Seeds.

Flowering & Fruiting : Flowering during July – August and fruiting in January – February.

Folk Use : Roots used as a tonic, sedative narcotic and hypnotic. Fruits considered as sedative. Seeds also accredited with hypnotic properties.

Active Constituents : Constituents characterized are anaferine, isopelletierine, tropine, pseudotropine, 3α-trigloyloxytropane, cuscohygrine, dl-isopelletierine, anahygrine, hygrine, withasomnine, withacoagin characterized as 5α, 20-dihydroxy-1-oxowitha-2, 6, 24-trienolide (*root*), and withanolide – WS 1 (*seeds*).

References : Ambasta (1986), Dhiman (2005), Bhattacharjee (2001), Gopal & Chandra (2003), Harborne & Baxter (2001), Hemlata *et al.* (2007), Kaushik & Dhiman (2000), Kirtikar & Basu (1984) Kshirsagar & Sagar (2007), Kumar (2006), Parrotta (2001); Prajapati *et al.* (2006), Ranjan (2000), Reddi *et al.* (2005), Samvatsar & Diwanji (2000), Sharma (2003, 2004), Singh (2001), Srivastava *et al.* (2003a), Thakur *et al.* (1989), Watt (1972).

984. *Woodfordia fruticosa* (L.) Kurz (Pl. 12D)

Syn.: Grislea punctata Sm.; *G. tomentosa* Roxb.; *Lythrum fruticosum* L.; *Woodflordia floribunda* Salisb.

Family: Lythraceae.

English, Hindi and Sanskrit Names: **Eng.** : Fire-flame Bush, Shiranjitia; **Hindi** : Dawi, Dhaula, Dhaura, Dhai, Santha, Thawi; **Sans.** : Agnijwala, Dhataki, Madaniyahetu.

Distribution : Grows wild up to 1,500m in the tropical deciduous forests.

Parts Used : Flower.

Flowering & Fruiting : Flowering during March – April and fruiting in June – July.

Folk Use : Dried flowers used as stimulant; also eaten as tonic by local people.

Active Constituents : Known to possess cyanidin-3,5-diglucoside, pelargonidin-3,5-diglucoside, myricetin-3-diglucoside, polystachoside, quercetin-3-rhamnoside, naringeuin-7-glucoside, Kaempferol-3-glucoside, octacosanol, chrysophanol-8-O-β-D-glucopyranoside, cyaniding-3,5-diglucoside, glucose, tannins, both pyrogallal type and hydrolysable type including the dimeric oenothenin B, woodfordine A, B and C, ellagic acid, myrictin-3-galactoside, pelargonidin-3,5-diglucoside and polysatachoside (*flower*).

References : Ambasta (1986), Kaushik & Dhiman (2000), Kothari & Rao (2000), Parrotta (2001), Prajapati *et al.* (2006), Reddi *et al.* (2005), Sharma (2004), Sharma & Sood (1997), Thakur *et al.* (1989), Watt (1972).

985. *Wrightia tinctoria* (Roxb.) R. Br.

Syn.: Nerium tinctorium Roxb.

Family: Apocynaceae.

English, Hindi and Sanskrit Names: **Eng. :** Pala Indigo Plant; **Hindi :** Dudhi, Khirni, Mitha; **Sans. :** Hayamaraka, Madhuindrayava, Svetakutaja.

Distribution : Rajputana, Central provinces, Deccan, Konkan, Circars, W. Ghats, Sri Lanka, Myanmar – Timor.

Parts Used : Whole Plant. Bark. Root.

Flowering & Fruiting : March – December.

Folk Use : Plant and bark used as tonic. Roots also possess tonic and narcotic properties.

Active Constituents : β-sitosterol, β-amyrin and its acetate and lupeol benzoate (*bark*).

References : Farooq (2005), Kirtikar & Basu (1984), Kothari & Londhe (2000), Parrotta (2001), Srivastava *et al.* (2003a), Subramaniam (2000), Thakur (2009).

986. *Xanthium strumarium* L. (Pl. 12E)

Syn.: X. discolor Wall.; *X. indicum* DC.; *X. orientale* Blume; *X. roxburghii* Wall.

Family: Asteraceae.

English, Hindi and Sanskrit Names: **Eng. :** Burweed, Cocklebur, Ditch Bur, Louse-bur; **Hindi :** Adhasisi, Banokra, Chota-dhatura, Chota-gokhru, Gokhru; **Sans. :** Arishta, Chanda, Itora, Shwetakusuma, Vanamalini.

Distribution : Native of America; naturalized throughout the hotter parts of India, ascending upto 1,828m in the Himalayas.

Parts Used : Whole Plant. Root. Bud. Flower. Fruit.

Flowering & Fruiting : Flowering and fruiting during rainy season.

Folk Use : Buds and roots valued as tonic. Plant and buds used as sedative. Fruits also possess both tonic and sedative properties. Flowers useful as tonic.

Active Constituents : These are stigmasterol, β-D-glucoside of β-sitosterol, xanthin & xanthinin (*fruit*), a sesquiterpenoid – xanthumin (*aerial parts*). heptacosanol, stigmasterol, β-sitosterol and its glucoside, 3, 4-dihydroxycinnamic acid, Kcl, KNO_3, and K_2SO_4 (*root*).

References : Ambasta (1986), Dhiman (2005), Bhalla *et al.* (1996), Farooq (2005), Kaushik & Dhiman (2000), Khan *et al.* (2003), Kirtikar & Basu (1984), Parrotta (2001), Prajapati *et al.* (2006), Sharma (2003), Siwakoti & Varma (1996), Viswanathan (2000).

987. *Xerospermum glabratum* (Kurz.) Radk.

Syn.: Cupanania glabrata Kurz.; *Xerospermum noronhianum* sensu Hiem.

Family: Sapindaceae.

Distribution : Found in the eastern Himalayas.

Parts Used : Fruit.

Folk Use : Ripe fruits eaten for improving memory among the Garos of Meghalaya.

Reference : Khatoon & Geol (2010).

988. *Zanthoxylum armatum* DC.

Syn.: Z. alatum Roxb.; *Z. planispinum* Sieb. & Zucc.

Family: Rutaceae.

English, Hindi and Sanskrit Names: **Eng. :** Toothache Tree; **Hindi :** Darmar, Nepali Dhaniya, Tej-phal, Tumru; **Sans. :** Dhiva, Gandhalu, Tumburu.

Distribution : Himalayas up to 2,100m and in Andhra Pradesh upto 1,350m.

Parts Used : Bark. Fruit. Seeds.

Flowering & Fruiting : April – June.

Folk Use : Fruits and seeds used as a tonic in fever and dyspepsia. Bark and fruits used as stimulant. Fruits, seeds and decoction of bark valued as tonic.

Active Constituents : Principle activities due to linalool, limonene, methyl cinnamate, cineole, arginine, glycine, histidine, threonine, tyrosine, aspartic acid, glutamic acid, monoterpenetriol-3, 7-dimethyl-1-octane-3,6,7-triol, trans-cinnamic acid, nevadensin, umbelliferone, β-sitosterol and its glucoside (*fruit*), 6-hydroxynonadec-4 (Z)-enoic, 8-hydroxypentadec-4 (Z)-enoic, 7-

hydroxyl-7-vinylhexadec-4 (Z)-enoic acid and hexadec-4 (Z)-enoic acids (*seeds*).

References : Ambasta (1986), Chhetri (2005), Gajurel *et al.* (2006), Kirtikar & Basu (1984), Prajapati *et al.* (2006), Ranjan (2000), Samant *et al.* (2001), Sharma (2003, 2009), Siwakoti & Varma (1996).

989. *Zanthoxylum limonella* (Dennst.) Alston

Syn.: Fagura budrunga Roxb.; *Zanthoxylum budrunga* Wall. ex DC.; *Z. rhetsa* DC.

Family: Rutaceae.

English, Hindi and Sanskrit Names: **Eng. :** Indian Prickly Ash; **Hindi :** Badrang; **Sans. :** Ashvaghra, Atitejani, Sutejasi.

Distribution : Konkan, Deccan, N. Kanara, W. Ghats in S. Kanara, Mysore, Malabar, Annamalais and Travancore at low elevations, Orissa, Sylhet, Khasia hills. Chittagong, Pegu & Martaban down to Tensserin.

Parts Used : Bark. Fruit.

Folk Use : Bark used as tonic and pericarp of fruits as stimulant.

Active Constituents : Known to possess rhetine, chelerythrine, rhetsine, evodiamine, rhetsinine, lupeol, rutaecarpine, xanthoxyletin, osthol and scopoletin (*bark*), zanthoxylin, arborine and dictamnine (*fruit*).

References : Ambasta (1986), Kirtikar & Basu (1984).

990. *Zanthoxylum nitidum* (Roxb.) DC.

Syn.: Z. hamiltonianum Wall. ex Hook. f.

Family: Rutaceae.

English, Hindi and Sanskrit Names: **Eng. :** Toothache Tree; **Hindi :** Tezbal; **Sans. :** Tumburuh.

Distribution : Bihar, eastwards to Sikkim and Assam.

Parts Used : Bark. Fruit.

Flowering & Fruiting : Flowering during March – May and fruiting in September – November.

Folk Use : Fruits and bark used as stimulant and tonic.

Active Constituents : Nitidine chloride and oxyterinamine (*bark*).

References : Ambasta (1986), Kirtikar & Basu (1984), Sharma (2004).

991. *Zanthoxylum oxyphyllum* Edgew.

Syn.: Z. violaceum Wall.

Family: Rutaceae.

English and Sanskrit Names: **Eng. :** Toothache Tree; **Sans. :** Tejowati.

Distribution : Throughout the hills of Kumaon: 2,000-3,000m, eastwards to Sikkim, and Bhutan, Khasia Hills, 1,350-2,000m.

Parts Used : Bark. Seeds.

Folk Use : Bark considered stimulant. Seeds and bark valued as tonic.

Active Constituents : Rhetsinine and zanoxyline (*bark*).

References : Ambasta (1986), Maity *et al.* (2003); Sharma (2004).

992. *Zataria multiflora* Boiss.

Family: Lamiaceae.

English Name: **Eng. :** Wild Thyme.

Distribution : Baluchistan, Afganistan, Persia.

Parts Used : Leaf. Stem.

Folk Use : Infusion leaves and stems used as a stimulant.

References : Ambasta (1986), Kirtikar & Basu (1984).

993. *Zehneria umbellata* Thw.

Syn.: Bryonia amplexicaulis Lamk.; *B. sinuosa* Wall.; *B. umbellate* Klein; *Harlandia bryonioides* Hance; *Karivia rheedii* Roem.; *K. umbellata* Roem.; *Momordica umbellate* Roxb.; *Z. hastata* Miq.

Family: Cucurbitaceae.

Hindi Names: **Hindi :** Amant-mul, Tarali.

Distribution : Throughout India and Ceylon, very common, Malaya, China, North Australia.

Parts Used : Root.

Folk Use : Roots used as stimulant.

Reference : Kapur (1996b).

994. *Zingiber cassumunar* Roxb.

Syn.: Zingiber cliffordii Andhr.; *Z. montanum* (Koen.) Thel.; *Z. purpureum* Rosc.

Family: Zingiberaceae.

English, Hindi and Sanskrit Names: **Eng. :** Cassumunar Ginger; **Hindi :** Banada; **Sans. :** Vanardraka.

Distribution : Throughout India, Sri Lanka, Malay Peninsula. Cultivated in tropical Asia.

Parts Used : Rhizome.

Folk Use : Rhizomes used as a stimulant.

Active Constituents : Two new pigments – cassumunaquinones I and II, three potent antioxidants – cassumunarins A, B and C, trans-3 (2, 4, 5-trimethoxyphenyl)-4 [(E) 3, 4-dimethoxystyryl] and cis-

1,2-bis [(E) 3, 4-dimethoxyphenyl] cyclobutane (*rhizome*) have been identified.

References : Ambasta (1986), Behura & Rout (2003), Rahman & Yusuf (1996), Tripathi & Goel (2001).

995. *Zingiber officinale* Rosc. (Pl. 12F)*

Syn.: Amomum zingiber L.

Family: Zingiberaceae.

English, Hindi and Sanskrit Names: **Eng.** : Ginger; **Hindi** : Ada, Adrak, Sonth (when dry); **Sans.** : Ardrakam, Gulmamula, Mahija, Shunthi.

Distribution : Cultivated in hot humid areas of South India, Assam and H.P.

Parts Used : Rhizome.

Flowering & Fruiting : Flowering in rainy season.

Folk Use : Rhizomes used as a stimulant.

Active Constituents : Active constituents are cumene, 3-carene, α-pinene, limonene, myrcene, p-cymene, methylheptenone, nonanal, decanol, neral, geraniol, linoloot, 2-nonanal, linolool, bornyl acetate, α-terpineol, sabinene, d-borneol and geroniol (*rhizome*).

References : Ambasta (1986), Bhattacharjee (2001), Farooq (2005), Kaushik & Dhiman (2000), Pal & Paul (2003), Prajapati *et al.* (2006), Rahman & Yusuf (1996), Ranjan (2000), Reddi *et al.* (2005), Saikia & Nath (2003b), Sharma (2004), Thakur *et al.* (1989).

996. *Zingiber zerumbet* Rosc. ex Sm.

Syn.: Amomum spurium Gmel.; *A. zerumbet* L.; *Zingiber spurium* Kon.

Family: Zingiberaceae.

English, Hindi and Sanskrit Names: **Eng.** : Wild Ginger, Zizania; **Hindi:** Mahabaribach, Narkachur; **Sans.:** Avanti, Kapooraharidra, Sthulagranthi.

Distribution : N.E. India (as an aquatic grass), common in moist forests, beach thickets, mangrove margins from sea – level to over 500m.

Parts Used : Rhizome.

Flowering & Fruiting : June – September.

Folk Use : Rhizome used as stimulant.

Active Constituents : Camphene, camphor, gingerol, zingeberol, zingerone, sesqui-terpenoids including zerumbone and zerumbone, epoxide, oxalic acid, kaempferol derivatives, afzelin, flavonoid glycosides, essential oils, chlorgenic acid and ferulic acid (*rhizome*).

Reference : Prajapati *et al.* (2006).

997. ***Ziziphus jujuba*** Mill.

Syn.: Z. sativa Gaertn.; *Z. vulgaris* Lam.

Family: Rhamnaceae.

English, Hindi and Sanskrit Names: **Eng.** : Chinese Date, Common Jujube, Indian Jujube, Jujube; **Hindi** : Ban Ber, Kandiari, Kandika, Pitni-ber, Singli; **Sans.** : Soubira.

Distribution : Native of North China; throughout India and in the outer Himalayas up to 1,350m.

Parts Used : Bark. Root. Fruit. Seeds.

Flowering & Fruiting : May – September.

Folk Use : Roots, bark and fruits used as a tonic. Seeds used as tonic to heart and brain, and also as a sedative.

Active Constituents : Stepharine, N-nornuciferine, asimilobine, cyclic AMP, zizyphus saponins I, II and III, jujuboside B (*fruit*).

References : Ambasta (1986), Kirtikar & Basu (1984), Thakur, Puri, Husain (1989).

998. ***Ziziphus mauritiana*** Lam.

Syn.: Rhamnus jujube L.; *Ziziphus jujuba* Lam., non Mill.

Family: Rhamnaceae.

English, Hindi and Sanskrit Names: **Eng.** : Common Jujube, Indian Jujube; **Hindi** : Baer, Ber; **Sans.** : Ajapriya, Badara, Karkandhu, Kuvala, Madhuraphala.

Distribution : Hotter parts of India; cultivated in gardens, villages and roadsides. Also, found wild in wasteplaces and tropical forests.

Parts Used : Root. Fruit. Seeds.

Flowering & Fruiting : Flowering during September – October and fruiting in October – March.

Folk Use : Fruits used as tonic. Roots and seeds valued as tonic and sedative. Roots considered as good stimulant.

Active Constituents : Saponins – jujubosides A and B and spinosin (*seeds*), three new aliphatic acid, p-coumarates (I, II and III), betulinic acid, aliphatic acid, new p-coumaroylates of maslinic acid-3-O-trans-p-coumaryl and 3-O-cis-p-coumaroyl maslinic acid, betulonic, oleanolic and maslinic acids and their methyl esters (*fruit*).

References : Ambasta (1986), Dhiman (2005), Farooq (2005), Gopal & Sinha (2000), Hebbar *et al.* (2003), Parrotta (2001), Prajapati *et al.* (2006), Sharma (2004).

999. *Ziziphus oenoplia* (L.) Mill.

Syn.: Rhamnus oenoplia L.; *Ziziphus albens* Roxb.; *Z. celtidifolia* DC.; *Z. rufula* Miq.; *Z. scandens* Roxb.

Family: Rhamnaceae.

English, Hindi and Sanskrit Names: **Eng. :** Jackal Jujube; **Hindi :** Makkay, Makoi; **Sans.:** Laghubadara, Karkandhauh, Srigalakoli.

Distribution : Common in village thicklets and hedges throughout the hotter parts of India.

Parts Used : Fruit. Stem Bark.

Flowering & Fruiting : April – July (flowering) and fruiting in winter.

Folk Use : Fruits and stem bark used as tonic.

Active Constituents : Cyclopetide alkaloids *viz.* abyssinenis A and B and ziziphines A – G (*stem bark*).

References : Dhiman (2005), Hebbar *et al.* (2003), Kaushik & Dhiman (2000), Parrotta (2001).

1000. *Zornia diphylla* (L.) Pers.

Syn.: Z. angustifolia Smith.; *Z. dictyocarpa* DC.; *Z. gibbosa* Spanoghe.

Family: Fabaceae.

Hindi Name: **Hindi :** Gewani.

Distribution : Himalayas – Sri Lanka and Myanmar, ascending to 1,350m in Kumaon.

Parts Used : Leaf.

Flowering & Fruiting : July – October.

Folk Use : Leaf powder mixed with cow's milk taken twice daily as tonic.

References : Singh (1999, 2003)

Conclusion and Prospects

The study has brought to light the use of 1,000 plant species belonging to 674 genera under 175 families for mood swings, tonic purposes and boosting memory **(Appendix I)**. Of these, 2 species (2 genera: *Cetraria, Parmelia*) belong to lichens; 19 species (14 genera: *Actinopteris, Adiantum, Asplenium, Ceratopteris, Cibotium, Drynaria, Helminthostachys, Lycopodium, Marsilea, Ophioglossum, Osmunda, Pteris, Selaginella, Tectoria*) to Pteridophytes; 11 species (9 genera: *Abies*, *Cedrus*, *Cycas*, *Ephedra*, *Ginkgo*, *Juniperus*, *Pinus*, *Taxus*, *Thuja*) to Gymnosperms; and the remaining 968 species (649 genera) to Angiosperms of which 127 species (83 genera) are the monocots **(Fig. 1; Appendix I)**. Relatively, the ratio of species, genera and families between monocots and dicots is 1:6.62; 1:6.81 and 1:5.69 **(Fig. 2)**. Evidently, the dicots predominate in their overall use.

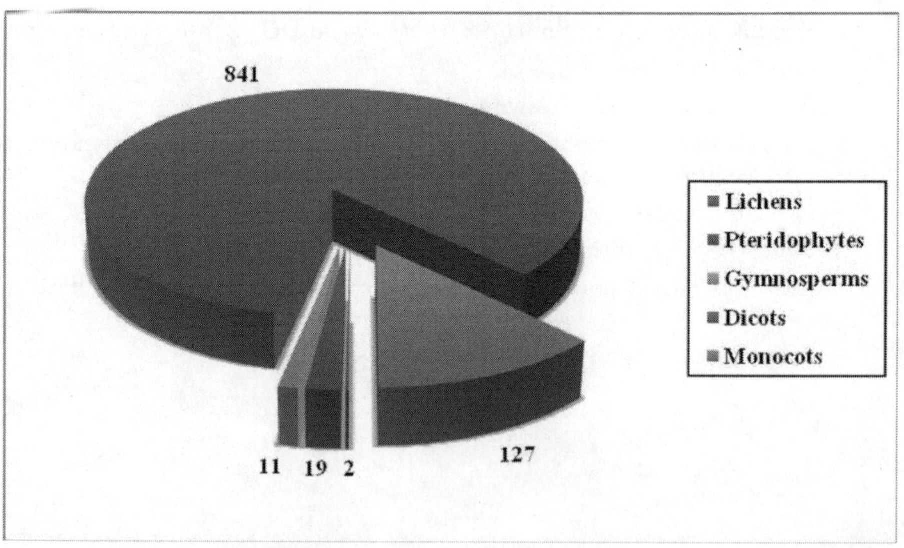

Fig. 1. Mood Swings, Tonic and Memory Boosters Under Various Divisions of Plant Kingdom.

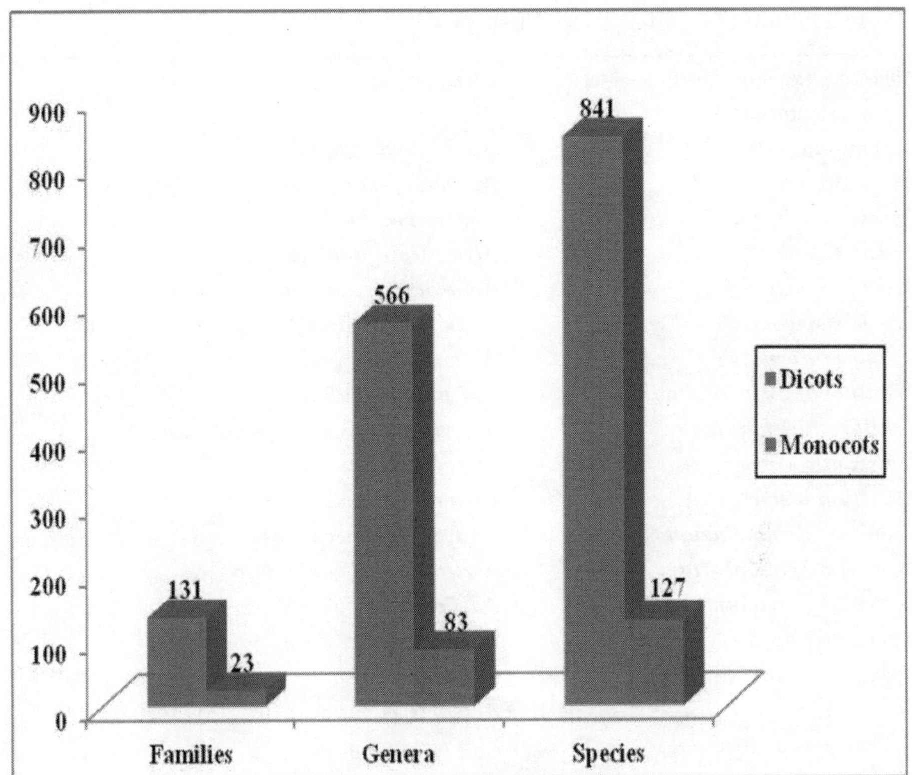

Fig. 2. Total Number of Plant Families, Genera and Species of Mood Swings, Tonic and Memory Boosters.

Of the estimated 1000 plant species, 796 species (560 genera, 159 families) are used for tonic purposes **(Table 1)**, 442 species (331 genera, 116 families) for mood swings **(Table 3)**, and 69 species (63 genera, 43 families) for boosting memory **(Table 2)**. Moreover, 221 species (188 genera) are used for both tonic and mood swings **(Table 4)**, 37 species (36 genera) as tonic and memory boosters **(Table 5)**, 20 species (*Acanthopanax trifoliatus, Achillea millefolium, Basella alba, Cinnamomum tamala, Citrus limon, Curcuma zedoaria, Embelia ribes, Emblica officinalis, Ferula narthex, Ginkgo biloba, Lagenaria siceraria, Majorana hortensis, Melissa officinalis, Mimusops elengi, Myrtus communis, Piper betle, Prunus amygdalus, Syzygium aromaticum, Terminalia bellirica, Ziziphus jujuba*) for tonic, mood swings and memory boosters **(Table 5)**, and 7 species (*Celastrus paniculatus, Coriandrum sativum, Dendrophthoe falcata, Ferula galbaniflua, Matricaria chamomilla, Passiflora foetida, Petroselinum crispum*) for mood swings and boosting memory **(Table 5)**. Apparently, the use of botanicals for tonic purposes is more prevalent than for mood swings and boosting memory in Indian societies.

Table 1. Plants Used for Tonic Purposes

Abelmoschus moschatus	*Aloe perryi*
Abies spectabilis	*Alpinia calcarata*
Abroma augusta	*Alpinia galanga*
Abrus precatorius	*Alpinia officinarum*
Abutilon indicum	*Alstonia scholaris*
Acacia catechu	*Alysicarpus pubescens*
Acacia nilotica	*Amaranthus spinosus*
Acacia senegal	*Amberboa divaricata*
Acampe praemorsa	*Ammi majus*
Acanthopanax trifoliatus	*Amomum subulatum*
Achillea millefolium	*Amorphophallus companulatus*
Achyranthes aspera	*Anacyclus pyrethrum*
Aconitum bisma	*Ananas comosus*
Aconitum deinorrhizum	*Andrographis echioides*
Aconitum heterophyllum	*Andrographis paniculata*
Aconitum koreanum	*Aneilema scapiflorum*
Aconitum napellus	*Anethum graveolens*
Aconitum violaceum	*Angelica glauca*
Acorus calamus	*Anisomeles indica*
Actinopteris australis	*Annona squamosa*
Adiantum capillus-veneris	*Anodendron manubrium*
Adiantum incisum	*Anogeissus latifolia*
Adiantum pedatum	*Anthemis nobilis*
Adiantum vensutum	*Anthocephalus chinensis*
Adonis aestivalis	*Antidesma ghaesembilla*
Aegle marmelos	*Apium graveolens*
Aerva lanata	*Aquilaria agallocha*
Aesculus hippocastanum	*Arachis hypogaea*
Ageratum conyzoides	*Ardisia solanacea*
Ailanthus excelsa	*Areca catechu*
Ailanthus triphysa	*Argemone mexicana*
Ajuga bracteosa	*Argyreia nervosa*
Alangium salviifolium	*Aristolochia indica*
Albizzia lebbeck	*Aristolochia saccata*
Alisma plantago-aquatica	*Aristolochia tagala*
Allium cepa	*Arnica montana*
Allium sativum	*Artemisia abrotanum*
Allophylus serratus	*Artemisia absinthium*
Aloe barbadensis	*Artemisia cina*

Artemisia indica
Artemisia maritima
Artemisia pallens
Artocarpus communis
Artocarpus heterophyllus
Asparagus adscendens
Asparagus racemosus
Asparagus scandens
Asperula odorata
Asplenium trichomanes
Atylosia scarabaeoides
Avena sativa
Averrhoa carambola
Azadirachta indica
Bacopa monnieri
Baliospermum montanum
Bambusa arundinacea
Barleria prionitis subsp. *dicantha*
Barringtonia acutangula ssp. *spicata*
Basella alba
Bauhinia tomentosa
Bauhinia vahlii
Bauhinia variegata
Benincasa hispida
Berberis aristata
Berberis chitra
Berberis lycium
Berberis ulicina
Berberis umbellata
Bergenia ciliata
Bergenia stracheyi
Beta vulgaris
Betula utilis
Bidens pilosa
Biophytum sensitivum
Bistorta affinis
Bixa orellana
Blepharis asperrima
Blepharis linariaefolia
Blepharis persica
Blumea lacera
Boerhavia diffusa
Bombax ceiba
Borassus flabelliffer
Boswellia serrata
Brassica juncea
Brassica oleracea
Brassica oleracea var. *botrytis*
Broussonetia papyrifera
Bryonopsis laciniosa
Buchanania lanzan
Bupleurum falcatum
Butea monosperma
Butea superba
Caccinea glauca
Caesalpinia crista
Caesalpinia pulcherrima
Calamintha clinopodium
Calamus rotang
Calendula officinalis
Callicarpa arborea
Calotropis gigantia
Calotropis procera ssp. *hamiltonii*
Calycopteris floribunda
Camellia sinensis
Canna edulis
Cannabis sativa
Canscora decussata
Canscora diffusa
Canthium parviflorum
Capparis deciduas
Capparis sepiaria
Capparis spinosa
Caralluma attenuata
Cardiospermum halicacabum
Careya arborea
Carica papaya
Carum carvi
Caryota urens
Casearia tomentosa
Cassia fistula
Cassia occidentalis

Cassytha filiformis
Castanea sativa
Cedrus deodara
Ceiba pentandra
Celastrus paniculatus
Celosia argentea
Centaurea cyanus
Centella asiatica
Centratherum anthelminticum
Cephalenthus occidentalis
Ceratopteris siliquosa
Ceropegia bulbosa
Ceropegia tuberosa
Cetraria islandica
Cheiranthus cheiri
Chelone glabra
Chenopodium album
Chimaphila maculata
Chlorophytum arundinaceum
Chlorophytum borivilianum
Chondrodendron tomentosum
Cibotium barometz
Cicer arietinum
Cichorium endivia
Cichorium intybus
Cinchona calisaya
Cinchona ledgeriana
Cinchona officinalis
Cinnamomum obtusifolium
Cinnamomum tamala
Cinnamomum zeylanicum
Cirsium arvense
Cirsium verutum
Cissampelos pareira
Citrullus vulgaris
Citrus aurantium
Citrus limon
Citrus medica
Clerodendrum indicum
Clerodendrum infortunatum
Clerodendrum multiflorum
Clerodendrum serratum
Clitoria ternatea
Cnicus benedictus
Coccinia grandis
Cocculus hirsutus
Cocculus pendulus
Cochlospermum religiosum
Cocos nucifera
Codonopsis clematidea
Codonopsis ovata
Coffea benghalensis
Coix lacryma-jobi
Cola nitida
Colebrookia oppositifolia
Coleus amboinicus
Commiphora mukul
Commiphora myrrha
Convallaria majalis
Convolvulus arvensis
Convolvulus microphyllus
Coptis teeta
Corchorus capsularis
Corchorus depressus
Corchorus fascicularis
Corchorus olitorius
Cordia dichotoma
Cordia macleodii
Corydalis flabellata
Corydalis govaniana
Corylus colurna
Coscinium fenestratum
Costus speciosus
Crataegus oxyacantha
Crataeva nurvala
Cressa cretica
Crinum asiaticum
Crocus sativus
Crotalaria albida
Cryptolepis buchananii
Cucumis melo var. *utilissimus*
Cucumis sativus

Cucurbita maxima
Curculigo orchioides
Curcuma angustifolia
Curcuma aromatica
Curcuma caesia
Curcuma domestica
Curcuma zedoaria
Cuscuta chinensis
Cuscuta reflexa
Cyamopsis tetragonoloba
Cycas revoluta
Cyclea arnottii
Cydonia oblonga
Cymbopogon martinii
Cynodon dactylon
Cyperus articulatus
Cyperus esculentus
Cyperus iria
Cyperus kyllingia
Cyperus longus
Cyperus rotundus
Cyperus scariosus
Cyperus serotinus var. *inundates*
Cyperus stoloniferus
Cytisus scoparius
Dactylorhiza hatagirea
Daphne cannabina
Daucus carota
Delphinium denudatum
Dendrobium alpestre
Dendrobium normale
Dendrocalamus strictus
Descurainea sophia
Desmodium gangeticum
Desmodium heterophyllum
Desmodium tiliaefolium
Desmodium triflorum
Desmotrichum fimbriatum
Dicentra canadensis
Dicliptera roxburghiana
Digitalis lanata
Digitalis purpurea
Dillenia indica
Dimocarpus longan
Dioclea javanica
Dioscorea alata
Dioscorea bulbifera
Dioscorea pentaphylla
Diplocyclos palmatus
Dipsacus japonicus
Dolichos biflorus
Doronicum falconeri
Doronicum hookeri
Doronicum pardalianches
Doronicum roylei
Drosera peltata
Drymaria cordata
Drynaria quercifolia
Drypetes roxburghii
Durio zibethinus
Echinops echinatus
Eclipta prostrata
Elaeagnus infundibularis
Elaeocarpus sphaericus
Elephantopus scaber
Elettaria cardamomum
Eleusine coracana
Embelia ribes
Emblica officinalis
Enicostemma littorale
Entada phaseoloides
Entada rheedii
Ephedra gerardiana
Eruca sativa
Eryngium billardieri
Eryngium caeruleum
Eryngium foetidum
Erythrina stricta
Erythroxylum coca
Erythroxylum monogynum
Eucalyptus globules
Eulophia campestris

Euodia officinalis
Eupatorium cannabinum
Eupatorium perfoliatum
Eupatorium purpureum
Eupatorium triplinerve
Euphorbia neriifolia
Euphorbia pilosa
Euphrasia officinalis
Euryale ferox
Evodia lunu-ankenda
Evolvulus alsinoides
Exacum tetragonum
Fagonia cretica
Fagonia indica
Fagraea racemosa
Feronia limonia
Ferula asafoetida
Ferula narthex
Ficus benghalensis
Ficus hispida
Ficus racemosa
Ficus religiosa
Ficus virens
Flacourtia indica
Foeniculum vulgare
Fraxinus excelsior
Fritillaria roylei
Fumaria officinalis
Galipea officinalis
Garcinia indica
Gardenia jasminoides
Gentiana algida
Gentiana kurroo
Gentiana lutea
Gentiana prostrata
Gentiana stracheyi
Gentiana villosa
Geum elatum
Gillenia trifoliata
Ginkgo biloba
Glechoma hederacea
Glinus oppositifolius
Gloriosa superba
Glycyrrhiza glabra
Gmelina arborea
Gossypium arboretum var. *obtusifolium*
Grewia microcos
Grewia tiliifolia
Gymnema sylvestre
Habenaria acuminata
Habenaria edgeworthii
Habenaria goodyeroides
Habenaria griffithi
Habenaria intermedia
Haematoxylon campechianum
Haldina cordifolia
Hedychium coronarium
Hedychium spicatum
Hedyotis corymbosa
Hedyotis herbacea
Heimia salicifolia
Helminthostachys zeylanica
Hemidesmus indicus
Hesperethusa crenulata
Heynea trijuga
Hibiscus rosa-sinensis
Hibiscus sabdariffa
Hippophae rhamnoides L. subsp. *turkestanica*
Hiptage benghalensis
Holarrhena antidysenterica
Holostemma ada-kodien
Hordeum vulgare
Houttuynia cordata
Humulus lupulus
Hybanthus enneaspermus
Hydnocarpus laurifolia
Hydrangea arborescens
Hydrastis canadensis
Hydrocotyle javanica
Hygrophila auriculata
Hygroryza aristata
Hypochoeris glabra

Hypoxis aurea
Hyptis capitata
Ichnocarpus frutescens
Impatiens balsamina
Imperata cylindrica
Indigofera articulata
Indigofera gerardiana
Indigofera glabra
Indigofera glandulosa
Indigofera tinctoria
Indigofera trita
Inula racemosa
Ipomoea batatas
Ipomoea digitata
Ipomoea pes – caprae
Ipomoea sepiaria
Ixora parviflora
Jasminum arborescens
Jasminum auriculatum
Jasminum grandiflorum
Jasminum multiflorum
Jateorhiza palmata
Jatropha curcas
Juglans cinerea
Juglans regia
Juniperus communis
Justicia adhatoda
Justicia beddomei
Kalanchoe integra
Krameria argentea
Krameria triandra
Lactuca indica
Lactuca serriola
Lagenaria siceraria
Lamium album
Lantana camara
Lawsonia inermis
Lens culinaris
Leonurus cardiaca
Leonurus japonicus
Lepidium sativum
Leptadenia reticulata
Leucas aspera
Lilium polyphyllum
Limnophila gratioloides
Linum usitatissimum
Liparis paradoxa
Liparis rostrata
Liquidambar orientalis
Liriope platyphylla
Litchi chinensis
Litsea glutinosa
Lodoicea maldivica
Luffa acutangula
Luffa cylindrica
Lycopersicon esculentum
Madhuca indica
Maerua arenaria
Magnolia grandiflora
Majorana hortensis
Malaxis acuminata
Malaxis muscifera
Mallotus philippensis
Malva parviflora
Mangifera indica
Manilkara hexandara
Manilkara kauki
Maranta arundinacea
Marrubium vulgare
Marsilea minuta
Matthiola flavida
Matthiola incana
Mazus japonicus
Meconopsis aculeata
Melia azedarach
Melia composite
Melissa officinalis
Melissa parviflora
Melothria maderaspatana
Memecylon augustifolium
Menispermum canadense
Mentha arvensis

Mentha pulegium
Menyanthes trifoliata
Meriandra bengalensis
Mesua ferrea
Michelia champaca
Mimosa hamata
Mimusops elengi
Momordica balsamina
Momordica charantia
Monochoria vaginalis
Morinda citrifolia
Morinda pubescens
Moringa oleifera
Mucuna monosperma
Mucuna prurita
Mundulea sericea
Murraya koenigii
Murraya paniculata
Musa paradisiaca
Mussaenda frondosa
Myrica esculenta
Myristica fragrans
Myrtus communis
Myxopyrum serratulum
Nardostachys jatamansi
Nasturtium officinale
Nelsonia campestris
Nelumbo nucifera
Nepata cataria
Nepeta discolor
Nepeta flocosa
Nepeta hindostana
Nerium indicum
Nervilia aragoana
Nicotiana tabacum
Nyctanthes arbor – tristis
Nymphaea nouchali
Ochna jabotapita
Ocimum americanum
Ocimum gratissimum
Ocimum sanctum
Onosma bracteatum
Ophioglossum reticulatum
Ophioglossum vulgatum
Ophiorrhiza mungos
Origanum vulgare
Ormocarpum cochinchinensis
Oroxylum indicum
Oryza sativa
Osmunda regalis
Ostrya virginiana
Ouratea serrata
Oxalis corniculata
Oxydendrum arboratum
Paederia foetida
Panax bipinnatifidus
Panax pseudo – ginseng
Pandanus odoratissimus
Papaver somniferum
Parthenium hysterophorus
Paspalum scrobiculatum
Pavetta indica ssp. tomentosa
Pedalium murex
Pennisetum americanum
Pennisetum flaccidum
Peperomia tetraphylla
Permelia perlata
Peucedanum grande
Peucedanum nagpurense
Phaseolus radiatus
Phoenix pusilla
Phoenix sylvestris
Pholidota articulata
Phyllostachys bambusoides
Physalis minima
Picrorhiza kurrooa
Pimenta dioica
Pinus roxburghii
Piper betle
Piper brachystachyum
Piper cubeba
Piper longum

Piper puberulum
Pistacia chinensis ssp. *integerrima*
Plantago major
Plantago ovata
Platycodon grandiflorum
Plectranthus mollis
Plumbago zeylanica
Podophyllum hexandrum
Polygonatum cirrhifolium
Polygonatum kingianum
Polygonatum verticillatum
Polygonum aviculare
Polygonum barbatum
Polygonum chinensis
Polygonum hydropiper
Polygonum multiflorium
Polygonum viviparum
Pongamia pinnata
Populus alba
Populus ciliata
Potentilla anserina
Potentilla argyrophylla
Potentilla sibbaldi
Potentilla supina
Prangos pabularia
Premna obtusifolia
Prunella vulgaris
Prunus amygdalus
Prunus armeniaca
Prunus cerasoides
Prunus cerasus
Prunus domestica
Pseudarthria viscida
Psidium guajava
Pteris vittata
Pterocarpus marsupium
Pterocarpus santalinus
Pterospermum acerifolium
Pueraria tuberosa
Pulsatilla koreana
Punica granatum
Pygmaeopremna herbacea
Quassia indica
Quercus velutina
Ranunculus laetus
Ranunculus sceleratus
Rauvolfia vomitora
Rehmania glutinosa
Rhamnus purpurea
Rhamnus triquetra
Rheum emodi
Rheum webbianum
Rhodiola imbricata
Rhododendron anthopogon
Rhododendron arboreum
Rhododendron campanulatum
Rhus succedanea
Rosa centifolia
Rosa damascena
Rosa laevigata
Roscoea alpina
Rosmarinus officinalis
Rourea minor
Roylea cinerea
Rubia cordifolia
Rubia tinctorum
Rubia wallichiana
Rubus coreanus
Ruellia patula
Rumex hymenosephalus
Rumex nepalensis
Ruta chalepensis
Ruta graveolens
Saccharum spontaneum
Salacia reticulata
Salix babylonica
Salix caprea
Salix elegans
Salvadora persica
Salvia officinalis
Samadera indica
Sandorium kortjape

Sansevieria trifasciata
Sansevieria zeylanica
Santalum album
Sapindus mukorossi
Sapindus trifoliatus
Sapium insigne
Sapium sebiferum
Saraca asoca
Sarcostemma acidum
Sassafras albidum
Satyrium nepalense
Saussurea lappa
Saussurea sacra
Saxifraga oppositifolia
Schinus terebinthifolius
Schleichera oleosa
Schweinfurthia sphaerocarpa
Scindapsus officinalis
Scutellaria lateriflora
Scutellaria prostrata
Securinega leucopyrus
Selaginella bryopteris
Semecarpus anacardium
Senecio jacquemontianus
Sesamum indicum
Sesbania grandiflora
Shefflera octophylla
Shorea robusta
Sida acuta
Sida cordata
Sida cordifolia
Sida ovata
Sida rhombifolia
Sida spinosa
Sissymbriun iria
Smilax china
Smithia conferta
Solanum indicum
Solanum melongena
Solanum nigrum
Solanum torvum
Solena amplexicaulis
Sonchus oleraceus
Soymida febrifuga
Sphaeranthus indicus
Spilanthes paniculata
Spondias pinnata
Sterculia urens
Stereospermum colais
Stereospermum suaveolens
Strobilanthes ciliates
Strychnos ignatii
Strychnos nux – vomica
Strychnos potatorum
Swertia angustifolia
Swertia chirayita
Swertia ciliata
Swertia cordata
Swertia petiolata
Swietenia mahagoni
Symplocos paniculata
Symplocos racemosa
Syzygium aromaticum
Syzygium cumini
Tabernaemontana divaricata
Tamarindus indica
Tanacetum vulgare
Taraxacum officinale
Tecoma stans
Tectoria cicutaria
Tephrosia purpurea
Tephrosia virginia
Teramnus labialis
Terminalia arjuna
Terminalia bellirica
Terminalia catappa
Terminalia chebula
Terminalia crenulata
Terminalia myriocarpa
Terminalia paniculata
Thalictrum foliolosum
Thalictrum pedunculatum

Theobroma cacao	*Ventilago madraspatana*
Thevetia peruviana	*Verbena officinalis*
Thuja orientalis	*Vernonia cinerea*
Thymus serpyllum	*Veronicastrum sibiricum*
Thymus vulgaris	*Vetiveria zizanioides*
Tinospora cordifolia	*Viburnum opulus*
Toddalia asiatica	*Viburnum stellulatum*
Toona ciliata	*Vigna mungo*
Trachyspermum ammi	*Vigna radiata*
Trachyspermum roxburghianum	*Vigna trilobata*
Trapa natans var. *bispinosa*	*Vigna unguiculata*
Tribulus terrestris	*Viola odorata*
Tricholepis glaberrima	*Viscum album*
Trichosanthes cucumerina	*Vitex negundo*
Trichosanthes lobata	*Vitex trifolia*
Tricodesma indicum	*Vitis vinifera*
Trigonella foenum – graecum	*Wagatea spicata*
Triticum aestivum	*Wedelia chinensis*
Triumfetta rhomboidea	*Withania coagulans*
Tropaeolum majus	*Withania somnifera*
Tulipa stellata	*Woodfordia fruticosa*
Tylophora indica	*Wrightia tinctoria*
Typha angustata	*Xanthium strumarium*
Uraria lagopodioides	*Zanthoxylum armatum*
Urginea indica	*Zanthoxylum limonella*
Urginea maritima	*Zanthoxylum nitidum*
Urtica dioica	*Zanthoxylum oxyphyllum*
Urtica hyperborea	*Ziziphus jujuba*
Urtica parviflora	*Ziziphus mauritiana*
Valeriana jatamansi	*Ziziphus oenoplia*
Vateria indica	*Zornia diphylla*

Table 2. Plants Used as Memory Boosters

Acanthopanax trifoliatus	*Celastrus paniculatus*
Achillea millefolium	*Centella asiatica*
Achyranthes aspera	*Cinnamomum tamala*
Acorus calamus	*Citrullus vulgaris*
Argyreia nervosa	*Citrus limon*
Bacopa monnieri	*Clitoria ternatea*
Basella alba	*Convallaria majalis*

Convolvulus arvensis
Convolvulus microphyllus
Cordia macleodii
Coriandrum sativum
Curcuma zedoaria
Dendrophthoe falcata
Desmodium tiliaefolium
Elephantopus scaber
Embelia ribes
Emblica officinalis
Evolvulus alsinoides
Ferula galbaniflua
Ferula narthex
Ficus benghalensis
Ginkgo biloba
Hedychium spicatum
Helminthostachys zeylanica
Hibiscus rosa-sinensis
Hydrocotyle sibthorpioides
Juglans regia
Lagenaria siceraria
Lantana camara
Lawsonia inermis
Limnophila gratioloides
Limnophila rugosa
Linum usitatissimum
Majorana hortensis
Matricaria chamomilla
Medicago sativa
Melia azedarach
Melissa officinalis
Melissa parviflora
Mimusops elengi
Myrtus communis
Nelumbo nucifera
Passiflora foetida
Petroselinum crispum
Picrorhiza kurrooa
Piper betle
Prunella vulgaris
Prunus amygdalus
Prunus cerasus
Rhodiola imbricate
Salvia officinalis
Semecarpus anacardium
Syzygium aromaticum
Tabernaemontana divaricata
Terminalia bellirica
Terminalia chebula
Tinospora cordifolia
Tricodesma indicum
Valeriana himalayana
Vitex trifolia
Xerospermum glabratum
Ziziphus jujuba

Table 3. Plants Used for Mood Swings

Abelmoschus esculantus
Abelmoschus moschatus
Acalypha indica
Acanthopanax trifoliatus
Achillea millefolium
Aconitum chasmanthum
Aconitum ferox
Aconitum napellus
Acorus gramineus
Actaea spicata
Adiantum pedatum
Aesculus hippocastanum
Ageratum conyzoides
Ajuga bracteosa
Allium cepa
Allium fistulosum
Allium sativum
Alpinia galanga
Alpinia officinarum
Alstonia scholaris

Amomum subulatum
Amomum xanthioides
Anacyclus pyrethrum
Anagallis arvensis
Anemone pulsatilla
Anethum graveolens
Angelica archangelica L. var. *himalaica*
Angelica glauca
Annona squamosa
Anthemis nobilis
Apium graveolens
Aquilaria agallocha
Aquilegia vulgaris
Argemone mexicana
Aristolochia saccata
Armoracia rusticana
Artemisia absinthium
Artemisia cina
Artemisia dracunculus
Artemisia maritima
Artocarpus heterophyllus
Asarum europaeum
Asparagus racemosus
Asplenium nidus
Atropa belladonna
Azadirachta indica
Baliospermum montanum
Bambusa arundinacea
Baptisia leucophaea
Barbarea vulgaris
Barringtonia acutangula ssp. *spicata*
Basella alba
Bidens pilosa
Bidens tripartita
Biophytum sensitivum
Blumea lacera
Bombax ceiba
Borassus flabelliffer
Borreria articularis
Bunium persicum
Butea superba
Caesalpinia crista
Caesalpinia pulcherrima
Calamintha clinopodium
Calendula officinalis
Camellia sinensis
Canna edulis
Cannabis sativa
Capparis deciduas
Capparis zeylanica
Capsella bursa-pastoris
Capsicum annum
Cardamine pratensis
Cardiospermum halicacabum
Carum carvi
Cassia absus
Catha edulis
Catharanthus roseus
Celastrus paniculatus
Centaurea cyanus
Centratherum anthelminticum
Cheiranthus cheiri
Chenopodium album
Chloranthus glaber
Chlorophytum arundinaceum
Cimicifuga foetida
Cimicifuga racemosa
Cinnamomum camphora
Cinnamomum tamala
Cinnamomum zeylanicum
Cissampelos pareira
Citrus aurantifolia
Citrus aurantium
Citrus limon
Citrus medica
Cleome viscosa
Clerodendrum serratum
Cocculus hirsutus
Cochlospermum religiosum
Codonopsis clematidea
Codonopsis ovata
Codonopsis rotundifolia

Coffea arabica
Coffea benghalensis
Cola acuminata
Cola nitida
Colocasia esculenta
Commiphora mukul
Commiphora myrrha
Conium maculatum
Corchorus capsularis
Corchorus fascicularis
Coriandrum sativum
Costus speciosus
Crataeva nurvala
Crocus sativus
Cucurbita maxima
Curcuma caesia
Curcuma domestica
Curcuma zedoaria
Cymbopogon citratus
Cymbopogon jwarancusa
Cymbopogon martinii
Cyperus articulatus
Cyperus esculentus
Cyperus iria
Cyperus rotundus
Cyperus serotinus var. *inundates*
Cyperus stoloniferus
Dalbergia sissoo
Datura innoxia
Datura metel
Datura sanguinea
Datura stramonium
Daucus carota
Delphinium denudatum
Dendrobium denudans
Dendrophthoe falcata
Derris trifoliata
Desmotrichum fimbriatum
Digitalis lanata
Digitalis purpurea
Dioclea javanica
Diospyros lotus
Dorema ammoniacum
Doronicum falconeri
Drypetes roxburghii
Duboisea myoporoides
Elaeagnus umbellata
Elaeocarpus sphaericus
Elettaria cardamomum
Embelia ribes
Emblica officinalis
Ephedra gerardiana
Eragrostis cynosuroides
Erigeron asteroides
Erigeron canadensis
Eriobotrya japonica
Eruca sativa
Erythrina variegata
Erythroxylum coca
Eucalyptus globules
Eupatorium purpureum
Eupatorium triplinerve
Euphorbia thymifolia
Evodia rutaecarpa
Fagonia cretica
Ferula asafoetida
Ferula galbaniflua
Ferula narthex
Ferula suaveolens
Foeniculum vulgare
Fritillaria roylei
Gardenia jasminoides
Gaultheria fragrantissima
Gaultheria trichophylla
Gelsemium sempervirens
Ginkgo biloba
Gloriosa superba
Grevillea robusta
Gymnema sylvestre
Hedera nepalensis
Hedychium coronarium
Hedyotis biflora

Hedyotis corymbosa
Heteropogon contortus
Hibiscus sabdariffa var. *sabdariffa*
Holostemma ada-kodien
Homalonema aromatica
Humulus lupulus
Hydrocotyle javanica
Hygrophila auriculata
Hyoscyamus muticus
Hyoscyamus niger
Hypecoum leptocarpum
Hypericum cernuum
Hypericum patulum
Hypericum perforatum
Hyptis capitata
Hyptis suaveolens
Hyssopus officinalis
Ilex paraguariensis
Illicium verum
Inula racemosa
Iochroma fuchsioides
Ipomea carnea ssp. *fistulosa*
Ipomoea digitata
Ipomoea violacea
Ixora coccinea
Jatropha curcas
Jatropha gossypifolia
Juniperus communis
Juniperus marcopoda
Jurinea dolomiaea
Kaempferia galanga
Lactuca serriola
Lagenaria siceraria
Lagerstroemia indica
Lagerstroemia speciosa
Lathyrus aphaca
Laurus nobilis
Leea hirta
Leonurus cardiaca
Lepidium sativum
Leptadenia reticulata
Leucas aspera
Leucas cephalotus
Leucas lavandulaefolia
Leucas stelligera
Ligusticum elatum
Liquidambar orientalis
Liriodendron tulipifera
Lithospermum officinale
Litsea monopetala
Lycopodium clavatum
Madhuca indica
Maerua arenaria
Magnolia grandiflora
Majorana hortensis
Mandragora autumnalis
Marsilea minuta
Matricaria chamomilla
Meconopsis aculeata
Melissa officinalis
Melodinus monogynus
Mentha aquatica
Mentha arvensis
Mentha longifolia
Mentha piperita
Mentha pulegium
Mentha spicata
Menyanthes trifoliata
Michelia champaca
Mimusops elengi
Moringa oleifera
Moschosma polystachyum
Mucuna monosperma
Mucuna prurita
Murraya koenigii
Murraya paniculata
Myrica esculenta
Myristica fragrans
Myroxylon balsamum
Myrtus communis
Nardostachys jatamansi
Nasturtium officinale

Nepeta cataria
Nepeta hindostana
Nicotiana tabacum
Nigella sativa
Nymphaea alba
Nymphaea stellata
Ocimum basilicum
Ocimum gratissimum
Ocimum sanctum
Oenothera biennis
Onosma bracteatum
Operculina turpethum
Ophiorrhiza mungos
Origanum vulgare
Ormocarpum cochinchinensis
Oryza sativa
Osmunda regalis
Ougeinia oojeinensis
Paeonia emodi
Panax pseudo – ginseng
Panax quinquefolium
Pandanus odoratissimus
Panicum miliaceum
Panicum sumatrense
Papaver rhoeas
Papaver somniferum
Parnassia palustris
Paspalum scrobiculatum
Passiflora foetida
Passiflora incarnata
Pedicularis pectinata
Peganum harmala
Perilla frutescens
Petroselinum crispum
Peucedanum grande
Phaseolus radiatus
Physochlaina praealta
Physostigma venenosum
Phytolacca acinosa
Pimenta dioica
Pimpinella anisum
Pinellia ternate
Pinus gerardiana
Pinus roxburghii
Piper betle
Piper cubeba
Piper longum
Piper nigrum
Piper retrofractum
Piscidia piscipula
Pistacia lentiscus
Pittosporum floribundum
Plantago major
Plumbago indica
Plumbago zeylanica
Plumeria acuminata
Podophyllum hexandrum
Pogostemon benghalense
Pogostemon parviflorus
Polemonium caeruleum
Polygonum amplexicaule
Polygonum hydropiper
Pongamia pinnata
Populus ciliata
Potentilla fruticosa
Prangos pabularia
Primula rosea
Primula veris
Prunus amygdalus
Prunus armeniaca
Prunus virginiana
Pterygota alata
Punica granatum
Pygmaeopremna herbacea
Pyrus communis
Randia spinosa
Ranunculus sceleratus
Rauvolfia serpentina
Rauvolfia vomitora
Rheum emodi
Rhododendron anthopogon
Rhus succedanea

Rorippa dufia
Rosmarinus officinalis
Rubus niveus
Ruta graveolens
Salvadora persica
Sambucus canadensis
Sambucus nigra
Sansevieria trifasciata
Sarcostemma acidum
Sarcostemma brunonianum
Sassafras albidum
Saussurea lappa
Saussurea sacra
Schinus terebinthifolius
Scindapsus officinalis
Scutellaria lateriflora
Selinum vaginatum
Selinum wallichianum
Serenoa serrulata
Sesbania roxburghii
Sesbania sesban
Seseli diffusum
Sissymbriun iria
Sium latijugum
Solanum dulcamara
Solanum melongena
Solanum nigrum
Solanum surattense
Solanum torvum
Solanum tuberosum
Solena amplexicaulis
Sonchus arvensis
Sonchus oleraceus
Spilanthes acmella
Stephania glabra
Streblus asper
Strychnos nux – vomica
Synadenium grantii
Syzygium aromaticum
Syzygium cumini
Tabebuia pentaphylla
Tagetes minuta
Taxus baccata
Tectona grandis
Tephrosia purpurea
Tephrosia virginia
Terminalia bellirica
Theobroma cacao
Thevetia peruviana
Thlaspi arvense
Thymus serpyllum
Tilia europaea
Toddalia asiatica
Trachyspermum ammi
Trachyspermum roxburghianum
Trifolium pratense
Tropaeolum majus
Turbina corymbosa
Tylophora indica
Typhonium trilobatum
Urginea indica
Urtica dioica
Vaccinium myrtillus
Valeriana fouriei
Valeriana hardwicikii
Valeriana jatamansi
Valeriana officinalis
Ventilago madraspatana
Verbascum coromandelianum
Verbascum thapsus
Verbena officinalis
Vetiveria zizanioides
Vigna mungo
Vigna trilobata
Vinca minor
Viscum album
Vitex negundo
Walsura trifoliata
Withania coagulans
Withania somnifera
Woodfordia fruticosa
Wrightia tinctoria

Xanthium strumarium
Zanthoxylum armatum
Zanthoxylum limonella
Zanthoxylum nitidum
Zanthoxylum oxyphyllum
Zataria multiflora
Zehneria umbellata
Zingiber cassumunar
Zingiber officinale
Zingiber zerumbet
Ziziphus jujuba
Ziziphus mauritiana

Table 4. Plants Possessing Tonic and Mood Swings Properties

Abelmoschus moschatus
Aconitum napellus
Adiantum pedatum
Aesculus hippocastanum
Ageratum conyzoides
Ajuga bracteosa
Allium cepa
Allium sativum
Alpinia galanga
Alpinia officinarum
Alstonia scholaris
Amomum subulatum
Anacyclus pyrethrum
Anethum graveolens
Angelica glauca
Annona squamosa
Anthemis nobilis
Apium graveolens
Aquilaria agallocha
Argemone mexicana
Aristolochia saccata
Artemisia absinthium
Artemisia cina
Artemisia maritima
Artocarpus heterophyllus
Asparagus racemosus
Azadirachta indica
Baliospermum montanum
Bambusa arundinacea
Barringtonia acutangula ssp. *spicata*
Bidens pilosa
Biophytum sensitivum
Blumea lacera
Bombax ceiba
Borassus flabelliffer
Butea superba
Caesalpinia crista
Caesalpinia pulcherrima
Calamintha clinopodium
Calendula officinalis
Camellia sinensis
Canna edulis
Cannabis sativa
Capparis deciduas
Cardiospermum halicacabum
Carum carvi
Celastrus paniculatus
Centaurea cyanus
Centratherum anthelminticum
Cheiranthus cheiri
Chenopodium album
Chlorophytum arundinaceum
Cinnamomum zeylanicum
Cissampelos pareira
Citrus aurantium
Citrus medica
Clerodendrum serratum
Cocculus hirsutus
Cochlospermum religiosum
Codonopsis clematidea
Codonopsis ovata
Coffea benghalensis
Cola nitida
Commiphora myrrha

Corchorus capsularis
Corchorus fascicularis
Costus speciosus
Crataeva nurvala
Crocus sativus
Cucurbita maxima
Curcuma caesia
Curcuma domestica
Cymbopogon martinii
Cyperus articulatus
Cyperus esculentus
Cyperus iria
Cyperus rotundus
Cyperus serotinus var. *inundates*
Cyperus stoloniferus
Daucus carota
Delphinium denudatum
Desmotrichum fimbriatum
Digitalis lanata
Digitalis purpurea
Dioclea javanica
Doronicum falconeri
Drypetes roxburghii
Elaeocarpus sphaericus
Elettaria cardamomum
Ephedra gerardiana
Eruca sativa
Erythroxylum coca
Eucalyptus globules
Eupatorium purpureum
Eupatorium triplinerve
Fagonia cretica
Ferula asafoetida
Foeniculum vulgare
Fritillaria roylei
Gardenia jasminoides
Gloriosa superba
Gymnema sylvestre
Hedychium coronarium
Hedyotis corymbosa
Holostemma ada-kodien
Humulus lupulus
Hydrocotyle javanica
Hygrophila auriculata
Hyptis capitata
Inula racemosa
Ipomoea digitata
Jatropha curcas
Juniperus communis
Lactuca serriola
Leonurus cardiaca
Lepidium sativum
Leptadenia reticulata
Leucas aspera
Liquidambar orientalis
Madhuca indica
Maerua arenaria
Magnolia grandiflora
Marsilea minuta
Meconopsis aculeata
Mentha arvensis
Mentha pulegium
Menyanthes trifoliata
Michelia champaca
Moringa oleifera
Mucuna monosperma
Mucuna prurita
Murraya koenigii
Murraya paniculata
Myrica esculenta
Myristica fragrans
Nardostachys jatamansi
Nasturtium officinale
Nepeta cataria
Nepeta hindostana
Nicotiana tabacum
Ocimum gratissimum
Ocimum sanctum
Onosma bracteatum
Ophiorrhiza mungos
Origanum vulgare
Ormocarpum cochinchinensis

Oryza sativa
Panax pseudo – ginseng
Pandanus odoratissimus
Papaver somniferum
Paspalum scrobiculatum
Peucedanum grande
Phaseolus radiatus
Pimenta dioica
Pinus roxburghii
Piper cubeba
Piper longum
Plantago major
Plumbago zeylanica
Podophyllum hexandrum
Polygonum hydropiper
Pongamia pinnata
Populus ciliata
Prangos pabularia
Prunus armeniaca
Punica granatum
Pygmaeopremna herbacea
Ranunculus sceleratus
Rauvolfia vomitora
Rheum emodi
Rhododendron anthopogon
Rhus succedanea
Rosmarinus officinalis
Ruta graveolens
Salvadora persica
Sansevieria trifasciata
Sarcostemma acidum
Sassafras albidum
Saussurea lappa
Saussurea sacra
Schinus terebinthifolius
Scindapsus officinalis
Scutellaria lateriflora
Sissymbriun iria
Solanum melongena
Solanum nigrum
Solanum torvum
Solena amplexicaulis
Sonchus oleraceus
Strychnos nux – vomica
Syzygium cumini
Tephrosia purpurea
Tephrosia virginia
Theobroma cacao
Thevetia peruviana
Thymus serpyllum
Toddalia asiatica
Trachyspermum ammi
Trachyspermum roxburghianum
Tropaeolum majus
Tylophora indica
Urginea indica
Urtica dioica
Valeriana jatamansi
Ventilago madraspatana
Verbena officinalis
Vetiveria zizanioides
Vigna mungo
Vigna trilobata
Viscum album
Vitex negundo
Withania coagulans
Withania somnifera
Woodfordia fruticosa
Wrightia tinctoria
Xanthium strumarium
Zanthoxylum armatum
Zanthoxylum limonella
Zanthoxylum nitidum
Zanthoxylum oxyphyllum
Ziziphus mauritiana

Table 5. Plants Used for Tonic and/or Boosting Memory, Mood Swings

Tonic and Memory Boosters	Tonic, Mood Swings and Memory Boosters	Mood Swings and Memory Boosters
Achyranthes aspera	*Acanthopanax trifoliatus*	*Celastrus paniculatus*
Acorus calamus	*Achillea millefolium*	*Coriandrum sativum*
Argyreia nervosa	*Basella alba*	*Dendrophthoe falcata*
Bacopa monnieri	*Cinnamomum tamala*	*Ferula galbaniflua*
Centella asiatica	*Citrus limon*	*Matricaria chamomilla*
Citrullus vulgaris	*Curcuma zedoaria*	*Passiflora foetida*
Clitoria ternatea	*Embelia ribes*	*Petroselinum crispum*
Convallaria majalis	*Emblica officinalis*	
Convolvulus arvensis	*Ferula narthex*	
Convolvulus microphyllus	*Ginkgo biloba*	
Cordia macleodii	*Lagenaria siceraria*	
Desmodium tiliaefolium	*Majorana hortensis*	
Elephantopus scaber	*Melissa officinalis*	
Evolvulus alsinoides	*Mimusops elengi*	
Ficus benghalensis	*Myrtus communis*	
Hedychium spicatum	*Piper betle*	
Helminthostachys zeylanica	*Prunus amygdalus*	
Hibiscus rosa-sinensis	*Syzygium aromaticum*	
Juglans regia	*Terminalia bellirica*	
Lantana camara	*Ziziphus jujuba*	
Lawsonia inermis		
Limnophila gratioloides		
Linum usitatissimum		
Melia azedarach		
Melissa parviflora		
Nelumbo nucifera		
Picrorhiza kurrooa		
Prunella vulgaris		
Prunus cerasus		
Rhodiola imbricate		
Salvia officinalis		
Semecarpus anacardium		
Tabernaemontana divaricata		
Terminalia chebula		
Tinospora cordifolia		
Tricodesma indicum		
Vitex trifolia		

From the distribution point of view, plant species having scattered or throughout distribution (558 species) predominate in usage for possessing tonic, mood swings and memory boosting properties, followed by N.W. & W. Himalaya (162 species), Peninsular India (Ghats, Coastal Plains, Deccan Plateau; 78 species), Entire Himalaya (77 species), Central Highlands & Plains of India (52 species), N.E. & E. Himalaya (33 species), Miscellaneous (24 species) and Tropical forests of India (16 species) **(Fig.3; Table 6)**. As a matter of fact, the use of botanicals for tonic, mood swings and boosting memory is prevalent in the rural populace/ethnic societies located in various physiographical regions of the country.

Table 6. Distribution Patterns of Indian Botanicals Possessing Tonic, Mood Swing and Memory Boosting Properties

Distribution	**No. of Species**
Throughout or Scattered Distribution	558
N.W. & W. Himalaya	162
Peninsular India (Ghats, Coastal Plains, Deccan Plateau)	78
Entire Himalaya	77
Central Highlands & Plains of India	52
N.E. & E. Himalaya	33
Miscellaneous	24
Tropical Forests of India	16

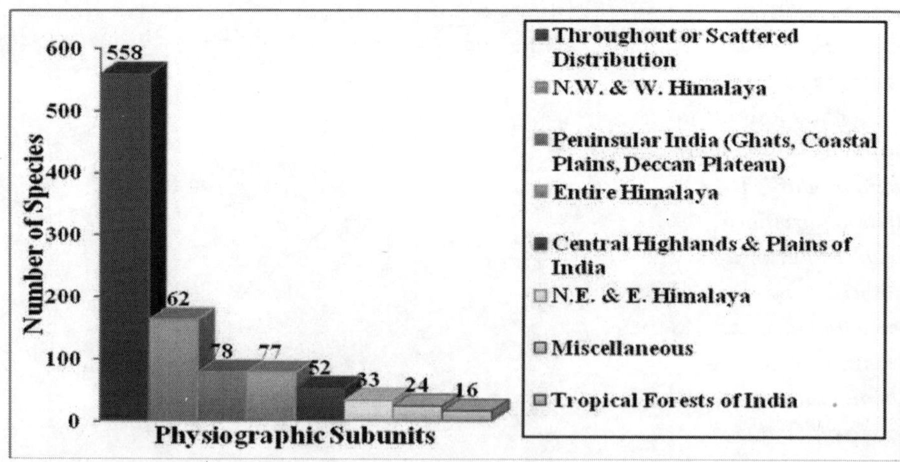

Fig. 3. Histogram Showing Distribution Patterns of Tonic, Mood Swings and Memory Boosting Plants.

Family-wise analysis **(Appendix II)** indicates that the highest number of species amongst the dicotyledonous families **(Fig. 4; Table 7)** belonged to Fabaceae (80 species), followed by Asteraceae (59 species), Lamiaceae (50 species), Apiaceae (35 species), Rubiaceae (27 species), Solanaceae (24 species), Ranunculaceae (23 species), Rosaceae, Rutaceae (22 species each), Cucurbitaceae (18 species), Brassicaceae (17 species), Gentianaceae (16 species), Convolvulaceae, Euphorbiaceae, Malvaceae, Scrophulariaceae (15 species each), Apocynaceae (14 species), Acanthaceae, Asclepiadaceae, Verbenaceae (13 each), Polygonaceae (12 species), Menispermaceae (10 species), etc. whereas these for monocots **(Fig. 5; Table 8)** are: Poaceae (23 species), Liliaceae (20 species), Orchidaceae (19 species), Zingiberaceae (18 species), Arecaceae, Cyperaceae (9 species each), Araceae (8 species), etc. Comparatively, relative % in taxa of the predominant dicotyledonous family, Fabaceae is 8% and for monocots (Poaceae) is 2.3% **(Tables 7, 8)**.

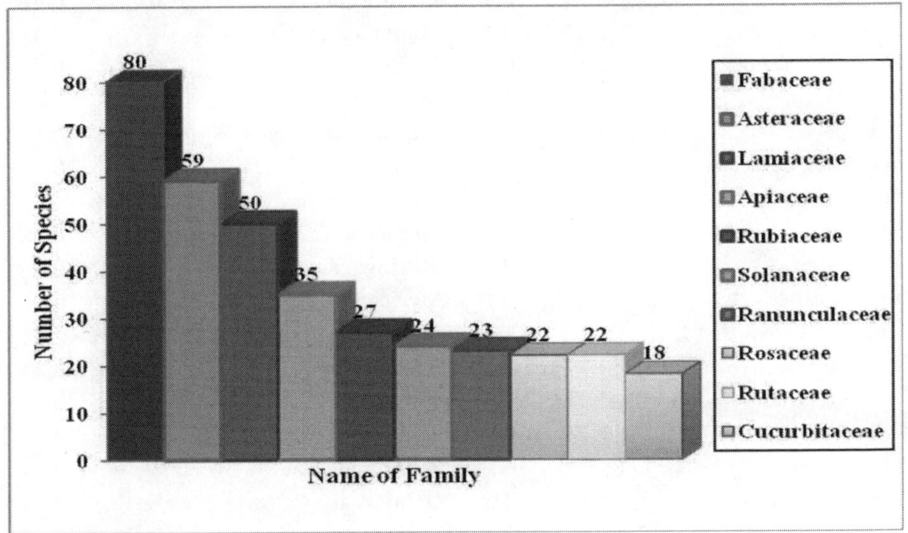

Fig. 4. Predominant Dicotyledonous Families of Mood Swings, Tonic and Memory Boosters.

Table 7. Predominant Dicot Families of Mood Swings, Tonic and Memory Boosters

Name of Family	Total No. of Species	%age of Taxa	Name of Family	Total No. of Species	%age of Taxa
Acanthaceae	13	1.3	Aristolochiaceae	4	0.4
Aizoaceae	1	0.1	Asclepiadaceae	13	1.3
Alangiaceae	1	0.1	Asteraceae	59	5.9
Amaranthaceae	4	0.4	Balsaminaceae	1	0.1
Anacardiaceae	8	0.8	Basellaceae	1	0.1
Annonaceae	1	0.1	Berberidaceae	6	0.6
Apiaceae	35	3.5	Betulaceae	3	0.3
Apocynaceae	14	1.4	Bignoniaceae	5	0.5
Aquifoliaceae	1	0.1	Bixaceae	1	0.1
Araliaceae	6	0.6	Bombacaceae	3	0.3

Name of Family	Total No. of Species	%age of Taxa	Name of Family	Total No. of Species	%age of Taxa
Boraginaceae	6	0.6	Lamiaceae	50	5
Brassicaceae	17	1.7	Lauraceae	9	0.9
Burseraceae	3	0.3	Lecythidaceae	1	0.1
Campanulaceae	4	0.4	Linaceae	1	0.1
Cannabinaceae	1	0.1	Loganiaceae	5	0.5
Capparaceae	7	0.7	Loranthaceae	1	0.1
Caprifoliaceae	4	0.4	Lythraceae	5	0.5
Caricaceae	1	0.1	Magnoliaceae	4	0.4
Caryophyllaceae	1	0.1	Malpighiaceae	1	0.1
Celastraceae	3	0.3	Malvaceae	15	1.5
Chenopodiaceae	2	0.2	Melastomataceae	1	0.1
Chloranthaceae	1	0.1	Meliaceae	9	0.9
Clusiaceae	1	0.1	Menispermaceae	10	1
Cochlospermaceae	1	0.1	Moraceae	9	0.9
Combretaceae	9	0.9	Moringaceae	1	0.1
Connaraceae	1	0.1	Myricaceae	1	0.1
Convolvulaceae	15	1.5	Myristicaceae	1	0.1
Crassulaceae	2	0.2	Myrsinaceae	3	0.3
Cucurbitaceae	18	1.8	Myrtaceae	7	0.7
Dilloniaceae	1	0.1	Nyctaginaceae	1	0.1
Dipsacaceae	1	0.1	Nymphaeaceae	5	0.5
Dipterocarpaceae	2	0.2	Ochnaceae	2	0.2
Droseraceae	1	0.1	Oleaceae	7	0.7
Ebenaceae	1	0.1	Onagraceae	1	0.1
Elaeagnaceae	3	0.3	Oxalidaceae	3	0.3
Elaeocarpaceae	1	0.1	Paeoniaceae	1	0.1
Ericaceae	7	0.7	Pandanaceae	1	0.1
Erythroxylaceae	2	0.2	Papaveraceae	8	0.8
Euphorbiaceae	15	1.5	Passifloraceae	2	0.2
Fabaceae	80	8	Pedaliaceae	2	0.2
Fagaceae	2	0.2	Phytolaccaceae	1	0.1
Flacourtiaceae	2	0.2	Piperaceae	8	0.8
Fumariaceae	1	0.1	Pittosporaceae	1	0.1
Gentianaceae	16	1.6	Plantaginaceae	2	0.2
Hamamelidaceae	1	0.1	Plumbaginaceae	2	0.2
Hippocastanaceae	1	0.1	Polemoniaceae	1	0.1
Hydrangeaceae	1	0.1	Polygonaceae	12	1.2
Hypericaceae	3	0.3	Primulaceae	2	0.2
Juglandaceae	2	0.2	Proteaceae	1	0.1
Krameriaceae	2	0.2	Punicaceae	1	0.1

Name of Family	Total No. of Species	%age of Taxa	Name of Family	Total No. of Species	%age of Taxa
Ranunculaceae	23	2.3	Sterculiaceae	6	0.6
Rhamnaceae	6	0.6	Symplocaeae	2	0.2
Rosaceae	22	2.2	Theaceae	1	0.1
Rubiaceae	27	2.7	Thymelaeaceae	2	0.2
Rutaceae	22	2.2	Tiliaceae	8	0.8
Salicaceae	5	0.5	Trapaceae	1	0.1
Salvadoraceae	1	0.1	Tropaeolaceae	1	0.1
Samydaceae	1	0.1	Urticaceae	4	0.4
Santalaceae	1	0.1	Vacciniaceae	1	0.1
Sapindaceae	8	0.8	Valerianaceae	6	0.6
Sapotaceae	4	0.4	Verbenaceae	13	1.3
Saururaceae	1	0.1	Violaceae	2	0.2
Saxifragaceae	4	0.4	Viscaceae	1	0.1
Scrophulariaceae	15	1.5	Vitaceae	2	0.2
Simaroubaceae	4	0.4	Zygophyllaceae	3	0.3
Solanaceae	24	2.4			

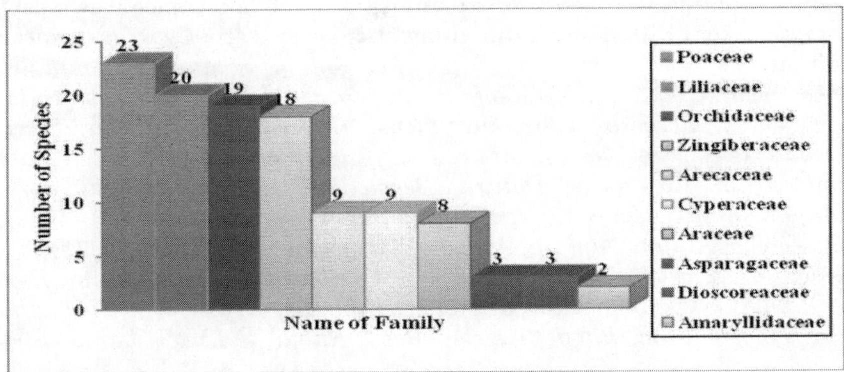

Fig. 5. Predominant Monocotyledonous Families of Mood Swings, Tonic and Memory Boosters.

Table 8. Predominant Monocot Families of Mood Swings, Tonic and Memory Boosters

Name of Family	Total No. of Species	%age of Taxa
Poaceae	23	2.3
Liliaceae	20	2
Orchidaceae	19	1.9
Zingiberaceae	18	1.8
Arecaceae	9	0.9
Cyperaceae	9	0.9

Name of Family	Total No. of Species	%age of Taxa
Araceae	8	0.8
Asparagaceae	3	0.3
Dioscoreaceae	3	0.3
Amaryllidaceae	2	0.2
Alismataceae	1	0.1
Bambusaceae	1	0.1
Bromeliaceae	1	0.1
Cannaceae	1	0.1
Commelinaceae	1	0.1
Convallariaceae	1	0.1
Costaceae	1	0.1
Hypoxidaceae	1	0.1
Iridaceae	1	0.1
Marantaceae	1	0.1
Musaceae	1	0.1
Pontederiaceae	1	0.1
Typhaceae	1	0.1

By and large, the most promising genera of tonic, mood swings and memory boosters with a maximum number of species are: *Cyperus* (9 species), *Aconitum* (8 species), *Artemisia*, *Piper*, *Polygonum*, *Solanum*, *Terminalia* (7 species each), *Gentiana*, *Indigofera*, *Ipomoea*, *Mentha*, *Prunus*, *Sida* (6 species each), *Berberis*, *Curcuma*, *Ficus*, *Habenaria*, *Potentilla*, *Swertia*, *Valeriana* (5 species each), *Adiantum*, *Capparis*, *Cinnamomum*, *Citrus*, *Clerodendrum*, *Corchorus*, *Datura*, *Desmodium*, *Doronicum*, *Eupatorium*, *Ferula*, *Jasminum*, *Leucas*, *Nepeta*, *Ocimum*, *Vigna*, *Zanthoxylum* (4 species each), *Acacia*, *Allium*, *Alpinia*, *Aristolochia*, *Asparagus*, *Bauhinia*, *Blepharis*, *Brassica*, *Cassia*, *Cinchona*, *Codonopsis*, *Cymbopogon*, *Dendrobium*, *Dioscorea*, *Eryngium*, *Euphorbia*, *Hedyotis*, *Hibiscus*, *Hypericum*, *Nymphaea*, *Panax*, *Polygonatum*, *Rhododendron*, *Rosa*, *Rubia*, *Salix*, *Sesbania*, *Strychnos*, *Urtica*, *Zingiber*, *Ziziphus* (3 species each) **(Fig 6; Table 9)**. All these species hold a great potential for drug discovery programme on the listed aspects.

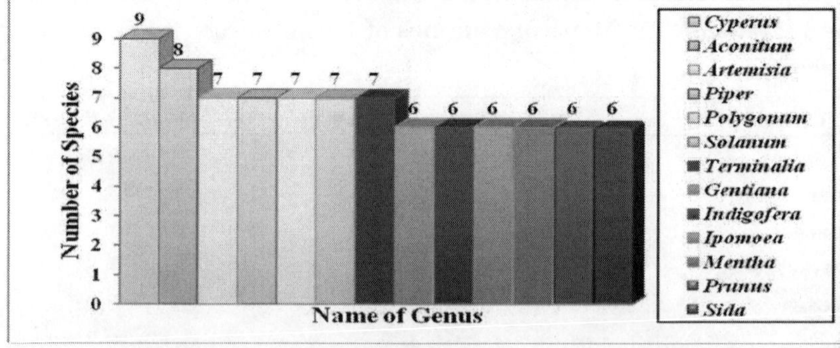

Fig. 6. Predominant Genera of Mood Swings, Tonic and Memory Boosters.

Table 9. Predominant Genera of Mood Swings, Tonic and Memory Boosters

Name of Genus	Number of Species	Name of Genus	Number of Species
Cyperus	9	*Ocimum*	4
Aconitum	8	*Vigna*	4
Artemisia	7	*Zanthoxylum*	4
Piper	7	*Acacia*	3
Polygonum	7	*Allium*	3
Solanum	7	*Alpinia*	3
Terminalia	7	*Aristolochia*	3
Gentiana	6	*Asparagus*	3
Indigofera	6	*Bauhinia*	3
Ipomea	6	*Blepharis*	3
Mentha	6	*Brassica*	3
Prunus	6	*Cassia*	3
Sida	6	*Cinchona*	3
Berberis	5	*Codonopsis*	3
Curcuma	5	*Cymbopogon*	3
Ficus	5	*Dendrobium*	3
Habenaria	5	*Dioscorea*	3
Potentilla	5	*Eryngium*	3
Swertia	5	*Euphorbia*	3
Valeriana	5	*Hedyotis*	3
Adiantum	4	*Hibiscus*	3
Capparis	4	*Hypericum*	3
Cinnamomum	4	*Nymphaea*	3
Citrus	4	*Panax*	3
Clerodendrum	4	*Polygonatum*	3
Corchorus	4	*Rhododendron*	3
Datura	4	*Rosa*	3
Desmodium	4	*Rubia*	3
Doronicum	4	*Salix*	3
Eupatorium	4	*Sesbania*	3
Ferula	4	*Strychnos*	3
Jasminum	4	*Urtica*	3
Leucas	4	*Zingiber*	3
Nepeta	4	*Ziziphus*	3

Organ-wise analysis revealed that the most predominantly employed uses of plant part/s for tonic purposes are met largely from roots (216 species), followed by whole plant (191 species), fruit (146 species), leaf (138 species), bark (112 species), stem (109 species), seed (105 species), inflorescence (64 species), aerial parts (19 species), wood (11 species) **(Fig.7; Appendix III)**; whereas those for mood swings are from whole plant (102 species), roots (95 species), leaf (83 species), seed (72 species), fruit (67 species), stem (57 species), inflorescence (41 species), bark (35 species), aerial

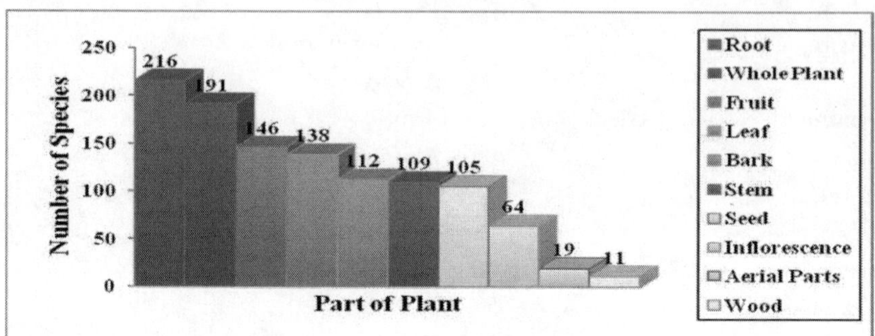

Fig. 7. Predominant Plant Parts Used for Tonic Purposes.

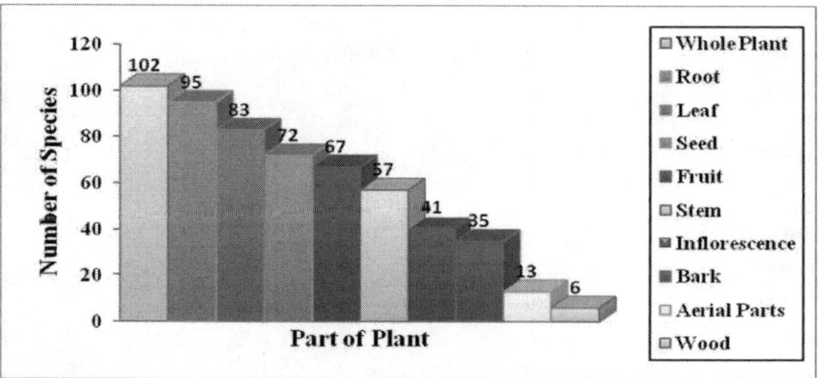

Fig. 8. Predominant Plant Parts Used for Mood Swings.

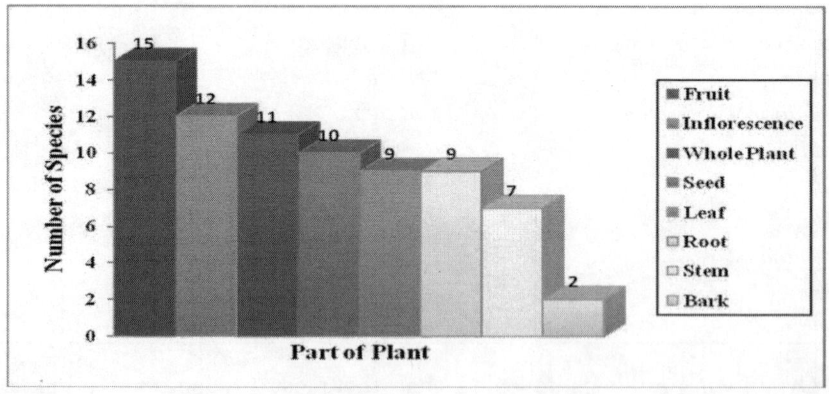

Fig. 9. Predominant Plant Parts Used for Boosting Memory.

parts (13 species), wood (6 species) **(Fig. 8; Appendix IV)**; and for boosting memory from fruits (15 species), inflorescence (12 species), whole plant (11 species), seed (10 species), leaf, root (9 species each), stem (7 species), bark (2 species) **(Fig. 9; Appendix V)**.

Hopefully, the generated data on botanicals for tonic, mood swings and memory boosting should be screened on scientific lines for necessary validation of the claims by the clinical trials so as to exploit the traditional knowledge for the ultimate benefit of living organism in particular and the society at large.

Literature Cited

Abraham, Z. 1997. Further ethnobotanical study of the Todas and Kotas of the Nilgiris: 249-254. In Jain, S. K. (*ed.*): *Contribution to Indian Ethnobotany*, Vol. 3. Scientific Publishers (India), Jodhpur.

Agarwal, V.S. 1985. *Drug Plants of Indian: Root Drugs*. Kalyani Publ., Delhi (India).

Agarwal, V.S. 1986. *Economic Plants of India*. Bishen Singh Mahendra Pal Singh, Dehradun (India).

Aggrawal, A. 1995. *Narcotic Drugs*. National Book Trust, New Delhi (India).

Ahuja, B.S. 1965. *Medicinal Plants of Saharanpur*. Survey of Medicinal Plants, Central Council of Ayurvedic Research Gurukul Kangri Vishwavidyalya, Hardwar.

Ainslie, W. 1813. *Materia Medica of Hindoostan*. Govt. Press Madras (India).

Ainslie, W. 1926. *Materia Medica*. Neeraj Publishing House, Delhi.

Alagesaboopathi, C., Dwarkan, P. & Balu, S. 2000. Plants used as medicine by tribals of Shevaroy Hill, Tamil Nadu: 391-393. In Maheshwari, J.K. (*eds.*): *Ethnobotany and Medicinal Plants of Indian Subcontinent*. Scientific Publishers (India), Jodhpur.

Aman, N. & Dar, G.H. 2003. Medicinal plants in Scrophulariaceae from Kashmir Himalaya: 914-921. In Singh, V. & Jain, A.P. (*eds.*): *Ethnobotany and Medicinal Plants of India and Nepal*, Vol. II. Scientific Publishers (India), Jodhpur.

Ambasta, S.P. 1986. *The Useful Plants of India*. C.S.I.R., New Delhi (India).

Ambasta, S.P., Ramachandran, K., Kashyapa, K. & Chand, R. 1992. *The Useful Plants of India*. P.I.D., C.S.I.R., New Delhi.

Anonymous, 1948-1976a. The Wealth of India (A Dictionary of Indian Raw Materials and Industrial Products), Vol. I-VIII, C.S.I.R., New Delhi (India).

Anonymous, 1976b. *Botanical Exploration Tawang*. R.R.L., Jorhat.

Anonymous, 1986. All India Cordinated Research Project Ethnobiology, Eastern Circle, Shillong, Annual Report. Ministry of Environment and Forests, India.

Anonymous, 1981. *Medicinal Plants of Kerala Forest*. Kerala Forest Research Institute, Peechi, Kerala.

Anonymous, 1999. *Medicinal Plants of Andhra Pradesh*. CCRUM, New Delhi.

Ansarali, K.C. & Sivdasan, M. 2009. Ethnobotanical investigations in Lakshadweep Islands, India. *Ethnobotany* ***21***: 18-24.

Arya, K.R. 2004. Traditionally potential plants of Western Himalaya for IRR value and utility of biotechnological tools for their conservation: 41-53. In Pande, P.C. & Joshi, N.K. (*eds.*): *Traditions and Traditional Knowledge of Central Himalayas*. Bishen Singh Mahendra Pal Singh, Dehradun (India).

Asolkar, L.V., Kakkar, K.K. & Chakre, O.J. 1992. *Second Supplement to Glossary of Indian Medicinal Plants with Active Principle*, Part I (A-K) (1965-81). C.S.I.R., New Delhi (India).

Aswal, B.S. & Mehrotra, B.N. 1994. *Flora of Lahaul Spiti (A Cold Desert in North West Himalaya)*. Bishen Singh Mahendra Pal Singh, Dehradun (India).

Aswal, B.S. 1996. Conservation of ethno-medicinal plants diversity of Garhwal Himalaya in India: 133-135. In Jain, S. K. (*ed.*): *Ethnobiology in Human Welfare*. Deep Publ., New Delhi.

Atal, C.K. 1982. *Cultivation and Utilization of Medicinal Plants*. R.R.L. (CSIR) Jammu- Tawi (India).

Atkinson, E.T. 1982. *Economic Botany of Himalayan Region*. Cosmos Publ., New Delhi.

Balasubramanian, P. & Prasad, S.N. 1996. Ethnobotany and conservation of medicinal plants by Irulas of Nilgiri Biosphere Reserve: 271-273. In Jain, S.K. (*ed.*): *Ethnobiology in Human Welfare*. Deep Publ., New Delhi.

Ballabh, B. & Chaurasia, O.P. 2006. Ethnobotanical studies on Boto tribe in Ladakh. *Ethnobotany **18***: 87-95.

Banerjee, A. 1996. Medicinal uses of some flowers by Santhals of West Bengal: 314-317. In Maheshwari, J.K. (*ed.*): *Ethnobotany in South Asia*. Scientific Publishers (India), Jodhpur.

Banerjee, A. 2000a. Ethnobotany of a few plant species in the eroded soil of Birbhum, West Bengal: 527-530. In Maheshwari, J.K. (*ed.*): *Ethnobotany and Medicinal Plants of Indian Subcontinent*. Scientific Publishers (India), Jodhpur.

Banerjee, A. 2000b. Ethnobotany of some trees in the Santhal villages of Birbhum, West Bengal: 531-534. In Maheshwari, J.K. (*ed.*): *Ethnobotany and Medicinal Plants of Indian Subcontinent*. Scientific Publishers (India), Jodhpur.

Banerjee, R.N. & Ghora, C. 1996. On the domestic use of some unreported plants of West Dinajpur district West Bengal: 325-328. In Maheshwari, J.K. (*ed.*): *Ethnobotany in South Asia*. Scientific Publishers (India), Jodhpur.

Bannerman, R. 1982. Traditional medicine in modern health care. *World Health Forum **3(1)***: 8-13.

Barua, P. & Sarma, G.C. 1984. Studies on the medicinal uses of plants by the Northeast tribes-III. *J. Econ. Taxon. Bot. **11***: 71-76.

Bedi, S.J. 1978. Ethnobotany of the Ratan Mahal Hills, Gujarat, India. *Econ. Bot. **32***: 278-284.

Behura, S. & Rout, N.C. 2003. Diversity of Zingiberaceae plants from Orissa: 546-550. In Singh, V. & Jain, A.P. (*eds.*): *Ethnobotany and Medicinal Plants of India and Nepal*, Vol. II. Scientific Publishers (India), Jodhpur.

Bhalla, N.P. & Bhalla, S. 2002. Ethnomedicinal and ecophysiological studies in Fabaceae (Leguminosae): 255-278. In Trivedi, P.C. (*ed.*): *Ethnobotany*. Aavishkar Publ., Distributors, Jaipur (India).

Bhalla, N.P., Sahu, T.R., Mishra, G.P. & Dakwale, R.N. 1982. Traditional plant medicines of Sagar Distt. Madhya Pradesh, India. *J. Econ. Taxon. Bot. **3***: 23-32.

Bhalla, S., Patel, J.R. & Bhalla, N.P. 1996. Ethnomedicinal observations on some Asteraceae of Bundelkhand region, Madhya Pradesh: 175-178. In Maheshwari, J.K. (*ed.*): *Ethnobotany in South Asia*. Scientific Publishers (India), Jodhpur.

Bhandary, M.J., Chandrashekar, K.R. & Kaverippa, K.M. 1996. Ethnobotany of Gowalis of Uttara Kannada district, Karnataka: 244-249. In Maheshwari, J.K. (*ed.*): *Ethnobotany in South Asia*. Scientific Publishers (India), Jodhpur.

Bharkatiya, M., Goyal, A., Gupta, G.D. & Gaud, R.S. 2006. Memory enhancers. *Indian Pharmacist.* ***5(8)***: 32-36.

Bhatt, D.C. 2002. Studies on some ethnobotanical plants from Saurashtra: 119-127. In Trivedi, P.C. (*ed.*): *Ethnobotany*. Aavishkar Publishers, Distributors, Jaipur (India).

Bhatt, D.C., Mehta, M.R., Mehta, S.K. & Parmar, R.P. 2002. Studies on some ethnomedicinal plants from Talja taluka of Bhavnagar district, Gujarat: 295-310. In Trivedi, P.C. (*ed.*): *Ethnobotany*. Aavishkar Publ., Jaipur (India).

Bhatt, D.C., Mehta, S.K., Nurani, M.K. & Baxi, U.S. 2003. Studies on some ethnobotanical plants of Barda hills from Saurashtra: 221-227. In Singh, V. & Jain, A. P. (*ed.*): *Ethnobotany and Medicinal Plants of India and Nepal*, Vol. I. Scientific Publishers (India), Jodhpur.

Bhatt, D.C. & Mitaliya, K. D. 1999. Ethnomedicinal plants of Victoria park (reserved forest) of Bhavnagar, Gujarat, India. *Ethnobotany* ***11***: 81-84.

Bhatt, U.D. & Jain, B.K. 2009. Ethnomedicinal uses of some important plants recorded from forest areas of Shoolpaneshwar Wildlife Sanctury, Narmada district (Gujarat). *Ethnobotany* ***21***: 70-74.

Bhattacharjee, S.K, 2001. *Handbook of Medicinal Plants.* Pointer Publ., Jaipur (India).

Bhattacharya, G. 1996. Medico-ethno-botanical value of Sourashtra weeds: 166-168. In Maheshwari, J.K. (*ed.*) *Ethnobotany in South Asia.* Scientific Publishers (India), Jodhpur.

Bhattacharya, S.K., Ray, A.B. & Dutta, S.C. 1975. Psychopharmacological investigations of the 4-methoxyindole alkaloids of *Alstonia venenata. Planta Med.* (Stultg). ***27***: 164-170.

Bhogaonkar, P.Y. & Devarkar, V.D. 2002. Some unique ethnomedicine plants of Korkus of Melghat Tiger Reserve (Maharashtra). *Ethnobotany* ***14***: 16-19.

Bhowmik, D., Chiranjib, Tiwari, P., Tripathi, K.K. & Sampath Kumar, K.P. 2010. Traditional Indian memory enhancer herbs and their medicinal importance. *Ann. Biol. Res.* ***1(1)***: 41-46.

Binu, S., Shanavaskhan, A.E., Santosh Kumar, E.S. & Pushpangadan, P. 2003. Plants used as medicine by the Irulas of Palghat district, Kerela, India: 808-814. In Singh, V. & Jain, A.P. (*eds.*): *Ethnobotany and Medicinal Plants of India and Nepal,* Vol. II. Scientific Publishers (India), Jodhpur.

Biswas, K. & Chopra, R.N. 1982. *Common Medicinal Plants of Darjeeling and Sikkim Himalayas.* Periodical Expert Book Agency, Delhi.

Bose, K.C. 1932. Pharmacopoeia Indica. Being A Collection of Vegetable, Mineral and Animal Drugs in Common Use in India. Bishen Singh Mahendra Pal Singh, Dehra Dun (India).

Brahmam, M. & Saxena, H.O. 1990. Ethnobotany of Gandhamardan Hills-some noteworthy folk-medicinal uses. *Ethnobotany* ***2***:71-79.

Caius, J.F. 2003 (ed.). The Medicinal and Aromatic Poisonous Plants of India. Scientific Publishers (India), Jodhpur.

Chatterjee, A. & Pakrashi, S.C. 1997. *The Treatise on Indian Medicinal Plants.* C.S.I.R., New Delhi (India).

Chaudhari, U.S. & Hutke, V. 2002. Ethno-medico-botanical information on some plants used by Melghat tribes of Amravati district, Maharashtra. *Ethnobotany* ***14***: 100-102.

Chaudhary, D. & Neogi, B. 2003. Ethnobotany of Khasi and Chakma tribes of North East India: 583-589. In Maheshwari, J.K. (*ed.*): *Ethnobotany and Medicinal Plants of Indian Subcontinent*, Vol. I. Scientific Publishers (India), Jodhpur.

Chauhan, D., Navneet & Prabhat. 2007. Medicinal plants of Rigveda: 281-298. In Chopra, A.K., Khanna, D.R., Prasad, G., Malik, D.S. & Bhutiani, R. (*eds.*): *Medicinal Plants: Conservation, Cultivation and Utilization*. Daya Publ. House, Delhi (India).

Chauhan, N.S. 1999. *Medicinal and Aromatic Plants of Himachal Pradesh*. Indus Publ. House, New Delhi.

Chauhan, N.S. 2001. Domestication of selected medicinal plants of Himachal Pradesh: 285-315. In Samant, S.S., Dhar, U. & Palni, M.S. (*eds.*): *Himalayan Medicinal Plants: Potential and Prospects*. Gyanodaya Prakashan, Nainital.

Chaurasia, O.P., Singh, B. & Attrey, D.P. 2003. Ethnobotany of Chongthang valley-Ladakh Himalaya: 108-113. In Singh, V. & Jain, A.P. (*eds.*): *Ethnobotany and Medicinal Plants of India and Nepal*, Vol. I. Scientific Publishers (India), Jodhpur.

Chaurasia, O.P., Singh, B. & Sareen, S.K. 2000. Ethno-medico-botanical survey of Nubra valley: 167-172. In Maheshwari, J.K. (*ed.*): *Ethnobotany and Medicinal Plants of Indian Subcontinent*. Scientific Publishers (India), Jodhpur.

Chhetri, D.R. 2005. Ethnomedicinal plants of Khangchendzonga National Park, Sikkim, India. *Ethnobotany* ***17***: 96-103.

Chhetri, R.B. 1994. Further observations of ethnomedicobotany of Khasi hills in Meghalaya, India: *Ethnobotany* ***6***: 33-36.

Chhetri, R.B. & Shrestha, R. 2004. Ethnobotany of some weeds of winter crops in Dhulikhel, Nepal. *Ethnobotany* ***16***: 108-112.

Chopra, I.C. 1960. *Medicinal Plants of Arid Zone*. Today & Tomorrow's Printers & Publ., Delhi.

Chopra, R.N. 1932. Medicinal and Economic Aspects of Some Indian Medicinal Plants. Patna Univ., Patna.

Chopra, R.N. 1933. *Chopra's Indigenous Drugs of India* (2nd ed. Revised and Enlarged by R. N. Chopra, I.C. Chopra, K.L. Handa and L.D. Kapur in 1958). Dhur & Sons Pvt. Ltd., Calcutta (India).

Chopra, R.N. and Chopra, I.C. 1955. *A Review of Work on Indian Medicinal Plants*. ICMR. New Delhi.

Chopra, R.N., Chopra, I.C. & Verma, B.S. 1974. *Supplement to Glossary of Indian Medicinal Plants*. Publications & Information Directorate Hillside Road, New Delhi.

Chopra, R.N., Nayar, S.L. & Chopra I.C. 1956. *Glossary of Indian Medicinal Plants*. C.S.I.R., New Delhi.

Dandiya, P.C. 1990. The pharmacological basis of herbal drug acting on CNS. *East Pharm.* ***33(396)***: 39-47.

Das, H.S., Panda, P.C. & Patnaik, S.N. 1996. Traditional uses of wetland plants of eastern Orissa: 306-313. In Maheshwari, J.K. (*ed.*): *Ethnobotany in South Asia*. Sci. Publ. Jodhpur, India.

Dastur, J.F. 1970. *The Medicinal Plants of India and Pakistan*. D.B. Toporowala & Sons, Bombay.

Datta, S.C. 1949. *Pharmacognosy of Indian Root and Rhizome Drugs*. Central Council for Research in Ayurveda and Siddha, New Delhi (India).

Datta, S.C. & Mukerji, B. 1952. *Pharmacognosy of Indian Leaf Drugs*. Ministry of Health, Govt. of India, Pharmacogo. Lab., Calcutta (India).

Datta, S.C., Bhattacharya, S.K. & Ray, A.B. 1976. Flower alkaloids of *Alstonia scholaris*. *Planta Medica* ***30***: 86-89.

Deshmukh, V.R., Muratkar, G.D. & Rothe, S.P. 2000. Preliminary observation on the medicinal and economically important leguminous plant species from Amravati Tehsil: 283-289. In Maheshwari, J.K. (*ed.*): *Ethnobotany and Medicinal Plants of Indian Subcontinent*. Scientific Publishers (India), Jodhpur.

Dey, K.L. 1896. *The Indigenous Drugs of India*. Thacker Spink, Calcutta (India).

Dhamija, H.K., Parashar, B. & Singh, J. 2011. Anti-depression potential of herbal drugs: an overview. *J. Chem. Pharm. Res.* ***3(5)***: 725-735.

Dhar, U., Samant, S.S., Rawal, R.S. & Sharma, S. 1997. Studies on biota and resource use patterns of the natives in Askot wildlife sanctuary of Kumaon Himalaya, India. *Tiger Paper* ***24***: 12-18.

Dhar, U., Samant, S.S., Rawal, R.S. & Sharma, S. 1998. Wild plant resources and inhabitants of Askot wildlife sanctuary of Kumaon in Indian Himalaya: Conservation Issues: 128-142. In Kotwal, P.C. & Banerjee, S. (*eds.*): *Biodiversity Conservation in Managed and Protected Areas*. Agro Botanica, Bikaner.

Dhasmana, M. & Rathore, R.K.S. 2007. Variability measurements in three wild collections of *Solanum nigrum* L. complex: 65-70. In Chopra, A.K., Khanna, D.R., Prasad, G., Malik, D.S. & Bhutiani, R. (*eds.*): *Medicinal Plants: Conservation, Cultivation and Utilization*. Daya Publ. House, Delhi.

Dhawan, B.N. 1986. *Current Research on Medicinal Plants*. Indian Natl. Sci. Acad., Delhi.

Dhiman, A.K. 2005. *Wild Medicinal Plants of India (With Ethnomedicinal Uses)*. Bishen Singh Mahendra Pal Singh, Dehra Dun (India).

Dhiman, A.K. 2007. Conservation of some useful medicinal plants of Haridwar district in Uttaranchal state: 147-166. In Chopra, A.K., Khanna, D.R., Prasad, G., Malik, D.S. & Bhutiani, R. (*eds.*): *Medicinal Plants: Conservation, Cultivation and Utilization*. Daya Publ. House, Delhi (India).

Dhingra, D. & Sharma, A. 2006. A review on antidepressant plants. *Natural Product Rad.* ***5(2)***: 144-152.

Dixit, R.D. & Kumar, R. 2003. Plants used by local people in human welfare: 53-59. In Singh, V. & Jain, A.P. (*eds.*): *Ethnobotany and Medicinal Plants of India and Nepal*. Vol. I. Scientific Publishers (India), Jodhpur.

Dubey, G.P., Pathak, S.R. & Gupta, B.S. 1994. Combined effect of Brahmi (*Bacopa monnieri*) and shankhpushpi (*Convolvulus pluricaulis*) on cognitive functions. *Pharmacopsychoecol* ***7(3)***: 249-251.

Dutt, N.B. 1928. *Commercial Drugs of India*. Bishen Singh Mahendra Pal Singh, Dehra Dun.

Dutta, A.C. 1985. Dictionary of Economic and Medicinal Plants.

Dutta, U.C. 1877. *The Materia Medica of the Hindus*. Thacker Spink, Calcutta (India).

Dwivedi, A.P. 2007. *Forests: The Non-Wood Resources*. Intl. Book Distributors, Dehradun, India.

Dwivedi, S.N. 2000. Traditional health care among tribals of Rewa district of Madhya Pradesh with special reference to conservation of endangered and vulnerable species: 315-320. In Maheshwari, J.K. (*ed.*): *Ethnobotany and Medicinal Plants of Indian Subcontinent*. Scientific Publishers (India), Jodhpur.

Dwivedi, S.N. 2003. Ethnobotanical studies and conservational strategies of wild and natural resources of Rewa district of Madhya Pradesh: 233-244. In Singh, V.P. & Jain, A.P. (*eds.*): *Ethnobotany and Medicinal Plants of India and Nepal,* Vol. 1. Scientific Publishers (India), Jodhpur.

Dymock, J.H., Warden, J.H. and Hooper, D. 1889-1893. *Pharmacographica Indica 2*: 423. Trubner Co. Ltd., London (U.K.).

Dymock, W. 1883. *The Vegetable Materia Medica of Western India.* Edu. Society's Press, Bombay (India).

Emboden, W.A. 1979. *Narcotic Plants* (Ed. 2). McMilan Publ. Co., New York.

Farnsworth, N.R. & Soejarto, D.D. 1991. Global importance of medicinal plants: 29-35. In Akrele, O., Heywood, V., Synge, H. (*eds.*): *Conservation of Medicinal Plants.* Cambridge Univ. Press, New York.

Farooq, S. 2005. *555 Medicinal Plants: Field and Laboratory Manual.* Intl. Book Distributors, Dehradun (India).

Farooqi, S.S. 1993. *Production Technology of Medicinal and Aromatic Crops.* Nat. Remedies Pvt. Ltd., Bangalore (India).

Flemming, J. 1810. A catalogue of Indian medicinal plants and drugs. *Asit. Res. 11*: 153-189.

Fluckiger, F.A. and Handburg, D. 1874. *Pharmacographia.* McMillan & Co. London.

Furst, P.T. (*ed.*) 1972. *Flesh of the God.* Praeger, New York.

Furst, P.T. 1976. *Hallucinogens and Culture.* Chandler & Sharp, San Francisco.

Gajurel, P.R., Rethy, P., Singh, B. & Angami, A. 2006. Ethnobotanical studies on Adi tribes in Dehang Debang Biosphere Reserve in Arunachal Pradesh, Eastern Himalaya. *Ethnobotany 18*: 114-118.

Gaur, R.D., Semwal, J.K. & Tiwari, J.K. 1983. A survey of high altitude medicinal plants of Garhwal Himalaya. *Bull. Medico-Ethnobot. Res. 4*: 102-116.

Gautam, A., Dalal, S. & Bisht, G.R.S. 2007. Studies on *in vitro* antimicrobial activity of essential oil of the *Nardostachys jatamansi* and *Zanthoxylum armatum*: 341-356. In Chopra, A.K., Khanna, D.R., Prasad, G., Malik, D.S. & Bhutiani, R. (*eds.*): *Medicinal Plants*: *Conservation, Cultivation and Utilization.* Daya Publ. House, Delhi (India).

Girach, R.D. & Aminuddin. 1995. Ethnomedicinal uses of plants among the tribals of Singhbum district, Bihar, India. *Ethnobotany 7*: 103-107.

Gogoi, R., Bokolial, D. & Hazarika, D. 2003. Preliminary observation of medicinal plants of Chandrapur area of Kamrup district, Assam: 410-415. In Singh, V. & Jain, A.P. (*eds.*): *Ethnobotany and Medicinal Plants of India and Nepal.* Vol. I. Scientific Publishers (India), Jodhpur, India.

Gogoi, R. & Das, M.K. 2003. Observations on some weeds of medicinal importance in Brahmputra valley of Assam: 434-441. In Singh, V. & Jain, A.P. (*eds.*): *Ethnobotany and Medicinal Plants of India and Nepal.* Vol. I. Scientific Publishers (India), Jodhpur.

Gopal, G.V. & Chandra, K.S.J. 2003. Ethnobotany of Soligas of Biligirirangana hills, Karnataka, India: 640-647. In Singh, V. & Jain, A.P. (*eds.*): *Ethnobotany and Medicinal Plants of India and Nepal,* Vol. II. Scientific Publishers (India), Jodhpur.

Gopal, G.V. & Sinha, M.P. 2000. Medicinal uses of various beads used in rosaries: 131-133. In Maheshwari, J.K. (*ed.*): *Ethnobotany and Medicinal Plants of Indian Subcontinent.* Scientific Publishers (India), Jodhpur.

Goyal, H.R. 1999. *Medicinal Plants of Kachchh (Gujarat).* CCRAS, New Delhi.

Gupta, R. 1997. Plants in folk medicine of the Himalaya: 55-60. In Jain, S.K. (*ed.*): *Contribution to Indian Ethnobotany*. Scientific Publishers (India), Jodhpur.

Hajra, P.K. & Baishya, A.K. 1981. Ethnobotanical notes on the Miris (Mishing) of Assam Plains: 161-169. In Jain S.K. (*ed.*): *Glimpses of Indian Ethnobotany*. Oxford and IBH Publ. Co., New Delhi, India.

Harborne, J.B. & Baxter, H. 2001. *Chemical Dictionary of Economic Plants*. John Wiley and Sons Ltd, Chichester, New York.

Haridason, K., Sukla, G.P. & Deori, M.L. 2001 Cultivation prospects of medicinal plants of Arunanchal Pradesh: a review: 329-343. In Samant, S.S., Dhar, U. & Palni, L.M.S. (*eds.*): *Himalayan Medicinal Plants: Potential and Prospects*. Gyanodaya Prakashan, Nainital (India).

Hebbar, S.S., Harsha. V.H., Shripathi, V. & Hegde, G.R. 2003. Wild edible fruits of Dharwad, Karnataka: 982-988. In Singh, V. & Jain, A.P. (*eds.*): *Ethnobotany and Medicinal Plants of India and Nepal*. Vol. II. Scientific Publishers (India), Jodhpur.

Hemlata, Kuniyal, C.P. & Pangtey, Y.P.S. 2007. Traditional healthcare in a remote area of district Chamoli (Garhwal), Uttaranchal: What could do with?: 1-10. In Chopra, A.K., Khanna, D.R., Prasad, G., Malik, D.S. & Bhutiani, R. (*eds.*): *Medicinal Plants: Conservation, Cultivation and Utilization*. Daya Publ. house, Delhi (India).

Hill, A.F. 1952. *Economic Botany*. Mc Graw-Hill Book Company, INC.

Hooker, J.D. 1872-1897. *The Flora of British India*. Vol. I-VIII. Lalit Mohan Basu, Allahabad (India).

Hosagoudar, V.B. & Henry, A.N. 1996. Ethnobotany of Soligas in Biligiri Rangana Betta, Karnataka, Southern India: 228-243. In Maheshwari, J.K. (*ed.*): *Ethnobotany in South Asia*. Scientific Publishers (India), Jodhpur.

Husain, G.M., Mishra, D., Singh, P.N., Rao, C.V. & Kumar, V. 2007. Ethnopharmacological review of native traditional medicinal plants for brain disorders. *Phcog. Rev.* ***1*:*1***: 20-29.

Hussain, A. 1992. *Dictionary of Indian Medicinal Plants*. Central Institute of Medicinal and Aromatic Plants, Lucknow.

Islam, M. 1983. Utilization of certain ferns and fern allies in the North-eastern region, India. *J. Econ. Taxon. Bot. 4*: 861-867.

Islam, M. A. & Jha, R.K. 2003. Ethnopharmacology of tribals of Ranchi district in Jharkhand: 300-310. In Singh, V. & Jain, A.P. (*eds.*): *Ethnobotany and Medicinal Plants of India and Nepal*, Vol. I. Scientific Publishers (India), Jodhpur.

Iyenger, M.A. 1977. *Bibliography on Investigated Indian Plants (1950-75)*. Kasturba Med. Coll., Manipur (India).

Jadeja, B.A. 1999. Plants used by the tribe Rabari in Barda Hills of Gujarat. *Ethnobotany* ***11***: 42-46.

Jain, S.K. 1987. *Manual of Ethnobotany*. Scientific Publishers (India), Jodhpur.

Jain, S.K. 1989 (*ed.*). *Methods and approaches in Ethnobotany*. Deep Pub., New Delhi, India.

Jain, S.K. 1991. Dictionary of Indian Folkmedicine and Ethnobotany. Deep Publ., New Delhi (India).

Jain, S.K. 2002. *Bibliography of Indian Ethnobotany*. Scientific Publishers (India), Jodhpur.

Jain, S.K. (ed.) 2009. Plants Affecting Human Mind (Psychoactive Plants). Deep Publ., New Delhi.

Jain, S.K. & Borthakur, S.K. 1980. Ethnobotany of Mikirs of India. *Econ. Bot.* ***34***: 264-272.

Jain, S.K. & De Filipps, R. 1991. *Medicinal Plants of India,* Vols. I, II. Reference Publ. Algonac Mich. U.S.A.

Jain, S.K. & Goel, A.K. 2005. Some Indian Plants in Tibetian traditional medicine-I. *Ethnobotany* ***17***: 127-136.

Jain, S.K. & Mudgal, V. 1999. *A Handbook of Ethnobotany*. Bishen Singh Mahendra Pal Singh, Dehra Dun.

Jain, S.K., Mudgal, V., Banerjee, D.K., Guha, A. & Das, D. 1984. *Bibliography of Ethnobotany*. B.S.I., Calcutta: 1-157.

Jain, S.K. & Sharma, P.P. 2000. A review of parallel or unique indigenous uses of twenty five plants in India and some African countries. *Ethnobotany* ***12***: 51-55.

Jain, S.K., Sinha, B.K. & Gupta, R.C. 1991. *Notable Plants in Ethnomedicine of India.* Deep Publ., New Delhi.

Jain, S.P. 1996. Ethno-medico-botanical survey of Chaibasa, Singhbhum district, Bihar: 403-408. In Maheshwari, J.K. (*ed.*): *Ethnobotany in South Asia*. Scientific Publishers (India), Jodhpur.

Jain, S.P., Singh, J. & Singh, S.C. 2003. Rare and endangered medicinal and aromatic plants of Madhya Pradesh: 925-932. In Singh, V. & Jain, A.P. (*eds.*): *Ethnobotany and Medicinal Plants of India and Nepal*, Vol. II. Scientific Publishers (India), Jodhpur.

Jain, S.P. & Singh, S.C. 1997. An ethnobotanical survey of Ambikapur district, M.P.: 83-91. In Jain, S.K. (*ed.*): *Contribution to Indian Ethnobotany*. Scientific Publishers (India), Jodhpur.

Jamir, N.S. 1997. Ethnobiology of Naga tribe in Nagaland: 1-Medicinal herbs. *Ethnobotany* ***9***: 101-104.

Jamir, N.S., Limasemba & Jamir, N. 2008. Ethnomedicinal plants used by Konyak Naga tribes of Mon district in Nagaland. *Ethnobotany* ***20***: 48-53.

Jeyapardha, D., Murugesan, P., Rajeshwar, G. & Karunakaran, K. 2011. Ethnopharmacological application of medicinal plants for skin ailments and cosmetics used by various tribal communities in Kolli hills of South India. *Intl. J. Pharma. Sci. Rev. & Res.* ***11(1)***: 102-104.

Jha, A.K., Suman, N.R. & Rathore, D.K. 2003. Medicinal pteridophytes of Bastar, Chattisgarh: 993-996. In Singh, V. & Jain, A.P. (*eds.*): *Ethnobotany and Medicinal Plants of India and Nepal*, Vol. II. Scientific Publishers (India), Jodhpur.

Jha, R.R. & Varma, S.K. 1996. Ethnobotany of Sauria Paharias of Santhal Pargana, Bihar: 1. Medicinal plants. *Ethnobotany* ***8***: 31-35.

Jones, W. 1799. Botanical Observations on Select Indian Plants, London.

Joshi, K. 2009. Indigenous uses of wetland plant diversity of two valleys (Kathmandu and Pokhara) in Nepal. *Ethnobotany* ***21***: 11-17.

Joshi, S. 2004. Plants of ayurvedic importance from Kumaon Himalaya: 73-85. In Pande, P.C. & Joshi, N.K. (*eds.*): *Traditions and Traditional Knowledge of Central Himalayas*. Bishen Singh Mahendra Pal Singh, Dehradun (India).

Joshi, S.G. 2000. *Medicinal Plants*. Oxford & IBH Publ., New Delhi.

Kalpesh, I., Rao, V.R., Parabia, F.M., Kothari, I.L. & Parabia, M.H. 2004. Ethnobotany of Little Rann of Kachchh, Gujarat. *Ethnobotany* ***16***: 40-43.

Kanwar, K.C. & Kanwar, U. 1992. Caffeine – a safe drug unhealthy addiction. *Everyman Sci.,* Feb. – Mar. 13-17. 1992.

Kapoor, S.L. 1979. *Herbal Drugs in Indian Pharmaceutical Industry*. N.B.R.I., Lucknow (India).

Kapoor, S.L. & Mitra, R. 1979. *Herbal Drugs Used in Pharmaceutical Industry*. N.B.R.I., Lucknow.

Kapur, S.K. 1996a. Traditionally important medicinal plants of Bhaderwah hills-Jammu Province-III: 62-69. In Maheshwari, J.K. (*ed.*): *Ethnobotany in South Asia*. Scientific Publishers (India), Jodhpur.

Kapur, S.K. 1996b. Traditionally important medicinal plants of Bhaderwah hills-Jammu Province-IV: 70-74. In Maheshwari, J.K. (*ed.*): *Ethnobotany in South Asia*. Scientific Publishers (India), Jodhpur.

Kapur, S.K. & Nanda, S. 1996. Traditionally important medicinal plants of Bhaderwah hills (Jammu Province) Part II: 56-61. In Maheshwari, J.K. (*ed.*): *Ethnobotany in South Asia*. Scientific Publishers (India), Jodhpur.

Kapur, S.K., Nanda, S. & Srivastava, T.N. 1996. Ethnobotanical uses of RRL-herbarium-III: 50-55. In Maheshwari, J.K. (*ed.*): *Ethnobotany in South Asia*. Scientific Publishers (India), Jodhpur.

Kapur, S.K. & Singh, P. 1996. Traditionally important medicinal plants of Udhampur district (Jammu Province) Part-I: 75-81. In Maheshwari, J.K. (*ed.*): *Ethnobotany in South Asia*. Scientific Publishers (India), Jodhpur.

Kapur, S.K. & Srivastava, T.N. 1996. Traditionally important medicinal plants of Udhampur district (Jammu Province) Part-II: 82-88. In Maheshwari, J.K. (*ed.*): *Ethnobotany in South Asia*. Scientific Publishers (India), Jodhpur.

Karnick, C.R., Tiwari, K.C., Majumdar, R. & Bhattacharjee, S. 1981. Newer ethnobotanical and folk-lore studies of some medicinal plants of Gauhati and surrounding areas. *Nagarjun* ***24***: 240-245.

Katark, N.N. 1903. *Materia Medica of Indian and Their Therapeutics*. Neeraj Publ. House, Delhi.

Katewa, S.S. & Arora, A. 1997. Some plants in folk medicines of Udaipur district (Rajasthan). *Ethnobotany* ***9***: 48-51.

Kaul, M.K. 1997. Medicinal Plants of Kashmir and Ladakh Temperate and Cold Arid Himalaya. Indus Publ. Co., New Delhi (India).

Kaushik, P. & Dhiman, A.K. 2000. *Medicinal Plants and Raw Drugs of India*. Bishen Singh Mahendra Pal Singh, Dehradun (India).

Kayal, R.N., Saha, S. & Ghoshal, P.P. 2003. A floristic account of Indian medicinal plants (originated, cultivated, naturalized or introduced) used in homoeopathic drug preparation: 60-67. In Singh, V. & Jain, A.P. (*eds.*): *Ethnobotany and Medicinal Plants of India and Nepal*, Vol. I. Scientific Publishers (India), Jodhpur.

Khan, A.A., Khan, S. & Singh, M.P. 2003. Weeds of soyabean [*Glycine max* (L.) Merrill] and their medicinal importance in Rewa division (M.P.) India: 245-248. In Singh, V. & Jain, A.P. (*eds.*): *Ethnobotany and Medicinal Plants of India and Nepal*, Vol. I. Scientific Publishers (India), Jodhpur.

Khanna, K.K. 2002. Ethnobotany of five districts of Terai region, Uttar Pradesh: 128-145. In Trivedi, P.C. (*ed.*): *Ethnobotany*. Aavishkar Publ., Distributors, Jaipur (India).

Khanna, K.K., Mudgal, V., Shukla, G. & Srivastava, P.K. 1996. Unreported ethnomedicinal uses of plants from Mirzapur district, Uttar Pradesh: 112-117. In Maheshwari, J.K. (*ed.*): *Ethnobotany in South Asia*. Scientific Publishers (India), Jodhpur.

Khatoon, S. & Goel, A.K. 2010. A brief review on ethno-medicines used in cognitive and brain disorders. *Ethnobotany* ***22***: 55-66.

Khory, R.N. 1887. *The Bombay Materia Medica and their Therapeutics*. Periodical Expert Book Agency, Delhi (India).

Kirn, H.S., Kapahi, B.K. & Srivastava, T.N. 2000. Taxo-ethnobotanical observations on the gymnosperms of Poonch district (J&K State) India: 155-160. In Maheshwari, J.K. (*ed.*): *Ethnobotany and Medicinal Plants of Indian Subcontinent*. Scientific Publishers (India), Jodhpur.

Kirtikar, K.R. & Basu, B.D. 1935. *Indian Medicinal Plants*, Vol. I-IV. Lalit Mohan Basu, Allahabad, India.

Kirtikar, K.R. & Basu, B.D. 1984. *Indian Medicinal Plants*. Vol. I-IV. Bishen Singh Mahendra Pal Singh, New Delhi (India).

Klemens, J. 2006. Herbs used for psychotropic or behaviour modifying activity. *JAAIM – online*: 1-9.

Kothari, M.J., Jadhav, C.R. & Singh, N.P. 2003. Ethnobotanical wealth of two sacred grooves in Junnar Taluka, Pune, Maharashtra: 585-591. In Singh, V. & Jain, A.P. (*eds.*): *Ethnobotany and Medicinal Plants of India and Nepal*, Vol. II. Sci. Publ., Jaipur (India).

Kothari, M.J. & Londhe, A.N. 2000. Ethnobotany in human health care of Chikhaldara, Amaravati district in Maharashtra state, India: 273-281. In Maheshwari, J.K. (*ed.*): *Ethnobotany and Medicinal Plants of Indian Subcontinent*. Scientific Publishers (India), Jodhpur.

Kothari, M.J. & Rao, K.M. 2000. Ethnobotanical studies of Thane district Maharashtra: 265-272. In Maheshwari, J.K. (*ed.*): *Ethnobotany and Medicinal Plants of Indian Subcontinent*. Scientific Publishers (India), Jodhpur.

Krishnamurthy, S. 1953. *Horticultural and Economical Plants of the Nilgiris*. Govt. Bot. Gdn. Centenary Committee, Udhagamandalam.

Kshirsagar, R.D. & Singh, N.P. 2000. Less – known ethnomedicinal uses of plants in Coorg district of Karnataka state, Southern India. *Ethnobotany* ***12***: 12-16.

Kshirsagar, R.D. & Singh, N.P. 2007. *Ethnobotany of Mysore and Coorg, Karnataka State*. Bishen Singh Mahendra Pal Singh, Dehradun (India).

Kshirsagar, S.R., Prabia, M.H. & Reddy, M.N. 2003. Ethnobotany of coastal areas in south Gujarat. *Ethnobotany* ***15***: 60-63.

Kumar, A. & Lal, J. 2003. Medicinal plant resources of the Kanha tiger reserve, Madhya Pradesh: 255-265. In Singh, V. & Jain, A.P. (*eds.*): *Ethnobotany and Medicinal plants of India and Nepal*, Vol. I. Scientific Publishers (India), Jodhpur.

Kumar, A., Tewari, D.D. & Pande, Y.N. 2003. Ethno-phytotherapeutics among Tharus of Beerpur-Semra forest range of Balrampur district. (U.P.): 839-844. In Singh, V. & Jain, A.P. (*eds.*): *Ethnobotany and Medicinal Plants of India and Nepal*, Vol. II. Scientific Publishers (India), Jodhpur.

Kumar, K. & Goel, A.K. 1998. Little known ethno-medicinal plants of Santhal and Paharia tribes in Santhal Paragana, Bihar, India. *Ethnobotany* ***10***: 66-69.

Kumar, P. 2006. Medicinal Plants in India Conservation and Sustainable Utilization in the Emerging Global Scenario. Bishen Singh Mahendra Pal Singh, Dehradun (India).

Kumar, R.V. & Pullaiah, T. 1998. Medicinal plants used by tribals of Prakasam district, Andhra Pradesh. *Ethnobotany* ***10***: 97-102.

Kumar, S. 1979. Marketing Dictionary of Medicinal Plants. CIMAP, Lucknow.

Kumar, S. 1999. *The Economic Plants of North-East India.* Scientific Publishers (India), Jodhpur.

Kumar, S. 2002. *The Medicinal Plants of North-East India.* Scientific Publishers (India), Jodhpur.

Kumar, S. 2011. *Centella asiatica*: a brain tonic at our feet. http://sanjeetbiotech.blogspot.in

Kumar, S. & Chauhan, A.K.S. 2005. Medicinal plants used by local inhabitants in Bharatpur district, Rajasthan. *Ethnobotany* ***17***: 179-183.

Kumar, S. & Goel, A.K. 2008. A preview of odoriferous plants found around Itchagarh hills in Ormanjhi block of Jharkhand. *Ethnobotany* ***20***: 135-137.

Kumar, T.D.C. & Pullaiah, T. 2000. Ethno-medicinal uses of some plants of Mahabubnagar district, Andhra Pradesh, India: 341-345. In Maheshwari, J.K. (*ed.*): *Ethnobotany and Medicinal Plants of Indian Subcontinent.* Scientific Publishers (India), Jodhpur.

Kumar, V. 2006. Potential medicinal plants for CNS disorders: an overview. *Phytother. Res.* ***20 (12)***: 1023-1035.

Kumar, V. & Rao, R.R. 2001. Some plant beverages used in traditional medicine. *Ethnobotany* ***13***: 36-39.

Kumar, Y., Haridasan, K. & Rao, R.R. 1980. Ethnobotanical notes on certain medicinal plants among some Garo people around Balphakram Sanctuary in Meghalaya. *Bull. Bot. Surv. India* ***22***: 161-165.

Lalramnghinglova, H. 2003. *Ethno-Medicinal Plants of Mizoram.* Bishen Singh Mahendra Pal Singh, Dehradun (India).

Lalramnghinglova, J.H. 1996. Ethnobotany of Mizoram – a proliminary survey: 439-459. In Maheshwari, J.K. (*ed.*): *Ethnobotany in South Asia.* Scientific Publishers (India), Jodhpur.

Levetin & McMohan 2003. Psychoactive plants: opium poppy, marijuana, and cocaine. Lecture outline prepared from *Plants and Society* (3rd edition). McGraw – Hil Publ. Co.

Mahato, R.B. & Chaudhary, R.P. 2005. Ethnomedicinal plants of Palpa district, Nepal. *Ethnobotany* ***17***: 152-163.

Maheshwari, J.K. (*ed.*) 1996. *Ethnobotany in South Asia.* Scientific Publishers (India), Jodhpur.

Maheshwari, J.K. (*ed.*) 2000. *Ethnobotany and Medicinal Plants of Indian Subcontinent.* Scientific Publishers (India), Jodhpur.

Maheshwari, J.K., Painuli, R.M. & Dwivedi, R.P. 1997. Notes on ethnobotany of the Oraon and Korwa tribes of Madhya Pradesh: 67-82. In Jain, S.K. (*ed.*): *Contribution to Indian Ethnobotany.* Scientific Publishers (India), Jodhpur.

Maheshwari, P. & Singh, U. 1965. *Dictionary of Economic Plants in India.* I.C.A.R., New Delhi.

Maikhuri, R.K., Nautiyal, S. & Rao, K.S. 2001. Medicinal plant cultivation practices of Bhotiyas in Nanda Devi biosphere reserve villages of Garhwal Himalaya: 317-328. In Samant, S.S., Dhar, U. & Palni, M.S. (*eds.*): *Himalayan Medicinal Plants: Potential and Prospects.* Gyanodaya Prakashan, Nainital.

Maity, D., Chauhan, A.S. & Maity, G.G. 2003. Ethnobotanical notes on some unexploited medicinal plants used by Lepchas and Nepalese communities of North Sikkim: 325-332. In Singh, V. & Jain, A.P. (*eds.*): *Ethnobotany and Medicinal Plants of India and Nepal*, Vol. I. Scientific Publishers (India), Jodhpur.

Malhotra, S.K. & Murthy, S. 1973. Some useful and medicinal plants of Chandrapur district (Maharashtra state). *Bull. Bot. Surv. India* ***15***: 13-21.

Malla, B. & Chhetri, R.B. 2009. Observations on some ethnomedicinal plants in Kavrepalanchowk district, Nepal. *Ethnobotany* ***21***: 41-45.

Manadhar, N.P. 2002. Ethnobotanical uses of 'Pakhanbed' (*Bergenia ciliata*) in Nepal. *Ethnobotany* ***14***: 13-15.

Manandhar, N.P. 2003. Ethnobotanical notes on the family Leguminosae in Nepal: 333-344. In Singh, V. & Jain, A.P. (*eds.*): *Ethnobotany and Medicinal Plants of India and Nepal*, Vol. I. Scientific Publishers (India), Jodhpur.

Mathur, A. 2003. Who owns traditional knowledge? Working Paper No. 96: 1-33. *Indian Council for Research on International Economic Relations.*

Maya, S., Prameela, K.S. & Menon, V.S. 2003. Ethnobotanical notes on the flora of sacred tanks of Kerela. *Ethnobotany 15*: 55-59.

Meena, S.L., Sharma, K.C. & Gopalan, R. 2003. Ethno-medicinal plants of Karauli district, Rajasthan: 177-180. In Singh, V. & Jain, A.P. (*eds.*): *Ethnobotany and Medicinal Plants of India and Nepal*, Vol. I. Scientific Publishers (India), Jodhpur.

Mehra, P.N., Bhatnagar, J.K. & Handa, S.S. 1969. Researches in pharmacognosy in India. *Res. Bull. Punjab Univ. 20*: 261-337.

Merlin, M.D. 1972. *Man and Marijuana.* Barnas & Co., New York.

Merlin, M.D. 1984. *On the Trail of Ancient Opium Poppy.* Assoc. Univ. Press, London & Toronto.

Milind, P. & Nirmal, S. 2004. Animal models for testing memory. *Asia Pac. J. Pharmacol.* ***16***: 101-120.

Mishra, D.P. & Sahu, T.R. 1984. Euphorbiaceous plants used in medicine by the tribals of Madhya Pradesh. *Ind. J. Econ. Taxon. Bot.* ***5***: 791-794.

Mishra, R.C. & Das, P. 2003. Wild poisonous seeds: some notable species from Gandhamarden hill ranges of Orissa: 513-518. In Singh, V. & Jain, A.P. (*eds.*): *Ethnobotany and Medicinal Plants of India and Nepal*, Vol. I. Scientific Publishers (India), Jodhpur.

Mishra, S. 2008. Ethnomedicinal studies of Kokru tribe, with Gotra names derived from plant names, from Khandwa district in Madhya Pradesh. *Ethnobotany* ***20***: 122-127.

Mitra, R. 1985. Bibliography on Pharmacognosy of Medicinal Plants. N.B.R.I. Lucknow (India).

Mitra, R. & Jain, S.K. 1991. Medicinal plant research in India. an overview. *Ethnobotany 3*: 65-77.

Mudgal, V., Pal, D.C., Kayal, R.N. & Saha, S. 1999. *Ethnobotany of Totopara.* Bishen Singh Mahendra Pal Singh, Dehradun (India).

Mughal, M.H., Saba, Srivastava, P.S. & Iqbal, M. 2000. Drumstick (*Moringa pterygosperma* Gaertn.): a unique source of food and medicine: 47-61. In Maheshwari, J.K. (*ed.*): *Ethnobotany and Medicinal Plants of Indian Subcontinent.* Scientific Publishers (India), Jodhpur.

Mukherjee, A., Mukherjee, S. & Mukherjee, M. 1986. Less known medicinal plants of Rubiaceae in Darjeeling Hills, West Bengal (India). *Indian J. Applied & Pure Biol.* ***1***: 86-91.

Mukherjee, A. & Namhata, D. 1990. Medicinal plantlore of tribals of Sundargarh district, Orissa. *Ethnobotany* ***2***: 57-60.

Mukherjee, A. & Roy, S.D. 1987. An account of Piperaceae and Saururaceae in the hills of Darjeeling district, West Bengal. *J. Econ. Taxon. Bot.* ***9***: 367-371.

Mukhopadhyaya, G. 1992. *History of Indian Medicine*. Munshiram Manoharlal Publ., Delhi (India).

Murty, K.S., Narayanappa, D. & Sharma, P.C. 1989. Some unknown narcotic folk claims from Orissa. *Bull. Med. Eth. Bot. Res.* ***10(3-4)***: 199-200.

Nadeem, M., Kumar, A., Nandi, S.K. & Palni, L.M.S. 2001. Tissue culture of medicinal plants with particular reference to Kumaon Himalaya: 231-268. In Samant, S.S., Dhar, U. & Palni, L.M.S. (*eds.*): *Himalayan Medicinal Plants - Potential and Prospects*. Gyanodaya Prakashan, Nainital (India).

Nadkarni, K.M. 1926. *Indian Materia Medica*. Popular Book Depot, Bombay (India).

Nahata, A., Patil, U.K. & Dixit, V.K. 2008. Effect of *Convolvulus pluricaulis* Choisy on learning behaviour and memory enhancement activity in rodents. *Nat. Prod. Res.* ***22***: 1472-1482.

Nambiar, V.P. 1985. *Studies on the Medicinal Plants of Kerala Forest*. Kerala Forest Res. Inst., Peechi, Kerala.

Nargas, J. & Trivedi, P.C. 2000. Traditional and medicinal importance of *Azadirachta indica* Juss. in India: 33-37. In Maheshwari, J.K. (*ed.*): *Ethnobotany and Medicinal Plants of Indian Subcontinent*. Scientific Publishers (India), Jodhpur.

Nath, K.K., Deka, P. & Borthakur, S.K. 2007. Ethnomedicinal aspects of some weeds from Darrang district of Assam. *Ethnobotany* ***19***: 82-87.

Nautiyal, S., Maikhuri, R.R., Rao, K.S. & Saxena, K.G. 2003. Ethnobotany of Tolccha Bhotiya tribe of the buffer zone villages in Nanda Devi Biosphere reserve, India: 119-142. In Jain, V. & Jain, A.P. (*eds.*): *Ethnobotany and Medicinal Plants of India and Nepal*, Vol. I. Scientific Publishers (India), Jodhpur.

Nayak, R.K., Nayak, P.K. & Choudhury, B.P. 2003. Some medicinal weeds of Mahanadi Delta: 533-538. In Singh, V. & Jain, A.P. (*eds.*): *Ethnobotany and Medicinal plants of India and Nepal*, Vol. II. Scientific Publishers (India), Jodhpur.

Nayar, M.P., Ramamurthy, K. & Agarwal, V.S. 1989. *Economic Plants of India*, Vols. I & II. Bot. Survey of India, Calcutta.

Negi, K.S., Gaur, R.D. & Tiwari, J.K. 1999. Ethnobotanical notes on the flora of Har-Ki-Doon (district Uttarkashi), Garhwal Himalaya, Uttar Pradesh, India. *Ethnobotany* ***11***: 9-17.

O' Shaughnessy, W.B. 1841. *The Bengal Dispensatory and Pharmacopoeia*. Thacker, Spink and Co., Calcutta (India).

Painuli, R.M. & Maheshwari, J.K. 1996. Some interesting ethnomedicinal plants used by Sahariya tribe of Madhya Pradesh: 179-185. In Maheshwari, J.K. (*ed.*): *Ethnobotany in South Asia*. Scientific Publishers (India), Jodhpur.

Pal, D.C. 2000. Some plants in traditional and homeopathic systems of medicine in India: 79-83. In Maheshwari, J.K. (*ed.*): *Ethnobotany and Medicinal Plants of Indian Subcontinent*. Scientific Publishers (India), Jodhpur.

Pal, D.C. & Jain, S.K. 1989. Notes on Lodha medicine in Midnapur district, West Bengal, India. *Econ. Bot.* ***43*(4)**: 464-470.

Pal, D.C. & Paul, C.R. 2003. Healthful culinary herbal species of India: 75-77. In Singh, V. & Jain, A.P. (*eds.*): *Ethnobotany and Medicinal Plants of India and Nepal*, Vol. I. Scientific Publishers (India), Jodhpur.

Panda, P.C. & Das, P. 2000. Medicinal plant-lore of the tribals of Baliguda sub-division, Phulbani district, Orissa: 515-521. In Maheshwari, J.K. (*ed.*):

Ethnobotany and Medicinal Plants of Indian Subcontinent. Scientific Publishers (India), Jodhpur.

Pande, P.C. & Joshi, G.C. 2001. Indigenous medicinal uses of cultivated plants in Kumaon Himalaya: 117-124. In Samant, S.S., Dhar, U. & Palni, L.M.S. (*eds.*) *Himalayan Medicinal Plants: Potential and Prospects.* Gyanodaya Prakashan, Nainital (India).

Pande, P.C., Joshi, G.C. & Kandpal, M.M. 1990. Ethnobotany of Kumaon Himalaya: 285-289. In Shah, N.K. *et al.* (*eds.*): *Himalaya: Environment, Resources and Development.* Shree Almora Book Depot, Almora.

Pandey, A.K. & Rout, S.D. 2006. Ethnobotanical uses of plants of Similipal Biosphere Reserve (Orissa). *Ethnobotany **18***: 102-106.

Pandey, G. 1995. *Medicinal Plants of Himalaya.* Sri Satguru Pub., Delhi.

Pandey, I.B. 2003. Some traditional herbal home remedies used in around Kanpur city (Uttar Pradesh), India. *Ethnobotany **15***: 129-131.

Panthi, M.P. & Chaudhary, R.P. 2003. Ethnomedicinal plant resources of Arghakhanchi district, West Nepal. *Ethnobotany **15***: 71-86.

Parle, M. & Vasudevan, M. 2006. Memory enhancing activity of Abana: an Indian ayurvedic poly-herbal formulation. *J. Health Sci. **53(1)***: 43-52.

Parrotta, J.A. 2001. *Healing Plants of Peninsular India.* CABI Publ., Wallingford (U.K.).

Pathak, S.K. & Pathak, S. 2007. Ethnomedicinal flora of West Nimar (Khargone) district, M.P., India: 207-211. In Chopra, A.K., Khanna, D.R., Prasad, G., Malik, D.S. & Bhutiani, R. (*eds.*): *Medicinal Plants: Conservation, Cultivation and Utilization.* Daya Publ. House, Delhi (India).

Patil, U.S. 2007. Ethnomedicinal plants in Melghat of Amravati district: a need for conservation: 53-63. In Chopra, A.K., Khanna, D.R., Prasad, G., Malik, D.S. & Bhutiani, R. (*eds.*) *Medicinal Plants: Conservation, Cultivation and Utilization.* Daya Publ. House, Delhi (India).

Perri, S. 2009. Tonic plants for pregnancy. http://www.healthychild.com

Perry, L.M. 1980. Medicinal Plants of East and South East Asia. MIT: Press, Mass. U.S.A.

Phurailatpam, K., Geetha, K.A., Gajbhiye, N.A., Saravanan, R. & Maiti, S. 2009. Comparative study of *Chlorophytum borivilianum* Sant. & Fernand and *C. arundinaceum* Baker – two 'safed musli' species used as vital tonics in Indian System of medicine. *Phytomorphology **59(1-2)***: 77-82.

Prajapati, N.D., Purohit, S.S., Sharma, A.K. & Kumar, T. 2006. *A Handbook of Medicinal Plants (A Complete Source Book).* Agrobios, Jodhpur.

Prakash, V. 1998. Indian medicinal plants: current status-1. *Ethnobotany 10*: 112-121.

Prakash, V. 2001. Indian medicinal plants: current status: 45-63. In Samant, S.S. Dhar, U. & Palni, L.M.S. (*eds.*) *Himalayan Medicinal Plants: Potential and Prospects.* Gyanodaya Prakashan, Nainital (India).

Prakash, V. & Mehrotra, B.N. 1991. Ethno-medicinal uses of some plants among Garos of Meghalaya. *Ethnobotany **3***: 41-45.

Prakash, V. & Mehrotra, B.N. 1996. Attributes on utility and scarcity of Valerianaceae in India: 318-322. In Jain, S.K. (*ed.*): *Ethnobiology in Human Welfare.* Deep Publ., New Delhi (India).

Prasad, P.N., Singh, A.J.A.R., Narayanan, L.M. & Natarajan, C.R. 1996. Ethnobotany of the Kanikkars of South Tamil Nadu-I: 292-298. In Maheshwari, J.K. (*ed.*): *Ethnobotany in South Asia.* Scientific Publishers (India), Jodhpur.

Prasad, V.K., Rajagopal, T. & Badarinath, K.V.S. 2002. Ethnobotany of Kondareddis of Rampa agency, Eastern Ghats, Andhra Pradesh: 205-220. In Trivedi, P.C. (*ed.*): *Ethnobotany*. Aavishkar Publishers, Distributors, Jaipur (India).

Prasad, V.K., Rajagopal, T. & Badarinath, K.V.S. 2003. Notes on economic importance of wild plants of Rampa agency – East Godavari district, Andhra Pradesh, India: 603-612. In Singh, V. & Jain, A.P. (*eds.*): *Ethnobotany and Medicinal plants of India and Nepal*, Vol. II. Scientific Publishers (India), Jodhpur.

Prusti, A.B. 2007. Plants used as ethnomedicine by Bondo tribe of Malkangiri district, Orissa. *Ethnobotany* ***19***: 105-110.

Punjani, B.L. 2002. Ethnobotanical aspects of some weeds from Gujarat. *Ethnobotany* ***14***: 78-80.

Punjani, B.L. 2003. Poisonous plants in the forest areas of Vijaynagar Taluka (North Gujarat): 196-204. In Singh, V. & Jain, A.P. (*eds.*): *Ethnobotany and Medicinal plants of India and Nepal*, Vol. I. Scientific Publishers (India), Jodhpur.

Punjani, B.L. 2006. Ethno-medicobotanical study of Kathodi tribe of Sabarkantha in Gujarat. *Ethnobotany* ***18***: 135-138.

Purohit, K. & Samant, S.S. 1995. *Fodder Trees and Shrubs of Central Himalaya*. Gyanodaya Prakashan, Nainital (U.P.), India.

Pushpagandhan, P. 1994. *Ethnobiology in India. A Staus Report*. Ministry of Environment and Forest, GOI, New Delhi.

Raghunathan, K. 1976. Medicinal Plants Growing in the Districts of Andhra Pradesh Through Which River Krishna Flows. Central Council for Research in Indian Medicine & Homeopathy, Delhi.

Rahman, M.A. & Yusuf, M. 1996. Diversity, ecology and ethnobotany of the *Zingiberaceae* of Bangladesh: 13-19. In Maheshwari, J.K. (*ed.*): *Ethnobotany in South Asia*. Scientific Publishers (India), Jodhpur.

Rai, K.S., Rao, M.S. & Karanth, K.S. 2000. *Clitorea ternatea* enhances learning and memory – an experimental study on rats. International congress on frontiers in pharmacology and therapeutics in 21st century. New Delhi, India. *Indian J. Pharmacol.* ***32***: 150.

Rai, L.K. 1994. Medicinal Plants of Sikkim Himalaya: Status, Usage and Potential. Bishen Singh Mahendra Pal Singh, Dehra Dun.

Rai, M.K. 1987. Ethno-medicinal studies of Patalkot and Tamiya (Distt. Chhindwara) M.P. – plants used as tonic. *Anc. Sci. Life* ***7(2)***: 119-127.

Rajan, G.B., Chezhiyan, N., Khan, I.A. & Khanum, A. 2003. Spices medicine: 115-173. In Khan, I.A. & Khanum, A. (*eds.*): *Role of Biotechnology in Medicinal and Aromatic Plants*, Vol VIII. Ukaaz Publ., Hyderabad (India).

Rajan, S., Gupta, H.C. & Kumar, S. 2000. Exotic medicinal plants of Lucknow: 205-222. In Maheshwari, J.K. (*ed.*): *Ethnobotany and Medicinal Plants of Indian Subcontinent*. Scientific Publishers (India), Jodhpur.

Rajendran, S.M. 2003. West Sikkim: a treasure house of medicinal plants and their utility by local tribals: 311-316. In Singh, V. & Jain, A.P. (*eds.*): *Ethnobotany and Medicinal Plants of India and Nepal*, Vol. I. Scientific Publishers (India), Jodhpur.

Rajendran, S.M. & Mehrotra, B.N. 1996. Unrecorded medicinal uses of plants among Parambikulam tribals, Kerala, India: 181-183. In Jain, S.K. (*ed.*): *Ethnobiology in Human Welfare*. Deep Publ., New Delhi (India).

Raju, R.A. 2000. *Wild Plants of Indian Sub-continent and Their Economic Use*. CBS Publishers & Distributors, New Delhi (India).

Rajwar, G.S. 1983. Low altitude medicinal plants of South Garhwal. *Bull. Medico-Ethno-Bot. Res.* ***4***: 14-28.

Ram, A.J., Raja, K., Reddy, K.E. & Raju, R.R.V. 2002. Medicinal plant lore of Sugalis of Gooty forests, Andhra Pradesh. *Ethnobotany* ***14***: 37-42.

Rana, T.S., Datt, B. & Rao, R.R. 1996. Strategies for sustainable utilization of plant resources by the tribals of the Tons Valley, Western Himalaya. *Ethnobotany* ***8***: 96-104.

Ranjan, P. 2000. A contribution to some of the medicinal plants of Indo-Nepal border area adjoining the districts of Madhubani and Sitamarhi: 651-659. In Maheshwari, J.K. (*ed.*): *Ethnobotany and Medicinal Plants of Indian Subcontinent*. Sci. Publ. Jodhpur.

Ranjan, P. 2003. A contribution to some ethnomedicinally important plants of Nepal: 345-353. In Singh, V. & Jain, A.P. (*eds.*): *Ethnobotany and Medicinal Plants of India and Nepal*, Vol. I. Scientific Publishers (India), Jodhpur.

Rao, N.R. & Henry, A.N. 1995. *The Ethnobotany of Eastern Ghats in Andhra Pradesh, India. 1996*. Botanical Survey of India, Calcutta.

Rao, R.R. 1996. Traditional knowledge and sustainable development: key role of ethnobiologists. *Ethnobotany* ***8***: 14-24.

Raskin, I., Ribnicky, Dm, Komarnytsky, S., Poulev, A., Borisjuk, N., Brinker, A., Moreno, Da, Ripoll, C., Yakoby, N., O'neal, J., Cornwell, T., Pastor, I., Fridlender, B. 2002. Plants and human health in the twenty-first century. *Trends in Biotech.* ***20(12)***: 522-553.

Rastogi, R.P. & Mehrotra, B.N. 1990. *Compendium of Medicinal Plants*. Vol. I. (1960-69). Central Drugs Research Institute, Lucknow and National Institute of Science Communication, New Delhi (India).

Rastogi, R.P. & Mehrotra, B.N. 1991. *Compendium of Medicinal Plants*. Vol. II. (1970-79). Central Drugs Research Institute, Lucknow and National Institute of Science Communication, New Delhi (India).

Rastogi, R.P. & Mehrotra, B.N. 1993. *Compendium of Medicinal Plants*. Vol. III. (1980-84). Central Drugs Research Institute, Lucknow and National Institute of Science Communication, New Delhi (India).

Rastogi, R.P. & Mehrotra, B.N. 1995a. *Compendium of Medicinal Plants*. Vol. IV. (1985-89). Central Drugs Research Institute, Lucknow and National Institute of Science Communication, New Delhi (India).

Rastogi, R.P. & Mehrotra, B.N. 1995b. *Compendium of Medicinal Plants*. Vol. V. (1990-94). Central Drugs Research Institute, Lucknow and National Institute of Science Communication, New Delhi (India).

Rawat, L., Vashistha, H.B., Khaliya, D. & Kamboj, S.K. 2007. Medicinal plant diversity in Pindari Glacier area of Nanda Devi biosphere reserve (NDBR), Uttaranchal: 167-178. In Chopra, A.K., Khanna, D.R., Prasad, G., Malik, D.S. & Bhutiani, R. (*eds.*): *Medicinal Plants: Conservation, Cultivation and Utilization*. Daya Publ. House, Delhi (India).

Rawat, M.S. & Chaudhary, S. 1998. *Ethno Medico Botany of Arunachal Pradesh (Nishi & Atapani Tribe)*. Bishen Singh Mahendra Pal Singh, Dehra Dun.

Rawat, R.B.S. 2002. Cultivation Practices of Some Commercially Important Medicinal Plants. Natl. Med. Plant Board, New Delhi (India).

Raychaudhuri, S.P. 1991. *Recent advances in Medicinal, Aromatic and Spice Crops*. Today and Tomorrow's Printers and Publ., Delhi (India).

Reddi, T.V.V.S., Prasanthi, S. & Naidu, B.V.A.R. 2005. Medicinal and aromatic plants of India: 1-146. In Khan, I.A. & Khanum, A. (*eds.*): *Role of Biotechnology in Medicinal and Aromatic Plants.* Vol XII. Ukaaz Publ., Hyderabad (India).

Reddy, M.M. 1996. Ethnobotany of *Semecarpus anacardium* (Anacardiaceae) – marking nut in Andhra Pradesh, India: 408-409. In Jain, S.K. (*ed.*): *Ethnobiology in Human Welfare.* Deep Publ., New Delhi (India).

Robert, K. 1932. *Vegetable Drugs of India.* New Delhi.

Rosakutty, P.J., Roslin, A.S. & Ignacimuthu, S. 2000. Some traditional folklore medicinal plants of Kanyakumari district (Tamil Nadu): 369-375. In Maheshwari, J.K. (*ed.*): *Ethnobotany and Medicinal Plants of Indian Subcontinent.* Scientific Publishers (India), Jodhpur.

Roxburgh, W. 1832. *Flora Indica.* Thakur, Spink & Co., Calcutta.

Royle, J.F. 1837. An Essay on Antiquity of Hindu Medicine. London.

Sahu, P. 2004. Medicinal utility of botanic garden's weeds: a case study of Sagar University botanical garden: 215-229. In Sahu, T.R. & Sahu, P.K. (*eds.*): *Biodiversity and Sustainable Utilization of Biological Resources.* Scientific Publishers (India), Jodhpur.

Saikia, N. & Nath, S.C. 2003a. Ethnomedicinal observations of some species of the genus *Curcuma* Linn. growing in Assam: 430-433. In Singh, V. & Jain, A.P. (*eds.*): *Ethnobotany and Medicinal Plants of India and Nepal*, Vol. I. Scientific Publishers (India), Jodhpur.

Saikia, N. & Nath, S.C. 2003b. Traditional use of some plants in religio-cultural purposes by 'Koch' community in Assam, India: 997-1001. In Singh, V. & Jain, A.P. (*eds.*): *Ethnobotany and Medicinal Plants of India and Nepal*, Vol. II. Scientific Publishers (India), Jodhpur.

Saini, D.C. 1996. Ethnobotany of Tharus of Basti district, Uttar Pradesh: 138-153. In Maheshwari, J.K. (*ed.*): *Ethnobotany in South Asia.* Scientific Publishers (India), Jodhpur.

Saini, D.C. 2008. Traditional uses of pteridophytes among Baiga tribes of Amarkantak, Anuppur district, M.P. *Ethnobotany* ***20***: 65-69.

Saklani, A. & Jain, S.K. 1994. *Cross-Cultural Ethnobotany of Northeast India.* Deep Publ., New Delhi.

Samant, S.S., Dhar, U. & Rawal, R.S. 1996. Natural resource use by natives within Nanda Devi Biosphere Reserve in Western Himalaya. *Ethnobotany* ***8***: 40-50.

Samant, S.S., Dhar, U. & Rawal, R.S. 2001. Diversity, distribution and indigenous uses of threatened medicinal plants of Askot wildlife sanctuary in West Himalaya; conservation and management perspectives: 167-184. In Samant, S.S., Dhar, U. & Palni, L.M.S. (*eds.*): *Himalayan Medicinal Plants: Potential and Prospects.* Gyanodaya Prakashan, Nainital (India).

Samvatsar, S. & Diwanji, V.B. 2000. Plants used for rheumatism by the tribals of western M.P.: 305-314. In Maheshwari, J.K. (*ed.*): *Ethnobotany and Medicinal Plants of Indian Subcontinent.* Scientific Publishers (India), Jodhpur.

Sanjay, Navneet, Manohar, M. & Prabhat 2007. Antimicrobial properties of herbal tooth powders: 303-317. In Chopra, A.K., Khanna, D.R., Prasad, G., Malik, D.S. & Bhutiani, R. (*eds.*): *Medicinal Plants: Conservation, Cultivation and Utilization.* Daya Publ. House, Delhi (India).

Sanyal, D. 1984. *Vegetable Drugs of India.* Bishen Singh Mahendra Pal Singh, Dehra Dun.

Sarahkhan, M.S. & Balick, M.J. 2001. Therapeutic plants of Ayurveda: a review of selected clinical and other studies for 106 species. *J. Alternative and Complementary Medicine* **7(5)**: 405-515.

Sarin, Y.K. 1990. Some less known but effective folk remedies in north-west Himalayan region. *Ethnobotany* **2**: 39-43.

Sarma, S.K., Bhattacharjya, D.K. & Devi, B. 2002. Traditional use of herbal medicines by Madahi tribe of Nalbari district of Assam. *Ethnobotany* **14**: 103-111.

Satapathy, K.B. & Brahmam, M. 1996. Some medicinal plants used by tribals of Sundargarh district, Orissa, India: 153-158. In Jain, S.K. *(ed.)*: *Ethnobiology in Human Welfare*. Deep Publ., New Delhi (India).

Satyavati, G.V. 1993. Leads from Ayurveda on medicinal plants acting on the nervous system. Proceedings of the Indo-US Symposium on Decade of the Brain, Bangalore: NIMHANS (in press).

Satyavati, G.V., Raina, M.R. and Sharma, M. (*eds.*) 1976. *Medicinal Plants of India*. Vol. I (A-G), ICMR, New Delhi.

Saxena, H.O. 1971. *Medicinal Plants of Patalkot*. SFRI, Jabalpur.

Schultes, R.E. 1960. Tapping our heritage of ethnobotanical lore. *Eco. Bot.* **14**: 257-262.

Schultes, R.E. 1993. The virgin field in psychoactive plant research. *Ethnobotany* **5**: 5-61.

Schultes, R.E. & Hofmann, A. 1980. *The Botany and Chemistry of Hallucinogens*. Charles C. Thomas. Illinois, U.S.A.

Schultes, R.E. & Hofmann, A. 1987. *Plants of the Gods*. Alfred Van Der Marck Edition, New York.

Sen, S. & Batra, A. 1997. Ethno-medico-botany of household remedies of Phagi tehsil of Jaipur district (Rajasthan). *Ethnobotany* **9**: 122-128.

Sethiya, N.K., Nahata, A., Mishra, S.H. & Dixit, V.K. 2009. An update on shankhpushpi, a cognition-boosting ayurvedic medicine. *J. Chin. Integr. Med.* ***7(11)***: 1001-1022.

Shah, G.L., Yadav, S.S. & Badri, N. 1983. Medicinal plants from Dahanu forest division in Maharashtra state. *J. Econ. Taxon. Bot.* **4**: 141-151.

Shah, N.C. 1995. *Rauvolfia serpentina*: from folk-medicine to modern medicine – some unknown aspects. *Ethnobotany* **7**: 109-112.

Shah, N.C. 2006. A historical and ethnobotanical study of *Nardostachys jatamansi*: an ancient incense & aromatic medicinal herb from Uttaranchal, India. *Ethnobotany* **18**: 37-45.

Shah, N.C. 2010. Ethnographical and ethnomedicinal studies of Ladakh. *Ethnobotany* **22**: 73-81.

Shah, N.C. & Joshi, M.C. 1971. An ethnobotanical study of Kumaon region, India. *Econ. Bot.* **25**: 414-422.

Shah, N.C. & Singh, S.C. 1990. Hitherto unreported phytotherapeutical uses from tribal pockets of Madhya Pradesh (India). *Ethnobotany* **2**: 91-95.

Shah, N.C. & Virmani, O.P. 1982. Directory of Crude Drugs and Aromatic Plants, Dealers, Producers and Exporters in India: 1-98. Centr. Inst. Med. Arom. Pl., Lucknow.

Sharma, B.D. 2009. *Life-sustaining Plants of The Himalayas*. Indus Publ. Co., New Delhi (India).

Sharma, B.D. & Rana, J.C. 2000. Traditional medicinal uses of plants of Himachal hills: 173-176. In Maheshwari, J.K. (*ed.*): *Ethnobotany and Medicinal Plants of Indian Subcontinent*. Scientific Publishers (India), Jodhpur.

Sharma, K.R. & Sood, M. 1997. *Important Medicinal Plants of Himachal Pradesh*. Dr. Y.S. Parmar Univ. of Hort. & Forestry, Nauni, H.P. (India).

Sharma, N. 1998. Myco-myths, mycetismus and medicines. *Ethnobotany **10***: 16-21.

Sharma, P.K., Chauhan, N.S. & Lal, B. 2003. Commercially important medicinal and aromatic plants of Parvati Valley, Himachal Pradesh: 937-942. In Singh, V. & Jain, A.P. (*eds.*): *Ethnobotany and Medicinal Plants of India and Nepal*, Vol. II. Scientific Publishers (India), Jodhpur.

Sharma, P.P. & Singh, N.P. 2001. Ethnomedicinal uses of some edible plants in Dadra Nagar Haveli and Daman (U.T.). *Ethnobotany **13***: 121-125.

Sharma, R. 2003. *Medicinal Plants of India – An Encyclopaedia*. Daya Publ. House, Delhi (India).

Sharma, U.K. 2004. *Medicinal Plants of Assam*. Bishen Singh Mahendra Pal Singh, Dehra Dun (India).

Sharma, U.K., Sharma, A.K. & Sharma, C.M. 2007. Clinical evaluation of the effect of *Centella asiatica* on cerebral higher functions: 357-367. In Chopra, A.K., Khanna, D.R., Prasad, G., Malik. D.S. & Bhutiani, R. (*eds.*): *Medicinal Plants: Conservation, cultivation and Utilisation*. Daya Publ. House, Delhi (India).

Shepard, G.H. Jr. 1998. Psychoactive plants and ethnopsychiatric medicines of the Matsigenka. *J. Psychoactive Drugs **30***: 321-332.

Shivakumar, L., Gauda, S.T., Venkat, R.N., Shalam & Verma, R. 2011. Evaluation of memory enhancing activity of SR-105 in experimental animals. *Int. J. Ayurved Pharm. **2(3)***: 973-977.

Shrivastava, A.K. & Singh, M. 2003. Medicinal potential of grasss in Rajasthan: 170-176. In Singh, V. & Jain, A.P. (*eds.*): *Ethnobotany and Medicinal Plants of India and Nepal*, Vol. I. Scientific Publishers (India), Jodhpur.

Shukla, G. & Verma, B.K. 1996. Roots – a vital plant part to cure body ailments among tribal/rural folklore of Western Bihar: 392-394. In Maheshwari, J.K. (*ed.*): *Ethnobotany in South Asia*. Scientific Publishers (India), Jodhpur.

Shukla, G., Verma, B.K. & Pandey, H.P. 2003. Folk-lore herbal therapy of Jalaun district, U.P., India: 143-147. In Singh, V. & Jain, A.P. (*eds.*): *Ethnobotany and Medicinal Plants of India and Nepal*, Vol. I. Scientific Publishers (India), Jodhpur.

Shukla, Rashmi & Jain, S.K. 1997. Some potential psychoactive plants of India. *Natl. Acad. Sci., Letters **20(3/4)***: 29-34.

Sikarwar, R.L.S. & Kaushik, J.P.1992. Some less known medicinal uses of trees among the Sahariyas of Morena district, M.P. India. *Ethnobotany **4**:* 71-74.

Singh, A.K. 2000a. A contribution to the ethnobotany of sub-himalayan region of eastern Uttar Pradesh: 237-246. In Maheshwari, J.K. (*ed.*): *Ethnobotany and Medicinal Plants of Indian Subcontinent*. Scientific Publishers (India), Jodhpur.

Singh, B., Chaurasia, O.P. & Jadhav, C.K.L. 1996. An ethnobotany study of Indus valley (Ladakh): 92-101. In Maheshwari, J.K. (*ed.*): *Ethnobotany and Medicinal Plants of Indian Subcontinent*. Scientific Publishers (India), Jodhpur.

Singh, G.S. 1999. A contribution to ethnomedicine of Alwar district of Rajasthan. *Ethnobotany **11**:* 97-99.

Singh, G.S. 2003. Ethnobotanical inventory of Sariska National Park in the Aravalli hills of eastern Rajasthan, India: 181-195. In Singh, V. & Jain, A.P. (*eds.*):

Ethnobotany and Medicinal Plants of India and Nepal, Vol. I. Scientific Publishers (India), Jodhpur.

Singh, H. 1985. *Medicinal Chemistry Research in India*. Central Drug Research Institute, Lucknow (India).

Singh, H.B. 2000b. Potential medicinal pteridophytes of India and their chemical constituents: 63-78. In Maheshwari, J.K. (*ed.*): *Ethnobotany and Medicinal Plants of Indian Subcontinent*. Scientific Publishers (India), Jodhpur.

Singh, H.B. 2000c. Alternate source for some conventional drug plants of India: 109-114. In Maheshwari, J.K. (*ed.*): *Ethnobotany and Medicinal Plants of Indian Subcontinent*. Scientific Publishers (India), Jodhpur.

Singh, H.B. & Viswanathan, M.V. 1996. Useful pteridophytes of India- A gift of nature to human beings: 24-36. In Maheshwari, J.K. (*ed.*): *Ethnobotany in South Asia*. Scientific Publishers (India), Jodhpur.

Singh, J & Trivedi, V.P. 1978. Vegetable psychotropic drugs of India. *Quart. J. Crude Drug Res.* ***16(1)***: 33-52.

Singh, K.K. & Kumar, K. 2000a. *Ethnobotanical Wisdom of Gaddi Tribe in Western Himalaya*. Bishen Singh Mahendra Pal Singh, Dehra Dun (India).

Singh, K.K. & Kumar, K. 2000b. Ethnotherapeutics of some medicinal plants used as antipyretic agents among the tribals of India: 135-141. In Maheshwari, J.K. (*eds.*): *Ethnobotany and Medicinal Plants of Indian Subcontinent*. Scientific Publishers (India), Jodhpur.

Singh, K.K. & Maheshwari, J.K. 1992. Folk medicinal uses of some plants among the Tharus of Gorakhpur district, Uttar Pradesh, India. *Ethnobotany* ***4***: 39-43.

Singh, P. Sisodia, S. & Shah, J. 2007. Angiospermous seeds of medicinal importance in Gujarat state: 17-24. In Chopra, A.K., Khanna, D.R., Prasad, G., Malik, D.S. & Bhutiani, R. (*eds.*): *Medicinal Plants: Conservation, Cultivation and Utilization*. Daya Publ. House, Delhi (India).

Singh, P.B. 2001. Medicinal plants of Kangra district in North-West Himalaya: 185-196. In Samant, S.S., Dhar, U. & Palni, M.S. (*eds.*): *Himalayan Medicinal Plants: Potential and Prospects*. Gyanodaya Prakashan, Nainital (India).

Singh, P.K., Elangbam, V.D. & Huidrom, B.K.S. 2000. Ethnomedicinal studies of some plants used to enhance vocalism by the traditional Meitei singers of Manipur: 629-635. In Maheshwari, J.K. (*ed.*): *Ethnobotany and Medicinal Plants of Indian Subcontinent*. Scientific Publishers (India), Jodhpur.

Singh, R., Dwivedi, S.K., Raut, B. & Mishra, S.N. 2003. Ethnobotany of *Hippophae* 'Seabuckthorn' in Ladakh. *Ethnobotany* ***15***: 1-5.

Singh, R.H., Shukla, S.P. & Mishra, B.K. 1981. The psychotropic effect of Medhya rasayana drug, Mandukaparni (*Hydrocotyle asiatica*): an experimental study, Part-II. *J. Res. Arya. Siddha* **2(1)**: 1-10.

Singh, S.C. & Srivastava, G.N. 2000. Exotic medicinal plants of Lucknow district, U.P., India: 223-235. In Maheshwari, J.K. (*ed.*): *Ethnobotany and Medicinal Plants of Indian Subcontinent*. Scientific Publishers (India), Jodhpur.

Singh, S.P., Tripathi, S. & Shukla, R.S. 2003. Ethnomedicinal heritage of bioprospecting and drug development in North-Eastern states of India: 384-395. In Singh, V. & Jain, A.P. (*eds.*): *Ethnobotany and Medicinal Plants of India and Nepal*, Vol. I. Scientific Publishers (India), Jodhpur.

Singh, U. 1996. *Dictionary of Economic Plants in India*. I.C.A.R., New Delhi.

Singh, U., Wadhwani, A.M. & John, B.M. 1983. *Dictionary of Economic Plants of India*. I.C.A.R., New Delhi.

Singh, V. 1996. Ethnomedicobotany of Dards tribe of Gurez valley in Kashmir Himalaya: 129-132. In Jain, S.K. *(ed.)*: *Ethnobiology in Human Welfare*. Deep Publ., New Delhi (India).

Singh, V. & Pandey, R.P. 1998. *Ethnobotany of Rajasthan, India*. Scientific Publishers (India), Jodhpur.

Singh, V.K. 1984. *Medicinal Plants of Gwalior Forest Division, Madhya Pradesh*. Central Council for Research in Unani Medicine, Delhi.

Sivakumar, A., Subramanian, M.S., Karunakaran, M. & Burkanudeen, A. 2003. Ethnobotany of Poliyars of Annamalai hills, Tamil Nadu: 679-685. In Singh, V. & Jain, A.P. (*eds.*): *Ethnobotany and Medicinal Plants of India and Nepal*, Vol. II. Scientific Publishers (India), Jodhpur.

Siwakoti, M. & Varma, S. K. 1996. Medicinal plants of the Terai of eastern Nepal: 423-438. In Maheshwari, J.K. (*ed.*): *Ethnobotany in South Asia*. Scientific Publishers (India), Jodhpur.

Sood, S.K., Bhardwaj, R. & Lakhanpal, T.N. 2005a. *Ethnic Indian Plants in Cure of Diabetes*. Scientific Publishers (India), Jodhpur.

Sood, S.K., Nath, R. & Kalia, D.C. 2001. *Ethnobotany of Cold Desert Tribes of Lahaul-Spiti (N.W. Himalaya)*. Deep Publ., New Delhi (India).

Sood, S.K., Parmar, S. & Lakhanpal, T.N. 2005b. *Ethnic Plants of India Used in Cancer Cure (A Compendium)*. Bishen Singh Mahendra Pal Singh, Dehradun.

Sood, S.K., Rana, S. & Lakhanpal, T.N. 2005c. *Ethnic Aphrodisiac Plants*. Scientific Publishers (India), Jodhpur.

Sood, S.K., Sharma, S.K., Kumar, N. & Kumar, H. 2009. *Ethnobotanical Studies on Trees, Shrubs and Climbers of Himalaya*. Satish Serial Publ. House, Delhi (India).

Sood, S.K. & Thakur, S. 2004. Ethnobotany of Rewalsar Himalaya (Dist. Mandi, Himachal Pradesh, India). Deep Publ., New Delhi (India).

Srivastava, R.C. 1989. *Drug Plant Resources of Central India*. Today and Tomorrow's Printers & Publ. New Delhi (India).

Srivastava, R.C., Singh, V.P. & Singh, M.K. 2003a. Medicinal plants of Jaunpur district (U.P.): 148-159. In Singh, V. & Jain, A.P. (*eds.*): *Ethnobotany and Medicinal Plants of India and Nepal*, Vol. I. Scientific Publishers (India), Jodhpur.

Srivastava, S.K., Mehrotra, S., Srivastava, G.K. & Chauhan, D.K. 2003b. Folk uses of herbs from Mahableshwar region of Sahyadri hills: 857-863. In Singh, V. & Jain, A.P. (*eds.*): *Ethnobotany and Medicinal Plants of India and Nepal*, Vol. II. Scientific Publishers (India), Jodhpur.

Suba, V., Murugesan, T., Rao, R.B., Pal, M., Mandal, S.C. & Saha, B.P. 2002. Neuropharmacological profile of *Barleria lupulina* Lindl. extract in animal models. *J. Ethnopharmacol.* ***81(2)***: 251-255.

Subramaniam, A. 2000. A survey of medicinal plants from Chitheri hills in Dharmapuri district, Tamil Nadu: 395-416. In Maheshwari, J.K. (*ed.*): *Ethnobotany and Medicinal Plants of Indian Subcontinent*. Scientific Publishers (India), Jodhpur.

Sudharani, D., Krishna, K.L., Deval, K., Safia, A.K. & Priya 2011. Pharmacological profiles of *Bacopa monnieri*: a review. *Intl. J. Pharm.* ***1(1)***: 15-23.

Sundararaj, D. 1959. *Guide to Economic Plants of South India*. Periodical Expert Book Agency, Delhi.

Sur, P.R. 1984. Narcotic Plants of India. *J. Econ. Taxon. Bot.* **5(1)**: 143-148.

Thakor, A.B. 2009. Ethnomedicinal plants of Vaarli tribe in Valsad district (Gujarat). *Ethnobotany* ***21***:116-120.

Thakur, R.S., Puri, H.S., & Husain, A. 1989. *Major Medicinal Plants of India*. Central Institute of Medicinal & Aromatic Plants, Lucknow, India.

Thomas, J. & Britto, A.J. D. 2000. Weeds of medicinal importance in Tirunelveli District in Tamilnadu: 363-367. In Maheshwari, J.K. (*ed.*): *Ethnobotany and Medicinal Plants of Indian Subcontinent*. Scientific Publishers (India), Jodhpur.

Tigga, M., Lakra, G.S. & Rao, P.S.N. 2003. Medicinal value of latex yielding plants from Andaman and Nicobar islands: 765-772. In Singh, V. & Jain, A.P. (*eds.*): *Ethnobotany and Medicinal Plants of India and Nepal*, Vol. II. Scientific Publishers (India), Jodhpur.

Tiwari, K.C. & Tiwari, V.P. 1996. Some important medicinal plants of the tropical, sub-tropical and temperate regions of Siang, Subansiri and Tirap districts of Arunanchal Pradesh: 359-363. In Maheshwari, J.K. (*ed.*): *Ethnobotany in South Asia*. Scientific Publishers (India), Jodhpur.

Tiwari, K.P., 1998. Medicinal Plants of Madhya Pradesh: Distribution, Cultivation and Trade. SFRI Jabalpur.

Tosh, J. 1996. Ethnobotanical study of Western Maharashtra: 169-174. In Maheshwari, J.K. (*ed.*): *Ethnobotany in South Asia*. Scientific Publishers (India), Jodhpur.

Tripathi, S. & Goel, A.K. 2001. Ethnobotanical diversity of Zingiberaceae in North-Eastern India. *Ethnobotany* ***13***: 67-79.

Trivedi, G.N., Kayal, R.N. & Rai Chaudhury, H.N. 1982. Some medicinal plants of Mayurbhanj (Orissa). *Bull. Bot. Surv. India* ***24***: 117-120.

Trivedi, P.C. 2002. Ethno-medicinal plants of Rajasthan state, India: 412-439. In Trivedi, P.C. (*ed.*): *Ethnobotany*. Aavishkar Publ., Distributors, Jaipur (India).

Uniyal, B.P. & Malhotra, C.L. 1989. Economic exploitation of rare North-Western plants: 221-225. In Paliwal, G.S. (*ed.*): *The Vegetation Wealth of The Himalayas*. Puja Publ., Pitampura, Delhi (India).

Uniyal, M.R. 1968. Medicinal plants of Bhagirathi valley lying in Uttar Kashi Forest division. *Indian For.* ***94***: 407-420.

Uniyal, M.R. 1989. *Medicinal Flora of Garhwal Himalayas*. Baidyanath Ayurvedic Research Centre, Patna.

Uniyal, M.R. 1997. *Medicinal Plants and Minerals of Uttarakhand Himalaya*. Baidyanath Ayurvedic Research Centre, Patna.

Upadhye, A.S. & Kumbhojkar, M.S. 1998. *Schleichera oleosa* (Lour.) Oken. tree: a boon for tribals. *Ethnobotany* ***10***: 38-41.

Upadhye, A.S., Vartak, V.D. & Kumbhojkar, M.S. 1994. Ethno-medico-botanical studies in Western Maharastra, India. *Ethnobotany* ***6***: 25-31.

Upadhyay, R. 2003. Ethnobotanical observations on some rare and endangered herbs of Singrauli in Madhya Pradesh: 989-992. In Singh, V. & Jain, A.P. (*eds.*): *Ethnobotany and Medicinal Plants of India and Nepal*, Vol. II. Scientific Publishers (India), Jodhpur.

Upadhyay, R. & Chauhan, S.V.S. 2003. Ethnobotanical uses of plant gums by the tribals: 601-602. In Singh, V. & Jain, A.P. (*eds.*): *Ethnobotany and Medicinal Plants of India and Nepal*, Vol. II. Scientific Publishers (India), Jodhpur.

Upadhyay, R. & Singh, J. 2005. Ethno-medicinal uses of plants from Tikri forest of Gonda district (U.P.). *Ethnobotany* ***17***: 167-170.

Upadhyay, V.P. 1986. Therapeutic role of Ayurvedic herb in mental disorders. *Vedic Path **69:1***: 24-29.

Uphof, J.C.T. 1986. *Dictionary of Economic Plants* (2nd *ed.*). Stechert-Hafner Service Agency, Inc., New York.

Vaidya, A.D. & Devasagayam, T.P. 2007. Current status of herbal drugs in India: an overview. *J. Clin. Biochem. Nutr **41(1)***: 1-11.

Varghese, E. 1996. Applied Ethnobotany - A Case Study Among the Kharias of Central India. Deep Publ., New Delhi (India).

Ved, D.K. 2000. Tropical Indian Medicinal Plants Propagation Methods. FRLHT, Bangalore (India).

Vedavathy, S., Mrudula, V. & Sudhakar, A. 1997. *Tribal Medicine of Chittore District, A.P.* Intl. Devlt. Res. Ctre., Canada & Sri Venkateshwara Arts College.

Verma, P., Khan, A.A. & Singh, K.K. 1995. Traditional phytotherapy among the Baiga tribe of Shahdol district of Madhya Pradesh, India. *Ethnobotany **7***: 69-73.

Verma, S. & Chauhan, N.S. 2006. Studies on ethno-medico-botany of Kunihar forest division, district Solan (H.P.). *Ethnobotany **18***: 160-165.

Vishwanathan, M.B., Kumar, E.H.P. & Ramesh, N. 2001. Ethnomedicines of Kanis in Kalakkad – Mundanthurai tiger reserve, Tamil Nadu. *Ethnobotany **13***: 60-66.

Vishwanathan, M.V. 2000. Edible and medicinal plants of Ladakh (J & K): 151-154. In Maheshwari, J.K. (*eds.*): *Ethnobotany and Medicinal Plants of Indian Subcontinent*. Scientific Publishers (India), Jodhpur.

Vishwanathan, M.V. & Singh, H.B. 1996. Plants used as household remedies in India: 105-107. In Jain, S.K. (*ed.*): *Ethnobiology in Human Welfare*. Deep Publ., New Delhi (India).

Warrier, P.K., Nambiar, V.P.K. & Raman, K. 1994-1996. *Indian Medicinal Plants – A Compendium of 500 Species*. Vol. I-V. Orient Longman Limited, Hyderabad.

Wasson, R.G. 1967. *Soma, Divine Mushroom of Immorality*. Harcourt, New York.

Watt, G. 1889-96. *A Dictionary of the Economic Products of India*, Vol. I-VI. Periodical Experts, New Delhi (India).

WHO, 2000. General Guidelines for Methodologies on Research and Evaluation of Traditional Medicine: 1-80. Geneva, Switzerland.

WHO, 2008. Congress on Traditional Medicine. Beijing, China.

Wilbert, J. 1988. *Tobacco and Shamanism in South America*. Yale Univ. Press, Haven.

Yasodamma, N., Mehar, S.K. & Paramageetham, C. 2009. Threat assessment (IUCN categorization) for ethnomedicinal plants used by Chenchu tribe of Gundlabramheswaram in Nallamalai hills in Andhra Pradesh. *Ethnobotany **21***: 51-60.

Appendix I-
Index to Families, Genera and Species of Plants

Division	No. of Families	Total No. of Genera/Species	Total No. of Species in a Genera
Lichens	1	2/2	*Cetraria* (1), *Parmelia* (1).
Pterido-phytes	14	14/19	*Actinopteris* (1), *Adiantum* (4), *Asplenium* (2), *Ceratopteris* (1), *Cibotium* (1), *Drynaria* (1), *Helminthostachys* (1), *Lycopodium* (1), *Marsilea* (1), *Ophioglossum* (2), *Osmunda* (1), *Pteris* (1), *Selaginella* (1), *Tectaria* (1).
Gymno-sperms	6	9/11	*Abies* (1), *Cedrus* (1), *Cycas* (1), *Ephedra* (1), *Ginkgo* (1), *Juniperus* (2), *Pinus* (2), *Taxus* (1), *Thuja* (1).
Dicots	131	566/841	*Abelmoschus* (2), *Abroma* (1), *Abrus* (1), *Abutilon* (1), *Acacia* (3), *Acalypha* (1), *Acanthopanax* (1), *Achillea* (1), *Achyranthes* (1), *Aconitum* (8), *Actaea* (1), *Adonis* (1), *Aegle* (1), *Aerva* (1), *Aesculus* (1), *Ageratum* (1), *Ailanthus* (2), *Ajuga* (1), *Alangium* (1), *Albizzia* (1), *Allophylus* (1), *Alstonia* (1), *Alysicarpus* (1), *Amaranthus* (1), *Amberboa* (1), *Ammi* (1), *Anacyclus* (1), *Anagallis* (1), *Andrographis* (2), *Anemone* (1), *Anethum* (1), *Angelica* (2), *Anisomeles* (1), *Annona* (1), *Anodendron* (1), *Anogeissus* (1), *Anthemis* (1), *Anthocephalus* (1), *Antidesma* (1), *Apium* (1), *Aquilaria* (1), *Aquilegia* (1), *Arachis* (1), *Ardisia* (1), *Argemone* (1), *Argyreia* (1), *Aristolochia* (3), *Armoracia* (1), *Arnica* (1), *Artemisia* (7), *Artocarpus* (2), *Asarum* (1), *Asperula* (1), *Atropa* (1), *Atylosia* (1), *Averrhoa* (1), *Azadirachta* (1), *Bacopa* (1),

Division	No. of Families	Total No. of Genera/Species	Total No. of Species in a Genera
			Baliospermum (1), *Baptisia* (1), *Barbarea* (1), *Barleria* (1), *Barringtonia* (1), *Basella* (1), *Bauhinia* (3), *Benincasa* (1), *Berberis* (5), *Bergenia* (2), *Beta* (1), *Betula* (1), *Bidens* (2), *Biophytum* (1), *Bistorta* (1), *Bixa* (1), *Blepharis* (3), *Blumea* (1), *Boerhavia* (1), *Bombax* (1), *Borreria* (1), *Boswellia* (1), *Brassica* (3), *Broussonetia* (1), *Bryonopsis* (1), *Buchanania* (1), *Bunium* (1), *Bupleurum* (1), *Butea* (2), *Caccinea* (1), *Caesalpinia* (2), *Calamintha* (1), *Calendula* (1), *Callicarpa* (1), *Calotropis* (2), *Calycopteris* (1), *Camellia* (1), *Cannabis* (1), *Canscora* (2), *Canthium* (1), *Capparis* (4), *Capsella* (1), *Capsicum* (1), *Caralluma* (1), *Cardamine* (1), *Cardiospermum* (1), *Careya* (1), *Carica* (1), *Carum* (1), *Casearia* (1), *Cassia* (3), *Cassytha* (1), *Castanea* (1), *Catha* (1), *Catharanthus* (1), *Ceiba* (1), *Celastrus* (1), *Celosia* (1), *Centaurea* (1), *Centella* (1), *Centratherum* (1), *Cephalenthus* (1), *Ceropegia* (2), *Cheiranthus* (1), *Chelone* (1), *Chenopodium* (1), *Chimaphila* (1), *Chloranthus* (1), *Chondrodendron* (1), *Cicer* (1), *Cichorium* (2), *Cimicifuga* (2), *Cinchona* (3), *Cinnamomum* (4), *Cirsium* (2), *Cissampelos* (1), *Citrullus* (1), *Citrus* (4), *Cleome* (1), *Clerodendrum* (4), *Clitoria* (1), *Cnicus* (1), *Coccinia* (1), *Cocculus* (2), *Cochlospermum* (1), *Codonopsis* (3), *Coffea* (2), *Cola* (2), *Colebrookia* (1), *Coleus* (1), *Commiphora* (2), *Conium* (1), *Convolvulus* (2), *Coptis* (1), *Corchorus* (4), *Cordia* (2), *Coriandrum* (1), *Corydalis* (2), *Corylus* (1), *Coscinium* (1), *Crataegus* (1), *Crataeva* (1), *Cressa* (1), *Crotalaria* (1), *Cryptolepis* (1), *Cucumis* (2), *Cucurbita* (1), *Cuscuta* (2), *Cyamopsis* (1), *Cyclea* (1), *Cydonia* (1), *Cytisus* (1), *Dalbergia* (1), *Daphne* (1), *Datura* (4), *Daucus* (1), *Delphinium* (1), *Dendrophthoe* (1), *Derris* (1), *Descurainea* (1), *Desmodium* (4), *Dicentra* (1), *Dicliptera* (1), *Digitalis* (2), *Dillenia* (1), *Dimocarpus* (1), *Dioclea* (1),

Division	No. of Families	Total No. of Genera/Species	Total No. of Species in a Genera
			Diospyros (1), *Diplocyclos* (1), *Dipsacus* (1), *Dolichos* (1), *Dorema* (1), *Doronicum* (4), *Drosera* (1), *Drymaria* (1), *Drypetes* (1), *Duboisea* (1), *Durio* (1), *Echinops* (1), *Eclipta* (1), *Elaeagnus* (2), *Elaeocarpus* (1), *Elephantopus* (1), *Embelia* (1), *Emblica* (1), *Enicostemma* (1), *Entada* (2), *Erigeron* (2), *Eriobotrya* (1), *Eruca* (1), *Eryngium* (3), *Erythrina* (2), *Erythroxylum* (2), *Eucalyptus* (1), *Euodia* (1), *Eupatorium* (4), *Euphorbia* (3), *Euphrasia* (1), *Euryale* (1), *Evodia* (2), *Evolvulus* (1), *Exacum* (1), *Fagonia* (2), *Fagraea* (1), *Feronia* (1), *Ferula* (4), *Ficus* (5), *Flacourtia* (1), *Foeniculum* (1), *Fraxinus* (1), *Fumaria* (1), *Galipea* (1), *Garcinia* (1), *Gardenia* (1), *Gaultheria* (2), *Gelsemium* (1), *Gentiana* (6), *Geum* (1), *Gillenia* (1), *Glechoma* (1), *Glinus* (1), *Glycyrrhiza* (1), *Gmelina* (1), *Gossypium* (1), *Grevillea* (1), *Grewia* (2), *Gymnema* (1), *Haematoxylon* (1), *Haldina* (1), *Hedera* (1), *Hedyotis* (3), *Heimia* (1), *Hemidesmus* (1), *Hesperethusa* (1), *Heynea* (1), *Hibiscus* (3), *Hippophae* (1), *Hiptage* (1), *Holarrhena* (1), *Holostemma* (1), *Houttuynia* (1), *Humulus* (1), *Hybanthus* (1), *Hydnocarpus* (1), *Hydrangea* (1), *Hydrastis* (1), *Hydrocotyle* (2), *Hygrophila* (1), *Hyoscyamus* (2), *Hypecoum* (1), *Hypericum* (3), *Hypochoeris* (1), *Hyptis* (2), *Hyssopus* (1), *Ichnocarpus* (1), *Ilex* (1), *Illicium* (1), *Impatiens* (1), *Indigofera* (6), *Inula* (1), *Iochroma* (1), *Ipomea* (6), *Ixora* (2), *Jasminum* (4), *Jateorhiza* (1), *Jatropha* (2), *Juglans* (2), *Jurinea* (1), *Justicia* (2), *Kalanchoe* (1), *Krameria* (2), *Lactuca* (2), *Lagenaria* (1), *Lagerstroemia* (2), *Lamium* (1), *Lantana* (1), *Lathyrus* (1), *Laurus* (1), *Lawsonia* (1), *Leea* (1), *Lens* (1), *Leonurus* (2), *Lepidium* (1), *Leptadenia* (1), *Leucas* (4), *Ligusticum* (1), *Limnophila* (2), *Linum* (1), *Liquidambar* (1), *Liriodendron* (1), *Litchi* (1), *Lithospermum* (1), *Litsea*

Division	No. of Families	Total No. of Genera/Species	Total No. of Species in a Genera
			(2), *Luffa* (2), *Lycopersicon* (1), *Madhuca* (1), *Maerua* (1), *Magnolia* (1), *Majorana* (1), *Mallotus* (1), *Malva* (1), *Mandragora* (1), *Mangifera* (1), *Manilkara* (2), *Marrubium* (1), *Matricaria* (1), *Matthiola* (2), *Mazus* (1), *Meconopsis* (1), *Medicago* (1), *Melia* (2), *Melissa* (2), *Melodinus* (1), *Melothria* (1), *Memecylon* (1), *Menispermum* (1), *Mentha* (6), *Menyanthes* (1), *Meriandra* (1), *Mesua* (1), *Michelia* (1), *Mimosa* (1), *Mimusops* (1), *Momordica* (2), *Morinda* (2), *Moringa* (1), *Moschosma* (1), *Mucuna* (2), *Mundulea* (1), *Murraya* (2), *Mussaenda* (1), *Myrica* (1), *Myristica* (1), *Myroxylon* (1), *Myrtus* (1), *Myxopyrum* (1), *Nardostachys* (1), *Nasturtium* (1), *Nelsonia* (1), *Nelumbo* (1), *Nepeta* (4), *Nerium* (1), *Nicotiana* (1), *Nigella* (1), *Nyctanthus* (1), *Nymphaea* (3), *Ochna* (1), *Ocimum* (4), *Oenothera* (1), *Onosma* (1), *Operculina* (1), *Ophiorrhiza* (1), *Origanum* (1), *Ormocarpum* (1), *Oroxylum* (1), *Ostrya* (1), *Ougeinia* (1), *Ouratea* (1), *Oxalis* (1), *Oxydendron* (1), *Paederia* (1), *Paeonia* (1), *Panax* (3), *Pandanus* (1), *Papaver* (2), *Parnassia* (1), *Parthenium* (1), *Passiflora* (2), *Pavetta* (1), *Pedalium* (1), *Pedicularis* (1), *Peganum* (1), *Peperomia* (1), *Perilla* (1), *Petroselinum* (1), *Peucedanum* (2), *Phaseolus* (1), *Physalis* (1), *Physochlaina* (1), *Physostigma* (1), *Phytolacca* (1), *Picrorhiza* (1), *Pimenta* (1), *Pimpinella* (1), *Piper* (7), *Piscidia* (1), *Pistacia* (2), *Pittosporum* (1), *Plantago* (2), *Platycodon* (1), *Plectranthus* (1), *Plumbago* (2), *Plumeria* (1), *Podophyllum* (1), *Pogostemon* (2), *Polemonium* (1), *Polygonum* (7), *Pongamia* (1), *Populus* (2), *Potentilla* (5), *Prangos* (1), *Premna* (1), *Primula* (2), *Prunella* (1), *Prunus* (6), *Pseudarthria* (1), *Psidium* (1), *Pterocarpus* (2), *Pterospermum* (1), *Pterygota* (1), *Pueraria* (1), *Pulsatilla* (1), *Punica* (1), *Pygmaeopremna* (1), *Pyrus* (1),

Division	No. of Families	Total No. of Genera/Species	Total No. of Species in a Genera
			Quassia (1), *Quercus* (1), *Randia* (1), *Ranunculus* (2), *Rauvolfia* (2), *Rehmania* (1), *Rhamnus* (2), *Rheum* (2), *Rhodiola* (1), *Rhododendron* (3), *Rhus* (1), *Rorippa* (1), *Rosa* (3), *Rosmarinus* (1), *Rourea* (1), *Roylea* (1), *Rubia* (3), *Rubus* (2), *Ruellia* (1), *Rumex* (2), *Ruta* (2), *Salacia* (1), *Salix* (3), *Salvadora* (1), *Salvia* (1), *Samadera* (1), *Sambucus* (2), *Sandorium* (1), *Santalum* (1), *Sapindus* (2), *Sapium* (2), *Saraca* (1), *Sarcostemma* (2), *Sassafras* (1), *Saussurea* (2), *Saxifraga* (1), *Shefflera* (1), *Schinus* (1), *Schleichera* (1), *Schweinfurthia* (1), *Scutellaria* (2), *Securinega* (1), *Selinum* (2), *Semecarpus* (1), *Senecio* (1), *Sesamum* (1), *Sesbania* (3), *Seseli* (1), *Shorea* (1), *Sida* (6), *Sissymbrium* (1), *Sium* (1), *Smithia* (1), *Solanum* (7), *Solena* (1), *Sonchus* (2), *Soymida* (1), *Sphaeranthus* (1), *Spilanthes* (2), *Spondias* (1), *Stephania* (1), *Sterculia* (1), *Stereospermum* (2), *Streblus* (1), *Strobilanthes* (1), *Strychnos* (3), *Swertia* (5), *Swietenia* (1), *Symplocos* (2), *Synadenium* (1), *Syzygium* (2), *Tabebuia* (1), *Tabernaemontana* (1), *Tagetes* (1), *Tamarindus* (1), *Tanacetum* (1), *Taraxacum* (1), *Tecoma* (1), *Tectona* (1), *Tephrosia* (2), *Teramnus* (1), *Terminalia* (7), *Thalictrum* (2), *Theobroma* (1), *Thevetia* (1), *Thlaspi* (1), *Thymus* (2), *Tilia* (1), *Tinospora* (1), *Toddalia* (1), *Toona* (1), *Trachyspermum* (2), *Trapa* (1), *Tribulus* (1), *Tricholepis* (1), *Trichosanthes* (2), *Tricodesma* (1), *Trifolium* (1), *Trigonella* (1), *Triumfetta* (1), *Tropaeolum* (1), *Turbina* (1), *Tylophora* (1), *Uraria* (1), *Urtica* (3), *Vaccinium* (1), *Valeriana* (5), *Vateria* (1), *Ventilago* (1), *Verbascum* (2), *Verbena* (1), *Vernonia* (1), *Veronicastrum* (1), *Viburnum* (2), *Vigna* (4), *Vinca* (1), *Viola* (1), *Viscum* (1), *Vitex* (2), *Vitis* (1), *Wagatea* (1), *Walsura* (1), *Wedelia* (1), *Withania* (2), *Woodfordia* (1), *Wrightia*

Division	No. of Families	Total No. of Genera/Species	Total No. of Species in a Genera
			(1), *Xanthium* (1), *Xerospermum* (1), *Zanthoxylum* (4), *Zataria* (1), *Zehneria* (1), *Ziziphus* (3), *Zornia* (1).
Monocots	23	83/127	Acampe (1), Acorus (2), Allium (3), Alisma (1), Aloe (2), Alpinia (3), Amomum (2), Amorphophallus (1), Ananas (1), Aneilema (1), Areca (1), Asparagus (3), Avena (1), Bambusa (1), Borassus (1), Calamus (1), Canna (1), Caryota (1), Chlorophytum (2), Cocos (1), Coix (1), Colocasia (1), Convallaria (1), Costus (1), Crinum (1), Crocus (1), Curculigo (1), Curcuma (5), Cymbopogon (3), Cynodon (1), Cyperus (9), Dactylorhiza (1), Dendrobium (3), Dendrocalamus (1), Desmotrichum (1), Dioscorea (3), Elettaria (1), Eleusine (1), Eragrostis (1), Eulophia (1), Fritillaria (1), Gloriosa (1), Habenaria (5), Hedychium (2), Heteropogon (1), Homalonema (1), Hordeum (1), Hygroryza (1), Hypoxis (1), Imperata (1), Kaempferia (1), Lilium (1), Liparis (2), Liriope (1), Lodoicea (1), Malaxis (2), Maranta (1), Monochoria (1), Musa (1), Nervilia (1), Oryza (1), Panicum (2), Paspalum (1), Pennisetum (2), Phoenix (2), Pholidota (1), Phyllostachys (1), Pinellia (1), Polygonatum (3), Roscoea (1), Saccharum (1), Sansevieria (2), Satyrium (1), Scindapsus (1), Serenoa (1), Smilax (1), Triticum (1), Tulipa (1), Typha (1), Typhonium (1), Urginea (2), Vetiveria (1), Zingiber (3).

Appendix II-
Index to Families

Name of Family	Genus/ Species	Total Number of Species under a Genus
Acanthaceae	9/13	*Andrographis* (2), *Barleria* (1), *Blepharis* (3), *Dicliptera* (1), *Hygrophila* (1), *Justicia* (2), *Nelsonia* (1), *Ruellia* (1), *Strobilanthes* (1).
Actinopteridaceae	1/1	*Actinopteris* (1).
Adiantaceae	1/4	*Adiantum* (4).
Aizoaceae	1/1	*Glinus* (1).
Alangiaceae	1/1	*Alangium* (1).
Alismataceae	1/1	*Alisma* (1).
Amaranthaceae	4/4	*Achyranthes* (1), *Aerva* (1), *Amaranthus* (1), *Celosia* (1).
Amaryllidaceae	2/2	*Crinum* (1), *Hypoxis* (1).
Anacardiaceae	7/8	*Buchanania* (1), *Mangifera* (1), *Pistacia* (2), *Rhus* (1), *Schinus* (1), *Semecarpus* (1), *Spondias* (1).
Annonaceae	1/1	*Annona* (1).
Apiaceae	25/35	*Ammi* (1), *Anethum* (1), *Angelica* (2), *Apium* (1), *Bunium* (1), *Bupleurum* (1), *Carum* (1), *Centella* (1), *Conium* (1), *Coriandrum* (1), *Daucus* (1), *Dorema* (1), *Eryngium* (3), *Ferula* (4), *Foeniculum* (1), *Hydrocotyle* (2), *Ligusticum* (1), *Petroselinum* (1), *Peucedanum* (2), *Pimpinella* (1), *Prangos* (1), *Selinum* (2), *Seseli* (1), *Sium* (1), *Trachyspermum* (2).
Apocynaceae	13/14	*Alstonia* (1), *Anodendron* (1), *Catharanthus* (1), *Holarrhena* (1), *Ichnocarpus* (1), *Melodinus* (1), *Nerium* (1), *Plumeria* (1), *Rauvolfia* (2), *Tabernaemontana* (1), *Thevetia* (1), *Vinca* (1), *Wrightia* (1).
Aquifoliaceae	1/1	*Ilex* (1).
Araceae	7/8	*Acorus* (2), *Amorphophallus* (1), *Colocasia* (1), *Homalonema* (1), *Pinellia* (1), *Scindapsus* (1), *Typhonium* (1).

Name of Family	Genus/ Species	Total Number of Species under a Genus
Araliaceae	4/6	*Acanthopanax* (1), *Hedera* (1), *Panax* (3), *Schefflera* (1).
Arecaceae	8/9	*Areca* (1), *Borassus* (1), *Calamus* (1), *Caryota* (1), *Cocos* (1), *Lodoicea* (1), *Phoenix* (2), *Serenoa* (1).
Aristolochiaceae	2/4	*Aristolochia* (3), *Asarum* (1).
Asclepiadaceae	10/13	*Calotropis* (2), *Caralluma* (1), *Ceropegia* (2), *Cryptolepis* (1), *Gymnema* (1), *Hemidesmus* (1), *Holostemma* (1), *Leptadenia* (1), *Sarcostemma* (2), *Tylophora* (1).
Asparagaceae	1/3	*Polygonatum* (3).
Aspleniaceae	1/2	*Asplenium* (2).
Asteraceae	39/59	*Achillea* (1), *Ageratum* (1), *Amberboa* (1), *Anacyclus* (1), *Anthemis* (1), *Arnica* (1), *Artemisia* (7), *Bidens* (2), *Blumea* (1), *Calendula* (1), *Centaurea* (1), *Centratherum* (1), *Cichorium* (2), *Cirsium* (2), *Cnicus* (1), *Doronicum* (4), *Echinops* (1), *Eclipta* (1), *Elephantopus* (1), *Erigeron* (2), *Eupatorium* (4), *Hypochoeris* (1), *Inula* (1), *Jurinea* (1), *Lactuca* (2), *Matricaria* (1), *Parthenium* (1), *Saussurea* (2), *Senecio* (1), *Sonchus* (2), *Sphaeranthus* (1), *Spilanthes* (2), *Tugetes* (1), *Tanacetum* (1), *Taraxacum* (1), *Tricholepis* (1), *Vernonia* (1), *Wedelia* (1), *Xanthium* (1).
Balsaminaceae	1/1	*Impatiens* (1).
Bambusaceae	1/1	*Bambusa* (1).
Basellaceae	1/1	*Basella* (1).
Berberidaceae	2/6	*Berberis* (5), *Podophyllum* (1).
Betulaceae	3/3	*Betula* (1), *Corylus* (1), *Ostrya* (1).
Bignoniaceae	4/5	*Stereospermum* (2), *Tabebuia* (1), *Tecoma* (1), *Oroxylum* (1).
Bixaceae	1/1	*Bixa* (1).
Bombacaceae	3/3	*Bombax* (1), *Ceiba* (1), *Durio* (1).
Boraginaceae	5/6	*Caccinea* (1), *Cordia* (2), *Lithospermum* (1), *Onosma* (1), *Tricodesma* (1).
Brassicaceae	14/17	*Armoracia* (1), *Barbarea* (1), *Brassica* (3), *Capsella* (1), *Cardamine* (1), *Cheiranthus* (1), *Descurainea* (1), *Eruca* (1), *Lepidium* (1), *Matthiola* (2), *Nasturtium* (1), *Rorippa* (1), *Sissymbrium* (1), *Thlaspi* (1).
Bromeliaceae	1/1	*Ananas* (1).
Burseraceae	2/3	*Boswellia* (1), *Commiphora* (2).
Campanulaceae	2/4	*Codonopsis* (3), *Platycodon* (1).

Name of Family	Genus/ Species	Total Number of Species under a Genus
Cannabinaceae	1/1	*Cannabis* (1).
Cannaceae	1/1	*Canna* (1).
Capparaceae	4/7	*Capparis* (4), *Cleome* (1), *Crataeva* (1), *Maerua* (1).
Caprifoliaceae	2/4	*Sambucus* (2), *Viburnum* (2).
Caricaceae	1/1	*Carica* (1).
Caryophyllaceae	1/1	*Drymaria* (1).
Celastraceae	3/3	*Catha* (1), *Celastrus* (1), *Salacia* (1).
Chenopodiaceae	2/2	*Beta* (1), *Chenopodium* (1).
Chloranthaceae	1/1	*Chloranthus* (1).
Clusiaceae	1/1	*Mesua* (1).
Cochlospermaceae	1/1	*Cochlospermum* (1).
Combretaceae	3/9	*Anogeissus* (1), *Calycopteris* (1), *Terminalia* (7).
Commelinaceae	1/1	*Aneilema* (1).
Connaraceae	1/1	*Rourea* (1).
Convallariaceae	1/1	*Convallaria* (1).
Convolvulaceae	8/15	*Argyreia* (1), *Convolvulus* (2), *Cressa* (1), *Cuscuta* (2), *Evolvulus* (1), *Ipomea* (6), *Operculina* (1), *Turbina* (1).
Costaceae	1/1	*Costus* (1).
Crassulaceae	2/2	*Kalanchoe* (1), *Rhodiola* (1).
Cucurbitaceae	14/18	*Benincasa* (1), *Bryonopsis* (1), *Citrullus* (1), *Coccinia* (1), *Cucumis* (2), *Cucurbita* (1), *Diplocyclos* (1), *Lagenaria* (1), *Luffa* (2), *Melothria* (1), *Momordica* (2), *Solena* (1), *Trichosanthes* (2), *Zehneria* (1).
Cupressaceae	2/3	*Juniperus* (2), *Thuja* (1).
Cycadaceae	1/1	*Cycas* (1).
Cyperaceae	1/9	*Cyperus* (9).
Dicksoniaceae	1/1	*Cibotium* (1).
Dilleniaceae	1/1	*Dillenia* (1).
Dioscoreaceae	1/3	*Dioscorea* (3).
Dipsacaceae	1/1	*Dipsacus* (1).
Dipterocarpaceae	2/2	*Shorea* (1), *Vateria* (1).
Droseraceae	1/1	*Drosera* (1).
Ebenaceae	1/1	*Diospyros* (1).
Elaeagnaceae	2/3	*Elaeagnus* (2), *Hippophae* (1).
Elaeocarpaceae	1/1	*Elaeocarpus* (1).
Ephedraceae	1/1	*Ephedra* (1).
Ericaceae	4/7	*Chimaphila* (1), *Gaultheria* (2), *Oxydendron* (1), *Rhododendron* (3).

Name of Family	Genus/ Species	Total Number of Species under a Genus
Erythroxylaceae	1/2	*Erythroxylum* (2).
Euphorbiaceae	11/15	*Acalypha* (1), *Antidesma* (1), *Baliospermum* (1), *Drypetes* (1), *Emblica* (1), *Euphorbia* (3), *Jatropha* (2), *Mallotus* (1), *Sapium* (2), *Securinega* (1), *Synadenium* (1).
Fabaceae	54/80	*Abrus* (1), *Acacia* (3), *Albizzia* (1), *Alysicarpus* (1), *Arachis* (1), *Atylosia* (1), *Baptisia* (1), *Bauhinia* (3), *Butea* (2), *Caesalpinia* (2), *Cassia* (3), *Cicer* (1), *Clitoria* (1), *Crotalaria* (1), *Cyamopsis* (1), *Cytisus* (1), *Dalbergia* (1), *Derris* (1), *Desmodium* (4), *Dioclea* (1), *Dolichos* (1), *Entada* (2), *Erythrina* (2), *Glycyrrhiza* (1), *Haematoxylon* (1), *Indigofera* (6), *Lathyrus* (1), *Lens* (1), *Medicago* (1), *Mimosa* (1), *Mucuna* (2), *Mundulea* (1), *Myroxylon* (1), *Ormocarpum* (1), *Ougeinia* (1), *Phaseolus* (1), *Physostigma* (1), *Piscidia* (1), *Pongamia* (1), *Pseudarthria* (1), *Pterocarpus* (2), *Pueraria* (1), *Saraca* (1), *Sesbania* (3), *Smithia* (1), *Tamarindus* (1), *Tephrosia* (2), *Teramnus* (1), *Trifolium* (1), *Trigonella* (1), *Uraria* (1), *Vigna* (4), *Wagatea* (1), *Zornia* (1).
Fagaceae	2/2	*Castanea* (1), *Quercus* (1).
Flacourtiaceae	2/2	*Flacourtia* (1), *Hydnocarpus* (1).
Fumariaceae	1/1	*Fumaria* (1).
Gentianaceae	6/16	*Canscora* (2), *Enicostemma* (1), *Exacum* (1), *Gentiana* (6), *Menyanthes* (1), *Swertia* (5).
Ginkgoaceae	1/1	*Ginkgo* (1).
Hamamelidaceae	1/1	*Liquidambar* (1).
Helminthostachyaceae	1/1	*Helminthostachys* (1).
Hippocastanaceae	1/1	*Aesculus* (1).
Hydrangeaceae	1/1	*Hydrangea* (1).
Hypericaceae	1/3	*Hypericum* (3).
Hypoxidaceae	1/1	*Curculigo* (1).
Iridaceae	1/1	*Crocus* (1).
Juglandaceae	1/2	*Juglans* (2).
Krameriaceae	1/2	*Krameria* (2).
Lamiaceae	30/50	*Ajuga* (1), *Anisomeles* (1), *Calamintha* (1), *Colebrookia* (1), *Coleus* (1), *Glechoma* (1), *Hyptis* (2), *Hyssopus* (1), *Lamium* (1), *Leonurus* (2), *Leucas* (4), *Majorana* (1), *Marrubium* (1), *Melissa* (2), *Mentha* (6), *Meriandra* (1), *Moschosma* (1), *Nepeta* (4), *Ocimum* (4), *Origanum* (1), *Perilla* (1),

Name of Family	Genus/ Species	Total Number of Species under a Genus
		Plectranthus (1), *Pogostemon* (2), *Prunella* (1), *Rosmarinus* (1), *Roylea* (1), *Salvia* (1), *Scutellaria* (2), *Thymus* (2), *Zataria* (1).
Lauraceae	5/9	*Cassytha* (1), *Cinnamomum* (4), *Laurus* (1), *Litsea* (2), *Sassafras* (1).
Lecythidaceae	1/1	*Barringtonia* (1).
Liliaceae	12/20	*Allium* (3), *Aloe* (2), *Asparagus* (3), *Chlorophytum* (2), *Fritillaria* (1), *Gloriosa* (1), *Lilium* (1), *Liriope* (1), *Sansevieria* (2), *Smilax* (1), *Tulipa* (1), *Urginea* (2).
Linaceae	1/1	*Linum* (1).
Loganiaceae	3/5	*Fagraea* (1), *Gelsemium* (1), *Strychnos* (3).
Loranthaceae	1/1	*Dendrophthoe* (1).
Lycopodiaceae	1/1	*Lycopodium* (1).
Lythraceae	4/5	*Heimia* (1), *Lagerstroemia* (2), *Lawsonia* (1), *Woodfordia* (1).
Magnoliaceae	4/4	*Illicium* (1), *Liriodendron* (1), *Magnolia* (1), *Michelia* (1).
Malpighiaceae	1/1	*Hiptage* (1).
Malvaceae	7/15	*Abelmoschus* (2), *Abutilon* (1), *Gossypium* (1), *Hibiscus* (3), *Malva* (1), *Pterospermum* (1), *Sida* (6).
Marantaceae	1/1	*Maranta* (1).
Marsileaceae	1/1	*Marsilea* (1).
Melastomataceae	1/1	*Memecylon* (1).
Meliaceae	8/9	*Azadirachta* (1), *Heynea* (1), *Melia* (2), *Sandorium* (1), *Soymida* (1), *Swietenia* (1), *Toona* (1), *Walsura* (1).
Menispermaceae	9/10	*Chondrodendron* (1), *Cissampelos* (1), *Cocculus* (2), *Coscinium* (1), *Cyclea* (1), *Jateorhiza* (1), *Menispermum* (1), *Stephania* (1), *Tinospora* (1).
Moraceae	4/9	*Artocarpus* (2), *Broussonetia* (1), *Ficus* (5), *Streblus* (1).
Moringaceae	1/1	*Moringa* (1).
Musaceae	1/1	*Musa* (1).
Myricaceae	1/1	*Myrica* (1).
Myristicaceae	1/1	*Myristica* (1).
Myrsinaceae	3/3	*Anagallis* (1), *Ardisia* (1), *Embelia* (1).
Myrtaceae	6/7	*Careya* (1), *Eucalyptus* (1), *Myrtus* (1), *Pimenta* (1), *Psidium* (1), *Syzygium* (2).
Nyctaginaceae	1/1	*Boerhavia* (1).
Nymphaeaceae	3/5	*Euryale* (1), *Nelumbo* (1), *Nymphaea* (3).

Name of Family	Genus/ Species	Total Number of Species under a Genus
Ochnaceae	2/2	*Ochna* (1), *Ouratea* (1).
Oleaceae	4/7	*Fraxinus* (1), *Jasminum* (4), *Myxopyrum* (1), *Nyctanthus* (1).
Onagraceae	1/1	*Oenothera* (1).
Ophioglossaceae	1/2	*Ophioglossum* (2).
Orchidaceae	11/19	*Acampe* (1), *Dactylorhiza* (1), *Dendrobium* (3), *Desmotrichum* (1), *Eulophia* (1), *Habenaria* (5), *Liparis* (2), *Malaxis* (2), *Nervilia* (1), *Pholidota* (1), *Satyrium* (1).
Osmundaceae	1/1	*Osmunda* (1).
Oxalidaceae	3/3	*Averrhoa* (1), *Biophytum* (1), *Oxalis* (1).
Paeoniaceae	1/1	*Paeonia* (1).
Pandanaceae	1/1	*Pandanus* (1).
Papaveraceae	6/8	*Argemone* (1), *Corydalis* (2), *Dicentra* (1), *Hypecoum* (1), *Meconopsis* (1), *Papaver* (2).
Parkeriaceae	1/1	*Ceratopteris* (1).
Parmeliaceae	2/2	*Cetraria* (1), *Parmelia* (1).
Passifloraceae	1/2	*Passiflora* (2).
Pedaliaceae	2/2	*Pedalium* (1), *Sesamum* (1).
Phytolaccaceae	1/1	*Phytolacca* (1).
Pinaceae	3/4	*Abies* (1), *Cedrus* (1), *Pinus* (2).
Piperaceae	2/8	*Peperomia* (1), *Piper* (7).
Pittosporaceae	1/1	*Pittosporum* (1).
Plantaginaceae	1/2	*Plantago* (2).
Plumbaginaceae	1/2	*Plumbago* (2).
Poaceae	19/23	*Avena* (1), *Coix* (1), *Cymbopogon* (3), *Cynodon* (1), *Dendrocalamus* (1), *Eleusine* (1), *Eragrostis* (1), *Heteropogon* (1), *Hordeum* (1), *Hygroryza* (1), *Imperata* (1), *Oryza* (1), *Panicum* (2), *Paspalum* (1), *Pennisetum* (2), *Phyllostachys* (1), *Saccharum* (1), *Triticum* (1), *Vetiveria* (1).
Polemoniaceae	1/1	*Polemonium* (1).
Polygonaceae	4/12	*Bistorta* (1), *Polygonum* (7), *Rheum* (2), *Rumex* (2).
Polypodiaceae	1/1	*Drynaria* (1).
Pontederiaceae	1/1	*Monochoria* (1).
Primulaceae	1/2	*Primula* (2).
Proteaceae	1/1	*Grevillea* (1).
Pteridaceae	1/1	*Pteris* (1).
Punicaceae	1/1	*Punica* (1).

Name of Family	Genus/ Species	Total Number of Species under a Genus
Ranunculaceae	13/23	*Aconitum* (8), *Actaea* (1), *Adonis* (1), *Anemone* (1), *Aquilegia* (1), *Cimicifuga* (2), *Coptis* (1), *Delphinium* (1), *Hydrastis* (1), *Nigella* (1), *Pulsatilla* (1), *Ranunculus* (2), *Thalictrum* (2).
Rhamnaceae	3/6	*Rhamnus* (2), *Ventilago* (1), *Ziziphus* (3).
Rosaceae	10/22	*Crataegus* (1), *Cydonia* (1), *Eriobotrya* (1), *Geum* (1), *Gillenia* (1), *Potentilla* (5), *Prunus* (6), *Pyrus* (1), *Rosa* (3), *Rubus* (2).
Rubiaceae	18/27	*Anthocephalus* (1), *Asperula* (1), *Borreria* (1), *Canthium* (1), *Cephalenthus* (1), *Cinchona* (3), *Coffea* (2), *Gardenia* (1), *Haldina* (1), *Hedyotis* (3), *Ixora* (2), *Morinda* (2), *Mussaenda* (1), *Ophiorrhiza* (1), *Paederia* (1), *Pavetta* (1), *Randia* (1), *Rubia* (3).
Rutaceae	13/22	*Aegle* (1), *Citrus* (4), *Euodia* (1), *Evodia* (2), *Feronia* (1), *Galipea* (1), *Garcinia* (1), *Hesperethusa* (1), *Murraya* (2), *Peganum* (1), *Ruta* (2), *Toddalia* (1), *Zanthoxylum* (4).
Salicaceae	2/5	*Populus* (2), *Salix* (3).
Salvadoraceae	1/1	*Salvadora* (1).
Samydaceae	1/1	*Casearia* (1).
Santalaceae	1/1	*Santalum* (1).
Sapindaceae	7/8	*Allophylus* (1), *Cardiospermum* (1), *Dimocarpus* (1), *Litchi* (1), *Sapindus* (2), *Schleichera* (1), *Xerospermum* (1).
Sapotaceae	3/4	*Madhuca* (1), *Manilkara* (2), *Mimusops* (1).
Saururaceae	1/1	*Houttuynia* (1).
Saxifragaceae	3/4	*Bergenia* (2), *Parnassia* (1), *Saxifraga* (1).
Scrophulariaceae	12/15	*Bacopa* (1), *Chelone* (1), *Digitalis* (2), *Euphrasia* (1), *Limnophila* (2), *Mazus* (1), *Pedicularis* (1), *Picrorhiza* (1), *Rehmania* (1), *Schweinfurthia* (1), *Verbascum* (2), *Veronicastrum* (1).
Selaginellaceae	1/1	*Selaginella* (1).
Simaroubaceae	3/4	*Ailanthus* (2), *Quassia* (1), *Samadera* (1).
Solanaceae	13/24	*Atropa* (1), *Capsicum* (1), *Datura* (4), *Duboisea* (1), *Hyoscyamus* (2), *Iochroma* (1), *Lycopersicon* (1), *Mandragora* (1), *Nicotiana* (1), *Physalis* (1), *Physochlaina* (1), *Solanum* (7), *Withania* (2).
Sterculiaceae	5/6	*Abroma* (1), *Cola* (2), *Pterygota* (1), *Sterculia* (1), *Theobroma* (1).
Symplocaceae	1/2	*Symplocos* (2).
Taxaceae	1/1	*Taxus* (1).
Tectariaceae	1/1	*Tectaria* (1).

Name of Family	Genus/ Species	Total Number of Species under a Genus
Theaceae	1/1	*Camellia* (1).
Thymelaeaceae	2/2	*Aquilaria* (1), *Daphne* (1).
Tiliaceae	4/8	*Corchorus* (4), *Grewia* (2), *Tilia* (1), *Triumfetta* (1).
Trapaceae	1/1	*Trapa* (1).
Tropaeolaceae	1/1	*Tropaeolum* (1).
Typhaceae	1/1	*Typha* (1).
Urticaceae	2/4	Humulus (1), Urtica (3).
Vacciniaceae	1/1	Vaccinium (1).
Valerianaceae	2/6	Nardostachys (1), Valeriana (5).
Verbenaceae	9/13	Callicarpa (1), Clerodendrum (4), Gmelina (1), Lantana (1), Premna (1), Pygmaeopremna (1), Tectona (1), Verbena (1), Vitex (2).
Violaceae	2/2	Hybanthus (1), Viola (1).
Viscaceae	1/1	Viscum (1).
Vitaceae	2/2	Leea (1),Vitis (1).
Zingiberaceae	8/18	Alpinia (3), Amomum (2), Curcuma (5), Elettaria (1), Hedychium (2), Kaempferia (1), Roscoea (1), Zingiber (3).
Zygophyllaceae	2/3	Fagonia (2), Tribulus (1).

Appendix III - Index to Plant Part/s Used for Tonic Purposes

Aerial Parts	**Aerial Parts**	*Artemisia abrotanum, Asperula odorata, Glechoma hederacea, Leonurus cardiaca, Meconopsis aculeata, Verbena officinalis.*
	Branch	*Anisomeles indica.*
	Fronds	*Adiantum capillus-veneris, Adiantum incisum, Adiantum vensutum, Ceratopteris siliquosa, Ophioglossum reticulatum, Ophioglossum vulgatum, Pteris vittata.*
	Shoot	*Bambusa arundinacea, Calamintha clinopodium, Scutellaria prostrata, Shorea robusta.*
	Twigs	*Cytisus scoparius.*
Bark	**Inner Bark**	*Juglans cinerea.*
	Root	*Abroma augusta, Acanthopanax trifoliatus, Aegle marmelos, Ailanthus excelsa, Ardisia solanacea, Asparagus adscendens, Azadirachta indica, Berberis aristata, Berberis umbellata, Capparis spinosa, Casearia tomentosa, Ceiba pentandra, Melia azedarach, Mundulea sericea, Oroxylum indicum, Salvadora persica, Sapium sebiferum, Schefflera octophylla, Sida spinosa, Stereospermum suaveolens, Tamarindus indica, Toddalia asiatica, Ventilago madraspatana.*
	Stem	*Acacia catechu, Acacia nilotica, Acanthopanax trifoliatus, Aesculus hippocastanum, Ailanthus excelsa, Ailanthus triphysa, Alangium salviifolium, Albizzia lebbeck, Alstonia scholaris, Anthocephalus chinensis, Antidesma ghaesembilla, Azadirachta indica, Bauhinia variegata, Berberis lycium, Berberis ulicina, Betula utilis, Bombax ceiba, Boswellia serrata,*

Butea monosperma, Caesalpinia pulcherrima, Callicarpa arborea, Capparis spinosa, Celastrus paniculatus, Cephalenthus occidentalis, Cinchona calisaya, Cinchona ledgeriana, Cinchona officinalis, Cinnamomum obtusifolium, Cinnamomum zeylanicum, Cordia dichotoma, Crataeva nurvala, Erythroxylum monogynum, Ficus benghalensis, Ficus racemosa, Flacourtia indica, Fraxinus excelsior, Galipea officinalis, Gmelina arborea, Grewia tiliifolia, Haldina cordifolia, Heynea trijuga, Hiptage benghalensis, Holarrhena antidysenterica, Ixora parviflora, Madhuca indica, Magnolia grandiflora, Manilkara hexandara, Memecylon augustifolium, Mimusops elengi, Murraya koenigii, Murraya paniculata, Myrica esculenta, Ochna jabotapita, Oroxylum indicum, Ostrya virginiana, Populus alba, Populus ciliata, Prunus cerasoides, Psidium guajava, Pterocarpus marsupium, Pterospermum acerifolium, Quercus velutina, Rhamnus purpurea, Rhamnus triquetra, Salix babylonica, Samadera indica, Saraca asoca, Schefflera octophylla, Schinus terebinthifolius, Sesbania grandiflora, Shorea robusta, Soymida febrifuga, Spondias pinnata, Stereospermum suaveolens, Swietenia mahagoni, Symplocos paniculata, Symplocos racemosa, Terminalia arjuna, Terminalia catappa, Terminalia chebula, Terminalia crenulata, Terminalia myriocarpa, Terminalia paniculata, Thevetia peruviana, Toona ciliata, Vateria indica, Viburnum opulus, Wrightia tinctoria, Zanthoxylum armatum, Zanthoxylum limonella, Zanthoxylum nitidum, Zanthoxylum oxyphyllum, Ziziphus jujuba, Ziziphus oenoplia.

Fruit, its Types and Rind	**Berry**	*Lycopersicon esculentum, Solanum nigrum, Viburnum opulus, Viscum album.*
	Drupe	*Cocos nucifera.*
	Endosperm	*Cocos nucifera.*
	Fruit	*Aegle marmelos, Alangium salviifolium, Allophylus serratus, Ammi majus, Ananas comosus, Anethum graveolens, Angelica glauca, Annona squamosa, Apium graveolens, Artocarpus communis, Artocarpus*

heterophyllus, Averrhoa carambola, Azadirachta indica, Barringtonia acutangula ssp. *spicata, Benincasa hispida, Berberis ulicina, Borassus flabelliffer, Buchanania lanzan, Caesalpinia crista, Careya arborea, Carica papaya, Chimaphila maculata, Citrullus vulgaris, Citrus aurantium, Citrus limon, Citrus medica, Coccinia grandis, Corchorus fascicularis, Cordia dichotoma, Crataegus oxyacantha, Cucumis melo* var. *utilissimus, Cucumis sativus, Cucurbita maxima, Cydonia oblonga, Dillenia indica, Dimocarpus longan, Dioscorea bulbifera, Drypetes roxburghii, Durio zibethinus, Elaeagnus infundibularis, Elettaria cardamomum, Embelia ribes, Emblica officinalis, Euodia officinalis, Feronia limonia, Ficus benghalensis, Ficus hispida, Ficus racemosa, Ficus virens, Foeniculum vulgare, Garcinia indica, Gardenia jasminoides, Gmelina arborea, Grewia microcos, Grewia tiliifolia, Hesperethusa crenulata, Juglans regia, Juniperus communis, Leptadenia reticulata, Litchi chinensis, Luffa acutangula, Luffa cylindrica, Madhuca indica, Mangifera indica, Manilkara hexandara, Melia azedarach, Melia composite, Melissa officinalis, Melissa parviflora, Michelia champaca, Mimusops elengi, Momordica charantia, Moringa oleifera, Musa paradisiaca, Myrtus communis, Nelumbo nucifera, Pedalium murex, Peucedanum grande, Phoenix pusilla, Phoenix sylvestris, Physalis minima, Pimenta dioica, Piper brachystachyum, Piper cubeba, Piper longum, Podophyllum hexandrum, Prangos pabularia, Prunus cerasus, Prunus domestica, Psidium guajava, Punica granatum, Rhus succedanea, Rosa laevigata, Rubus coreanus, Sapindus mukorossi, Sapindus trifoliatus, Schleichera oleosa, Schweinfurthia sphaerocarpa, Scindapsus officinalis, Semecarpus anacardium, Shorea robusta, Solanum melongena, Solanum torvum, Spondias pinnata, Syzygium cumini, Tamarindus indica, Teramnus labialis, Terminalia arjuna, Terminalia bellirica, Terminalia chebula, Thevetia peruviana, Tinospora cordifolia, Trachyspermum ammi,*

		Trachyspermum roxburghianum, Trapa natans var. *bispinosa, Tribulus terrestris, Viburnum stellulatum, Vitex negundo, Vitex trifolia, Vitis vinifera, Xanthium strumarium, Zanthoxylum armatum, Zanthoxylum nitidum, Ziziphus jujuba, Ziziphus mauritiana, Ziziphus oenoplia.*
	Grain	*Eleusine coracana, Hordeum vulgare, Oryza sativa, Paspalum scrobiculatum, Pennisetum americanum, Triticum aestivum.*
	Kernel	*Lodoicea maldivica, Prunus cerasus, Semecarpus anacardium.*
	Nut	*Areca catechu, Caryota urens, Corylus colurna.*
	Pod	*Cyamopsis tetragonoloba.*
	Pulp	*Aegle marmelos, Cassia fistula, Hippophae rhamnoides* subsp. *turkestanica, Sapindus trifoliatus.*
	Rind	*Mangifera indica, Pongamia pinnata.*
Inflorescence, Bud, Flower and Pollen Grains	**Calyx**	*Hibiscus sabdariffa.*
	Carpel	*Nelumbo nucifera.*
	Florets	*Centaurea cyanus.*
	Flower	*Acacia nilotica, Aesculus hippocastanum, Albizzia lebbeck, Anthemis nobilis, Azadirachta indica, Brassica oleracea* var. *botrytis, Butea monosperma, Calotropis gigantia, Calotropis procera* ssp. *hamiltonii, Careya arborea, Convolvulus arvensis, Convolvulus microphyllus, Crocus sativus, Elaeagnus infundibularis, Erythrina stricta, Evodia lunu-ankenda, Evolvulus alsinoides, Gossypium arboretum* var. *obtusifolium, Hibiscus rosa-sinensis, Hiptage benghalensis, Humulus lupulus, Impatiens balsamina, Indigofera gerardiana, Jasminum auriculatum, Jasminum multiflorum, Justicia adhatoda, Lamium album, Lawsonia inermis, Linum usitatissimum, Madhuca indica, Mesua ferrea, Michelia champaca, Mimusops elengi, Moringa oleifera, Nelumbo nucifera, Nymphaea nouchali, Ocimum americanum, Onosma bracteatum, Pavetta indica* ssp. *tomentosa, Pterospermum acerifolium, Rhododendron arboreum, Rhododendron campanulatum, Rosa centifolia, Rosa damascena, Semecarpus anacardium, Shorea robusta, Sphaeranthus indicus,*

		Stereospermum colais, Vitex negundo, Vitis vinifera, Woodfordia fruticosa, Xanthium strumarium.
	Flower Bud	*Hibiscus sabdariffa, Phoenix sylvestris, Saraca asoca, Syzygium aromaticum, Xanthium strumarium.*
	Flowering Tops	*Achillea millefolium, Arnica Montana, Nepata cataria, Spilanthes paniculata, Tanacetum vulgare.*
	Inflorescence	*Urtica parviflora.*
	Pollen Grains	*Typha angustata.*
	Strobilus	*Humulus lupulus.*
Leaf	**Bud**	*Ficus religiosa.*
	Leaf	*Abies spectabilis, Abrus precatorius, Achillea millefolium, Aconitum deinorrhizum, Ageratum conyzoides, Ailanthus excelsa, Ajuga bracteosa, Albizzia lebbeck, Aloe perryi, Amaranthus spinosus, Andrographis paniculata, Annona squamosa, Artemisia cina, Artemisia indica, Artemisia maritima, Asparagus racemosus, Asparagus scandens, Bambusa arundinacea, Barringtonia acutangula* ssp. *spicata, Bergenia ciliata, Beta vulgaris, Blepharis asperrima, Blepharis persica, Brassica oleracea, Buchanania lanzan, Butea monosperma, Caesalpinia pulcherrima, Calycopteris floribunda, Camellia sinensis, Cannabis sativa, Cardiospermum halicacabum, Castanea sativa, Celastrus paniculatus, Chelone glabra, Chimaphila maculata, Cinnamomum tamala, Clerodendrum indicum, Clerodendrum infortunatum, Clerodendrum multiflorum, Clerodendrum serratum, Colebrookia oppositifolia, Coleus amboinicus, Corchorus capsularis, Corchorus depressus, Corchorus fascicularis, Corchorus olitorius, Crotalaria albida, Cryptolepis buchananii, Cynodon dactylon, Daphne cannabina, Digitalis lanata, Digitalis purpurea, Drosera peltata, Drymaria cordata, Eclipta prostrata, Erythroxylum coca, Euphorbia neriifolia, Euphorbia pilosa, Euryale ferox, Evodia lunu-ankenda, Feronia limonia, Ficus religiosa, Heimia salicifolia, Hemidesmus indicus, Heynea trijuga, Hiptage benghalensis, Holarrhena antidysenterica,*

		Houttuynia cordata, Indigofera articulata, Indigofera glabra, Indigofera tinctoria, Ipomoea pes – caprae, Jasminum arborescens, Juglans regia, Justicia adhatoda, Kalanchoe integra, Lactuca indica, Lactuca serriola, Lantana camara, Lawsonia inermis, Malva parviflora, Melissa officinalis, Mentha arvensis, Menyanthes trifoliata, Meriandra bengalensis, Michelia champaca, Monochoria vaginalis, Morinda citrifolia, Morinda pubescens, Mucuna prurita, Murraya koenigii, Murraya paniculata, Myxopyrum serratulum, Nepata cataria, Nerium indicum, Nicotiana tabacum, Nyctanthes arbor – tristis, Ocimum americanum, Ocimum sanctum, Onosma bracteatum, Ouratea serrata, Oxalis corniculata, Oxydendrum arboratum, Paederia foetida, Pandanus odoratissimus, Pedalium murex, Piper betle, Piper puberulum, Pterospermum acerifolium, Ranunculus laetus, Rauvolfia vomitora, Rhododendron anthopogon, Rhododendron campanulatum, Rosmarinus officinalis, Roylea cinerea, Rubia cordifolia, Salix babylonica, Salix caprea, Salix elegans, Salvadora persica, Salvia officinalis, Sapium insigne, Schweinfurthia sphaerocarpa, Sesbania grandiflora, Sonchus oleraceus, Sphaeranthus indicus, Tanacetum vulgare, Thuja orientalis, Thymus serpyllum, Tribulus terrestris, Urtica parviflora, Verbena officinalis, Vigna trilobata, Vitex negundo, Wedelia chinensis, Withania coagulans, Zornia diphylla.
	Oil	*Eucalyptus globules, Salix caprea.*
Root, its Modification and Exudates	**Aerial Roots**	*Ficus benghalensis.*
	Exudates	*Ferula asafoetida.*
	Root	*Abroma augusta, Abrus precatorius, Abutilon indicum, Aconitum bisma, Aconitum deinorrhizum, Aconitum heterophyllum, Aconitum napellus, Aconitum violaceum, Aerva lanata, Alisma plantago-aquatica, Amaranthus spinosus, Anacyclus pyrethrum, Andrographis paniculata, Aneilema scapiflorum, Angelica glauca, Anodendron manubrium, Argyreia nervosa, Aristolochia indica, Aristolochia saccata, Aristolochia tagala, Artemisia indica, Asparagus*

adscendens, Asparagus racemosus, Baliospermum montanum, Bambusa arundinacea, Barringtonia acutangula ssp. *spicata, Berberis chitra, Berberis lycium, Berberis ulicina, Bergenia ciliata, Beta vulgaris, Bistorta affinis, Boerhavia diffusa, Bombax ceiba, Bupleurum falcatum, Butea superba, Calamus rotang, Calotropis gigantia, Canna edulis, Canthium parviflorum, Capparis deciduas, Capparis spinosa, Cassia fistula, Ceiba pentandra, Ceropegia bulbosa, Chlorophytum borivilianum, Chondrodendron tomentosum, Cibotium barometz, Cichorium endivia, Cichorium intybus, Cirsium verutum, Cissampelos pareira, Clerodendrum multiflorum, Clitoria ternatea, Cocculus hirsutus, Cocculus pendulus, Codonopsis ovata, Coptis teeta, Corydalis govaniana, Coscinium fenestratum, Costus speciosus, Curculigo orchioides, Dactylorhiza hatagirea, Delphinium denudatum, Desmodium gangeticum, Desmodium heterophyllum, Desmodium tiliaefolium, Desmodium triflorum, Dipsacus japonicus, Doronicum falconeri, Doronicum hookeri, Doronicum pardalianches, Doronicum roylei, Drynaria quercifolia, Echinops echinatus, Eclipta prostrata, Ephedra gerardiana, Eryngium billardieri, Eryngium caeruleum, Eupatorium purpureum, Fagraea racemosa, Ficus racemosa, Gentiana algida, Gentiana kurroo, Gentiana lutea, Gentiana stracheyi, Gentiana villosa, Gillenia trifoliata, Gloriosa superba, Glycyrrhiza glabra, Gmelina arborea, Hedychium spicatum, Hemidesmus indicus, Holostemma ada-kodien, Hydrangea arborescens, Hydrastis canadensis, Hydrocotyle javanica, Hygrophila auriculata, Hygroryza aristata, Hypochoeris glabra, Hypoxis aurea, Ichnocarpus frutescens, Indigofera articulata, Indigofera tinctoria, Inula racemosa, Ipomoea batatas, Jateorhiza palmata, Justicia adhatoda, Krameria argentea, Litsea glutinosa, Maerua arenaria, Malaxis muscifera, Mallotus philippinensis, Melia azedarach, Menispermum canadense, Menyanthes trifoliata, Monochoria vaginalis, Morinda pubescens, Moringa oleifera, Mucuna prurita,*

Murraya koenigii, Murraya paniculata, Musa paradisiaca, Nardostachys jatamansi, Nelsonia campestris, Nerium indicum, Ophiorrhiza mungos, Ormocarpum cochinchinensis, Osmunda regalis, Ouratea serrata, Panax bipinnatifidus, Panax pseudo – ginseng, Pandanus odoratissimus, Parthenium hysterophorus, Pavetta indica ssp. *tomentosa, Peucedanum nagpurense, Phyllostachys bambusoides, Picrorhiza kurrooa, Platycodon grandiflorum, Plumbago zeylanica, Podophyllum hexandrum, Polygonatum cirrhifolium, Polygonatum verticillatum, Polygonum hydropiper, Polygonum multiflorium, Polygonum viviparum, Potentilla argyrophylla, Potentilla sibbaldi, Potentilla supina, Premna obtusifolia, Pseudarthria viscida, Pueraria tuberosa, Pulsatilla koreana, Pygmaeopremna herbacea, Rauvolfia vomitora, Rehmania glutinosa, Rheum emodi, Rheum webbianum, Rhodiola imbricata, Rourea minor, Rubia cordifolia, Rubia tinctorum, Ruellia patula, Rumex hymenosephalus, Rumex nepalensis, Saccharum spontaneum, Salacia reticulata, Sandorium kortjape, Sansevieria trifasciata, Sansevieria zeylanica, Sassafras albidum, Satyrium nepalense, Saussurea lappa, Senecio jacquemontianus, Sida acuta, Sida cordifolia, Sida ovata, Sida rhombifolia, Sida spinosa, Solanum indicum, Sonchus oleraceus, Stereospermum colais, Strobilanthes ciliates, Strychnos nux – vomica, Tabernaemontana divaricata, Taraxacum officinale, Tecoma stans, Tephrosia purpurea, Tephrosia virginia, Thalictrum foliolosum, Thalictrum pedunculatum, Tribulus terrestris, Trichosanthes cucumerina, Triumfetta rhomboidea, Tylophora indica, Urtica dioica, Valeriana jatamansi, Veronicastrum sibiricum, Vetiveria zizanioides, Vigna radiata, Viola odorata, Vitex negundo, Vitex trifolia, Withania somnifera, Wrightia tinctoria, Xanthium strumarium, Ziziphus jujuba, Ziziphus mauritiana.*

Root Tuber *Chlorophytum arundinaceum, Dioscorea bulbifera, Ipomoea digitata, Roscoea alpina, Solena amplexicaulis, Wagatea spicata.*

Seed and its Parts	**Cotyledon**	*Cola nitida.*
	Kernel	*Elaeocarpus sphaericus.*
	Nutmeg	*Myristica fragrans.*
	Oil	*Celastrus paniculatus, Cucurbita maxima, Origanum vulgare, Prunus armeniaca.*
	Seed	*Abelmoschus moschatus, Abrus precatorius, Achyranthes aspera, Alangium salviifolium, Albizzia lebbeck, Amomum subulatum, Apium graveolens, Arachis hypogaea, Avena sativa, Bauhinia tomentosa, Bauhinia vahlii, Bixa orellana, Blepharis asperrima, Boerhavia diffusa, Brassica juncea, Broussonetia papyrifera, Butea monosperma, Caesalpinia crista, Cardiospermum halicacabum, Cassia fistula, Celastrus paniculatus, Celosia argentea, Centratherum anthelminticum, Cheiranthus cheiri, Cicer arietinum, Citrullus vulgaris, Coffea benghalensis, Coix lacryma-jobi, Cola nitida, Corchorus depressus, Cordia macleodii, Cucumis sativus, Cucurbita maxima, Cuscuta chinensis, Cuscuta reflexa, Daucus carota, Delphinium denudatum, Descurainea Sophia, Dioclea javanica, Dolichos biflorus, Drypetes roxburghii, Echinops echinatus, Elettaria cardamomum, Entada phaseoloides, Entada rheedii, Eruca sativa, Euryale ferox, Ficus benghalensis, Gossypium arboretum* var. *obtusifolium, Holarrhena antidysenterica, Hydnocarpus laurifolia, Hygrophila auriculata, Indigofera glandulosa, Indigofera trita, Jatropha curcas, Lactuca serriola, Lagenaria siceraria, Lens culinaris, Lepidium sativum, Linum usitatissimum, Manilkara kauki, Matthiola incana, Melia azedarach, Mimosa hamata, Mimusops elengi, Mucuna monosperma, Mucuna prurita, Ocimum americanum, Papaver somniferum, Phaseolus radiatus, Plantago major, Plantago ovata, Plectranthus mollis, Polygonum barbatum, Pongamia pinnata, Prunus amygdalus, Prunus armeniaca, Punica granatum, Ranunculus sceleratus, Salvadora persica, Schleichera oleosa, Scutellaria lateriflora, Sesamum indicum, Shorea robusta, Sida cordifolia, Sissymbriun iria, Strychnos ignatii, Strychnos nux – vomica, Strychnos potatorum,*

		Tamarindus indica, Tephrosia purpurea, Theobroma cacao, Thuja orientalis, Tinospora cordifolia, Trigonella foenum – graecum, Vigna mungo, Vigna unguiculata, Zanthoxylum armatum, Zanthoxylum oxyphyllum, Ziziphus jujuba, Ziziphus mauritiana.
Stem, its Modification and Inclusions	**Bulb**	*Allium cepa, Allium sativum, Buchanania lanzan, Crinum asiaticum, Fritillaria roylei, Lilium polyphyllum, Liparis rostrata, Tulipa stellata, Urginea indica, Urginea maritima.*
	Corm	*Amorphophallus companulatus.*
	Culm	*Bambusa arundinacea, Dendrocalamus strictus.*
	Gall	*Pistacia chinensis* ssp. *integerrima, Rhus succedanea.*
	Gum	*Acacia Senegal, Anogeissus latifolia, Azadirachta indica, Bauhinia variegata, Bombax ceiba, Buchanania lanzan, Butea monosperma, Cochlospermum religiosum, Commiphora mukul, Commiphora myrrha, Sterculia urens.*
	Resin	*Vateria indica.*
	Rhizome	*Aconitum koreanum, Acorus calamus, Actinopteris australis, Adiantum pedatum, Alpinia galanga, Alpinia officinarum, Aristolochia indica, Bergenia stracheyi, Coptis teeta, Costus speciosus, Curculigo orchioides, Curcuma angustifolia, Curcuma aromatica, Curcuma caesia, Curcuma domestica, Curcuma zedoaria, Cyperus kyllingia, Cyperus longus, Drynaria quercifolia, Eulophia campestris, Gentiana kurroo, Gentiana villosa, Glycyrrhiza glabra, Hedychium coronarium, Hedychium spicatum, Helminthostachys zeylanica, Imperata cylindrica, Maranta arundinacea, Nardostachys jatamansi, Nymphaea nouchali, Picrorhiza kurrooa, Polygonatum kingianum, Polygonatum verticillatum, Rheum emodi, Rheum webbianum, Smilax china, Taraxacum officinale, Tectaria cicutaria, Thalictrum foliolosum, Valeriana jatamansi.*
	Siliceous Matter	*Dendrocalamus strictus.*
	Stem	*Basella alba, Berberis aristata, Berberis chitra, Borassus flabelliffer, Caralluma attenuata,*

		Chondrodendron tomentosum, Cocculus hirsutus, Ferula narthex, Indigofera tinctoria, Musa paradisiaca, Pedalium murex, Sarcostemma acidum, Schweinfurthia sphaerocarpa, Sonchus oleraceus, Tinospora cordifolia.
	Stem Base	*Malaxis acuminata, Malaxis muscifera.*
	Straw	*Avena sativa.*
	Tuber	*Asparagus racemosus, Ceropegia bulbosa, Ceropegia tuberosa, Cycas revoluta, Cyclea arnottii, Cyperus articulatus, Cyperus esculentus, Cyperus rotundus, Cyperus scariosus, Cyperus serotinus* var. *inundates, Cyperus stoloniferus, Dactylorhiza hatagirea, Dicentra canadensis, Dioscorea alata, Dioscorea pentaphylla, Gloriosa superba, Habenaria acuminata, Habenaria edgeworthii, Habenaria goodyeroides, Habenaria griffithi, Habenaria intermedia, Liriope platyphylla, Pueraria tuberosa.*
Whole Plant		*Acampe praemorsa, Achillea millefolium, Aconitum napellus, Adiantum capillus-veneris, Adiantum vensutum, Adonis aestivalis, Aerva lanata, Ageratum conyzoides, Ajuga bracteosa, Aloe barbadensis, Alpinia calcarata, Alysicarpus pubescens, Amaranthus spinosus, Amberboa divaricata, Andrographis echioides, Andrographis paniculata, Anisomeles indica, Aquilaria agallocha, Areca catechu, Argemone mexicana, Artemisia absinthium, Artemisia pallens, Asplenium trichomanes, Atylosia scarabaeoides, Azadirachta indica, Bacopa monnieri, Barleria prionitis* subsp. *dicantha, Basella alba, Bidens pilosa, Biophytum sensitivum, Blepharis linariaefolia, Blumea lacera, Boerhavia diffusa, Bombax ceiba, Boswellia serrata, Bryonopsis laciniosa, Caccinea glauca, Calamintha clinopodium, Calendula officinalis, Calotropis procera* ssp. *hamiltonii, Cannabis sativa, Canscora decussata, Canscora diffusa, Capparis deciduas, Capparis sepiaria, Carum carvi, Cassia occidentalis, Cassytha filiformis, Centella asiatica, Cetraria islandica, Chelone glabra, Chenopodium album, Cichorium intybus, Cirsium arvense, Citrus medica, Cnicus benedictus, Codonopsis clematidea,*

Convallaria majalis, Convolvulus microphyllus, Corchorus depressus, Corchorus olitorius, Corydalis flabellata, Cressa cretica, Cuscuta reflexa, Cycas revoluta, Cymbopogon martinii, Cynodon dactylon, Cyperus iria, Dendrobium alpestre, Dendrobium normale, Desmodium gangeticum, Desmotrichum fimbriatum, Dicliptera roxburghiana, Diplocyclos palmatus, Drynaria quercifolia, Echinops echinatus, Eclipta prostrata, Elephantopus scaber, Enicostemma littorale, Ephedra gerardiana, Eryngium foetidum, Eupatorium cannabinum, Eupatorium perfoliatum, Eupatorium triplinerve, Euphrasia officinalis, Evodia lunu-ankenda, Evolvulus alsinoides, Exacum tetragonum, Fagonia cretica, Fagonia indica, Fumaria officinalis, Gentiana kurroo, Gentiana lutea, Gentiana prostrata, Geum elatum, Ginkgo biloba, Glinus oppositifolius, Gymnema sylvestre, Hedyotis corymbosa, Hedyotis herbacea, Houttuynia cordata, Hybanthus enneaspermus, Hygrophila auriculata, Hyptis capitata, Indigofera trita, Ipomoea pes – caprae, Ipomoea sepiaria, Jasminum grandiflorum, Justicia beddomei, Krameria triandra, Lantana camara, Leonurus japonicus, Leptadenia reticulata, Leucas aspera, Limnophila gratioloides, Liparis paradoxa, Liquidambar orientalis, Majorana hortensis, Marrubium vulgare, Marsilea minuta, Matthiola flavida, Mazus japonicus, Melothria maderaspatana, Mentha pulegium, Menyanthes trifoliata, Momordica balsamina, Murraya koenigii, Mussaenda frondosa, Nasturtium officinale, Nepeta flocosa, Nepeta hindostana, Nervilia aragoana, Ocimum gratissimum, Onosma bracteatum, Origanum vulgare, Osmunda regalis, Oxalis corniculata, Papaver somniferum, Parthenium hysterophorus, Paspalum scrobiculatum, Pedalium murex, Pennisetum flaccidum, Peperomia tetraphylla, Permelia perlata, Peucedanum grande, Phaseolus radiatus, Pholidota articulata, Physalis minima, Plantago major, Polygonum aviculare, Polygonum chinensis, Polygonum hydropiper, Potentilla anserina, Prunella vulgaris,

		Rosmarinus officinalis, Rubia tinctorum, Ruta chalepensis, Ruta graveolens, Salvia officinalis, Saussurea sacra, Saxifraga oppositifolia, Securinega leucopyrus, Selaginella bryopteris, Sesamum indicum, Sida cordata, Sida cordifolia, Smithia conferta, Solanum indicum, Solanum nigrum, Sphaeranthus indicus, Swertia angustifolia, Swertia chirayita, Swertia ciliata, Swertia cordata, Swertia petiolata, Tephrosia purpurea, Tephrosia virginia, Terminalia bellirica, Thymus serpyllum, Thymus vulgaris, Tribulus terrestris, Tricholepis glaberrima, Trichosanthes cucumerina, Trichosanthes lobata, Tricodesma indicum, Tropaeolum majus, Uraria lagopodioides, Urtica hyperborea, Vernonia cinerea, Viscum album, Wrightia tinctoria.
Wood	**Heart Wood**	Acacia catechu, Haematoxylon campechianum, Prunus cerasoides, Santalum album.
	Wood	Aquilaria agallocha, Pinus roxburghii, Pterocarpus santalinus, Quassia indica, Samadera indica, Strychnos nux – vomica.
	Wood Oil	Cedrus deodara.

Appendix IV - Index to Plant Part/s Used for Mood Swings

Aerial Parts	**Aerial Parts**	*Citrus aurantifolia, Grevillea robusta, Ipomea carnea* ssp. *fistulosa, Jatropha curcas, Leonurus cardiaca, Melissa officinalis, Nepeta cataria, Punica granatum, Verbena officinalis.*
	Branches	*Solanum dulcamara.*
	Shoot	*Bambusa arundinacea, Calamintha clinopodium, Myrtus communis.*
Bark	**Root**	*Acanthopanax trifoliatus, Capparis zeylanica, Liriodendron tulipifera, Piscidia piscipula, Salvadora persica, Toddalia asiatica, Ventilago madraspatana.*
	Stem	*Acanthopanax trifoliatus, Aesculus hippocastanum, Alstonia scholaris, Barringtonia acutangula* ssp. *spicata, Caesalpinia pulcherrima, Cinnamomum tamala, Cinnamomum zeylanicum, Crataeva nurvala, Dendrophthoe falcata, Iochroma fuchsioides, Lagerstroemia indica, Litsea monopetala, Magnolia grandiflora, Michelia champaca, Murraya koenigii, Oenothera biennis, Ougeinia oojeinensis, Piscidia piscipula, Pittosporum floribundum, Plumeria acuminata, Pongamia pinnata, Populus ciliata, Prunus virginiana, Randia spinosa, Sambucus nigra, Schinus terebinthifolius, Walsura trifoliata, Zanthoxylum armatum, Zanthoxylum nitidum, Zanthoxylum oxyphyllum.*
Fruit, its Types and Parts	**Berry**	*Hedera nepalensis.*
	Capsules	*Papaver somniferum.*
	Fruit	*Anethum graveolens, Angelica archangelica* L. var. *himalaica, Annona squamosa, Apium*

		graveolens, Borassus flabelliffer, Bunium persicum, Caesalpinia crista, Capsicum annum, Carum carvi, Citrus limon, Citrus medica, Conium maculatum, Coriandrum sativum, Cucurbita maxima, Datura stramonium, Drypetes roxburghii, Elaeocarpus sphaericus, Elettaria cardamomum, Embelia ribes, Emblica officinalis, Eriobotrya japonica, Evodia rutaecarpa, Foeniculum vulgare, Gardenia jasminoides, Gaultheria trichophylla, Illicium verum, Juniperus communis, Juniperus marcopoda, Laurus nobilis, Ligusticum elatum, Meconopsis aculeata, Michelia champaca, Moringa oleifera, Myrica esculenta, Passiflora incarnata, Peucedanum grande, Pimenta dioica, Pimpinella anisum, Piper cubeba, Piper nigrum, Piper retrofractum, Prangos pabularia, Pyrus communis, Rhus succedanea, Sarcostemma brunonianum, Scindapsus officinalis, Serenoa serrulata, Seseli diffusum, Solanum melongena, Solanum surattense, Syzygium cumini, Taxus baccata, Trachyspermum ammi, Trachyspermum roxburghianum, Vaccinium myrtillus, Withania coagulans, Withania somnifera, Xanthium strumarium, Zanthoxylum armatum, Zanthoxylum nitidum.
	Grain	*Oryza sativa, Paspalum scrobiculatum.*
	Kernel	*Terminalia bellirica.*
	Oil	*Illicium verum.*
	Peel	*Tabebuia pentaphylla.*
	Pericarp	*Zanthoxylum limonella.*
Inflorescence, Bud and Flower	**Capitula**	*Spilanthes acmella, Tagetes minuta.*
	Florets	*Centaurea cyanus.*
	Flower	*Alstonia scholaris, Artemisia absinthium, Cheiranthus cheiri, Citrus aurantium, Citrus medica, Cochlospermum religiosum, Elaeagnus umbellata, Madhuca indica, Matricaria chamomilla, Mentha longifolia, Michelia champaca, Mimusops elengi, Moringa oleifera, Nymphaea stellata, Ocimum basilicum, Oenothera biennis, Onosma bracteatum, Passiflora incarnata, Primula veris, Rosmarinus officinalis, Sambucus nigra, Solanum surattense, Tilia europaea, Trifolium*

		pratense, Woodfordia fruticosa.
	Flower Bud	*Citrus medica, Syzygium aromaticum, Xanthium strumarium.*
	Flowering Tops	*Angelica archangelica* var. *himalaica, Anthemis nobilis, Cannabis sativa, Datura stramonium, Hyoscyamus niger, Nepeta hindostana.*
	Inflorescence	*Pedicularis pectinata, Scindapsus officinalis.*
	Petals	*Papaver rhoeas.*
	Stigma & Styles	*Crocus sativus.*
	Strobiles	*Humulus lupulus.*
Leaf	**Bracts**	*Pandanus odoratissimus.*
	Leaf	*Achillea millefolium, Ageratum conyzoides, Ajuga bracteosa, Angelica archangelica* var. *himalaica, Artemisia absinthium, Artemisia cina, Artemisia dracunculus, Artemisia maritima, Atropa belladonna, Baliospermum montanum, Bambusa arundinacea, Baptisia leucophaea, Blumea lacera, Caesalpinia pulcherrima, Camellia sinensis, Catha edulis, Cinnamomum tamala, Cochlospermum religiosum, Colocasia esculenta, Corchorus capsularis, Corchorus fascicularis, Cymbopogon citratus, Cymbopogon jwarancusa, Dalbergia sissoo, Datura innoxia, Datura metel, Datura stramonium, Digitalis lanata, Digitalis purpurea, Duboisea myoporoides, Eruca sativa, Erythrina variegata, Erythroxylum coca, Euphorbia thymifolia, Gymnema sylvestre, Hedera nepalensis, Hibiscus sabdariffa* var. *sabdariffa, Hygrophila auriculata, Hyoscyamus muticus, Hyoscyamus niger, Hyssopus officinalis, Ilex paraguariensis, Iochroma fuchsioides, Lactuca serriola, Laurus nobilis, Lepidium sativum, Leucas lavandulaefolia, Lithospermum officinale, Marsilea minuta, Melodinus monogynus, Mentha arvensis, Mentha longifolia, Menyanthes trifoliata, Moschosma polystachyum, Murraya koenigii, Murraya paniculata, Myrtus communis, Nasturtium officinale, Nepeta hindostana, Nicotiana tabacum, Ocimum sanctum, Oenothera biennis, Physochlaina praealta, Piper betle, Piper nigrum, Pogostemon parviflorus, Potentilla fruticosa, Primula veris, Rauvolfia vomitora,*

		Rhododendron anthopogon, Rosmarinus officinalis, Salvadora persica, Solanum melongena, Taxus baccata, Thevetia peruviana, Verbascum coromandelianum, Vigna trilobata, Vitex negundo, Zataria multiflora.
	Oil	*Eucalyptus globules, Gaultheria fragrantissima, Gaultheria trichophylla.*
Root, its Modification and Exudates	**Oil**	*Sassafras albidum, Vetiveria zizanioides.*
	Oleoresin	*Ferula asafoetida.*
	Root	*Aconitum ferox, Aconitum napellus, Actaea spicata, Anemone pulsatilla, Angelica glauca, Aristolochia saccata, Armoracia rusticana, Asparagus racemosus, Atropa belladonna, Barringtonia acutangula* ssp. *spicata, Bombax ceiba, Borassus flabelliffer, Capparis deciduas, Catharanthus roseus, Cimicifuga foetida, Cissampelos pareira, Cleome viscosa, Clerodendrum serratum, Cocculus hirsutus, Codonopsis ovata, Codonopsis rotundifolia, Costus speciosus, Datura metel, Delphinium denudatum, Doronicum falconeri, Eupatorium purpureum, Fritillaria roylei, Gelsemium sempervirens, Heteropogon contortus, Holostemma ada-kodien, Inula racemosa, Ixora coccinea, Jurinea dolomiaea, Lagerstroemia speciosa, Leea hirta, Maerua arenaria, Mandragora autumnalis, Matricaria chamomilla, Meconopsis aculeata, Melodinus monogynus, Moringa oleifera, Mucuna prurita, Murraya koenigii, Nardostachys jatamansi, Operculina turpethum, Ophiorrhiza mungos, Ormocarpum cochinchinensis, Osmunda regalis, Panax pseudo – ginseng, Panax quinquefolium, Petroselinum crispum, Phaseolus radiatus, Pinus roxburghii, Piper longum, Plumbago indica, Plumbago zeylanica, Podophyllum hexandrum, Pogostemon benghalense, Pogostemon parviflorus, Polemonium caeruleum, Polygonum amplexicaule, Polygonum hydropiper, Pygmaeopremna herbacea, Rauvolfia serpentina, Rauvolfia vomitora, Rheum emodi, Rubus niveus, Sansevieria trifasciata, Sassafras albidum, Saussurea lappa, Selinum vaginatum, Selinum wallichianum, Sium latijugum, Solanum melongena, Tephrosia virginia, Tylophora*

		indica, Urtica dioica, Valeriana fouriei, Valeriana hardwicikii, Valeriana jatamansi, Valeriana officinalis, Vigna mungo, Vitex negundo, Withania somnifera, Wrightia tinctoria, Zehneria umbellata, Ziziphus mauritiana.
	Root Tuber	*Aconitum chasmanthum, Anacyclus pyrethrum, Chlorophytum arundinaceum, Ipomoea digitata, Solena amplexicaulis.*
	Tap Root	*Daucus carota.*
Seed and its Parts	**Kernel**	*Elaeocarpus sphaericus.*
	Nutmeg	*Myristica fragrans.*
	Oil	*Apium graveolens, Celastrus paniculatus, Prunus amygdalus.*
	Seed	*Abelmoschus esculantus, Abelmoschus moschatus, Amomum subulatum, Amomum xanthioides, Anethum graveolens, Argemone mexicana, Baliospermum montanum, Borreria articularis, Bunium persicum, Butea superba, Caesalpinia crista, Capsella bursa-pastoris, Cardamine pratensis, Cassia absus, Celastrus paniculatus, Centratherum anthelminticum, Cleome viscosa, Coffea arabica, Coffea benghalensis, Cola acuminata, Cola nitida, Datura innoxia, Datura metel, Datura sanguinea, Datura stramonium, Daucus carota, Delphinium denudatum, Dioclea javanica, Diospyros lotus, Drypetes roxburghii, Elaeagnus umbellata, Elettaria cardamomum, Euphorbia thymifolia, Ginkgo biloba, Hyoscyamus niger, Hypericum cernuum, Hypericum patulum, Ipomoea violacea, Lactuca serriola, Lagenaria siceraria, Lagerstroemia indica, Lagerstroemia speciosa, Lathyrus aphaca, Meconopsis aculeata, Mucuna monosperma, Nigella sativa, Panicum miliaceum, Peganum harmala, Physochlaina praealta, Physostigma venenosum, Pinus gerardiana, Plantago major, Prunus armeniaca, Pterygota alata, Salvadora persica, Scutellaria lateriflora, Sesbania roxburghii, Sesbania sesban, Sissymbriun iria, Solanum melongena, Strychnos nux – vomica, Theobroma cacao, Thlaspi arvense, Turbina corymbosa, Verbascum thapsus, Withania somnifera, Ziziphus jujuba, Ziziphus*

		mauritiana.
Stem, its Modification and Inclusions	**Bulb**	*Allium cepa, Allium fistulosum, Allium sativum, Urginea indica.*
	Corm	*Pinellia ternate.*
	Culm	*Bambusa arundinacea, Eragrostis cynosuroides.*
	Gall	*Rhus succedanea.*
	Gum	*Azadirachta indica, Cochlospermum religiosum, Commiphora mukul, Commiphora myrrha.*
	Latex	*Streblus asper.*
	Oil	*Myroxylon balsamum, Origanum vulgare.*
	Rhizome	*Acorus gramineus, Adiantum pedatum, Alpinia galanga, Alpinia officinarum, Angelica archangelica* var. *himalaica, Cimicifuga racemosa, Costus speciosus, Curcuma caesia, Curcuma domestica, Curcuma zedoaria, Ferula narthex, Ferula suaveolens, Gelsemium sempervirens, Hedychium coronarium, Homalonema aromatica, Kaempferia galanga, Menyanthes trifoliata, Nardostachys jatamansi, Nymphaea alba, Podophyllum hexandrum, Rheum emodi, Valeriana jatamansi, Valeriana officinalis, Zingiber cassumunar, Zingiber officinale, Zingiber zerumbet.*
	Stem	*Basella alba, Cissampelos pareira, Cocculus hirsutus, Leea hirta, Pedicularis pectinata, Sonchus oleraceus, Zataria multiflora.*
	Tuber	*Cyperus articulatus, Cyperus esculentus, Cyperus rotundus, Cyperus serotinus* var. *inundates, Cyperus stoloniferus, Gloriosa superba, Solanum tuberosum, Stephania glabra, Typhonium trilobatum.*
Whole Plant		*Acalypha indica, Achillea millefolium, Aconitum napellus, Anagallis arvensis, Aquilaria agallocha, Aquilegia vulgaris, Argemone mexicana, Asarum europaeum, Asplenium nidus, Barbarea vulgaris, Basella alba, Bidens pilosa, Bidens tripartita, Biophytum sensitivum, Blumea lacera, Calendula officinalis, Canna edulis, Cardamine pratensis, Cardiospermum halicacabum, Chenopodium album, Chloranthus glaber, Cinnamomum camphora, Codonopsis clematidea, Cymbopogon martinii,*

		Cyperus iria, Dendrobium denudans, Dendrophthoe falcata, Derris trifoliata, Desmotrichum fimbriatum, Dorema ammoniacum, Ephedra gerardiana, Erigeron asteroides, Erigeron canadensis, Eupatorium triplinerve, Fagonia cretica, Ferula galbaniflua, Gaultheria fragrantissima, Hedyotis biflora, Hedyotis corymbosa, Hydrocotyle javanica, Hyoscyamus muticus, Hypecoum leptocarpum, Hypericum perforatum, Hyptis capitata, Hyptis suaveolens, Hyssopus officinalis, Jatropha gossypifolia, Kaempferia galanga, Lactuca serriola, Leptadenia reticulata, Leucas aspera, Leucas cephalotus, Leucas stelligera, Liquidambar orientalis, Lycopodium clavatum, Majorana hortensis, Marsilea minuta, Matricaria chamomilla, Melissa officinalis, Mentha aquatica, Mentha arvensis, Mentha longifolia, Mentha piperita, Mentha pulegium, Mentha spicata, Menyanthes trifoliata, Ocimum basilicum, Ocimum gratissimum, Origanum vulgare, Paeonia emodi, Panicum sumatrense, Papaver somniferum, Parnassia palustris, Passiflora foetida, Pedicularis pectinata, Perilla frutescens, Peucedanum grande, Phytolacca acinosa, Pistacia lentiscus, Plantago major, Polygonum hydropiper, Prangos pabularia, Primula rosea, Ranunculus sceleratus, Rorippa dufia, Ruta graveolens, Sambucus canadensis, Sarcostemma acidum, Saussurea sacra, Seseli diffusum, Solanum nigrum, Solanum surattense, Solanum torvum, Sonchus arvensis, Synadenium grantii, Tephrosia purpurea, Tephrosia virginia, Thymus serpyllum, Tropaeolum majus, Vinca minor, Viscum album, Xanthium strumarium.
Wood	**Agar**	*Aquilaria agallocha.*
	Wood	*Aquilegia vulgaris, Artocarpus heterophyllus, Melodinus monogynus, Pinus roxburghii, Tectona grandis.*

Appendix V - Index to Plant Part/s Used As Memory Boosters

Bark	**Root**	*Acanthopanax trifoliatus.*
	Stem	*Acanthopanax trifoliatus, Celastrus paniculatus.*
Fruit		*Citrus limon, Convolvulus microphyllus, Coriandrum sativum, Embelia ribes, Emblica officinalis, Juglans regia, Melissa officinalis, Melissa parviflora, Myrtus communis, Prunus cerasus, Syzygium aromaticum, Terminalia bellirica, Terminalia chebula, Xerospermum glabratum.*
	Kernel	*Semecarpus anacardium.*
Flower		*Achillea millefolium, Convallaria majalis, Convolvulus arvensis, Evolvulus alsinoides, Hibiscus rosa-sinensis, Lawsonia inermis, Linum usitatissimum, Matricaria chamomilla, Mimusops elengi, Nelumbo nucifera, Salvia officinalis, Semecarpus anacardium.*
Leaf		*Bacopa monnieri, Basella alba, Centella asiatica, Cinnamomum tamala, Elephantopus scaber, Lantana camara, Passiflora foetida, Piper betle, Vitex trifolia.*
Root		*Argyreia nervosa, Desmodium tiliaefolium, Ferula galbaniflua, Petroselinum crispum, Picrorhiza kurrooa, Rhodiola imbricata, Tabernaemontana divaricata, Valeriana himalayana.*
	Prop Root	*Ficus benghalensis.*
Seed		*Achyranthes aspera, Celastrus paniculatus, Citrullus vulgaris, Clitoria ternatea, Cordia macleodii, Lagenaria siceraria, Melia azedarach, Prunus amygdalus, Tinospora cordifolia, Ziziphus jujube.*
Stem		*Achyranthes aspera, Basella alba, Ferula narthex.*
	Rhizome	*Acorus calamus, Curcuma zedoaria, Hedychium spicatum, Helminthostachys zeylanica.*
Whole Plant		*Convolvulus microphyllus, Dendrophthoe falcata, Evolvulus alsinoides, Ginkgo biloba, Hydrocotyle sibthorpioides, Limnophila gratioloides, Limnophila rugosa, Majorana hortensis, Medicago sativa, Prunella vulgaris, Tricodesma indicum.*

Appendix VI - Index to Botanical Names

Appendix VII - Index to English Names

Appendix VIII - Index to Hindi Names

Appendix IX - Index to Sanskrit Names

Plate - 1

Abelmoschus esculentus

Abrus precatorius

Acacia catechu

Adiantum capillus-veneris

Allium sativum

Asparagus racemosus

Plate - 2

A: *Avena sativa*

B: *Azadirachta indica*

C: *Bauhinia vahlii*

D: *Berberis aristata*

E: *Bergenia ciliata*

F: *Broussonetia papyrifera*

Plate - 3

Calendula officinalis

Cannabis sativa

Capsicum annum

Carica papaya

Cassia fistula

Catharanthus roseus

Plate - 4

Cedrus deodara

Centella asiatica

Citrus aurantium

Colebrookia oppositifolia

Colocasia esculenta

Coriandrum sativum

Plate - 5

Curcuma domestica

Cynodon dactylon

Datura metel

Dioscorea bulbifera

Eucalyptus globules

Ficus benghalensis

Plate - 6

Ficus hispida

Ficus religiosa

Foeniculum vulgare

Ginkgo biloba

Grevillea robusta

Hedera nepalensis

Plate - 7

Hibiscus rosa-sinensis

Hypericum patulum

Jasminum multiflorum

Justicia adhatoda

Lantana camara

Lycopersicon esculentum

Plate - 8

Magnolia grandiflora

Mallotus philippinensis

Mangifera indica

Musa paradisiaca

Nerium indicum

Nicotiana tabacum

Plate - 9

Oxalis corniculata

Papaver somniferum

Parthenium hysterophorus

Psidium guajava

Rhododendron arboreum

Rubus niveus

Plate - 10

Rumex nepalensis

Sida rhombifolia

Solanum melongena

Solanum nigrum

Solanum surattense

Solanum tuberosum

Plate - 11

Taraxacum officinale

Tecoma stans

Terminalia bellirica

Thuja orientalis

Toona ciliata

Triticum aestivum

Plate - 12

Urtica dioica

B

Viola odorata

Withania somnifera

Woodfordia fruticosa

Xanthium strumarium

Zingiber officinale